ACTA PHYSICA AUSTRIACA / SUPPLEMENTUM VII

DEVELOPMENTS IN HIGH ENERGY PHYSICS

PROCEEDINGS OF THE
IX. INTERNATIONALE UNIVERSITÄTSWOCHEN
FÜR KERNPHYSIK 1970 DER KARL-FRANZENS-UNIVERSITÄT
GRAZ, AT SCHLADMING (STEIERMARK, AUSTRIA)
23rd FEBRUARY—7th MARCH 1970

SPONSORED BY
BUNDESMINISTERIUM FÜR UNTERRICHT
THE INTERNATIONAL ATOMIC ENERGY AGENCY
STEIERMÄRKISCHE LANDESREGIERUNG AND
KAMMER DER GEWERBLICHEN WIRTSCHAFT FÜR STEIERMARK

EDITED BY
PAUL URBAN
GRAZ

WITH 134 FIGURES

1970

SPRINGER-VERLAG / WIEN · NEW YORK

Organizing Committee:

Chairman: Prof. Dr. PAUL URBAN
Vorstand des Institutes für Theoretische Physik, Universität Graz

Committee Members: Prof. Dr. P. URBAN
Dr. R. BAIER
Dr. H. KÜHNELT

Proceedings: Dr. H. J. FAUSTMANN
Dr. P. PESEC

Secretary: M. PAIL

Acta Physica Austriaca / Supplementum I
Weak Interactions and Higher Symmetries
published in 1964

Acta Physica Austriaca / Supplementum II
Quantum Electrodynamics
published in 1965

Acta Physica Austriaca / Supplementum III
Elementary Particle Theories
published in 1966

Acta Physica Austriaca / Supplementum IV
Special Problems in High Energy Physics
published in 1967

Acta Physica Austriaca / Supplementum V
Particles, Currents, Symmetries
published in 1968

Acta Physica Austriaca / Supplementum VI
Particle Physics
published in 1969

Softcover reprint of the hardcover 1st edition 1970

Library of Congress Catalog Card Number 77-133409

ISBN 978-3-7091-5837-1 ISBN 978-3-7091-5835-7 (eBook)
DOI 10.1007/ 978-3-7091-5835-7

Title-No. 9278

Contents

Ladies and Gentlemen, dear Colleagues!

With great pleasure I welcome you most cordially to our ninth Internationale Universitätswochen für Kernphysik. The presence of about 200 participants from 17 countries gives me the hope that this winterschool will maintain its international reputation among high energy physicists.

So, first of all, thank you for coming. I am also very grateful for the cooperation with all our lecturers.

It gives me great satisfaction to see among our guests of honour many distinguished members of our public authorities showing their interest in our meeting.

It is a great honour for me to welcome Prof. Dr. A. SALAM, the representative of the International Atomic Energy Agency; Prof. Dr. GÜNTER B. FETTWEIS, the Chancellor of the University of Mining, Metallurgy and Material Science, Leoben; the Dean Prof. Dr. G. POROD, the representative of the Chancellor of the University of Graz; Dr. TÖPFNER, the representative of the Chamber of Commerce of Styria; the representative of the Mayor of the city of Schladming and the members of the local council; Director LAURICH, who had sponsored all the preceding meetings; and Mr. FRANZ ANGERER, Manager of the Fremdenverkehrsverein in Schladming.

I would like to emphasize especially financial supports from the Ministry of Education, the provincial government of Styria, as well as the IAEA and the Chamber of Commerce which made the organisation of this school possible. Last not least we have to thank our host, the city of Schladming, for its hospitality and help in organising the meeting.

May I continue with a few short remarks about this year's scientific program. As a consistent continuation of the topics of our last school we will try to discuss the latest developments of the concept of duality and the new attempts in Regge theory. In addition we will have lectures on general aspects and non-linear techniques in S-matrix theory. A survey of the status of chiral symmetries and related topics should round off the present situation in strong interaction physics. Taking into account the increasing interest in general relativity, we included lectures on some relevant mathematical techniques and their application in relativity. Finally, let me mention the lecture on the construction of physical states in quantum field theory, which will give us some insight into the exciting recent developments of this branch of physics.

I am sure that this year's meeting also will bring us a small step further in the understanding of high energy phenomena, and in opening this school on "Developments in High Energy Physics" I wish you a profitable and successful two weeks.

PAUL URBAN

Acta Physica Austriaca, Suppl. VII, 1—31 (1970)

NON-POLYNOMIAL LAGRANGIAN THEORIES*

BY

ABDUS SALAM**

International Centre for Theoretical Physics,
Miramare - Trieste, Italy

1. INTRODUCTION

Barring lepton electrodynamics, most Lagrangians of physical interest are "non-renormalizable", the apparent non-renormalizability arising either from their non-polynomial nature or from higher spins. Typical non-polynomial cases are the chiral SU(2) × SU(2) Lagrangian

$$L = \frac{(\partial_\mu \phi)^2}{(1+f\phi^2)^2} \qquad (1.1)$$

in Weinberg's representation or the gravitational Lagrangian

$$L = \frac{1}{K^2} \sqrt{-g}\, g^{\mu\nu}(\Gamma^\lambda_{\mu\rho}\, \Gamma^\rho_{\nu\lambda} - \Gamma^\lambda_{\mu\nu}\, \Gamma^\rho_{\lambda\rho}) \qquad (1.2)$$

where

$$\Gamma^\lambda_{\mu\nu} = \frac{1}{2}\, g^{\lambda\rho}(\partial_\mu\, g_{\nu\rho} + \partial_\nu\, g_{\mu\rho} - \partial_\rho\, g_{\mu\nu}) .$$

The components $g_{\mu\nu}$ which enter the expression for $g=\det g_{\alpha\beta}$ are a ratio of two polynomials in $g^{\mu\nu}$. A typical example

*Presented at the Coral Gables Conference, Miami, January 23-25,1970,and the IX. Internationale Universitätswochen für Kernphysik, Schladming, February 23 - March 7, 1970.

**On leave of absence from Imperial College, London,England.

of a higher spin case is the intermediate-boson mediated weak Lagrangian, e.g., the neutral vector W_μ interacting with quarks Q ,

$$L_{int} = f \bar{Q} \gamma_\mu (1+\gamma_5) Q W_\mu . \qquad (1.3)$$

So far as non-renormalizability is concerned, this is manifested most simply by transforming (1.3) into a non-polynomial form. In Stückelberg variables ($Q' = \exp\{-i\gamma_5 f\frac{B}{\kappa}\} Q$ $W_\mu = A_\mu + \frac{1}{\kappa} \partial_\mu B$) an equivalent interaction is given by

$$L_{int} = f \bar{Q}'\gamma_\mu (1+\gamma_5) Q' A_\mu + m \bar{Q}'(\exp\{i\gamma_5 \frac{f}{\kappa} B\}-1) Q' \qquad (1.4)$$

It is clear therefore that if Lagrangian theory is to play any direct role in particle physics beyond that for electrodynamics, methods must be developed to extract numbers from non-polynomial theories. Basically any such methods must ensure the resolution of the two distinct difficulties of non-renormalizable theories, i.e., an infinite number of distinct infinity types and a high-energy behaviour which violates Froissart-like bounds.

Problems with conventional treatment of non-renormalizable theories

1. An infinite number of infinity types:

Ignoring derivatives for the moment, one may write L_{int} in the typical form

$$L_{int} = G \sum \frac{v(n)}{n!} (\phi)^n$$

where v(n) contain powers of f (typically $v(n) \propto f^n$). (We shall call f the minor coupling constant.) A perturbation expansion may be written to any given order N in the major coupling constant G and to any desired n order in the minor coupling f. In this linearized form all contributions of $f^n \phi^n$ interactions with $n > 4$ give rise to non-renormalizable infinities. To remove these in the conventional manner, one would need more and more counter-terms in each order, reducing very considerably the predictive power of the theory.

2. Unacceptable high-energy behaviour

The high-energy dependence of individual graphs in all theories with $L \propto f^n\phi^n$ $(n > 4)$ increases (unacceptably) as the order increases and is not polynomially bounded. (One aspect of this is that the counter-terms needed to cancel infinities must contain arbitrarily high-order derivatives of field variables, making the counter-Lagrangians non-local.)

To my knowledge the first acceptable treatment of problem 1) was given by S. Okubo [1] as early as 1954 in a paper which was apparently overlooked by others who subsequently worked on different aspects of this problem. These include Arnowitt and Deser, Fradkin, Efimov, Feinberg and Pais, Güttinger, Volkov, Fried, Lee and Zumino, Fivel and Mitter in addition to Delbourgo, Strathdee, Boyce and Sultoon, and Koller, Hunt and Shafi [2]. I shall review the earlier results and also state some new ones particularly relating to renormalization constants. These are joint work of Trieste and London groups.

The basic idea in dealing with problem 1) is that for a fixed order in the major coupling constant G^N one can Borel-sum the entire perturbation series to all orders in f^n. Formally this is an asymptotic series with each term given by an infinite expression. These Borel sums have the remarkable property that the summation automatically quenches most of the infinities. (This is perhaps not too unexpected a result when one considers that the Lagrangians of the type

$$\frac{1}{1 + f \phi^2}$$

visibly appear to possess a built-in damping factor for higher frequencies.) For some Lagrangians this quenching is so strong that all matrix elements are rendered finite, offering thus the possibility of computing even self-masses and self-charges. For others some few infinities still survive and these need renormalizing.

There are a number of different formulations of the summation procedure - several variants - which fall basically into two classes: the x-space methods and the p-space methods. The results obtained using either method are equivalent. The chief problem is to ensure that the Borel sums i) possess the requisite analyticity properties in p-space, ii) satisfy unitarity and iii) are unique. Since good reviews [3] of these methods exist, I shall not attempt to make this report comprehensive; I shall confine myself to a statement of results. In respect of problem 1), these are: i) the requirements of analyticity and unitarity are most likely met by these asymptotic sums, though uniqueness seems to need additional criteria; ii) for a large class of non-polynomial Lagrangians, a consistent renormalization programme can be devised where all infinities can be incorporated into acceptable counter-Lagrangians.

Regarding problem 2), which concerns the high-energy behaviour of Borel sums in the minor coupling constant, we obtain a perfectly acceptable behaviour for space-like momenta. For time-like momenta the cross-sections computed to order G^2 in the major coupling constant increase unacceptably fast with energy. It appears, however, that a further summation, this time in the major coupling constant G, of sets of chain-graphs alters this, just as is the case in conventional theory where, for example, a summation of ladder type perturbation diagrams produces Regge asymptotic behaviour.

2. A RAPID EXPOSE OF THE METHODS

The basic ideas of the summation methods can perhaps be rapidly illustrated by considering [4]

$$L_{int}(\phi) = G \frac{1}{1+f\phi} \quad .$$

a) The formal series expansion for amplitudes

Formally an expectation value like

$$F(\Delta) = \langle L_{int}(\phi(x_1)), L_{int}(\phi(x_2))\rangle$$

equals the asymptotic series:

$$G^2 \sum_{n=0}^{\infty} n! \, f^{2n} \, \Delta_F^n(x_1-x_2) \quad . \qquad (2.1)$$

Each term is infinite. Indeed as n increases, the singularity of $\Delta^n(x) \propto (1/x^2)^n$ gets worse and worse. We shall use the Borel method to sum the series. Ultimately we are interested in the Fourier transform of this sum:

$$\tilde{F}(p^2) = \int F(\Delta) \, e^{ipx} \, d^4x \quad . \qquad (2.2)$$

The criterion for an acceptable summation technique is that $\tilde{F}(p^2)$ should exhibit conventional p-space analyticity.

b) The euclidicity postulate

To guarantee this consider the Symanzik region in p-space ($p^2 < 0$). (When more than one external momentum p_i is involved, the Symanzik region is the region for which $p_i^2 \leq 0$, $p_i p_j \leq 0$. Certain other restrictions on momenta are also placed but the heart of the matter is that all momenta can be simultaneously chosen such that $p_{i0}=0$.) For $p^2 < 0$, choose the frame where $p_0=0$. Clearly we may make a Wick rotation $x_0 \to ix_4$ without altering the value of $\tilde{F}$. Thus for the Symanzik region of p-space one needs to consider $\Delta^n(x)$ for euclidean vectors x^2 only. (For a zero-mass field $\Delta(x) = -1/4\pi^2x^2$, where $x^2 = -x_4^2 - \underline{x}^2$ and is real and positive.) For p-space regions outside the Symanzik region we analytically continue (2.2). (It cannot be emphasized strongly enough that for divergent series of the type (2.1) one is not starting by "proving" the validity of the Wick rotation. Rather, euclidicity is a basic postulate - part of the process of defining the theory. One accepts it for the Symanzik region in p-space: outside this region one makes an analytic continuation in the momenta.)

c) Borel summation

To give meaning to the divergent sum $F(\Delta)$, use Borel transforms and write:

$$F(\Delta) = \sum_{n=0}^{\infty} \int_0^{\infty} e^{-\zeta}(f^2 \zeta\Delta)^n \quad . \tag{2.3}$$

using the identity:

$$n! = \int_0^{\infty} \zeta^n e^{-\zeta} d\zeta \quad .$$

d) The x-space method

The x-space method consists of inverting integration and summation in (2.3) and writing it as

$$F(\Delta) = \int_0^{\infty} d\zeta \; e^{-\zeta}(1-\zeta f^2\Delta)^{-1} \quad . \tag{2.4}$$

The expression (2.4) defines the amplitude $F(\Delta)$. For zero-mass particles (m=0) this equals:

$$\int_0^\infty \frac{r^2}{r^2 - \zeta f^2} e^{-\zeta} d\zeta \ .$$

Notice that as $r \to 0$, $F(\Delta)$ is perfectly well behaved. The Borel summation has quenched the ultraviolet infinities. (Fradkin and Efimov have given explicit expressions of the type (2.4) for Borel sums to all orders G^N where this quenching effect can be explicitly seen.)

e) At this stage we encounter our first problem in the x-space method: the integrand has a pole on the integration path at $\zeta = 1/f^2\Delta$ which equals $4\pi^2 r^2/f^2$ when $m = 0$ $(r^2 = \underline{x}^2 + x_4^2)$.

We must define how to go round this singularity, the final objective being that the Fourier transform should be an analytic function which when continued to positive p^2 (outside the Symanzik region) has the unitarity cut from $p^2 = 0$ to ∞ .

One answer is: take the principal value. This is because, from (2.3), $F(\Delta)$ must be real. The p.v. prescription for the integral representation (2.4) of $F(\Delta)$ will guarantee this [1]. The Fourier transform of (2.4) when m=0 can be explicitly evaluated and a continuation to time-like values of p^2 carried out to demonstrate explicitly that $\tilde{F}(p^2)$ possesses the correct analyticity structure in the p^2-plane. The asymptotic behaviour of $\tilde{F}(p^2)$ is:

$$\begin{aligned} \tilde{F}(s) &\to \frac{1}{(f^2 s)^3} && s \to -\infty \\ &\to \pm i\pi \exp(f^2 s) && s \to +\infty \pm i0 \end{aligned} \tag{2.5}$$

where $s = p^2$. [2]

f) The p-space method

One can use an alternative method which works directly in p-space. It depends on Volkov's observation of the power of the Gel'fand-Shilov investigation of the Fourier transform of the generalized function $(\Delta(m=0))^z = r^{-2z}$ in the range $0 < \mathrm{Re}\, z < 2$.

The crucial formula is

$$\Delta^z(x) = \frac{1}{(2\pi)^4}\int d^4p\, e^{-ipx}\, \frac{(-p^2)^{z-2}\pi(4\pi)^{2-2z}}{\sin\pi z\Gamma(z)\Gamma(z-1)}$$

$$0 < \mathrm{Re}\, z < 2 \quad . \tag{2.6}$$

To use this formula go back to the Borel sum (2.3) and employ a Sommerfeld-Watson transformation to convert the series into a formal integral of the form

$$F(\Delta) = \frac{i}{2}\int_\Gamma \frac{dz}{\sin\pi z}\int d\zeta\, e^{-\zeta}(-\zeta f^2\Delta)^z \tag{2.7}$$

with the contour Γ enclosing the positive real axis in the z-plane. Straighten the contour to lie along the imaginary axis with Re z constrained to lie in the range $0 < \mathrm{Re}\, z < 2$. Using Gel'fand-Shilov's formula we obtain

$$\tilde{F}(p^2) = \frac{i}{2}\int_{\alpha-i\infty}^{\alpha+i\infty} \frac{dz}{\sin\pi z}\,\frac{(-f^2)^z(-p^2)^{z-2}\Gamma(z+1)}{\sin\pi z\Gamma(z)\;\Gamma(z-1)} + (2\pi)^4\,\delta(p) \tag{2.8}$$

where $0 < \alpha < 2$. (The term $\delta(p)$ corresponds to a graph which contains no internal line.)

f) Formula (2.8) is the master formula. By closing the contour along the left, one can immediately obtain the asymptotic behaviour of $\tilde{F}(p^2)$ for $p^2 \to -\infty$ and the result (2.5). As in Regge pole theory, the right-most pole of the integrand gives the leading contribution to the asymptotic behaviour; in this case the right-most pole lies[3] at z=-1 , giving the asymptotic expression $\approx 1/(f^2p^2)^3$ as before in (2.5).

3. HIGHER ORDERS

a) Super-graphs

Consider

$$L_{int}(\phi) = G \sum_{n=D_o}^{\infty} \frac{v(n)}{n!} (\phi)^n$$

($v(n)$ contains the minor coupling parameter f^n).

It is easy to verify that the G^N contribution to an amplitude $F(x_1,\ldots,x_N)$ with E external lines can be written as a sum of contributions from a set of super-graphs constructed as follows:

a) Take N points $x_1, x_2,\ldots,x_N$.

b) Join all points pair-wise with just one super-line joining two distinct points (x_i,x_j); associate with this line a positive integer n_{ij} .

c) For each line write the factor $\frac{1}{n_{ij}!}[\Delta_F(x_i-x_j)]^{n_{ij}}$.

d) For each point x_i write a vertex factor $v(\sum_j n_{ij}+m_i)$. Here m_i is the number of external lines impinging on the point x_i .

e) The contribution of the super-graph to the amplitude equals

$$F_{m_1 m_2 \ldots}(x_1,\ldots,x_N) =$$

$$= G \sum_{n_{ij}} \prod_i v(\sum_j n_{ij} + m_i) \prod_{i<j} \frac{(\Delta_F(x_i-x_j))^{n_{ij}}}{n_{ij}!} \quad (3.1)$$

The limits of the n_{ij} are given by

$$L_i = \sum_j n_{ij} + m_i \geq D_o \quad .$$

f) To get the total contribution in order G^N, sum over all configurations of the external lines with the m_i lines at the i-th vertex distributed over the various vertices, such that

$$\sum m_i = E \quad .$$

b) Super-graphs in momentum space

The great beauty of the p-space method lies in the similarity of the p-space expressions for super-graphs and normal Feynman diagrams.

One can introduce Feynman's auxiliary parameters and carry out the loop integrations. The result is an elegant expression for the super-graph contribution as a weighted average integral of contributions of conventional graphs. The utility of such an expression is two-fold.

i) The sums of super-graphs in different orders of G closely resemble the sums for conventional graphs and the methods previously discussed by Polkinghorne, Federbush [5] and others for carrying through the summation can be taken over.

ii) The discontinuity formulae of Cutkosky - and the proof of the unitary relations using such formulae - follow the conventional lines.

For the zero mass case, the integral expression for the N-th order super-graph is the following: (We consider here the simple case $D_0=0$.)

Associate with each super-line a four-momentum vector q_{ij}. The Sommerfeld-Watson transform of (3.1) in p-space equals:

$$\tilde{F}(p_i) = G^N \prod_{i<j} \int dz_{ij}\ \rho(z_{ij}) \int d^4q_{ij}(-q_{ij}^2)^{z_{ij}-2}\,\delta^4(\Sigma p_i+\Sigma q_{ij}) \,. \tag{3.2}$$

Here $\rho(z_{ij})$ is the product of the vertex factors $v(\sum_{i\neq j} z_{ij}+m_i$ the factors $1/\sin\pi z_{ij}$ (or more generally $(1+b\cos\pi z_{ij})/\sin\pi z_{ij}$) and the factors $1/(\sin\pi z_{ij}\Gamma(z_{ij})\Gamma(z_{ij}-1))$ for each super-line. The p_i's are the momenta carried by the external lines at the i-th vertex and the δ-functions express conservation of energy and momentum. The contour in each z_{ij}-plane for the case $D_0=0$ lies along the imaginary axis for each z_{ij} . (We consider later the location of these contours when $D_0 \neq 0$. The problem of any surviving infinities in the theory is bound up with the location of these contours.)

Introduce Feynman's auxiliary parameters, using the integral representation[4]

$$(-q^2)^{z-2} = \frac{1}{\pi\Gamma(2-z)} \int_0^\infty d\alpha\ \alpha^{1-z}\ e^{\alpha q^2} \ . \qquad (3.3)$$

One may now carry through the d^4q integrations in the subsidiary integral I defined by

$$I(p_i,\alpha_{ij}) = \int (\exp \sum \alpha_{ij}\ q_{ij}^2)[\delta^4(\sum p_i + \sum q_{ij})]^N \times$$

$$\times \prod d^4 q_{ij} \qquad (3.4)$$

The result is identical to the case of conventional Feynman graphs with $F = (N(N-1))/2$ internal lines. (This is because $I(p_i,\alpha_{ij})$ is not z_{ij}-dependent.) The evaluation of $I(p_i,\alpha_{ij})$ can easily be carried through using the methods of Chisholm [5]; the final expression for the amplitude $F(p_i)$ reads:

$$\tilde{F}(p_i) = \prod_{i<j} \int dz_{ij}\ \rho'(z_{ij}) \int d\alpha_{ij}\ \alpha_{ij}^{1-z_{ij}}\ I(p_i,\alpha_{ij}) \qquad (3.5)$$

where ρ' differs from ρ by the factors

$$\prod_{i,j} \frac{1}{\pi\Gamma(2-z_{ij})}$$

The result for the N-point function evaluated in order G^N can therefore be stated thus:

Draw a Feynman graph with internal lines joining all the N-points pair-wise. Introduce Feynman parameters; the result of performing loop integrations is the standard Chisholm expression $I(p_i,\alpha_{ij})$. Multiply this by the factors $(\alpha_{ij})^{1-z_{ij}}$ and the weight function $\rho'(z_{ij})$; integrate over Feynman parameters ρ_{ij} and the Sommerfeld-Watson parameters z_{ij} . This gives the super-graph contribution.

4. SUPER-GRAPHS

Infinities and renormalization

1. Using super-graphs one can investigate quite simply the possible infinities of non-polynomial theories. Among these are theories with no infinities whatsoever. The physically interesting cases, however, are of mixed theories where polynomial and non-polynomial Lagrangians both occur together. Such, for example, is the case for chiral Lagrangians (with nucleons interacting with pions for example) or weak Lagrangians (where the Stückelberg-B-field occurs non-polynomially while the A-field interacts polynomially (see (1.4))). Not all these mixed theories are renormalizable. *By renormalizable we shall mean theories where all infinities can be absorbed in a finite set of counter-terms. (Naturally the counter-terms must NOT contain arbitrarily high-order derivatives of field variables* $\partial_\mu\phi$, for example in non-polynomial combinations like $1/(1+f(\partial\phi)^2)$, otherwise the counter-terms would represent non-local additions to the original Lagrangian.)

2. Before we proceed, it is important to remark that, for non-polynomial Lagrangians with multitudes of external lines coming out of single vertices, the familiar statement of graphs getting less and less singular as the number of their external lines increases needs revision. The worst offenders in this respect are graphs with only two vertices. Here we have the surprising result: $S_{m,0} \approx S_{0,0}$. To see this, consider the simple case

$$L_{int}(\phi) = \sum_{n\geq 1} v(n) \frac{\phi^n}{n!}$$

$$S_{m,0} = \sum_{n\geq 1} \frac{v(n+m)v(n)}{n!} \Delta^n(x) .$$

In momentum space

$$S_{m0}(p^2) = \int dz \frac{1}{\sin\pi z} \frac{v(z+m)v(z)}{\Gamma(z+1)} \frac{(p^2)^{z-2}\Gamma(1-z)}{\Gamma(z-1)} ,$$

where the contour lies parallel to the imaginary axis along $\mathrm{Re}\, z = 1$. For $p^2 < 0$ the high-energy behaviour is given by the first pole of the integrand on the left of $\mathrm{Re}\, z = 1$. Clearly this lies to the left of $z = 0$ (it would come from the factor $v(z)$) irrespective of what the value of m is. (Barring special cases, the corresponding singularity of $v(m+z)$ is still further left since $m > 0$.)

One can easily prove the following results which give the precise connection between the singularities of graphs with different number of external lines.

Theorem: If N is the total number of vertices:

For $N = 2$

$$S_{m_1,m_2}(\Delta_{12}(x)) = \left(\frac{\partial}{\partial\Delta_{12}}\right)^{m_2} S_{m_1-m_2,0}(\Delta_{12}) , \quad m_1 > m_2 \tag{4.1}$$

For $N = 3$ or greater,

$$S_{2,0,0}(\Delta_{12},\Delta_{23},\Delta_{13}) = \left[\int_{-\infty}^{\Delta_{23}} d\Delta_{23} \frac{\partial}{\partial\Delta_{12}} \frac{\partial}{\partial\Delta_{13}}\right] S_{0,0,0} \tag{4.2}$$

Thus, for $N \geq 3$, all $S_{m_1,m_2,m_3,\ldots}$ can be related by repeated differentiations (A) or by operations of the type (B) above to the amplitude $S_{0,0,0,\ldots}$, or at worst to $S_{1,0,0,\ldots}$. Roughly this states that if in momentum space $\tilde{S}_{0,0,0,\ldots}(p_i)$ behaves like $\underset{M\to\infty}{M^\alpha}$, as external momenta p_i go to infinity $\tilde{S}_{m_1,m_2,\ldots}$ behaves like $M^{\alpha-\Sigma m}$.

3. A rough estimate for finiteness of super-graphs may be stated at this stage. The total number of super-lines F in a super-graph where all vertices are connected to each other is given by $F = \frac{N(N-1)}{2}$ while the number of loops equals (F-N+1) . Thus the convergence of an integral

$$\tilde{S} = \int \frac{(d^4k)\ \text{loops}}{(k^\alpha)^F}$$

requires that the factor associated with each super-line $1/k^\alpha$ must be such that

$$k^{(4-\alpha)F-4N+4} \tag{4.3}$$

does not increase with N. Clearly $\alpha \geq 4$ is sufficient to ensure this. Roughly, then, each super-propagator should behave for large k like a dipole $1/k^4$ for finiteness. Later we make this criterion more precise.

4. Although the considerations of this section are really more general, to simplify discussion consider interactions of the type:

$$L = \frac{\phi^{D_0}}{1+f\phi^{D_0-D}}$$

where D_0 and D are integers.
Note that

$$L \underset{\phi\to\infty}{\longrightarrow} \phi^D$$

The index D is the Dyson index which for conventional polynomial theories determines the possible infinities of the theory and if the theory is renormalizable. For example, for theories with D=3, the second-order vacuum graph (with no external lines E=0) is quadratically infinite ($\underset{M\to\infty}{\approx} M^2$) while second-order self-energy (E=2) is logarithmically infinite ($\approx\log M$) . For D=4, all vacuum graphs (E=0) behave like M^4, self-energy graphs (E=2) like M^2 and scattering graphs (E=4) like log M .

For non-polynomial cases the infinities and renormalizability again depend on the index D but in a more subtle manner. Our tentative results are:

A) When D < 2 there are no infinities.

B) For D = 2 , the only graphs possibly infinite are the star-fish graphs which modify the fundamental super-vertex

to

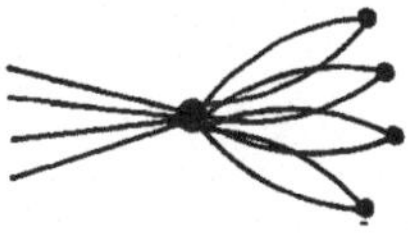

(with arbitrary numbers of external lines and arbitrary numbers of stars). Counter-terms (independent of field-derivatives) can be introduced to absorb these.

C) For D = 3 , the only infinities again come from modifications of the fundamental super-vertex. These are of two types:

1) Tadpole modifications:

(4.4)

2) Proper self-energy-like modifications:

(4.5)

Plus what we shall call repetitions of these patterns; for example tadpole repetitions

(4.6)

Similarly the self-energy-like repetitions:

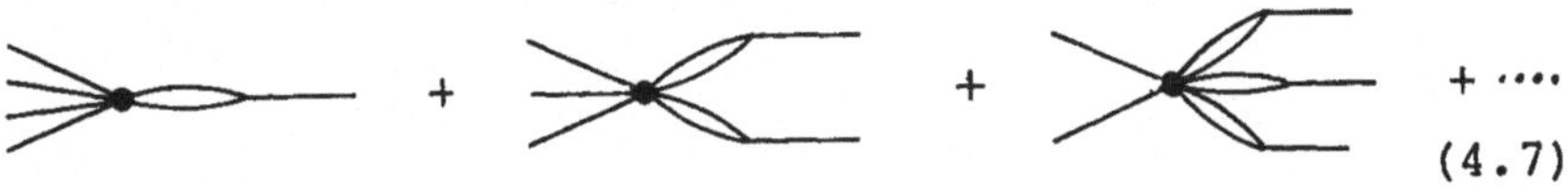

(4.7)

These infinities can be absorbed in non-derivative

counter-terms which renormalize the basic super-vertex. (These counter-terms are exhibited in the next section.)

D) For $D = 4$, the infinities come from the star-fish modifications of the basic super-vertex

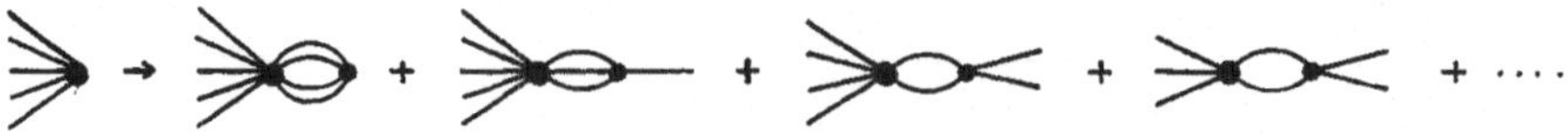

Type I Type II Type III Type IV

One may write counter-terms to absorb these, but this time there are an infinity of distinct types of counter-terms and these also contain derivatives of field-variables to arbitrary high orders. Even if this fundamental super-vertex is made finite, new infinities arise when graphs with two and more super-vertices are considered. Thus a non-polynomial theory like $\phi^5/(1+f\phi)$ is <u>non-renormalizable</u> .

E) To complete the statement of renormalizable theories, it appears that one may introduce as many as four derivatives without affecting renormalizability, though this conclusion is as yet tentative. Symbolically, for $L = \partial^{\alpha}\phi^{D}$ with $D \leq 3$, $D+\alpha \leq 4$, it is likely that no new problems arise, but this needs further examination.

5. To prove these results, consider the basic x-space expression:

$$\sum_{n_{ij}} \ldots v(m_i + \sum_j n_{ij}) \ldots \frac{\Delta^{n_{ij}}(x_i - x_j)}{n_{ij}!} \ldots$$

where all $n_{ij} \geq 0$ and are subject to the restrictions

$$L_i = \sum_j n_{ij} + m_i \geq D_0 \quad .$$

(L_i is the total number of lines at the vertex i.)

A Sommerfeld-Watson complexification of n_{ij} gives

$$\int \frac{1}{\sin\pi z_{ij}} \cdots \frac{v(m_i + \sum_j z_{ij})}{\Gamma(z_{ij}+1)} \cdots (-\Delta)^{z_{ij}} \cdots$$

with the contours in z_{ij} planes encircling the real axes

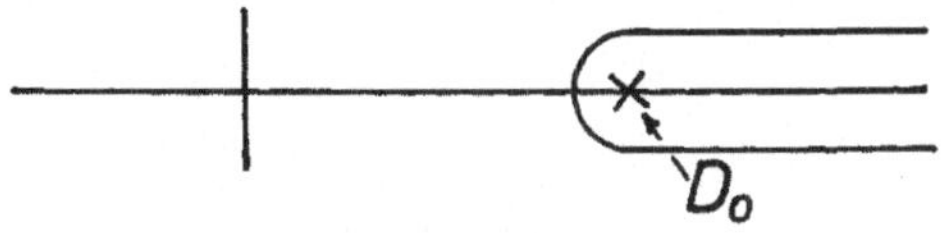

subject to the restrictions:

$$\text{Re } L_i = \sum_j \text{Re } z_{ij} + m_i \geq D_0 \; ; \qquad \text{Re } z_{ij} \geq 0 \; .$$

Now, from the Gel'fand-Shilov theorem we know that the Fourier transform of Δ^z exists provided $0 < \text{Re } z < 2$. Our first task is to shift the contours so that $\text{Re } z < 2$; in the process we shall pick up infinite modifications of super-vertices which will need renormalizing. A second minor task will be to shift the contours still further down to $0 < \text{Re } z < 1$ to get all superpropagators to behave like $1/k^4$ for space-like k^2 . This will give rise to some completely harmless tadpoles, the finiteness of the theory being maintained.

This double shifting task is facilitated by expressing the limits

$$\sum_{L_i \geq D}$$

in the form

$$\sum_{\substack{L_i \geq 0 \\ \text{all } i}} - \sum_{\substack{L_i \geq 0 \\ \text{one } L_i = 0,1,2,\ldots,D_0}} - \sum_{\substack{L_i \geq 0 \\ \text{two } L_i = 0\ 1,2,\ldots,D_0}} \cdots\cdots -$$

$$\cdots - \sum_{\text{all } L_i = 0,1,2,\ldots,D_0} \tag{4.8}$$

We shall call these subtracted terms the "ghost terms". There is a very simple graphical representation of these. Write:

$$L = \frac{\phi^{D_0}}{1+f\phi^{D_0-D}} = \sum_{0}^{D} a_r \phi^r + \sum_i \frac{\alpha_i}{1+\beta_i\phi} = L_P + L_{N\,P} \quad , \quad (4.9)$$

i.e., as a sum of a polynomial Lagrangian L_P and non-polynomial Lagrangians $L_{N\,P}$. For the latter the relevant index $D_0=0$. The important point to stress is that in the expansion (4.9) the highest polynomial term has the index D and not D_0 . We shall assume henceforth that $D \leq 4$ - if we do not make this assumption the polynomial part of L is non-renormalizable from the start. In terms of this split of the Lagrangian (4.9), the expansion (4.8) of the matrix element has the following meaning. The first term in the sum $\sum_{L_i \geq 0}$ corresponds to the contribution from $(L_{N\,P})^N$, the first ghost term to $(L_{N\,P})^{N-1}(L_P)$ and so on, the last ghost term corresponding to $(L_P)^N$. (As an illustration consider

$$L = \frac{\lambda+\phi^3+f\phi^4}{1+f\phi} = \phi^3 + \frac{\lambda}{1+f\phi} \; .$$

The third-order vacuum graphs are

Thick lines represent super-lines. Thin lines are ordinary lines with propagators Δ_F).

Consider the graph which consists of super-lines only. There is no difficulty in going over to Fourier space; this is because $L_i \geq 0$ implies $n_{ij} \geq 0$, so that the z_{ij} contours can be Sommerfeld-Watson rotated to lie between $0 < \text{Re } z_{ij} < 1$,

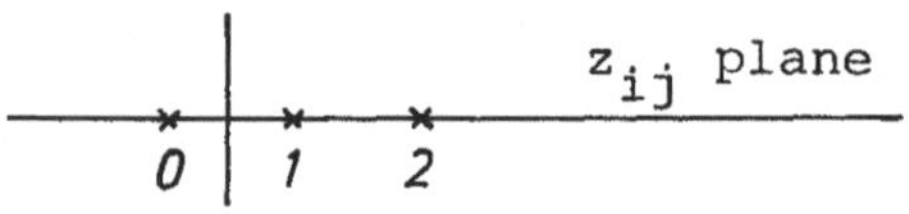

To estimate the high-energy behaviour of these graphs we need knowledge of the left-most singularity of the integrand. Barring some exceptional cases, this will lie to the left of $z_{ij} = 0$, giving at least a factor $(k_{ij}^2)^{z_{ij}-2} = (k_{ij}^2)^{-2}$ for each super-line. This guarantees the finiteness of all such graphs.

Consider now the cases D=1,2,3,4 individually for singularities of ghost graphs. The sub-graphs which are entirely made from the polynomial Lagrangian ϕ^r, $r \leq D$ may have their own singularities; these will need the conventional counter-terms and we shall assume that these have been introduced. We have only to consider mixed graphs and, in particular, star-fish modifications of super-vertices.

a) D = 1

The star-fish consists of spokes with a factor $\Delta(p=0,m^2) = (1/m^2)$. These tadpoles are harmless so far as infinities are concerned.

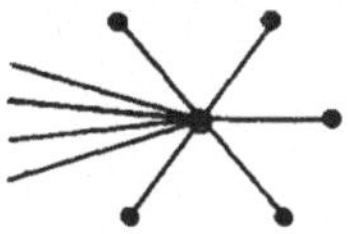

b) D = 2

The star-fish consists of the basic star

plus repetitions:

The full contribution of these diagrams is

$$\left[c_0\left(\frac{\partial}{\partial \Phi}\right)^2 + c_0'\left(\frac{\partial}{\partial \Phi}\right)^4 + c_0''\left(\frac{\partial}{\partial \Phi}\right)^6 + \ldots\right] L_{N\,P}(\phi) \quad (4.10)$$

where

$$c_0 = \int \Delta^2(x)\, d^4x \quad , \; c_0' = \frac{1}{2!}\, c_0^2 \; , \; c_0'' = \frac{1}{3!} c_0^3 \; \ldots \quad .$$

Clearly (and not unexpectedly) the series (4.10) sums to an exponential

$$\left([\exp(c_0 \frac{\partial^2}{\partial\phi^2})]-1\right) L_{N\,P} \quad .$$

If we had started with the Lagrangian

$$(L+\delta L)_{NP} = \exp[-c_0(\frac{\partial^2}{\partial\phi^2})]\, L_{N\,P}(\phi)$$

instead of $L_{N\,P}$ there would be no star-fish infinities.

c) D = 3

In this case there are two types of modifications:

Type I - tadpole-like modifications:

+ repetitions of these, like

The modified Lagrangian reads:

$$\exp(-[c_1(\frac{\partial^2}{\partial\phi^2}) + c_2(\frac{\partial^3}{\partial\phi^3}) + c_3(\frac{\partial^4}{\partial\phi^4}) + \ldots])\, L_{N\,P}(\phi) \quad .$$

$c_1, c_2, c_3, \ldots,$ are the (infinite) contributions from the basic graphs. The repetitions are all taken care of by the exponential.

Type II - The second category of infinities arises from self-energy-like graphs

These are taken care of by starting the theory with the modified Lagrangian

$$\exp(-d_1 \frac{\partial^2}{\partial\phi^2}\, \phi) L_{NP} \quad .$$

One final modification. Replace $L_{NP}(\phi)$ in the formula above by $L(\phi)$. This takes care now of the self-energy infinity  as well .

d) D = 4

The tadpole-like infinities present no difficulty, but

the infinities of Type II ,

being quadratic (M^2), quartic (M^4),..., now need a modified Lagrangian:

$$\exp[(d_0\phi + d_0'\partial^2\phi)(\frac{\partial}{\partial\phi})^3 + \ldots]L_{NP}(\phi)$$

i.e., the modified Lagrangian contains derivatives of field variables to any arbitrarily high order. Clearly, this - according to the criterion stated earlier - is a non-renormalizable situation .

Having taken care of vertex modifications of Type I and Type II for D=1,2,3 cases, we now need to see if there is any possibility of new infinities arising from joining pure L_P sub-graphs with pure L_{NP} graphs. For D=2, the proof that none so arise is trivial. One can get, at worst, situations like

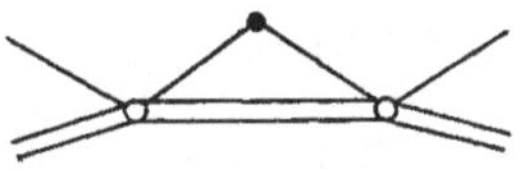

which are finite if one remembers that the super-line gives a factor like $1/k^4$. A slightly more complicated argument is necessary for D=3 . Basically, the proof needs the result (4.2) stated earlier, viz., a super-graph $S_{m_1,m_2,\ldots,}(k_i)$, with $\sum m_i$ external lines, decreases faster by a factor $1/(k^{\Sigma m_i - 1})$ than $S_{1,0,0,\ldots}$. Consider a super-graph connected with m internal lines with a graph made of ϕ^3 vertices only. The worst case for infinities is when neither the supergraph nor the pure ϕ^3 graph has any external lines. From the well-known Dyson count, the pure ϕ^3 graph contributes a factor

$$\frac{S_{0,0,0,\ldots}(k)}{k^{\Sigma m}} \approx \frac{1}{k^{\Sigma m-2}} \quad .$$

From (4.2) the super-graph contributes

$$\frac{S'_{0,0,0\ldots}(k)}{k^{\Sigma m-1}} \quad ;$$

the m-lines give $1/k^{2\Sigma m}$. There are at most $(\sum m-1)$ new loops made by these m connecting lines, so that the over-all behaviour of the mixed graph is finite and given by:

$$\underset{M\to\infty}{\ell t}\int k^{-4\Sigma m+3}\,(d^4k)^{\Sigma m-1} \approx \frac{1}{M}$$

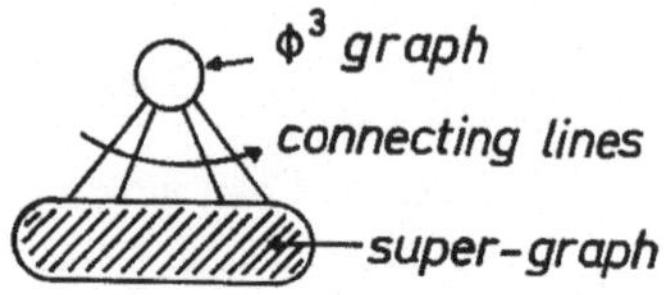

(Here we have assumed (see (4.3)) that

$$S'_{0,0,0\ldots}(k) \approx \frac{k^{4F_s-4N_s+4}}{k^{4F_s}} \approx k^{-4N_s+4} \quad .)$$

Note that our final result regarding high-energy behaviour for the total vacuum contribution of a non-polynomial ϕ^D Lagrangian is

$$\tilde{S} \approx \log M \quad \text{for } D = 2$$

$$\tilde{S} \approx M^2 \quad \text{for } D = 3 \quad .$$

For D = 3 , for example, although the pure super-graphs gave a finite contribution, the pure polynomial ϕ^3 graphs give the well-known behaviour M^2 of the ϕ^3-theory.

6. This is perhaps the stage at which one might remark on derivative couplings. Notice that when we power-count for super-graphs, in the naive count

$$\int \frac{k^{4F_s-4N_s+4}}{k^{4F_s}}$$

there appears the factor k^{-4N_s} . If each super-vertex carried derivatives up to fourth, the extra contribution would not exceed k^{4N_s} . Thus, provided that all sub-graph infinities (like those of star-fish graphs) could be consistently removed, up to fourth-order derivatives might be acceptable (with Lagrangians of the type $\partial^\alpha \phi^D$, $\alpha+D\leq 4$.)

7. The ghost-diagrams which have played such an essential role in the above analysis can always be associated with ghost-Lagrangians whenever we can write L in the form

$$\sum a_r \phi^r + \frac{\alpha_r}{1+\beta_r \phi} \quad .$$

When I-spin or unitary spin is present and terms like $(\partial_\mu \Phi . \Phi)$ are involved this is clearly impossible. We believe one can still carry through the ideas of ghost-graphs without writing corresponding ghost-Lagrangians; we hope to discuss this in detail elsewhere.

8. Let us now finally turn to weak interactions. This is the case of mixed fields and, as we shall see, here new types of infinities will arise and will need renormalizing. But before considering this difficult case, take a simple example of a mixed theory with two fields ϕ and B with L_{int} of the type:

$$L_{int} = B^p \frac{\phi^{D_o}}{1+f\phi^{D_o-D}} \quad .$$

At each of the N vertices there are p B-lines and (N-1) superlines. A necessary condition for renormalizability is clearly $p+D\leq 4$. (This is easily seen by writing

$$\frac{\phi^{D_o}}{1+f\phi^{D_o-D}} = \sum a_r \phi^r + \frac{\alpha_i}{1+f_i\phi} \quad .)$$

In what follows we concentrate on the case D=0, $p\leq 4$. All infinities come from the vacuum, self-energy and scattering graphs of the B-field, embedded inside super-graphs. For example, for $L = B^4 f(\phi)$ where $f(\phi) \approx \phi^o$,

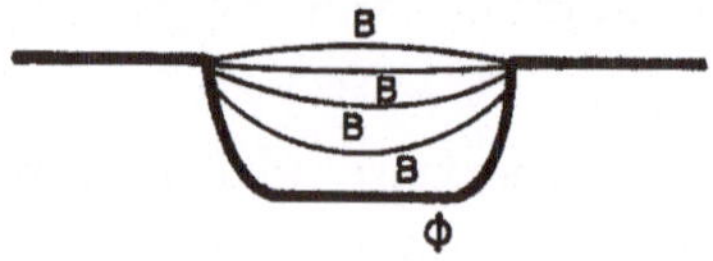

one may expect from the graph shown in the figure an overall singularity at worst of the type

$$\int (k^4) \times \frac{d^4k}{k^4} \approx M^4$$

(The factor k^4 is the contribution of the B-field vacuum graph; the factor $1/k^4$ is for the ϕ super-propagator.) This infinity needs a new variety of counter-term of the form

$$\delta L(x) = \int f(\phi(y))d^4y[c_o\delta(x-y)+c_1\partial^2\delta(x-y)+c_2\partial^4\ \delta(x-y)]f(\phi(x))$$

where c_o, c_1, c_2 are <u>infinite constants</u> of order M^4, M^2, log M , respectively. The important point is that the counter-Lagrangian which has the form

$$\delta L = d_1(\phi^o)+d_2\partial^2(\phi^o)+d_3\partial^4(\phi^o)$$

may itself produce singularities in its turn and, to absorb these, an exponential form of counter-term discussed before will be needed, but any counter-terms needed at any stage appear to fall within what we have called the renormalizable class.

The situation above is typically the weak interaction situation. Using a formalism involving the intermediate boson W , write the <u>minimal</u> weak Lagrangian:

$$L_W = J^+_\mu W^-_\mu + \text{h.c.}$$
$$L_{EM} = [J^{EM}_\mu + J^{EM}_\mu(W)]\ A^o_\mu$$

Here $J^\pm_\mu$ are the charged weak currents; $W^\pm_\mu$ are charged intermediate bosons. For fixing ideas one may assume that $J^\pm_\mu$ and J^{EM}_μ are currents made up of quarks (Q) and lepton fields (ℓ) and are of order M^3 .

For the W-fields themselves it is not essential, but it makes things very much easier, if we consider not just the two charged fields but a self-interacting triplet of Yang-Mills fields

$$L_{YM} = \hat{\underline{W}}_{\mu\nu} \cdot \hat{\underline{W}}_{\mu\nu} + m^2 \underline{W}_\mu^2$$

where

$$\hat{\underline{W}}_{\mu\nu} = (\partial_\nu W_\mu - \partial_\mu W_\nu + if\, \underline{W}_\mu \times \underline{W}_\nu) .$$

We now make a non-linear Stückelberg transformation on the $\underline{W}_\mu$ field variables; write $W_\mu = \underline{W}_\mu \cdot \underline{\tau}$ and introduce two fields A_μ and B by the relation

$$W_\mu = \Omega(B)\ A_\mu\ \Omega^{-1}(B) + \frac{i}{f}\, \Omega(B)\partial_\mu\ \Omega^{-1}(B) .$$

Here $\Omega(B)$ is a unitary matrix; (in Weinberg's representation one may write $\Omega(B)$ in the form $\frac{1-if\underline{\tau}\cdot\underline{B}}{1+if\underline{\tau}\cdot\underline{B}}$; $\Omega(B) \approx M^0$). Writing

$$\frac{im}{f}\, \Omega(B)\partial_\mu\ \Omega^{-1}(B) = X_\mu$$

the net effect on L_{YM} is to transform it to the form

$$L_{YM} = \hat{\underline{A}}_{\mu\nu} \cdot \hat{A}_{\mu\nu} + m^2\underline{A}_\mu^2 + 2m\underline{A}_\mu \cdot \underline{X}_\mu + \underline{X}_\mu \cdot \underline{X}_\mu .$$

There are corresponding changes in the interaction Lagrangians for both weak and E.M. cases. For example the new weak Lagrangian has the form

$$L = F^{(1)}(Q,\ell,A_\mu)\ f^{(1)}(B) + F_\mu^{(2)}(Q,\ell,A_\mu)\ f_\mu^{(2)}(B)$$
$$+ X_\mu(B)\ X_\mu(B)$$

$F^{(1)}$ is at most of order M^4 with $f^{(1)}(B) \approx B^0$ while $f_\mu^{(2)}(B)$ is at most of order $\partial_\mu(B)$ with $F_\mu^{(2)}\ f_\mu^{(2)}(B) \approx M^4$.

Now comes the important point. Boulware [6] has shown

that this non-linear Stückelberg analysis gives, for the A and B fields, propagators which are perfectly normal (i.e. are no more singular than $\Delta(x)$) and the S-matrix is unitary provided the Lagrangian is supplemented by an additional term of the form $\bar{\underline{F}} \times \partial_\mu \underline{F} \cdot A_\mu$ where the triplets of $\underline{F}$ represent "fictitious" particles first introduced by Feynman. From the point of view of renormalizability all we need to know is that in our power counts, $A \sim M$ and $B \sim M$, while $\Omega \gtrsim M^0$ and $X_\mu \gtrsim \partial_\mu (M^{-1})$. Clearly L_W falls within the category of renormalizable interactions tabulated in this discussion, and so does the additional Lagrangian for the fictitious particles.

We have not written out in detail all the counter-terms, nor is it interesting for anyone undertaking any practical calculations. As in most renormalization theory, what is important is the existence theorem - the statement that it can be done. We expect the practical rules for writing S-matrix elements to be:

1) Replace W_μ by the Stückelberg field A_μ.
2) For closed loops of W_μ fields, introduce Feynman's fictitious particles to preserve unitarity.
3) Add to the contributions above, super-graph contributions involving B-particles. These will need renormalizing. (In practice, knowing that these super-graphs, after renormalization, are finite, one may as a first approximation neglect these B-particle contributions. One may be certain that unitarity is preserved with just the contributions 1) and 2) .)

All this was on the assumption that we do not wish to modify the basic weak Lagrangian but wish to start with what we have called the minimal Lagrangian. There is no reason why one may not modify the weak Lagrangian itself such that its Dyson index is less than two and it produces no infinities whatsoever. This is what Mitter and Fivel [2] have done.

5. SUMMARY

We summarize the situation regarding non-polynomial Lagrangians: I should make the qualification that an enormous amount of verification is needed before the problems of renormalizability are all sorted out, but one may tentatively state:

1) All matrix elements are finite for theories where the Dyson index D is less than two.

2) For the cases when D=2 or 3, counter-terms have been explicitly written which absorb all infinities and the theories are renormalizable.

3) Mixed theories of polynomial and non-polynomial fields appear to be renormalizable provided the Dyson indices separately and jointly fulfill renormalizability criteria. We believe that weak interactions, chiral Lagrangians and Yang-Mills theory fall into this class though detailed proofs have not yet been constructed.

4) It seems likely that to each order in the major coupling (and to all orders in the minor coupling) the S-matrix elements, as computed by methods outlined, satisfy the necessary unitarity and analyticity requirements.

5) The real parts of the physical amplitudes (the parts not restricted by unitarity) are non-unique. This appears to be similar to the type of non-uniqueness one meets in conventional renormalization theory for polynomial cases, i.e. arbitrariness up to finite renormalizations. Unitarity requirements restrict this lack of uniqueness though they do not completely eliminate it. If one imposes on the theory the criterion that all such extra terms must be represented by (finite) modifications to the starting Lagrangian - and with no derivatives of arbitrarily high order appearing - no arbitrariness remains.

6) Non-polynomial theories give perfectly acceptable high-energy behaviour in the Symanzik region and where external momenta are space-like or on the mass shells. For time-like momenta, however, the lowest order in the major coupling constant gives cross-sections increasing arbitrarily fast with energy. If now a simple chain diagram is summed, or a Regge ladder summation carried out, the results alter drastically. Alternatively and perhaps equivalently, if in the Symanzik region one computed K-matrix elements and used the expression $S=(1-iK)/(1+iK)$ to continue to time-like momenta, the exponential growth would not survive. (It is important to realize that in the Symanzik region T_N (the T-matrix element to order G^N in the major coupling constant) equals K_N . This very powerful method will be elaborated on elsewhere.) This appears in line with a general result recently claimed by Fradkin and Feinberg (unpublished) where they assert axiomatic CPT, spin and statistics and polynomial boundedness in energy for theories of the type we have considered [7]. If this result holds and if a reliable summation technique in the major coupling can be devised, the last major objection to these theories would disappear. This is because if one extrapolates the results of Jaffe, Glimm, Hepp and others for polynomial Lagrangians in two dimensions to those in four dimensions there is no hope of obtaining finite self-masses and finite self-charges. We must turn to Lagrangians described in this paper if we are ever to compute these constants finitely and to make acceptable statements regarding phenomena like Goldstone bosons and symmetry breaking through the graphs for vacuum expectation values of scalar fields.

REFERENCES

1) S. Okubo, Progr. Theoret. Phys. (Kyoto) 11, 80 (1954).

2) R. Arnowitt and S. Deser, Phys. Rev. 100, 349 (1955);
E. S. Fradkin, Nucl. Phys. 49, 624 (1963);
G. V. Efimov, Soviet Phys. - JETP 17, 1417 (1963);
G. Feinberg and A. Pais, Phys. Rev. 131, 2724 (1963);
W. Güttinger, Fortschr. Phys. 14, 483 (1966);
M. K. Volkov, Ann. Phys. (N.Y.) 49, 202 (1968);
H. M. Fried, Nuovo Cimento 52A, 1333 (1967);
B. W. Lee and B. Zumino, CERN, preprint TH.1053 (1969);
D. I. Fivel and P. K. Mitter, "A theory of weak interactions without divergences", Univ. of Maryland preprint UMD-70-029 (1969);
R. Delbourgo, Abdus Salam and J. Strathdee, ICTP, Trieste, preprint IC/69/17, to appear in Phys. Rev. (and on renormalizability, in preparation).
J. Boyce and J. Sultoon, "Form factors in non-polynomial theories", Imperial College (in preparation).
K. Koller, A. P. Hunt and Q. Shafi, "Self-masses in chiral Lagrangian theories", Imperial College (in preparation).

3) G. V. Efimov, CERN, preprint 1087, Oct. 1969;
Abdus Salam and J. Strathdee, ICTP, Trieste, preprint IC/69/120, to appear in Phys. Rev.

4) S. Fells (UCLA preprint) has shown that the vacuum self-energy of ϕ^3-like non-polynomial Lagrangians which are odd in powers of ϕ (e.g., $\phi/(1+f^2\phi^2)$) do not possess a lower bound. Presumably ϕ^4 - like even theories do not suffer from this objection. Our use of a mixed power theory $1/(1+f\phi)$ is purely to illustrate the necessary techniques. Physical theories (like the chiral theory) are even in field variables.

5) J. C. Polkinghorne, J. Math. Phys. 4, 503 (1963);
P. G. Federbush and M. T. Grisaru, Ann. Phys. (N.Y.)

22, 263, 299 (1963);
J. S. R. Chisholm, Proc. Cambridge Phil. Soc. 48, 300 (1952)

6) D. Boulware, Seattle preprint (1969).

7) While this was being printed, Dr. O. V. Steinmann (Zürich) has sent a preprint which shows that, for non-polynomial theories, the scattering matrix exists in an axiomatic formulation and also that the LSZ reduction procedure can be carried through even though, for rational non-polynomial theories, strict local commutativity may possibly not hold. It appears that the quiet revolution which has been taking place with non-polynomial physics may acquire respectability yet.

FOOTNOTES

1 Ambiguities arise if instead of the p.v. we consider the more general real combination

$$(\frac{1}{2} + ib)F(\Delta, f^2+i\varepsilon) + (\frac{1}{2} - ib)F(\Delta, f^2-i\varepsilon) \ .$$

The result differs from the principal value integral by a purely real term of the form $b \exp(1/f^2\Delta)$ which possesses everywhere a zero expansion around $\Delta=0$, and which, when added to the p.v., does not affect its perturbation representation: $\sum_{n=0}^{\infty} n!(f^2\Delta)^n$.

The Fourier transform of this additional term is analytic in the entire p^2-plane so that in this order unitarity places no restriction on it. Higher-order unitarity, however, does seem to restrict such ambiguous terms. In Ref.3 it is argued that once the constant b is defined in the second-order super-propagator (see also footnote 3) the same constant or its multiples appear in all higher orders.

[2] The basic reason why the Fourier transforms of the Borel sums possess the correct unitarity cuts has been spelled out by Lee and Zumino. While the infinities come from small r values of $(\Delta(r) \underset{r\to o}{\approx} 1/r^2)$ of the propagators, the unitarity (singularity and threshold) structure arises from large values of r

$$\left(\Delta(r) \underset{r \to \infty}{\approx} e^{-mr}/r^{3/2}\right) .$$

Once it can be shown that (2.4) is an asymptotic representation of the perturbation expression (2.3) for large r, the correct unitarity behaviour of (2.4) is guaranteed.

[3] The principal value ambiguity of the x-space method noted in footnote 1 has a counterpart when we take into account the appearance of the (-) sign in front of Δ in $(-\Delta)^z$ in the Sommerfeld-Watson transform. To see this more explicitly, introduce a multiplier λ in front of Δ ; thus

$$F(\lambda\Delta) = \frac{i}{2} \int \frac{dz}{\sin\pi z} \int d\zeta\, e^{-\zeta} (-\zeta\lambda f^2\Delta)^z .$$

We must interpret the limit $\lambda \to +1$ by a <u>real</u> average of the values $(-\lambda)^z = e^{i\pi z}$ and $(-\lambda)^z = e^{-i\pi z}$, obtaining in general:

$$F(\Delta) = \int dz \left(\frac{1}{\tan\pi z} + b\right) \Gamma(z+1) (f^2\Delta)^z$$

with b an arbitrary real constant. This ambiguity of the constant b parallels the ambiguity noted in footnote 1. As noted in footnote 1, from unitarity one can show that all ambiguous constants arising in higher orders are multiples of this second-order b .

[4] For z = 1 we recover Feynman's formula for normal propagators.

Acta Physica Austriaca, Suppl. VII, 32–70 (1970)

INTRODUCTION TO THE USE OF NON-LINEAR TECHNIQUES IN S-MATRIX THEORY[*]

BY

D. ATKINSON

CERN-Geneva

CONTENTS

1. INTRODUCTION

I am going to explain to you how one can tackle certain problems in S-matrix theory that involve non-linear functional equations. A physicist's usual reaction to a non-linear equation of this kind would be either to try to get an approximate solution by iteration, or to introduce a linearization, perhaps in the neighbourhood

---

[*] Lecture given at IX. Internationale Universitätswochen für Kernphysik, Schladming, February 23 - March 7,1970

of a known approximate solution. I will introduce some concepts of Banach space analysis [1], which will enable us to put these intuitive ideas on a rigorous basis. The advantage is that one can sometimes prove the existence of solutions of the exact equations, without any approximations. For almost all of these talks, I will limit myself to the Contraction Mapping Principle, which is perhaps the simplest technique available, and corresponds precisely to trying to find a function, say $\Phi(x,y,\ldots)$, that satisfies a non-linear functional equation,

$$\Phi(x,y,\ldots) = P[\Phi;x,y,\ldots] \qquad (1.1)$$

by means of an iteration

$$\Phi_{n+1}(x,y,\ldots) = P[\Phi_n;x,y,\ldots] \quad . \qquad (1.2)$$

I will develop the proof that, when certain "contraction" conditions are observed, then, not only does the iteration (1.2) converge in a well-defined sense, but the limit-function satisfies the exact equation (1.1) . Moreover, we will be able to estimate the error involved in stopping the iteration after N steps, which will clearly be very useful, since in practice one always has to truncate an iteration, if only because someone else wants to use the computer.

I will first of all apply the technique to a problem that has been considered by Martin [2], namely, under what circumstances does a measurement of the differential cross-section, in the elastic region, serve to determine the phase-shifts uniquely, via the elastic unitarity condition. It turns out that, if the modulus of the amplitude satisfies a certain explicit condition, then its phase is uniquely determined (except for an overall sign). I will give a slightly simplified version of part of Martin's proof.

In this application, one can use a Banach space of

continuous functions; but when the non-linear equations also involve principal-value integrals, one needs to use a space of Hölder-continuous functions. I will introduce this space and apply it to the pion-pion equations in the Shirkov [3] approximation. My proof will be similar in some ways to Warnock's [4] treatment of the Low equation, except that I will consider the equation with no subtractions, since this simplifies matters. In particular, I will need no cut-off.

Then I will consider the exact Mandelstam [5] equations for pion-pion scattering; and I will explain the simplest version of the existence proof I [6] worked out two years ago. Again Hölder-continuous functions are used, but this time with respect to two variables.

Lastly, I propose to sketch the progress that has been made with generalizations of this last proof, in particular the introduction of subtractions [7] and CDD poles [8]. I will also mention the possible use of the Newton-Kantorovich method, which is the rigorous way to linearize in the vicinity of an approximate solution. I will finish by indicating a few outstanding problems, which may, or may not be tractable.

2. CONTRACTION MAPPING PRINCIPLE

A Banach space is a complete, normed, linear metric space. We will always be talking about Banach spaces in which the elements or "points" are functions, either of one or of two variables. That is, each of our spaces will consist in a set of functions that satisfy certain specific properties that are characteristic of the space in question. The norm of a function, $\Phi(x)$, is a number that is associated with $\Phi(x)$, and is written $\|\Phi\|$. This assignment of numbers (norms) to the functions cannot be done in a completely arbitrary way, but must be such that, for any Φ and Ψ belonging to the space, the following properties hold good:

$$||\Phi|| \geq 0 \tag{2.1}$$

$$||\Phi|| = 0 \quad \text{if and only if } \Phi(x) \equiv 0 \tag{2.2}$$

$$||\Phi+\Psi|| \leq ||\Phi|| + ||\Psi|| \tag{2.3}$$

The requirement of linearity means that, for any real (or complex) number, λ,

$$||\lambda \ \Phi|| = |\lambda| \ ||\Phi|| \tag{2.4}$$

To say that the normed space is complete is to assert that every Cauchy sequence of functions in the space converges to a function that belongs to the space. To say that $\{\Phi_n\}$ is a Cauchy sequence means that, for any $\varepsilon>0$, one can find an N such that

$$||\Phi_{p+n} - \Phi_p|| < \varepsilon \tag{2.5}$$

for any $p \geq N$, and $n=1,2,3,\ldots$. If the space is complete, then there necessarily exists a limit function, Φ, which belongs to the space. That is, there is a Φ such that

$$||\Phi - \Phi_p|| < \varepsilon \tag{2.6}$$

for any $p \geq N$. It is these two properties of having a linear norm structure, and completeness that make the use of Banach spaces indispensable in functional analysis.

The contraction mapping principle, specialized to a Banach space, can be stated as follows: Suppose that the non-linear operator, P, maps a complete set in the space into itself, and that Φ and Ψ are any two "points" belonging to this set, with

$$\Phi' = P(\Phi) \tag{2.7}$$

$$\psi' = P(\psi) \tag{2.8}$$

If

$$\|\Phi' - \Psi'\| \leq k\|\Phi - \Psi\| \tag{2.9}$$

where $k<1$, then the equation

$$\Phi = P(\Phi) \tag{2.10}$$

has a unique solution in the set in question, which may be obtained by iteration

$$\Phi_{n+1} = P(\Phi_n) \tag{2.11}$$

so long as Φ_o belongs to the set.

This principle, which I will prove in a moment, is just the common sense statement that an iteration converges if successive steps get smaller and smaller. I want to draw your attention to the fact, however, that it is crucial that the set be complete in the first place. The technique of proof is in fact to show that the sequence $\{\Phi_n\}$, defined by eq. (2.11), is Cauchy. For

$$\|\Phi_{n+p} - \Phi_p\| \leq \sum_{m=p}^{n+p-1} \|\Phi_{m+1} - \Phi_m\| \tag{2.12}$$

by eq. (2.3). Now from eq. (2.11) and (2.7) - (2.9),

$$\|\Phi_{m+1} - \Phi_m\| \leq k\,\|\Phi_m - \Phi_{m-1}\| \tag{2.13}$$

so that, by iteration,

$$\|\Phi_{m+1} - \Phi_m\| \leq k^m\,\|\Phi_1 - \Phi_o\| \ . \tag{2.14}$$

Hence, from eq. (2.12),

$$\|\Phi_{n+p} - \Phi_p\| \leq \|\Phi_1 - \Phi_o\| \sum_{m=p}^{n+p-1} k^m \leq \|\Phi_1 - \Phi_o\| \frac{k^p}{1-k} \ . \tag{2.15}$$

Since $k<1$, it follows that, for any pre-assigned $\varepsilon>0$, one can certainly choose p so that the right-hand side of eq. (2.15) is smaller than ε, so that $\{\Phi_n\}$ is a Cauchy sequence, and hence has a limit, say Φ. It is easy to see, by letting $n\to\infty$ in eq. (2.15), that

$$\|\Phi - \Phi_p\| \leq \|\Phi_1 - \Phi_o\| \frac{k^p}{1-k} \ . \tag{2.16}$$

This is a useful inequality, since it gives a bound on the error committed by stopping the iteration after p steps.

In order to finish the proof in the tidy way that mathematicians like, we should show (a) that the limit function, Φ, really satisfies eq. (2.10), and (b) that it is the only function (within the complete set in question) that does so. In view of eq. (2.16), given any $\varepsilon > 0$, we can certainly find a p such that

$$\|\Phi - \Phi_m\| < \varepsilon/2 \qquad (2.17)$$

for all $m \geq p$. Then

$$\|P(\Phi)-\Phi\| \leq \|P(\Phi)-P(\Phi_p)\| + \|P(\Phi_p)-\Phi\| \leq k\|\Phi-\Phi_p\| + $$
$$+ \|\Phi_{p+1} - \Phi\| < k\,\frac{\varepsilon}{2} + \frac{\varepsilon}{2} < \varepsilon \quad . \qquad (2.18)$$

Since ε can be as small as one likes, one must have

$$\|P(\Phi) - \Phi\| = 0 \qquad (2.19)$$

from which eq. (2.10) follows, by virtue of property (2.2).

Lastly, one can prove, by reductio ad absurdum, that Φ is locally unique. For suppose, on the contrary, that there were two different functions, Φ and Ψ, belonging to the complete set, such that

$$\Phi = P(\Phi) \qquad (2.20)$$

and

$$\Psi = P(\Psi) \quad . \qquad (2.21)$$

According to the contraction condition, eq. (2.9),

$$\|\Phi-\Psi\| \leq k\,\|\Phi-\Psi\| \quad . \qquad (2.22)$$

Since $\Phi-\Psi$ is not identically zero, it follows from eq.(2.1) and (2.2) that

$$\|\Phi - \Psi\| > 0 \quad . \qquad (2.23)$$

Hence eq. (2.22) implies

$$k \geq 1 \quad . \tag{2.24}$$

This is absurd, since one knows that $k < 1$.

3. SPACE OF CONTINUOUS FUNCTIONS

A simple, and often very useful Banach space is the set of all continuous functions, $\Psi(x)$, $-1\leq x\leq 1$, with the norm

$$||\Psi|| = \sup_{-1\leq x\leq 1} |\Psi(x)| \quad . \tag{3.1}$$

It is easy to check eq. (2.1) - (2.4). To show that the space is complete, let $\{\Psi_n(x)\}$ be a Cauchy sequence in the space. For a given, fixed x, one knows, by the Bolzano-Weierstrass theorem, that $\Psi_n(x)$ tends to a limit, that may be called $\Psi(x)$, as $n\to\infty$. It has to be shown that $\Psi(x)$ is continuous, and so belongs to the space. Now, for any n,

$$|\Psi(x_1)-\Psi(x_2)| \leq |\Psi(x_1)-\Psi_n(x_1)| + |\Psi(x_2)-\Psi_n(x_2)| + |\Psi_n(x_1)-\Psi_n(x_2)| \quad . \tag{3.2}$$

Given any $\varepsilon>0$, one can certainly choose n so large that the first two terms on the right-hand side of eq. (3.2) are each less than $\varepsilon/3$. One can find a δ so small that

$$|\Psi_n(x_1) - \Psi_n(x_2)| < \varepsilon/3 \tag{3.3}$$

for all $|x_1-x_2| \leq \delta$, since $\Psi_n(x)$ is continuous. Hence

$$|\Psi(x_1) - \Psi(x_2)| < \varepsilon \tag{3.4}$$

for all $|x_1-x_2| \leq \delta$, which means that $\Psi(x)$ is continuous.

This space has been used by Martin to tackle the following problem: Suppose that you know the modulus, B , of a two-particle elastic scattering amplitude, in the elastic region (for example from a measurement of the differential scattering cross-section). Under what circumstances does the elastic unitarity condition serve to define uniquely the phase, Φ , of the scattering amplitude, and hence to determine uniquely the phase -shifts? The unitarity condition can be written

$$B(z)\ \sin\Phi(z) = \tag{3.5}$$
$$= \frac{1}{4\pi}\int_{-1}^{1} dz_1 \int_{0}^{2\pi} d\phi_1\ B(z_1)B(z_2)\exp\{i[\Phi(z_1)-\Phi(z_2)]\}$$

where z is the cosine of the scattering angle, and where the dependence on the energy has been suppressed. In eq. (3.5), z_2 is to be considered as a function of z, z_1 and ϕ_1, according to

$$z_2 = z\ z_1 + (1-z^2)^{\frac{1}{2}}(1-z_1^2)^{\frac{1}{2}} \cos\ \phi_1 \quad . \tag{3.6}$$

Since B(z) is known, eq. (3.5) is to be regarded as an equation for the unknown $\Phi(z)$. Under what conditions on B is there a unique solution? We will apply the contraction mapping principle in the space of continuous functions on the domain $-1 \leq z \leq 1$.

The equation (3.5) can be rewritten

$$\Phi(z) \equiv P[\Phi;z] = \tag{3.7}$$
$$= \sin^{-1}\{\int d\Omega_1\ H(z,z_1,\phi_1)\cos[\Phi(z_1)-\Phi(z_2)]\}$$

where

$$H(z,z_1,\phi_1) = \frac{B(z_1)B(z_2)}{4\pi\ B(z)} \tag{3.8}$$

and where the symmetry of the integral (3.5) has been used to eliminate the imaginary part. Suppose that H is such that

$$\int d\Omega_1\ H(z,z_1,\phi_1) \leq \sin\mu < 1 \tag{3.9}$$

where $0<\mu<\frac{\pi}{2}$. This imposes a restriction on B(z). From eq. (3.7), it follows that

$$|\Phi(z)| \leq \sin^{-1}\{\int d\Omega\ H(z,z_1,\phi_1)\} \leq \mu\ . \qquad (3.10)$$

Let Φ_{max} and Φ_{min} be the maximum and minimum values of $\Phi(z)$. If we define $\Phi(o)$ to lie between $-\frac{\pi}{2}$ and $\frac{\pi}{2}$, then, since we are looking for a solution of eq. (3.7) that is continuous, it follows from (3.10) that

$$-\frac{\pi}{2} < \Phi_{min} \leq \Phi_{max} < \frac{\pi}{2} \qquad (3.11)$$

Hence

$$0 \leq \Phi_{max} - \Phi_{min} < \pi \qquad (3.12)$$

I will now show, following Martin, that in fact eq. (3.10) can be strengthened to

$$0 \leq \Phi(z) \leq \mu\ . \qquad (3.13)$$

Consider two cases: either

$$0 \leq \Phi_{max} - \Phi_{min} \leq \frac{\pi}{2}\ , \qquad (3.14)$$

in which case $\cos[\Phi(z_1)-\Phi(z_2)]$ in eq. (3.7) can never be negative, so that $\Phi(z) \geq o$ for all z. On the other hand, if

$$\frac{\pi}{2} < \Phi_{max} - \Phi_{min} < \pi\ , \qquad (3.15)$$

then $\cos[\Phi(z_1)-\Phi(z_2)]$ could apparently be negative for some values of z_1 and z_2, but eq. (3.7) implies that

$$\sin\Phi_{min} \geq \int d\Omega_1\ H(z,z_1,\phi_1)\cos[\Phi_{max} - \Phi_{min}] \geq \qquad (3.16)$$

$$\geq \sin\mu[\cos\Phi_{max}\cos\Phi_{min} + \sin\Phi_{max}\sin\Phi_{min}]$$

since $\cos[\Phi_{max} - \Phi_{min}]$ is negative, by assumption (3.15). Hence

$$\sin\Phi_{min} \geq \frac{\sin\mu\ \cos\Phi_{max}\cos\Phi_{min}}{1 - \sin\mu\ \sin\Phi_{max}} \geq \frac{\sin\mu\ \cos^2\mu}{1 - \sin^2\mu} = \sin\mu \qquad (3.17)$$

This contradicts eq. (3.10), so in fact the apparent alternative (3.15) is disallowed.

This result can be rephrased as follows: Consider

the mapping

$$\Phi'(z) = P[\Phi(z);z] \tag{3.18}$$

where P is defined in eq. (3.7). What has been shown is that if eq. (3.9) is satisfied, then the set in the Banach space of continuous functions that is defined by

$$0 \leq \Phi(z) \leq \mu \tag{3.19}$$

is mapped into itself by P . In technical terms, this set would be described as the intersection of the ball

$$||\Phi|| \leq \mu \tag{3.20}$$

with the norm of eq. (3.1), and the cone

$$\Phi(z) \geq 0 \quad . \tag{3.21}$$

A cone in a Banach space is a set such that if Φ belongs to it, then so does $c\Phi$, where c is any nonnegative real number.

If one were to iterate eq. (3.18), according to

$$\Phi_{n+1} = P[\Phi_n] \tag{3.22}$$

for $n = 0,1,2,\ldots$, with $\Phi_o(z)$ satisfying (3.19), it is clear that, for any $\sin\mu < 1$, the infinite set of iterates all satisfy eq. (3.19). Martin has shown that these iterates have at least one limit-point in the set (3.19), but the proof involves the Schauder principle, which I do not intend to explain in this course. If $\sin\mu$ is substantially smaller than unity, one can use the contraction mapping principle to show that there is one, and only one limit-point in the space of continuous functions. I will explain this to you.

Suppose that $\Phi^a(z)$ and $\Phi^b(z)$ are any two continuous functions that satisfy the inequality (3.19). Then it follows from eq. (3.18), by a series of elementary trigonometric manipulations, that

$$\sin \tfrac{1}{2}[\Phi'^{a}(z)-\Phi'^{b}(z)] = 2 \sec \tfrac{1}{2}[\Phi'^{a}(z)-\Phi'^{b}(z)] \times$$

$$\times \int d\Omega_1 \; H(z,z_1,\phi_1) \sin \tfrac{1}{2}[\Phi^{a}(z_1)-\Phi^{b}(z_1)] \times$$

$$\times \quad \cos \tfrac{1}{2}[\Phi^{a}(z_2)-\Phi^{b}(z_2)] \sin\tfrac{1}{2}[\Phi^{a}(z_1) +$$

$$+ \quad \Phi^{b}(z_1)-\Phi^{a}(z_2)-\Phi^{b}(z_2)] \tag{3.23}$$

The inequality (3.19), which has been shown to hold also for Φ', implies

$$|\sin\tfrac{1}{2}[\Phi'^{a}(z)-\Phi'^{b}(z)]| \le \frac{2\sin^2\mu}{\cos\mu} \sup_{-1\le z_1\le 1} [\sin\tfrac{1}{2} \Phi^{a}(z_1)- -\Phi^{b}(z_1)]| \tag{3.24}$$

Since

$$x\ge \sin x \ge \frac{2x}{\pi} \tag{3.25}$$

for $0\le x\le \frac{\pi}{2}$, it follows from eq. (3.24) that

$$|\Phi'^{a}(z)-\Phi'^{b}(z)| \le \frac{\pi\sin^2\mu}{\cos\mu} \sup_{-1\le z\le 1} |\Phi^{a}(z)-\Phi^{b}(z)| \tag{3.26}$$

or

$$||\Phi'^{a}-\Phi'^{b}|| \le \frac{\pi\sin^2\mu}{\cos\mu} ||\Phi^{a} - \Phi^{b}|| \tag{3.27}$$

The condition for a contraction mapping is accordingly

$$\frac{\pi\sin^2\mu}{\cos\mu} < 1 \tag{3.28}$$

or

$$\sin\mu < [\frac{2}{1+(1+4\pi^2)^{1/2}}]^{1/2} \tag{3.29}$$

The bound (3.29) imposes a restriction on the magnitude of B(z), the modulus of the amplitude, for the applicability of the above contraction mapping proof of the existence and uniqueness of a solution of the equation (3.7). One can extend the domain of the proof by remarking that

$$||\Phi||_1 \equiv \sup_{-1\le z\le 1} |\sin\tfrac{1}{2}\Phi(z)| \tag{3.30}$$

can be used as an alternative norm. One has to check the triangle inequality, eq. (2.3), but this is easily done (exercise). With this norm, one has, directly from eq. (3.24),

$$\|\phi'^{a}-\phi'^{b}\|_1 \leq \frac{2\sin^2\mu}{\cos\mu}\,\|\phi^{a}-\phi^{b}\|_1 \qquad (3.31)$$

and this leads to the requirement

$$\sin\mu < \left[\frac{2}{1+(17)^{1/2}}\right]^{1/2} \approx 0.62 \qquad (3.32)$$

which is an improvement.

By including in longer trigonometrical manipulations, Martin has managed to make the uniqueness proof work for

$$\sin\mu < 0.79 \quad . \qquad (3.33)$$

I will not go into his proof, which would take us too far afield, without introducing any new point of principle. I can refer you to his paper if you are interested. Incidentally, there is an outstanding problem: there is some reason to expect that one should have uniqueness for any

$$\sin\mu < 1 \qquad (3.34)$$

but no-one has been able to bridge the gap between 0.79 and 1.00 . I leave it as an exercise for the student to extend the proof to 1.00, or to find a counter-example. In either case, tell André Martin immediately!

4. PRINCIPAL-VALUE INTEGRALS

When one has to deal with mappings that involve principal-value integrals, one can no longer use the space of continuous functions, because the principal-value integral of a continuous function is not necessarily continuous. I will show you how to prove that the principal-value integral of a Hölder-continuous function is itself Hölder continuous. Then we will construct a Banach space of Hölder-continuous functions, in which we will use the contraction mapping theorem again.

The theorem I will prove is as follows: Suppose that

$$f(x) = \frac{P}{\pi}\int_0^1 \frac{dx'\ \sigma(x')}{x'-x} \quad , \tag{4.1}$$

where $\sigma(x)$ satisfies

$$\sigma(o) = 0 = \sigma(1) \tag{4.2}$$

and

$$|\sigma(x_1)-\sigma(x_2)| \le \xi |x_1-x_2|^\mu \tag{4.3}$$

for any x_1, x_2 in the interval [0,1] where ξ is constant, and where μ satisfies $0<\mu<1$. Eq. (4.3) is the statement of Hölder continuity. Then we will prove that

$$|f(x_1)-f(x_2)| \le c\xi |x_1-x_2|^\mu \tag{4.4}$$

for any x_1, x_2 in [0,1], where c depends only on the Hölder index, μ.

One has to be a little bit careful about the end points of the integration range, in order to avoid logarithm singularities. It is for this reason that one needs eq. (4.2). One can work the proof most elegantly by extending formally the integration range in eq. (4.1) to

$$f(x) = \frac{P}{\pi}\int_{-2}^{2} \frac{dx'\ \sigma(x')}{x'-x} \tag{4.5}$$

by defining $\sigma(x')=0$ for $-2\le x'\le 0$ and $1\le x'\le 2$. Because of eq. (4.2), one can extend the Hölder-continuity (4.3) over the whole range $-2\le x_1,x_2\le 2$. For example, suppose $0\le x_1\le 1$ and $-2\le x_2\le 0$. Then

$$\begin{aligned} |\sigma(x_1)-\sigma(x_2)| &= |\sigma(x_1)-0| = \\ &= |\sigma(x_1)-\sigma(o)| \le \xi|x_1-0|^\mu \le \xi|x_1-x_2|^\mu \ , \end{aligned} \tag{4.6}$$

and similarly for the other possibilities. Note the use of eq. (4.2) in the second line. Although x' now ranges over [-2,2] , x is still restricted to the range [0,1] in eq. (4.5).

One has

$$f(x) = \frac{1}{\pi}\int_{-2}^{2} dx' \, \frac{\sigma(x')-\sigma(x)}{x'-x} + \frac{\sigma(x)}{\pi} \, P\int_{-2}^{2} \frac{dx'}{x'-x} \tag{4.7}$$

so that

$$|f(x_1)-f(x_2)| \le B_1 + B_2 \tag{4.8}$$

where

$$B_1 = \frac{1}{\pi}\left|\int_{-2}^{2} dx'\left\{\frac{\sigma(x')-\sigma(x_1)}{x'-x_1} - \frac{\sigma(x')-\sigma(x_2)}{x'-x_2}\right\}\right| \tag{4.9}$$

and

$$B_2 = \frac{1}{\pi}\left|\sigma(x_1)\log\frac{2-x_1}{2+x_1} - \sigma(x_2)\log\frac{2-x_2}{2+x_2}\right| . \tag{4.10}$$

We have to tackle B_1 and B_2 in turn, and show that each is less than $|x_1-x_2|^{\mu}$, multiplied by a constant. Consider B_1 first, and suppose $x_2 \ge x_1$ for definiteness. Define $\theta = x_2-x_1$. We have a delicate piece of engineering to do. We will divide the integral (4.9) into the interval $x_1-2\theta \le x' \le x_1+2\theta$, which we might call Ω , and the rest, called $\bar{\Omega}$. I will leave you to check that $x_1+2\theta \ge x_2$ so that we have both "Cauchy points", $x'=x_1$ and $x'=x_2$, inside the interval Ω . Also you can check that $x_1-2\theta \ge -2$ and $x_1+2\theta \le 2$, which is why I extended the integration range as far as I did. Write

$$B_1 \le B_{11} + B_{12} + B_{13} \tag{4.11}$$

where

$$B_{11} = \frac{1}{\pi}\int_{\Omega} dx'\left\{\left|\frac{\sigma(x')-\sigma(x_1)}{x'-x_1}\right| + \left|\frac{\sigma(x')-\sigma(x_2)}{x'-x_2}\right|\right\} \tag{4.12}$$

$$B_{12} = \frac{1}{\pi}\left|\int_{\bar{\Omega}} dx'\{\sigma(x')-\sigma(x_1)\}\left\{\frac{1}{x'-x_1} - \frac{1}{x'-x_2}\right\}\right| \tag{4.13}$$

and

$$B_{13} = \frac{1}{\pi}\left|\int_{\bar{\Omega}} \frac{dx'}{x'-x_2}\{[\sigma(x')-\sigma(x_1)]-[\sigma(x')-\sigma(x_2)]\}\right| . \tag{4.14}$$

So far as B_{11} is concerned, we use the Hölder continuity directly to yield

$$B_{11} \le \frac{\xi}{\pi} \int_{\Omega} dx' \{ |x'-x_1|^{-1+\mu} + |x'-x_2|^{-1+\mu} \} =$$

$$= \frac{\xi}{\pi\mu} [1+2^{1+\mu} + 3^{\mu}] |x_1-x_2|^{\mu} \quad , \tag{4.15}$$

so that piece has the right form. For B_{12} one can be a bit more brutal, because the integrand has no singularity:

$$\begin{aligned} B_{12} &\le \frac{1}{\pi} \int_{\Omega} dx' |\sigma(x')-\sigma(x_1)| \left| \frac{x_1-x_2}{(x'-x_1)(x'-x_2)} \right| \le \\ &\le \frac{|x_1-x_2|\xi}{\pi} \int_{\bar{\Omega}} dx' |x'-x_1|^{-1+\mu} |x'-x_2|^{-1} \le \\ &\le \frac{2^{\mu}\,\xi}{\pi(1-\mu)} |x_1-x_2|^{\mu} \quad , \end{aligned} \tag{4.16}$$

which is nice again. Finally, B_{13} is easy:

$$B_{13} \le \frac{1}{\pi} |\sigma(x_1)-\sigma(x_2)| \left| \int_{\bar{\Omega}} \frac{dx'}{x'-x_2} \right| \le \frac{\xi}{\pi} \log 3 |x_1-x_2|^{\mu} \quad , \tag{4.17}$$

thus completing the hat-trick (for those of you conversant with gaming, or cricket jargon).

We still have to perform upon B_2, but this presents no difficulty. From eq. (4.10),

$$B_2 \le B_{21} + B_{22} \quad , \tag{4.18}$$

where

$$B_{21} = \frac{1}{\pi} |\sigma(x_1)-\sigma(x_2)| \log \frac{2+x_1}{2-x_1} \quad , \tag{4.19}$$

and

$$B_{22} = \frac{1}{\pi} |\sigma(x_2)| \{ \log(1+\frac{x_2-x_1}{2+x_1}) + \log(1+\frac{x_2-x_1}{2-x_2}) \} \; . \tag{4.20}$$

Clearly

$$B_{21} \le \frac{\xi}{\pi} \log 3 |x_1-x_2|^{\mu} \tag{4.21}$$

since log 3 is the largest value $\log(2+x_1/2-x_1)$ can have, for $0 \le x_1 \le 1$. I extended the range of x' up to two precisely to avoid the logarithmic divergence that we would have had otherwise. From eq. (4.2) and (4.3),

$$|\sigma(x_2)| = |\sigma(x_2)-\sigma(o)|$$
$$\leq \xi\, x_2^{\mu}$$
$$\leq \xi \qquad ; \tag{4.22}$$

and since

$$\log(1 + A) \leq A , \tag{4.23}$$

for $A \geq 0$, it follows that

$$B_{22} \leq \frac{3\xi}{2\pi}|x_1-x_2|^{\mu} \tag{4.24}$$

Thus one has proved eq. (4.4), with the explicit estimate

$$c = \frac{1}{\pi}\left\{\frac{1}{\mu}(1+2^{1+\mu}+ 3^{\mu}) + \frac{2^{\mu}}{1-\mu} + 2 \log 3 + \frac{3}{2}\right\} \tag{4.25}$$

Notice that this explodes as $\mu \to o$ or $\mu \to 1$.

5. SPACE OF HÖLDER-CONTINUOUS FUNCTIONS

I have shown that the property of Hölder-continuity is transmitted through a principal-value integration, as it were. Now, I will now show how one can construct a complete space of Hölder-continuous functions; and then we will be ready to tackle a non-linear, singular integral equation.

Consider the space of all functions, $\sigma(x)$, defined for $0 \leq x \leq 1$, for which

$$\sigma(o) = 0 \quad , \tag{5.1}$$

and for which the condition of Hölder-continuity, eq. (4.3), is satisfied for some ξ. Consider the following norm

$$||\sigma|| \equiv \sup_{0 \leq x_1, x_2 \leq 1} \frac{|\sigma(x_1)-\sigma(x_2)|}{|x_1 - x_2|^{\mu}} \quad . \tag{5.2}$$

One can easily check that the conditions (2.1) - (2.4) are satisfied, so that the set of all functions satisfying (5.1), and with a norm (5.2), constitute a linear, normed space.

I will now show that this space is complete, that is, it is a Banach space. The proof follows the lines of the corresponding proof for the space of continuous functions with the norm (3.1), but is a little more involved.

Let $\{\sigma_n\}$ be a uniformly bounded Cauchy sequence in the space, i.e.

$$\|\sigma_n\| \leq B \tag{5.3}$$

for all n; and, given any $\varepsilon > o$, there exists an N such that

$$\|\sigma_{N+p} - \sigma_N\| < \varepsilon , \tag{5.4}$$

for p=1,2,3,... . It has to be shown that

$$\sigma_n \to \sigma^* , \tag{5.5}$$

where the limit-function, σ^*, must belong to the space (i.e. it must be Hölder-continuous). It will be shown in fact that

$$\|\sigma^*\| \leq B . \tag{5.6}$$

First of all, it follows from the Bolzano-Weierstrass theorem that $\sigma^*(x)$ exists, if the limit (5.5) is understood in terms of the "usual" topology, i.e. for any $\bar{\varepsilon}>o$, there exists an n_o such that

$$|\sigma_n(x) - \sigma^*(x)| < \bar{\varepsilon} \tag{5.7}$$

for all $n>n_o$. For any N and p, it follows directly from the triangle inequality that

$$\frac{|\sigma^*(x_1)-\sigma^*(x_2)|}{|x_1 - x_2|^\mu} \leq \frac{|\sigma^*(x_1)-\sigma_{N+p}(x_1)|}{|x_1 - x_2|^\mu} + \frac{|\sigma_{N+p}(x_2)-\sigma^*(x_2)|}{|x_1 - x_2|^\mu}$$

$$+ \frac{|[\sigma_{N+p}(x_1)-\sigma_N(x_1)]-[\sigma_{N+p}(x_2)-\sigma_N(x_2)]|}{|x_1 - x_2|^\mu} + \frac{|\sigma_N(x_1)-\sigma_N(x_2)|}{|x_1 - x_2|^\mu} \tag{5.8}$$

The last term here is not greater than $\|\sigma_N\|$, whereas the penultimate term is bounded by $\|\sigma_{N+p} - \sigma_N\|$. Hence one can choose N such that the sum of these two terms is not greater than $B+\varepsilon$, while p is still completely free. Lastly, for any given x_1 and x_2, $x_1 \neq x_2$ one choose p to be so great that

$$|\sigma^*(x_1)-\sigma_{N+p}(x_1)| < \varepsilon |x_1-x_2|^\mu \qquad (5.9)$$

and

$$|\sigma^*(x_2)-\sigma_{N+p}(x_2)| < \varepsilon |x_1-x_2|^\mu \qquad (5.9)$$

This is certainly possible, according to eq. (5.7), if one sets $\bar{\varepsilon} = \varepsilon |x_1-x_2|^\mu$. Hence eq. (5.8) reduces to

$$\frac{|\sigma^*(x_1)-\sigma^*(x_2)|}{|x_1 - x_2|^\mu} < B + 3\varepsilon \quad . \qquad (5.10)$$

Since ε may be as small as one likes, one may drop it from eq. (5.10), if $<$ is replaced by $\leq$. Moreover, for any $\varepsilon > o$,

$$|\sigma^*(o)| = |\sigma^*(o) - \sigma_N(o)| < \varepsilon \qquad (5.11)$$

for N large enough. But since ε can be made indefinitely small, it follows that $\sigma^*(o)$ must vanish. Hence eq. (5.6) has been demonstrated, and with it the completeness of the space.

6. APPLICATION TO THE SHIRKOV EQUATIONS

I will first apply these results on Hölder-continuous functions to the Shirkov pion-pion equations, in the SP approximation. Let $F^I(s,t)$ be the total pion-pion scattering amplitude, the superscript, I=0,1,2, being the isospin. One can write a dispersion relation for the forward amplitude:

$$F(s,o) = \frac{1}{\pi} \int_4^\infty \frac{ds'}{s'-s} \text{Im } F(s',o) + \frac{1}{\pi} \int_{-\infty}^0 \frac{ds'}{s'-s} \text{Im } F(s',o) \; . \qquad (6.1)$$

Change the integration variable in the second integral from s' to 4-s', and use the crossing relation

$$F(s,o) = \eta\beta\eta\, F(4-s,o) \qquad (6.2)$$

where the crossing matrices are

$$\beta = \frac{1}{6}\begin{pmatrix} 2 & 6 & 10 \\ 2 & 3 & -5 \\ 2 & -3 & 1 \end{pmatrix} \qquad (6.3)$$

$$\eta = \begin{pmatrix} 1 & 0 & 0 \\ 0 & -1 & 0 \\ 0 & 0 & 1 \end{pmatrix} \qquad (6.4)$$

to obtain

$$F(s,o) = \frac{1}{\pi}\int_4^\infty ds'\left[\frac{1}{s'-s} - \frac{\eta\beta\eta}{s'+s-4}\right]\mathrm{Im}\, F(s',o) \quad . \qquad (6.5)$$

Now introduce the approximation of retaining only S and P waves, so that

$$F^I(s,o) = f^I(s) \qquad (6.6)$$

for I=0,2,, , where f^0 and f^2 are the S-wave amplitudes, and

$$F^1(s,o) = 3f^1(s) \qquad (6.7)$$

where f^1 is the P-wave amplitude. The real part of eq.(6.5) becomes

$$\mathrm{Re}\, f(s) = \frac{P}{\pi}\int_4^\infty ds'\left[\frac{1}{s'-s} + \frac{\gamma}{s'+s-4}\right]\mathrm{Im}\, f(s') \qquad (6.8)$$

where

$$\gamma = \begin{pmatrix} \frac{1}{3} & -3 & \frac{5}{3} \\ -\frac{1}{9} & \frac{1}{2} & \frac{5}{18} \\ \frac{1}{3} & \frac{3}{2} & \frac{1}{6} \end{pmatrix} \qquad (6.9)$$

The unitarity relation connects the real and imaginary parts of f(s) according to

$$\mathrm{Im}\, f^I(s) = \left(\frac{s-4}{s}\right)^{1/2}\left\{[\mathrm{Re}\, f^I(s)]^2 + [\mathrm{Im}\, f^I(s)]^2\right\} + v^I(s) \qquad (6.10)$$

where v(s) is the contribution from the inelastic channels. It must vanish below the inelastic threshold, s=16 . In terms of the usual elasticity function, $\eta^{I}(s)$, one has

$$v^{I}(s) = \frac{1}{4}\{1-[\eta^{I}(s)]^2\} . \qquad (6.11)$$

It will be assumed that $\eta(s)$, or equivalently $v(s)$, is known, and the problem is to construct solutions, $f(s)$, of the non-linear singular system (6.8) and (6.10) .

The equation (6.10), with Re $f(s)$ regarded as being defined in terms of Im $f(s)$ by (6.8), is a non-linear expression for Im $f(s)$ in terms of itself, which may be summarized

$$\text{Im } f(s) = P[\text{Im } f;s] . \qquad (6.12)$$

We will seek to find a solution in the space of Hölder-continuous functions, since eq. (6.8) involves a principal value integral.

A minor difficulty is that the integral in eq. (6.8) is over the infinite range $4 \le s' < \infty$, whereas, in Section 4, the proof was given for a finite domain, $0 \le x' \le 1$. It would be possible to transform eq. (6.8) according to $x=\frac{4}{s}$ and $x'=\frac{4}{s'}$, but it is neater simply to translate the theorem of Section 4 by the same transformation, read backwards. This, however, involves one significant change. Instead of eq. (4.1), consider

$$\bar{f}(x) = \frac{P}{\pi}\int_0^1 dx'\,\bar{\sigma}(x')\left[\frac{1}{x-x'} + \frac{1}{x'}\right] , \qquad (6.13)$$

since the above transformation takes this into

$$\bar{f}\left(\frac{4}{s}\right) = \frac{P}{\pi}\int_4^\infty \frac{ds'}{s'-s}\,\bar{\sigma}\left(\frac{4}{s'}\right) . \qquad (6.14)$$

Suppose that $\bar{\sigma}(x')$ satisfies the same conditions (4.2) and (4.3) as did $\sigma(x')$. Then $\bar{f}(x)$ will satisfy the condition of Hölder-continuity, eq. (4.4), since the extra term $1/x'$ in eq. (6.13) has no effect on this proof. More than this,

$$\bar{f}(0) = 0 \qquad (6.15)$$

due to cancellation between the two terms in the square

parentheses in eq. (6.13). I leave you to check this rigorously, given that $\bar{\sigma}(x')$ is Hölder-continuous. Set

$$\bar{f}(\tfrac{4}{s}) = F(s) \quad ; \quad \bar{\sigma}(\tfrac{4}{s}) = \Psi(s) \quad . \tag{6.16}$$

The theorem may be re-phrased as follows: If

$$F(s) = \frac{P}{\pi}\int_4^\infty \frac{ds'}{s'-s}\,\Psi(s') \quad , \tag{6.17}$$

where

$$\Psi(4) = 0 = \Psi(\infty) \ , \tag{6.18}$$

and

$$|\Psi(s_1) - \Psi(s_2)| \le b\left|\frac{s_1-s_2}{s_1 s_2}\right|^\mu \quad , \tag{6.19}$$

for $4 \le s_1, s_2 < \infty$, then

$$|F(s_1)-F(s_2)| \le Cb\left|\frac{s_1-s_2}{s_1 s_2}\right|^\mu \tag{6.20}$$

for $4 \le s_1, s_2 < \infty$, where C only depends on μ , and

$$F(\infty) = 0 \ , \tag{6.21}$$

this equation being the translation of eq. (6.15). Note that in general $F(4) \neq 0$.

A solution of the system (6.12) will be looked for in the Banach space of Hölder-continuous functions, which, in terms of the variable s, may be described as the set of all functions $\Psi(s)$, defined on $4 \le s \le \infty$, for which

$$\Psi(\infty) = 0 \ , \tag{6.22}$$

and which have a finite norm

$$\|\Psi\| = \sup_{4 \le s_1, s_2 < \infty} \frac{|\Psi(s_1)-\Psi(s_2)|}{\left|\frac{s_1-s_2}{s_1 s_2}\right|^\mu} \quad . \tag{6.23}$$

We have now set up the apparatus with which to probe eq. (6.12). Consider the mapping

$$\text{Im } f'(s) = P[\text{Im } f;s] \quad . \qquad (6.24)$$

Suppose that Im f(s) belongs to the space defined by eqs. (6.22) and (6.23), and in fact satisfies

$$\|\text{Im } f(s)\| \leq b \quad , \qquad (6.25)$$

and also

$$\text{Im } f(4) = 0 \ . \qquad (6.26)$$

Suppose that the known function, v(s), belongs to the ball

$$\|v(s)\| \leq B \quad , \qquad (6.27)$$

and to the cone defined by

$$v(s) = 0 \quad , \qquad (6.28)$$

for $4 \leq s \leq 16$, and

$$v(s) \geq 0 \qquad (6.29)$$

for $s > 16$.

It follows from eq. (6.8), the properties (6.25) and (6.26), and the theorem embodied in eqs.(6.17) - (6.21), that

$$\|\text{Re } f\| \leq C_1 b \quad , \qquad (6.30)$$

where C_1 is a quantity that depends only on the Hölder index, μ . I leave you to show that there is no difficulty in handling the second piece of eq. (6.8), which only involves a vulgar non-singular integral.

From eq. (6.10), rewritten with a prime on the left-hand side, one sees that

$$\|\text{Im } f'(s)\| \leq [C_1^2+1]b^2+B \quad . \qquad (6.31)$$

Moreover, because of the phase-space factor in eq. (6.10), one sees that

$$\text{Im } f'(4) = 0 \quad . \qquad (6.32)$$

Hence, if one can find values of b and B such that

$$\Gamma b^2 + B \leq b \ , \qquad (6.33)$$

where

$$\Gamma = C_1^2 + 1 \ , \qquad (6.34)$$

then one will have shown that P has mapped the set (6.25), (6.26) into itself.

The inequality (6.33) is equivalent to

$$(b-b_+)(b-b_-) \leq 0 \quad , \tag{6.35}$$

where

$$b_\pm = \frac{1 \pm (1-4\Gamma B)^{1/2}}{2\Gamma} \quad , \tag{6.36}$$

so that if

$$B \leq (4\Gamma)^{-1} \tag{6.37}$$

then the roots $b_\pm$ are real, and then if b satisfies

$$b_- \leq b \leq b_+ \quad , \tag{6.38}$$

it follows that the inequality (6.35) is indeed observed, so that the set has been mapped into itself.

To complete the contraction mapping proof, one has to consider any two functions, $\text{Im } f^{(1)}(s)$ and $\text{Im } f^{(2)}(s)$, that belong to the set (6.25), (6.26). It follows immediately from eq. (6.8) that

$$\| \text{Re } f^{(1)}(s) - \text{Re } f^{(2)}(s) \| \leq C_1 \| \text{Im } f^{(1)}(s) - \text{Im } f^{(2)}(s) \| \, . \tag{6.39}$$

From eq. (6.10), one can write

$$\text{Im } f'^{I(1)}(s) - \text{Im } f'^{I(2)}(s) = \tag{6.40}$$

$$= \left(\frac{s-4}{s}\right)^{1/2} \{ [\text{Re } f^{I(1)}(s) + \text{Re } f^{I(2)}(s)][\text{Re } f^{I(1)}(s) - \text{Re } f^{I(2)}(s)]$$

$$+ \quad [\text{Im } f^{I(1)}(s) + \text{Im } f^{I(2)}(s)][\text{Im } f^{I(1)}(s) - \text{Im } f^{I(2)}(s)] \}$$

Hence

$$\| \text{Im } f'^{(1)} - \text{Im } f'^{(2)} \| \leq 2\Gamma b \| \text{Im } f^{(1)} - \text{Im } f^{(2)} \| , \tag{6.41}$$

so that the contraction condition is

$$b < (2\Gamma)^{-1} \, . \tag{6.42}$$

This condition is only consistent with (6.38) if (6.37) is weakened to

$$B < (4\Gamma)^{-1} \, , \tag{6.43}$$

for then $$b_- < (2\Gamma)^{-1} \quad , \tag{6.44}$$

as can be seen from eq. (6.36).

The conclusion is that if v(s) is such that (6.43) is satisfied, then one has a contraction mapping for any b that satisfies

$$b_- \leq b < (2\Gamma)^{-1} \quad . \tag{6.45}$$

Hence it follows easily that the equations (6.8) and (6.10) have one, and only one solution in the ball

$$b \leq b_- \quad , \tag{6.46}$$

and no solutions in the annulus between this ball and the ball

$$b < (2\Gamma)^{-1} \quad . \tag{6.47}$$

Thus each allowed inelastic input, v(s), generates a locally unique Im f(s) . It will be shown that two different driving terms, $v_1(s)$ and $v_2(s)$, generate two different solutions, Im $f_1(s)$ and Im $f_2(s)$. For suppose the converse, namely that

$$\mathrm{Im}\ f_1(s) \equiv \mathrm{Im}\ f_2(s) \quad . \tag{6.48}$$

Then eq. (6.8) would mean that

$$\mathrm{Re}\ f_1(s) \equiv \mathrm{Re}\ f_2(s) \quad , \tag{6.49}$$

and so eq. (6.10) implies

$$v_1(s) \equiv v_2(s) \quad , \tag{6.50}$$

in contradiction to the supposition that v_1 and v_2 were different. This means that two different v's cannot generate the same Im f .

7. THE MANDELSTAM EQUATIONS

The Shirkov equations may be regarded as an approximation to the exact equations that were developed by Mandelstam in 1958 . We now turn to these equations, and we will develop a contraction mapping proof that involves no approximation, either of crossing symmetry, or of unitarity.

Part of the proof is closely parallel to that of the previous section. We will again look at unsubtracted equations, but now we have two variables, s and t, a decidedly non-trivial complication. The total pion-pion amplitude F(s,t), that is to be constructed, will have an unsubtracted Mandelstam representation,

$$F(s,t) = A(t,u)+\beta A(s,u)+\eta\beta\eta A(t,s) \quad , \qquad (7.1)$$

where

$$A(t,u) = \frac{\beta}{\pi^2}\int_4^\infty dt' \int_4^\infty du' \frac{\rho(t,u')}{(t'-t)(u'-u)} \quad , (7.2)$$

and the spectral-function is crossing-symmetric:

$$\rho(x,y) = \beta\rho(y,x) \quad . \qquad (7.3)$$

The isospin matrices have already been given in eqs.(6.3) and (6.4), and u=4-s-t .

As Mandelstam showed, the elastic unitarity relation, eq. (3.5), will be satisfied for $4 \leq s \leq 16$ if

$$\rho(s,t) = \rho^{el}(s,t) + \beta\rho^{el}(t,s) \quad , \qquad (7.4)$$

where

$$\rho^{elI}(s,t) = \int_4^{g(s;t,4)} dt_1 \int_4^{g(s;t,t_1)} dt_2 \times \qquad (7.5)$$

$$\times K(s;t,t_1,t_2) D^{I*}(s,t_1) D^{I}(s,t_2) \quad ,$$

with

$$D(s,t) = \frac{1}{\pi}\int_4^\infty ds'[\frac{1}{s'-s} + \frac{\eta\beta\eta}{s'-u}]\rho(s',t) \quad , \qquad (7.6)$$

$$K(s;t,t_1,t_2) = \qquad (7.7)$$

$$= \frac{4}{\pi}[s(s-4)]^{-1/2}[t^2+t_1^2+t_2^2-2t\,t_1-2t_1t_2-2t_2t-\frac{4tt_1t_2}{s-4}]^{-1/2}$$

$$g(s;t,t_1) = t+t_1+ \frac{2tt_1}{s-4} - 2[tt_1(1+ \frac{t}{s-4})(1+ \frac{t_1}{s-4})]^{1/2} \quad (7.8)$$

The form of eq. (7.4) guarantees that $\rho(s,t)$ satisfies the crossing symmetry, eq. (7.3). Moreover, one sees from eq. (7.5) that $\rho^{el}(s,t)$ vanishes when

$$g(s,t,4) \leq 4 \quad (7.9)$$

that is, when

$$t \leq \frac{16s}{s-4} \quad (7.10)$$

Hence, for any $s \geq 4$,

$$\rho^{el}(s,t) = 0 \quad (7.11)$$

for $t \leq 16$. Hence eq. (7.4) means that

$$\rho(s,t) = \rho^{el}(s,t) \quad (7.12)$$

for $s \leq 16$, i.e. elastic unitarity is exactly satisfied for $s \leq 16$, as should be the case. Above s=16, there is an inelastic contribution, $\beta\rho^{el}(t,s)$. In fact, one is free to add any crossing symmetric contribution that vanishes for $s \leq 16$, so as to preserve elastic unitarity. We will use this freedom to ensure that the inelastic unitarity constraints are not violated. In fact, in order to do this, it will prove necessary to re-cast the equations (7.4) - (7.6), and write them, not for $\rho^{el}(s,t)$ directly, but rather for

$$\bar{\rho}(s,t) \equiv \beta\rho^{el}(s,t) \quad . \quad (7.13)$$

Instead of eq. (7.4) one has

$$\rho(s,t) = \beta[\bar{\rho}(s,t)+v(s,t)] +[\bar{\rho}(t,s)+v(t,s)] , \quad (7.14)$$

where the identity $\beta^2= 1$ has been used, and where $v(s,t)$ is an inelastic generating function, which is constrained to vanish for $s \leq 16$ and $t \leq 16$, and which will be chosen in such a way that the inelastic inequalities are observed for $s>16$. Eq. (7.5) will be rewritten

$$\bar{\rho}^{I}(s,t) = \sum_{J,M,N} \beta_{IJ}\beta_{JM}\beta_{JN}\iint dt_1 dt_2 K(s,t,t_1,t_2) d^{M*}(s,t_1) \times$$

$$\times\ d^N(s,t_2) \tag{7.15}$$

where the summations are over the values 0,1,2, where the integration limits are as in eq. (7.5), and where

$$d(s,t) = \beta D(s,t) \quad . \tag{7.16}$$

Since $\eta\beta\eta = \beta\eta\beta$, and $\rho(s,t)$ satisfies eq. (7.3), it follows from eq. (7.6) that

$$d(s,t) = \frac{1}{\pi}\int_4^\infty ds'\left[\frac{1}{s'-s} + \frac{\eta}{s'-u}\right]\rho(t,s') \tag{7.17}$$

The reason that eqs.(7.14) - (7.17) are better than eqs. (7.4) - (7.6) will be explained later. One can regard eq. (7.15), with d(s,t) defined by eq. (7.17), and $\rho(s,t)$ by eq. (7.14), as a non-linear equation for $\bar{\rho}(s,t)$ in terms of itself:

$$\bar{\rho}(s,t) = P[\bar{\rho};s,t] \quad . \tag{7.18}$$

The idea will be to show that, for a suitable, given generating function, v(s,t), the equation (7.18) defines a contraction mapping in a suitable space.

The real part of eq. (7.17) involves a principal value integral in the variable s, while eq. (7.14) contains an exchange of s and t. So, at the very least, our experience with the Shirkov equation would lead us to require double Hölder-continuity, with respect both to s and with respect to t . A new feature is that the behaviour with respect to t has to be preserved under the integration (7.15). As I will explain in a moment, it turns out that a simple power behaviour $t^{-\mu}$ is not so preserved, but the form $t^{-\mu}(\log t)^{-1-\varepsilon}$, $\varepsilon>0$, is preserved, if $0<\mu<\frac{1}{2}$.

Thus one is led to the following generalization of the space of eqs. (6.22) - (6.23) : The set of all functions $\bar{\rho}(s,t)$ defined for $4\leq s,\ t<\infty$, for which

$$\bar{\rho}(s,\infty) = 0 = \bar{\rho}(\infty,t) \quad , \tag{7.19}$$

and for which there exists a norm

$$\|\bar{\rho}\| = \sup_{4\le s_1,s_2,t_1,t_2<\infty} \frac{|\rho(s_1,t_1)-\rho(s_2,t_2)|\,(\log\bar{s}\log\bar{t})^{1+\varepsilon}}{\left|\dfrac{s_1-s_2}{s_1s_2\bar{t}}\right|^{\mu}+\left|\dfrac{t_1-t_2}{t_1t_2\bar{s}}\right|^{\mu}} \qquad (7.20)$$

where $\bar{s} = \min(s_1,s_2)$, $\bar{t} = \min(t_1,t_2)$, and the Hölder-index satisfies $0<\mu<\frac{1}{2}$. Thus, in particular

$$|\bar{\rho}(s_1,t)-\bar{\rho}(s_2,t)| \le \|\bar{\rho}\| \left|\frac{s_1-s_2}{s_1s_2}\right|^{\mu} (\log\bar{s})^{-1-\varepsilon}\, t^{-\mu}(\log t)^{-1-\varepsilon} \qquad (7.21)$$

I will first indicate the outline of the existence proof, and then I will sketch in some of the difficult points in the algebra, which I do not have time to give in full detail.

Consider the mapping

$$\bar{\rho}'(s,t) = P[\bar{\rho};s,t] \quad . \qquad (7.22)$$

Let $\bar{\rho}(s,t)$ belong to the ball

$$\|\bar{\rho}\| \le \bar{b} \quad , \qquad (7.23)$$

and to the cone

$$\bar{\rho}(s,t) = 0 \qquad (7.24)$$

for $t\le \frac{16s}{s-4}$. Let the known function belong to the ball

$$\|v\| \le B \qquad (7.25)$$

and to the cone

$$v(s,t) = 0 \quad , \qquad (7.26)$$

for $s\le 16$ and $t\le 16$.

First of all, it can be shown from eq. (7.17) and (7.14), much as in the one-dimensional case, that

$$|d(s_1,t)-d(s_2,t)| \le (\bar{b}+B)C_1\left|\frac{s_1-s_2}{s_1s_2t}\right|^{\mu} (\log\bar{s}\,\log t)^{-1-\varepsilon} \qquad (7.27)$$

where C_1 depends only on μ. One has to carry the extra factor $(\log s)^{-1-\varepsilon}$ through the proof, and there is the extra t-dependence in the denominator $s'-u=s+s+t-4$ in eq. (7.17) to worry about, but this is not too hard. Next, one can use (7.27) to show from eq. (7.15), with a prime on the left-hand side, that

$$\|\bar{\rho}'\| \leq \Gamma(\bar{b}+B)^2 \tag{7.28}$$

where Γ only depends on the Hölder index, μ. I will sketch the algebra leading to this result in a moment. The condition that the ball (7.23) be mapped into itself is then that

$$\Gamma(\bar{b}+B)^2 \leq \bar{b} \tag{7.29}$$

or, if one defines

$$b = \bar{b} + B \quad , \tag{7.30}$$

the condition is

$$\Gamma b^2 + B \leq b \quad . \tag{7.31}$$

This inequality is exactly the same as eq. (6.33), so the solution is the same, viz.

$$B \leq (4\Gamma)^{-1} \tag{7.32}$$

and

$$b_- \leq b \leq b_+ \tag{7.33}$$

with $b_\pm$ defined in eq. (6.36).

Consider now two functions, $\bar{\rho}^{(1)}(s,t)$ and $\bar{\rho}^{(2)}(s,t)$ each of which belongs to the set (7.23), (7.24). One proves that

$$\|\bar{\rho}'^{(1)} - \bar{\rho}'^{(2)}\| \leq 2\Gamma b \|\bar{\rho}^{(1)} - \bar{\rho}^{(2)}\| \quad , \tag{7.34}$$

so that, as in eqs. (6.41) - (6.45), the conditions for a contraction mapping are

$$B < (4\Gamma)^{-1} \tag{7.35}$$

$$b_- \leq b < (2\Gamma)^{-1} \quad . \tag{7.36}$$

One has again that a locally unique solution, in this case $\bar{\rho}(s,t)$, is generated by each $v(s,t)$ for which eq. (7.35) holds; and that different generating functions, $v(s,t)$, give necessarily different solutions, $\rho(s,t)$.

I will now give some details of the derivation of

eq. (7.28) from eq. (7.27), and we will incidentally see why the factor $(\log t)^{-1-\varepsilon}$ is needed. I will in fact explain in detail only the simpler problem of showing that

$$|\bar{\rho}'(s,t)| \leq \Gamma b^2 (s\ t)^{-\mu} (\log s \log t)^{-1-\varepsilon} , \tag{7.37}$$

given

$$|d(s,t)| \leq Cb(s\ t)^{-\mu} (\log s \log t)^{-1-\varepsilon} . \tag{7.38}$$

One has, from eq. (7.15), that

$$|\bar{\rho}'(s,t)| \leq \frac{4}{\pi} C^2 b^2 (s-4)^{-1/2} s^{-1/2-2\mu} (\log s)^{-2-2\varepsilon} \times$$

$$\times \int_4^{g(s;t,4)} dt_1 t_1^{-\mu} (\log t_1)^{-1-\varepsilon} y(t,t_1) \tag{7.39}$$

where

$$y(t,t_1) = \int_4^{g(s;t,t_1)} dt_2 [h(s;t,t_1)-t_2]^{-1/2} [g(s;t,t_1)-$$

$$-t_2]^{-1/2} t_2^{-\mu} (\log t_2)^{-1-\varepsilon} \tag{7.40}$$

with

$$h(s;t,t_1) = g(s;t,t_1) + 4t^{1/2} t_1^{1/2} (1+\frac{t}{s-4})^{1/2} (1+\frac{t_1}{s-4})^{1/2} \tag{7.41}$$

Now $[h(s;t,t_1)-t_2]^{-1/2}$ can be majorized in eq. (7.40) by

$$[h(s;t,t_1)-g(s;t,t_1)]^{-1/2} = \tag{7.42}$$

$$= \frac{1}{2} t^{-1/4} t_1^{-1/4} (1+\frac{t}{s-4})^{1/4} (1+\frac{t_1}{s-4})^{1/4} \leq \frac{1}{2} (\frac{s-4}{tt_1})^{1/2} .$$

The factor $t_2^{-\mu} (\log t_2)^{-1-\varepsilon}$ can be written

$$t_2^{-1/2} t^{1/2-\mu} (\log t_2)^{-1-\varepsilon}$$

and it may be shown that

$$t_2^{1/2-\mu} (\log t_2)^{-1-\varepsilon}$$

is majorized by

$$C_2 [(s-4)\frac{t}{t_1}]^{1/2-\mu} (\frac{s}{s-4})^{\mu} [\log (\frac{4t}{t_1})]^{-1-\varepsilon} \tag{7.43}$$

where C_2 is constant. Hence

$$y(t,t_1) \le \frac{C_2}{2} s^{\mu}(s-4)^{1-2\mu} t^{-\mu} t_1^{-1+\mu} (\log\frac{4t}{t_1})^{-1-\varepsilon} \times$$

$$\times \int_4^g \frac{dt_2}{(g-t_2)^{1/2} t_2^{1/2}} \tag{7.44}$$

The integral that is left here may be shown to be less than 4, so that one finds

$$|\bar{\rho}'(s,t)| \le \frac{8}{\pi} c^2 b^2 C_2 s^{-1/2-\mu} (s-4)^{1/2-2\mu} (\log s)^{-2-2\varepsilon} t^{-\mu}$$

$$\int_4^t \frac{dt_1}{t_1} (\log t_1)^{-1-\varepsilon} (\log\frac{4t}{t_1})^{-1-\varepsilon} . \tag{7.45}$$

The integral here may be divided into two pieces, the first being

$$\int_4^{t^{1/2}} \frac{dt_1}{t_1} (\log t_1)^{-1-\varepsilon} (\log\frac{4t}{t_1})^{-1-\varepsilon} \le$$

$$\le [\log(4t^{1/2})]^{-1-\varepsilon} \int_4^{t^{1/2}} \frac{dt_1}{t_1} (\log t_1)^{-1-\varepsilon} \le$$

$$\le 2^{1+\varepsilon} (\log t)^{-1-\varepsilon} [\frac{(\log t_1)^{-\varepsilon}}{-\varepsilon}]_4^{t^{1/2}} \le$$

$$\le \frac{2}{\varepsilon} (\log 2)^{-\varepsilon} (\log t)^{-1-\varepsilon} . \tag{7.46}$$

Notice how crucial the power $(-1-\varepsilon)$ was in the second and third lines here. The other piece of the integral in eq. (7.45), from $t^{1/2}$ to t, may be shown to have a similar bound, with the help of the transformation $t_1 \to t/t_1$. On gathering together the pieces, one finds a majorant of the form (7.37).

One now has to show that $\bar{\rho}'(s,t)$ is Hölder-continuous with respect to s and with respect to t. These two pieces of the proof are best treated separately. The Hölder-continuity with respect to t can simply be derived from the bound (7.38), whereas one needs eq. (7.27) to show that $\bar{\rho}'(s,t)$ is Hölder-continuous with respect to s . One has to break up the differences of the double integrals, evaluated at different points, into lots of little pieces, and work very patiently. The work follows the lines of the above proof of eq. (7.37), but is more complicated. I will simply refer you to the original references.

8. INELASTIC UNITARITY

It has been shown that one can construct solutions, in fact an infinite number of solutions, of the non-linear equation (7.18). Each solution satisfies crossing-symmetry and, for $4 \leq s \leq 16$, exact elastic unitarity. In general, the inelastic inequalities would be violated for s>16, but we will now show that, if some extra constraints are imposed on v(s,t), then we can arrange that these inequalities are safe.

At the fixed-point of the mapping (7.22), we know that the amplitude, F(s,t), has the unsubtracted Mandelstam representation, eqs.(7.1), (7.2). On combining this with the partial-wave projection

$$f_\ell(s) = \frac{1}{s-4} \int_{4-s}^{0} dt\, P_\ell\left(1+\frac{2t}{s-4}\right) F(s,t), \qquad (8.1)$$

we find the Froissart-Gribov form

$$f_\ell(s) = \frac{1}{s-4} \int_{4}^{\infty} dt\, Q_\ell\left(1+\frac{2t}{s-4}\right) D(s,t) , \qquad (8.2)$$

where D was defined in eq. (7.6) and is related to d by eq. (7.16). The imaginary part of (8.2), for $s \geq 4$, is

$$\operatorname{Im} f_\ell(s) = \frac{1}{s-4} \int_{4}^{\infty} dt\, Q_\ell\left(1+\frac{2t}{s-4}\right) \rho(s,t) . \qquad (8.3)$$

Now $\rho(s,t)$ was written in eq. (7.14) as the sum of four parts, the first part being

$$\beta\bar{\rho}(s,t) = \rho^{el}(s,t) \quad . \tag{8.4}$$

This part must just yield the elastic contribution to $\mathrm{Im}\, f_\ell(s)$. That is, we must have

$$\frac{1}{s-4}\int_4^\infty dt\ Q_\ell(1+\frac{2t}{s-4})\rho^{el}(s,t) = (\frac{s-4}{s})^{1/2}|f_\ell(s)|^2 \tag{8.5}$$

This is in fact true, and it follows from the identity

$$(s-4)\int_0^\infty dt\ Q_\ell(1+\frac{2t}{s-4})K(s;t,t_1,t_2) = \\ = Q_\ell(1+\frac{2t_1}{s-4})Q_\ell(1+\frac{2t_2}{s-4}) \quad . \tag{8.6}$$

If we take the contribution (8.5) on to the left-hand side of eq. (8.3), what remains is

$$\mathrm{Im}\, f_\ell(s) - (\frac{s-4}{s})^{1/2}|f_\ell(s)|^2 = \tag{8.7}$$

$$= \frac{1}{s-4}\int_{16}^\infty dt\ Q_\ell(1+\frac{2t}{s-4})[\bar{\rho}(t,s)+\beta v(s,t)+v(t,s)]$$

The inelastic unitarity constraint is that the left-hand side of this equation should be non-negative for $s \geq 16$. I will show that one can arrange that

$$[\bar{\rho}(t,s) + \beta v(s,t) + v(t,s)] \geq 0 \quad , \tag{8.8}$$

everywhere. Now we can prove, simply by glancing at the Laplace representation

$$Q_\ell(z) = \int_0^\infty du[z+\cosh u(z^2-1)^{1/2}]^{-\ell-1} \quad , \tag{8.9}$$

that

$$Q_\ell(1+\frac{2t}{s-4}) \geq 0 \quad , \tag{8.10}$$

for $s>4$, $t>0$. Hence eqs. (8.8) and (8.10) imply that

$$\mathrm{Im}\, f_\ell(s) - (\frac{s-4}{s})^{1/2}|f_\ell(s)|^2 \geq 0 \quad , \tag{8.11}$$

for $s \geq 16$, directly from eq. (8.7).

I have, then, to demonstrate eq. (8.8). It is convenient to divide the s-t plane into four pieces:

$$\begin{array}{llll} \text{I.} & 4\leq s\leq 20 & , & 4\leq t\leq 20 \\ \text{II.} & 4\leq s\leq 20 & , & t>20 \\ \text{III.} & s>20 & , & 4\leq t\leq 20 \\ \text{IV.} & s>20 & , & t>20 \end{array} \tag{8.12}$$

It will be supposed that, in addition to the requirements (7.25) and (7.26), v(s,t) also satisfies

$$\beta v(s,t)+v(t,s) \geq \gamma B[\frac{(s-20)(t-20)}{s^2t^2}]^{\mu} (\text{logs logt})^{-1-\varepsilon} \tag{8.13}$$

for s and t in IV, and also

$$g(s,t) \geq \gamma B[\frac{t-20}{t^2}]^{\mu} (\text{logt})^{-1-\varepsilon} \quad , \tag{8.14}$$

and

$$\beta g(s,t) \geq \gamma B[\frac{t-20}{t^2}]^{\mu} (\log)^{-1-\varepsilon} \quad , \tag{8.15}$$

both for s and t in II, where

$$g(s,t) = P \int_{16}^{\infty} ds'[\frac{1}{s'-s} + \frac{\eta}{s'-u}][v(s',t)+ +\beta v(t,s')] \tag{8.16}$$

In eqs. (8.13) - (8.15), the number γ is to satisfy

$$0 < \gamma < 1 \tag{8.17}$$

The basis of the proof is that while $\beta v(s,t)+v(t,s)$ in IV, and $g(s,t)$ and $\beta g(s,t)$ in II, are of the order B and <u>positive</u>, $\bar{\rho}(s,t)$ is only of order B^2. This can be seen from eqs. (7.28) and (7.30), which imply that, at the fixed-point,

$$\begin{aligned} ||\bar{\rho}|| &\leq \Gamma(b\_)^2 = \\ &= \Gamma[\frac{2B}{1+(1-4\Gamma B)^{1/2}}]^2 \\ &\leq 4\Gamma B^2 \quad . \end{aligned} \tag{8.18}$$

Hence, in IV the inequality can certainly be arranged, simply by choosing B so small that $\beta v(s,t)+v(t,s)$, which is positive, swamps $\bar{\rho}(t,s)$, which could be negative. The little tail of $\bar{\rho}(t,s)$ in $16<s<20$, $t>20$, likewise presents no difficulty, but what we have to do is to show that $\bar{\rho}(t,s)$ can be made non-negative in III, or, what is equivalent, that $\bar{\rho}(s,t)$ can be made non-negative in II.

This will be done by requiring that $\bar{\rho}(s,t)$ belongs to the cone

$$\bar{\rho}(s,t) \geq 0 \tag{8.19}$$

for $4 \leq s \leq 20$, $t > 20$ and by showing that this restricted set is still mapped into itself. This will then complete the proof. The proof of eq. (8.19) is somewhat complicated by the fact that some of the elements of the crossing matrices, β and η, are negative.

I think it might be clearer where the proof is going if I give it backwards. In eq. (7.15) the symmetry between M and N, and between t_1 and t_2, may be exploited to replace $d^{M*}(s,t_1)d^N(s,t_2)$ by

$$\mathrm{Re}\, d^M(s,t_1)\,\mathrm{Re}\, d^N(s,t_2) + \rho^M(t_1,s)\rho^N(t_2,s) \; . \tag{8.20}$$

It will be shown that Re d(s,t) is non-negative in I and II, for all isospin states. Thus (8.20) can be made non-negative throughout I and II, because in I $\rho(t,s)$ vanishes, while in II it is of a higher order in B than is Re d(s,t). This is enough to prove the positivity of $\bar{\rho}(s,t)$ in II, because the kernel, K, in eq. (7.15), is positive, and moreover

$$\sum_J \beta_{IJ}\, \beta_{JM}\, \beta_{JN} \geq 0 \tag{8.21}$$

for all I, M. N, even though β_{IJ} itself has negative elements. I leave you to check this important result.

So, how do we prove that Re d(s,t) is non-negative in I and II ? In II it follows from (8.14) and the "order of B" argument. In I it follows because we can prove that

$$\rho(t,s) = \beta\rho(s,t) \geq 0 \tag{8.22}$$

for $s \geq 80\,\frac{t+20}{t-4}$, and this positive contribution to d(s,t) can be shown to more than compensate the contribution from $\frac{16t}{t-4} \leq s \leq 80\,\frac{t+20}{t-4}$. Finally, eq. (8.22), or equivalently

$$\rho(s,t) \geq 0 \tag{8.23}$$

for $t \geq 80\,\frac{s+20}{s-4}$, is demonstrated by using eq. (8.16), and the fact that, for $t \gtrsim 80\,\frac{s+20}{s-4}$, the integrand in eq. (7.5) contains some positive contribution, which can be relied upon to force (8.23). If you would like to see this proof in full detail, I refer you to the original paper.

9. PROBLEMS AND PROSPECTS

The above existence proof has been generalized by the introduction of subtractions [7] and CDD poles [8]. The way that one does this is to subtract out from eq. (7.17) a finite number of partial waves, which must then be treated separately from the double spectral function equations. For example, if $\sigma(t)$ is the absorptive part of the S-wave in the t-channel, then eq. (7.17) can be rewritten

$$d(s,t) = \sigma(t) + \frac{1}{\pi}\int_4^\infty ds'\left[\frac{1}{s'-s} + \frac{\eta}{s'-u} - \frac{1+\eta}{t-4}\log\left(1+\frac{t-4}{s'}\right)\right]\times$$
$$\times\ \rho(t',s) \qquad (9.1)$$

Now $\sigma(t)$ can be determined from an S-wave dispersion relation, in which the left-hand discontinuity is given exactly in terms of $d(s,t)$. A double mapping $(\bar{\rho},\sigma)\to(\bar{\rho}',\sigma')$ is involved, and consequently a double contraction mapping. One can have double-spectral functions that diverge now as $s\to\infty$, so that the Mandelstam representation needs subtractions. So far, it is only known how to construct elastic spectral functions that need one subtraction, although the inelastic generating function may need more subtractions [7]. The fiercest divergence that it has proved possible to allow so far [9] is

$$|F(s,t)| \lesssim \text{const}\ t(\log t)^{-2-\varepsilon} \qquad (9.2)$$

as $t\to\infty$, for $0\leq s\leq 16$. This would allow happily for the ρ-trajectory, but not for the Pomeranchuk. It is not yet known how to go beyond (9.2), without spoiling the in -

elastic unitarity bounds. It is possible to resolve the S-wave part of eq. (9.1) by N/D equations instead of straightforward dispersion relations, and then one can add CDD poles [8] thus further enlarging the set of crossing-symmetric, unitary functions.

To conclude, I want to mention the Newton-Kantorovich method, and show how it might be used as a computer algorithm for proceeding from small to large values of the coupling. Define the operator $\Phi \equiv 1-P$ by

$$\Phi[\bar\rho;s,t] = \bar\rho(s,t) - \iint dt_1 dt_2\, K(s;t,t_1t_2)d^*(s,t_1)d(s,t_2) \tag{9.3}$$

where it is understood that d(s,t) is defined in terms of $\bar\rho(s,t)$ by (7.17) and (7.14). The first Fréchet derivative of $\Phi[\bar\rho]$ with respect to $\bar\rho$ may be defined to be the <u>linear operator</u>, $\Phi'[\bar\rho]$, if it exists, such that if h(x,y) is any function belonging to the Banach space, then

$$\lim_{\lambda\to 0} \left\| \frac{\Phi[\bar\rho+\lambda h]-\Phi[\bar\rho]}{\lambda} - \Phi'[\bar\rho]h \right\| = 0\,, \tag{9.4}$$

where λ is a real number. We can get Φ' by differentiating eq. (9.3) with respect to $\bar\rho$ at fixed v. The result is

$$\Phi'[\bar\rho]h(s,t) = h(s,t)-2\mathrm{Re}\iint dt_1 dt_2 K(s;t,t_1t_2)d^*(s,t_1)g(s,t_2) \tag{9.5}$$

where

$$g(s,t) = \frac{1}{\pi}\int_4^\infty ds'\left[\frac{1}{s'-s} + \frac{\eta}{s'-u}\right][h(s',t)+\beta h(t,s')]\,. \tag{9.6}$$

The second Fréchet derivative is defined analogously, and it may be written

$$\Phi'' h_1 h_2(s,t) = -2\mathrm{Re}\iint dt_1 dt_2 K(s;t,t_1t_2)\, g_1^*(s,t_1)g_2(s,t_2) \tag{9.7}$$

where the connection between g_1 and h_1, and between g_2 and h_2, is the same as that between g and h. Note that Φ'' is a constant operator, that is, it does not depend on

$\bar{\rho}$. This simplifies the application of the Newton-Kantorovich method, since one can immediately obtain a numerical bound on $\|\Phi''\|$.

Let us consider the so-called "modified Newton iteration", namely

$$\bar{\rho}_{n+1} = \bar{\rho}_n - [\Phi'(\bar{\rho}_o)]^{-1} \Phi(\bar{\rho}_n) \ . \qquad (9.8)$$

A sufficient condition for the convergence of this iteration is that $\Phi'(\bar{\rho}_o)$ have an inverse, and that $\bar{\rho}_1$ be so close to $\bar{\rho}_o$ that

$$\|\bar{\rho}_1 - \bar{\rho}_o\| \cdot \|[\Phi'(\bar{\rho}_o)]^{-1}\| \cdot \|\Phi''\| \leq \frac{1}{2} \ . \qquad (9.9)$$

One could use this technique to proceed from small to large values of $\|v\|$. For example, let us replace $v(s,t)$ by $\lambda v(s,t)$, where λ is a number. Then we have already proved that there is a contraction mapping solution for λ small enough, say $\lambda<\lambda_c$. One could attempt to get a solution at a point outside the contraction circle, say at $\lambda=\lambda_c+\varepsilon$, by taking, as the starting point, $\bar{\rho}_o$, of the Newton-Kantorovich iteration, the known contraction-mapping solution at the point $\lambda=\lambda_c-\varepsilon$. Having obtained a new solution at $\lambda=\lambda_c+\varepsilon$, one could then use it as the starting point for an iteration at the point $\lambda=\lambda_c+2\varepsilon$, and so on.

The difficulty with this technique is that $\Phi'(\bar{\rho}_o)$ could fail to have an inverse outside the contraction circle. One could always ask the computer to work out $\|\Phi'(\rho_o)\|$, which in practice means the determinant of a matrix, as a preparatory step. If this is very small, it means that one is near a singular point of the iteration. One possible solution is to circumnavigate the bad point in the complex λ-plane. If, when one regains the real axis, the spectral-function is again real, and the inelastic inequalities are still safe (!), one can proceed to still larger values of λ, as if nothing has happened. However, so far no-one has proved that one definitely can escape from the contraction circle, but it would seem natural to expect that one could proceed at least some distance along the real λ-axis, before getting into trouble.

REFERENCES

1. M. A. Krasnosel'skii, "Topological Methods in the Theory of Nonlinear Integral Equations", McMillan, London and New York (1964).
2. A. Martin, Nuovo Cimento 59A, 131 (1969).
3. V. V. Serebryakov and D. V. Shirkov, Fortschr. d. Phys. 13, 227 (1965).
4. R. L. Warnock, Phys. Rev. 170, 1323 (1968).
5. S. Mandelstam, Phys. Rev. 112, 1344 (1958); 115, 1741, 1752 (1959).
6. D. Atkinson, Nucl. Phys. B7, 375 (1968); B8, 377 (1968).
7. D. Atkinson, Nucl. Phys. B13, 415 (1969). J. Kupsch, Nucl. Phys. B11, 573 (1969); B12, 155 (1969).
8. D. Atkinson and R. L. Warnock, to be published in Phys. Rev.
9. D. Atkinson, CERN preprint (in preparation)

Acta Physica Austriaca, Suppl. VII, 71–90 (1970)

SOME EXACT RESULTS ON $\pi\pi$ SCATTERING*

BY

A. MARTIN
CERN - Geneva

1. INTRODUCTION

In present day physics, at least in the field of strongly interacting particles the product of the mathematical rigor times the predictive power or the ability to produce measurable numbers, is approximately a constant. So the few people who want to obtain physical results, by purely deductive means, starting from postulates which are not obviously inconsistent with one another have a very hard time. One may be sometimes tempted to think that they are not really making physics progress. On the other hand, we see the flexibility of certain "physical" theories, when new unexpected experimental facts come in, which shows that these theories are not as predictive as they looked. So it might not be too bad if a minority of theoreticians tries to remain pure, with, in mind, the purpose of helping those who try phenomenological attempts, by producing some limitations which any good theory should fulfill.

The $\pi\pi$ system was the first to which the considerations of unitarity and analyticity were applied by Chew and Mandelstam [1] because it was recognized that this

* Lecture given at IX. Internationale Universitätswochen für Kernphysik, Schladming, February 23 - March 7,1970.

system has great simplicity features:

(i) complete crossing symmetry

(ii) stability of the particles and absence of unphysical cuts and anomalous thresholds, due to the small mass of the pion.

For rigorous studies, exactly the same is true; the $\pi\pi \to \pi\pi$ scattering amplitude is, undoubtedly, one on which most results are obtained, for the same reasons. First we would like to describe the basic axiomatic results on the $\pi\pi$ amplitude. Some of these results are in fact more general.

2. BASIC AXIOMATIC RESULTS

We know that the $\pi\pi \to \pi\pi$ can be described by three functions [1]:

$$\begin{cases} A(s,t,u) \\ B(s,t,u) \\ C(s,t,u) \quad , \text{ with } s+t+u = 4m_\pi^2 \end{cases} \tag{1}$$

B and C are obtained from A by circular permutation; A is symmetric in t,u , B in s,u , C in s,t , and the isospin combinations in the channel where s is the square of the energy are

$$\frac{1}{3}(I=0) + \frac{2}{3}(I=2) = A + B + C$$

$$(I=1) = B - C$$

$$(I=2) = B + C \tag{2}$$

The most remarkable combination is A+B+C, the $\pi_o\pi_o \to \pi_o\pi_o$ amplitude completely symmetric in s,t,u .

The pure isospin states possess the unitarity property; they can be expanded in partial waves for s>4 and the expansion is convergent (Lehmann ellipse [2]) :

$$F^I(s,t,u) = \Sigma(2\ell+1)\; f^I_\ell(s)\; P_\ell(\cos\theta_s)$$

with $\cos\theta_s = 1 + \frac{2t}{s-4}$ (3)

and $\mathrm{Im}\, f^I_\ell(s) \geq |f^I_\ell(s)|^2 \frac{2k}{\sqrt{s}}$ for all $s>4$

and $\mathrm{Im}\, f^I_\ell(s) = |f^I_\ell(s)|^2 \frac{2k}{\sqrt{s}}$ for $16>s>4$

(with $k = \sqrt{(s-4)}/4$) (4)

One of the most important aspects of these unitarity conditions is that they imply the positivity properties which means that the absorptive part of F^I, F^I_s is a function of positive type. This property can be proved directly if one does not like to go through the partial wave expansion. This was done by Glaser [3]. Important consequences of this positivity are the following: the absorptive part and all its derivatives with respect to $\cos\theta$ are positive for $\cos\theta = +1$. Further the modulus of any derivative for $-1<\cos\theta<+1$ is less than its values for $\cos\theta = 1$.

Of course these positivity properties are valid for a sum of isospin amplitudes with positive coefficients for instance the $\pi_o\pi_o\to\pi_o\pi_o$ amplitude. They have a rather important consequence which is the following:

We know, from local field theory in any version, including the theory of local observables and the Jaffé fields [4] that the $\pi\pi$ amplitude satisfies dispersion relations for $-28\mu^2 < t \leq 0$, i.e. t "physical" since $t=-2k^2(1-\cos\theta)$. We also know that the $\pi\pi$ amplitude is analytic for fixed energy in the Lehmann ellipse [2] with foci ±1 and extremities $\cos\theta_o = [1+\frac{64}{k^2 s}]^{1/2}$.

I proved in 1966 that the combination of these "classical" results of field theory with the positivity properties of the absorptive part imply much stronger analyticity properties [5], namely the validity of dispersion relations for

$$|t| < 4$$

Dispersion relations hold also for fixed u , $|u|<4$ and fixed s , $|s|<4$. Such an analyticity domain in two variables is not a natural domain of holomorphy and can be considerably extended. I will not describe here the whole results but just quote the only information which will be used later on, namely that the partial wave amplitudes are analytic in a domain like this at least:

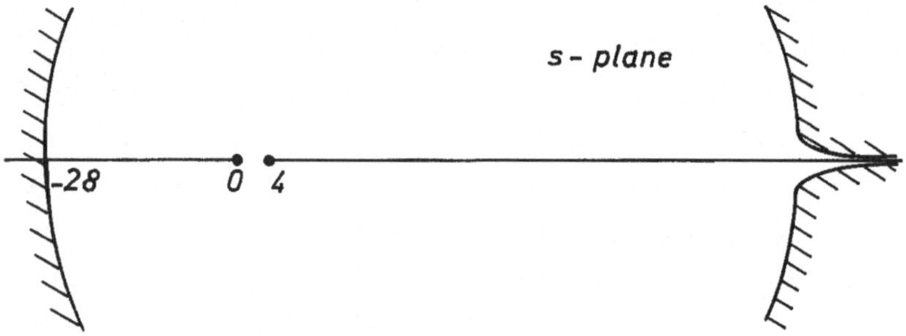

The importance of the fixed, positive t dispersion relations lies in the fact that they imply the Froissart bound [6] for the forward amplitude: The fixed t dispersion relations with a finite number of subtractions N, equal, if it is even, to the number of subtractions for t=0, imply that in the mean $F_s(s,t,u)$ is bounded by s^N. Then we have

$$s^N > \Sigma(2\ell+1) \text{ Im } f_\ell(s) \; P_\ell(1+\frac{2t}{s-4})$$

where the r.h.s. is a sum of positive terms, so that each of them is separately bounded by s^N. On the other hand we have

$$|F(s,t=0)| < \Sigma(2\ell+1)|f_\ell(s)|$$

If you combine this with

$$\text{Im } f_\ell(s) \geq |f_\ell(s)|^2$$

it is an elementary variational problem to show that

$$|F(s,t=0)| < C\, s \log^2 s$$

where C depends on N.

At this point you notice that N, the number of subtractions for t=0 (and also for t<0) is less or equal to 2. This remains so when one continues dispersion relations to t=4 and allows to sharpen further the Froissart bound by a new application of unitarity. In the end one finds

$$\sigma_t(s) < \frac{\pi}{m_\pi^2}(\log s)^2 \qquad (5)$$

This, however, is an asymptotic statement and one could argue that it is useless. In fact this is not true. A striking way of showing this has been invented by Yndurain [7]: he shows that one can write

$$\frac{1}{(s-4)^2}\int_4^s (s'-4)\sigma_t(s')ds' < C_1(a_2)(\log \frac{s}{\mu})^2 + C_2(a_2) \qquad (6)$$

where C_1 and C_2 can be calculated from the D-wave scattering length or from some bounds on the amplitude in the unphysical region obtained by Lukasuk and myself a few years ago and on which we shall come back. This is based on the equation

$$a_2 = c\int \frac{F_s(s',4)ds'}{s'^3}$$

where c is a numerical constant.

3. ABSOLUTE LIMITS TO THE $\pi\pi$ AMPLITUDE

Here I shall be largely repeating what I told in Trieste in 1965 [8]. The point is the following. We have seen that

$$\frac{1}{s-4}\int_4^s \sigma_t(s')ds'$$

can be bounded by a function of

$$\int_4^\infty \frac{F_s(s',t)}{s'^3}\,ds'$$

with $0<t<4$ (For simplicity we work first with the $\pi_o\pi_o\to\pi_o\pi_o$ amplitude).

Now the question is whether it is not possible to find a strict limit on

$$\int_4^\infty \frac{F_s(s',t)}{s'^3}\,ds'$$

The reason why one can hope to do such a thing is crossing symmetry. For simplicity I shall consider a simplified situation where you do not need any subtraction in dispersion relations for $0\le t\le 4$ (instead we may need two subtractions). Then take some value of t, say t=2 .

$$F(s=2,t=2) = \frac{1}{\pi}\int F_s(s',t=2)\left[\frac{1}{s'-2}+\frac{1}{s'}\right]ds' \qquad (7)$$

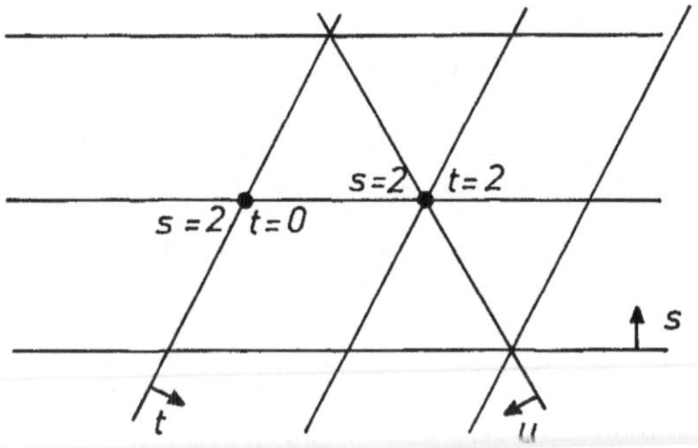

When F(2,2) is fixed, we have a limitation on the forward amplitude through unitarity. We have

$$F(s',t=0) = \Sigma(2\ell+1)\ f_\ell(s')$$

$$F_s(s',t=2) = \Sigma(2\ell+1)\ \mathrm{Im}\ f_\ell(s')\ P_\ell\left(1+\frac{4}{s'-4}\right)$$

Then we can make use of Schwarz inequality:

$$|\Sigma(2\ell+1)\ f_\ell(s')|^2 \le \Sigma(2\ell+1)\,|f_\ell(s')|^2 P_\ell\left(1+\frac{4}{s'-4}\right) \times$$

$$\times\ \Sigma(2\ell+1)\ \frac{1}{P_\ell\left(1+\frac{4}{s'-4}\right)} \qquad (8)$$

We call $\sum \frac{(2\ell+1)}{P_\ell(x)} = \phi(x)$.

All you need to know is that $\phi(x)$ exists for $x>1$ because $P_\ell(x)$ increases exponentially with ℓ; we can prove that

$$\phi(x) < \frac{\text{const}}{x-1} \quad \text{for } x \to 1 \quad .$$

Using now the unitarity condition (4) we get

$$|F(s',0)|^2 < \frac{\sqrt{s'}}{2k'} F_s(s',2)\ \phi(1+\frac{4}{s'-4}) \tag{9}$$

At this point we are almost ready to close the circle. So far we have used unitarity only. Now let us use crossing symmetry: we have $F(s=2,\ t=2) \equiv F(s=2,\ t=0)$. Consider now the function

$$G(s) = \frac{F(s,\ t=0)}{[1+\sqrt{s(4-s)}]^P} \tag{10}$$

where P will be adjusted later. The trick is that the denominator has <u>no singularity</u> on the physical sheet. Now

$$G(2) = \frac{F(s=2,t=0)}{3^P} = \frac{2}{\pi}\int \frac{\text{Im}\ G(s')ds'}{s'-2} \leq \frac{2}{\pi}\int_4^\infty \frac{|F(s',t=0)|ds'}{[1+s'(s'-4)]^{P/2}(s'-2)} \tag{11}$$

Now we are ready to close the circle: by Schwarz inequality we can write

$$\frac{|F(2,0)|^2}{3^{2P}} \leq \frac{1}{\pi}\int_4^\infty |F(s',t=0)|^2[\frac{1}{s'} + \frac{1}{s'-2}]\ \frac{1}{\phi(1+\frac{4}{s'-4})} \cdot$$

$$\frac{2k'}{\sqrt{s'}}\ ds' \times \frac{4}{\pi}\int_4^\infty \frac{\phi(1+\frac{4}{s'-4})\ \frac{\sqrt{s'}}{2k'}\ \ ds'}{(s'-2)^2[1+s'(s'-4)]^P\ [\frac{1}{s'} + \frac{1}{s'-2}]} \tag{12}$$

In the first integral of the right hand side of (12) we recognize, because of inequality (9) a lower bound of $F(2,2)$ as given by (7); but from crossing symmetry this is nothing but $F(2,0)$. If we divide out $F(2,0)$ (it is positive if there are no subtractions), we get

$$|F(2,0)| < 3^{2P} \times \frac{4}{\pi} \int_4^\infty \frac{\frac{\sqrt{s}}{2k}\,\phi(1+\frac{4}{s-4})\, ds}{(s-2)^2[1+s(s-4)]^P\,[\frac{1}{s}+\frac{1}{s-2}]} \qquad (13)$$

P has to be chosen in such a way that the integral of the r.h.s. converges. We see that since $\phi \sim s$, we can take P=1.

The remarkable character of this equation is that everything is known and therefore for s=2, t=0 you have a numerical bound for the amplitude. What we have done could be called antibootstrap: if F(s=2,t=2) gets big then F(s=2,t=0) can also get large but, however, not as fast because of unitarity which is responsible for inequality (9), and since by crossing symmetry of $\pi_o\pi_o \to \pi_o\pi_o$ amplitude these two quantities are equal there is a maximum allowed value. This can be extended to all points such that either $|s|<4$ or $|t|<4$ or $|u|<4$ lying outside the cuts. In particular for s=4/3, t=4/3 and u=4/3, we find $|\lambda|<0.6$, as defined by Chew and Mandelstam [1].

The next problem is to extend these considerations to the case where the amplitude has two subtractions which is what we expect from a physical point of view. Here you have to eliminate the subtraction constant (there is only one because of the symmetry) from the dispersion relation. So you start for instance from two values of s with the same t

$$F(s_1,t)-F(s_2,t) = \frac{1}{\pi}\int F_s(s',t)\left[\frac{1}{s'-s_1}+\frac{1}{s'-u_1}-\frac{1}{s'-s_2}-\frac{1}{s'-u_2}\right]ds' \qquad (14)$$

It is advantageous to choose s_2 and t such that $u_2=0$; then the same kind of argument can be applied to connect $F(s_2,0)$ to $F(s_1,t)-F(s_2,t)$, using the t=0 dispersion relation and unitarity. It is somewhat more complicated to connect $F(s_1,4-s_1-t)$, to $F(s_1,t)-F(s_2,t)$; if one takes $s_1>s_2$, $4-s_1-t$ is negative and one uses dispersion relations for $t<0$. Some complications arise from the unphysical cut $4<s<s_1+t$ but positivity helps to overcome them. In the end Lukasuk and I found for instance

$$-2.6 < \lambda < 17 \tag{15}$$

again following the definition of Chew and Mandelstam.

Now we come to the question raised at the end of section 2: In terms of $\int \frac{F_s(s',t)}{s'^3}\, ds'$

one can write a bound for

$$\frac{1}{(s-4)^2} \int_4^s \sigma_t(s')\ (s'-4)\ ds'$$

Now if we look at equation (14) in which we get bounds for $F(s_1,t)$ and $F(s_2,t)$ we see that the integrand behaves precisely like $F_s(s',t)/s'^3$ and we get therefore a bound for

$$\int_4^\infty \frac{F_s(s',t)}{s'^3}\ ds'$$

Therefore, it is possible to find a strict numerical bound for

$$\frac{1}{(s-4)^2} \int_4^s \sigma(s')(s'-4)ds'$$

Even if this bound is extremely bad it has the merit of existing and that is a very remarkable fact that microcausality, crossing symmetry and the existence of a minimum mass allow this.

Perhaps this is even more striking in a scalar theory. You look at the $\sigma\sigma \rightarrow \sigma\sigma$ amplitude (of course my σ's have to be stable unfortunately). This amplitude has a pole at $t=m_\sigma^2$ since the coupling $\sigma\sigma\sigma$ is not forbidden. (Naturally to make the theory nice you can think that it has also a σ^4 coupling to make the spectrum positive). V. F. Müller [9] has been able to calculate, following exactly the same ideas, a bound for the renormalized coupling constant of the trilinear term. I won't tell you what it was, because you would jump into the air, but again it exists.

To finish this let me say that these results exhibit a basic stability of the theory, because they hold irrespective of the presence of other couplings, other particles (if they are heavy enough!). It seems to me that the only crucial information which has been used is that the $\pi\pi$ (or the $\sigma\sigma$) system has no bound state. The reason why the coupling cannot be too strong is that otherwise you would have a bound state.

An important by-product of these bounds is the existence of lower bounds for the $\pi\pi$ scattering lengths, as obtained first by Vinh Mau and Bonnier [10] and later by Common. First notice in the $\pi_o\pi_o$ case that from the positivity of $F_s(s,0)$ it is easy to show that $F(s=4,t=0) > F(s=2,t=0)$. Since we can get a bound for $|F(2,0)|$ we can get a lower bound for $F(4,0)$. But the technique of Vinh Mau and Bonnier is more involved than that. I have no time to describe it completely. Let me say this:

1) They extend the bounds to complex s,t,u to be able to get a bound on the S-wave for complex s, as defined by

$$\frac{2}{s-4} \int_{\frac{4-s}{2}}^{0} F(s,t)\, dt$$

This can be done for $|s-4|<8$ and in fact in a larger domain obtained by analytic completion [11].

2) One applies a kind of Poisson formula in the following domain to $S(s) = 1+2i \frac{2k}{\sqrt{s}} f_o(s)$

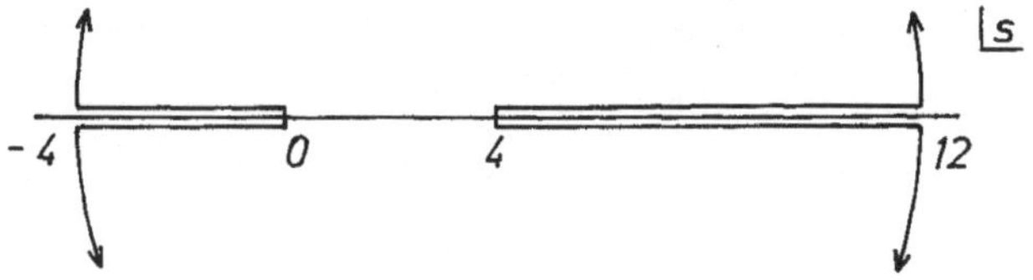

and use is made of the fact that $|S(s)| \leq 1$ on the right hand cut and that the sign of Re S(s) on the left hand cut from 0 to -4 is known. Somehow unitarity is re-injected at the last moment in this calculation. The result is surprisingly good

$$a_{\pi_0\pi_0\to\pi_0\pi_0} > -3m_\pi^{-1} \qquad (16)$$

Of course you would like to see a number 10 times smaller but this is not bad.

Common has extended these calculations to $a_{T=0}$ and $a_{T=2}$ separately [10]. You may say that these lower limits are very low. However, let me mention that I saw once a paper by Donnachie on unitarity corrections to the Weinberg calculation of the $\pi\pi$ scattering lengths. One of his solutions had an arbitrarily large negative scattering length [12]. We see that such a thing is impossible and that somewhere crossing symmetry was badly mistreated.

4. POSITIVITY AND LINEAR CONSTRAINTS ON THE PARTIAL WAVES

All of what I have said in section 3 is certainly important from a fundamental point of view. However, it has little pradical importance because of the very large numerical value of the limits obtained. What I want to discuss now might have more practical importance. It is the problem of imposing crossing symmetry on partial wave amplitudes. More exactly we shall discuss first crossing symmetry alone and next crossing symmetry + positivity. To begin the discussion will be limited to $\pi_0\pi_0\to\pi_0\pi_0$ and later we shall indicate how to extend this to other isospin combinations.

You all know that historically theoreticians have first looked at $\pi\pi$ partial wave amplitude. In spite of the work of David Atkinson, who works with the full amplitude, reported here at this school, this is still largely true right now. People like partial waves because unitarity is easy to impose, but then crossing symmetry is hard to impose because this requirement involves the full amplitude. An attempt to impose conditions from crossing symmetry on partial wave amplitude was done by Balachandran and Nuyts [13] and, in fact, contained the basic ideas. However, their paper was obscured by a fancy group theoretical sauce and as a result nobody paid much attention to it, until Roskies [14], with the help of Nussinov, and later Morel, Cohen-Tannoudji and Basdevant [15] understood how simple things were. Well, to make a long story short, take the "Euclidean region" $s>0$, $t>0$, $u>0$, and suppose you have a completely symmetric function $F(s,t,u)$, the $\pi_0\pi_0\to\pi_0\pi_0$ amplitude. Consider the integral

$$\iint d\sigma\,(s-u)^{2N+1}\,F(s,t,u)\,t^P \qquad (17)$$

where $d\sigma$ is the surface element and the integral extends to the triangle $s>0$, $t>0$, $u>0$. P and N are integers. This integral is <u>zero</u>. Now insert the partial wave series in the s-channel: the integral becomes

$$0 = \int_0^4 ds \int_{-1}^{+1} (4-s)\,d\cos\theta\,\left(s-\frac{4-s}{2}(1+\cos\theta)\right)^{2N+1}\left(\frac{4-s}{2}\right)^P \times$$
$$\times(1-\cos\theta)^P\,\Sigma(2\ell+1)\;f_\ell(s)\;P_\ell(\cos\theta)$$

By orthogonality, in the $\cos\theta$ integration only a finite number of partial waves will survive and we shall be left with a condition

$$\int_0^4 ds \sum_0^{2N+P+1} c_\ell^{2N+1,P}\;f_\ell(s) = 0 \qquad (18)$$

This condition is necessary. However, the complete system of conditions (18), applied to a function which has already $t \rightleftarrows u$ crossing symmetry built in by taking only ℓ even, is also sufficient, if the f_ℓ's are known to decrease exponentially for $0<s\leq 4$ and well behaved near $s=0$ and $s=4$. This is because the polynomials $(s-u)^{2N+1}t^P$ form a complete set of functions antisymmetric in $s \rightleftarrows u$.

If you look at the simplest conditions you see that they are such that except for the S-wave ($\ell=0$) you will never be able to get a constraint on the ℓth wave alone. Constraints involve the ℓth wave and all the lower waves. Let us write the lowest constraints:

$-\ell = 0$: we have to take $N=0$, $P=0$, we get

$$\int_0^4 ds(4-s)(3s-4)\ f_0(s) = 0 \tag{19}$$

$$\left.\begin{array}{l} -\ell = 2 \\ +\ell = 0 \end{array}\right\} \text{ we can take } \begin{array}{l} N=0,\ P=1 \\ N=0,\ P=2 \\ N=1,\ P=0 \end{array} \tag{20}$$

we have therefore three equations linking f_0 and f_2. Of course these equations apply to the unphysical interval $0<s<4$. They are therefore difficult to confront with "experimental" $\pi\pi$ phase shifts but can be confronted with models. For instance the Brown-Gobel model [14] did not pass the test. The Padé models of the Saclay group [15] passed the test.

Now you may wonder whether there is anything else to do since the system is complete. Well this is an illusion, because in practice you have only the lowest partial waves at your disposal and you want to concentrate as much information as possible on these partial waves. If you want to do that you have to use another ingredient which is positivity. It is obvious that this plays a role because for instance there are combinations of the $\ell=0$, $\ell=2$ constraints which contain an integral over the D-wave with a <u>positive</u> weight function. Now from our "first principles" quoted in section 2, we know that for $0<s<4$ we have dispersion relations with at most two subtractions. Therefore the Froissart - Gribov representation is valid for $\ell \geq 2$

$$f_\ell(s) = \frac{4}{\pi(4-s)} \int_4^\infty Q_\ell\left(\frac{2t}{4-s} - 1\right) F_t(s,t)\, dt \tag{21}$$

(where $u \rightleftarrows t$ crossing is taken care of). F_t is the absorptive part in the t channel for $0<s<4$. In this interval, the partial wave expansion

$$F_t = \Sigma(2\ell+1)\ \mathrm{Im}\ f_\ell(t)\ P_\ell(\cos\theta_t)$$

converges, and is a sum of <u>positive</u> terms from unitarity. Therefore, $f_\ell(s) > 0$ for $\ell > 0$, $0<s<4$ and hence that Roskies equation involving $\ell=0$ and $\ell=2$ with the coefficient of $\ell=2$ positive can be turned into an <u>inequality</u> involving the $\ell=0$ wave only.

There are essentially two approaches to accumulate conditions (all of which are inequalities). One followed by Roskies [16] and independently Piguet and Wanders [17] consists in looking for conditions on <u>moments of the</u> S-wave for $0<s<4$. The other, followed by Auberson, Brander, Mahoux and myself consists in looking for inequalities involving values of $f_0(s)$ at various points in the interval $0<s<4$.

Let me show the principle of this for the $\pi^0\pi^0$ case [18]. You write

$$F(s,t,u) = \sum_0^{L-2} (2\ell+1)\ f_\ell(s)\ P_\ell(\cos\theta_s) + R_L(s,\cos\theta_s) \tag{22}$$

where R_L is the rest of the series. R_L has a simple expression:

$$R_L = \frac{2L}{\pi} \int_{z_0}^\infty \frac{dz[zQ_{L-1}(z)P_L(\cos\theta) - \cos\theta P_\ell(\cos\theta)Q_L(z)]F_t(s,t')}{z^2 - \cos^2\theta} \tag{23}$$

where F_t is the absorptive part in the t-channel. It must be understood that $z = \frac{2t'}{4-s} - 1$. We already see that if we can show that the bracket in front of F_t has a constant sign, we get the sign of $R_L(s,\cos\theta_s)$. It is also allowed to write a partial wave expansion in the t channel

$$F = \sum_{0}^{L'-2} (2\ell+1)\ f_\ell(t)\ P_\ell(\cos\theta_t) + R_{L'-2}(t,\cos\theta_t)$$

These two expressions should agree. It is now clear that a sufficient condition to obtain an inequality involving only a finite number of partial waves is to find values of s and t such that

$$R_L(s,\cos\theta_s)\ R_{L'-2}(t,\cos\theta_t) < 0 \qquad (24)$$

Then one of the two partial sums is less than F, the other larger than F. So all we have to do is to make a study of the sign of

$$z\ Q_{L-1}(z)\ P_L(\cos\theta_s) - Q_L(z)\ \cos\theta_s\ P_{L-1}(\cos\theta_s)$$

$$\text{for } z_0(s) < z < \infty$$

It is possible to show that if the signs at $z=z_0$ and $z=\infty$ agree this quantity has a constant sign. With $z_0(s) = \frac{4+s}{4-s}$ we see that the critical values are given by $P_L(\cos\theta_s)=0$ and

$$z_0\ Q_{L-1}(z_0)\ P_L(\cos\theta_s) - Q_L(z_0)\ \cos\theta_s\ P_{L-1}(\cos\theta_s) = 0$$

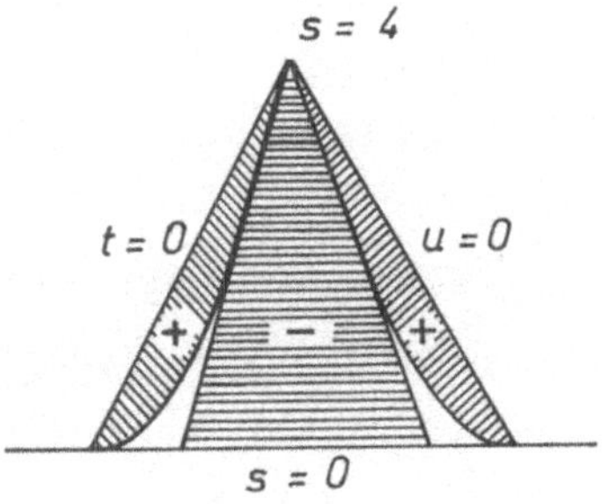

For instance for L=2 we find for the sign of $R_2(s,\cos\theta_s)$

for L=4 we find for the sign of $R_4(s,\cos\theta_s)$

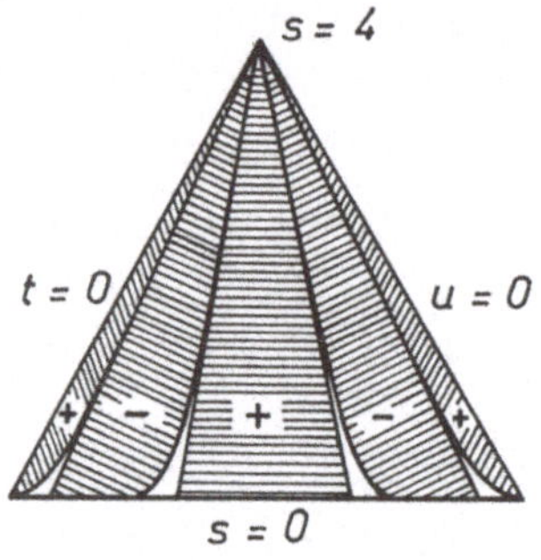

To get inequalities all we have to do is to superpose the picture giving the sign of $R_L(s,\cos\theta_s)$ and that giving the sign of $R_{L'}(t,\cos\theta_t)$.

In this way you get for instance

$$f_o(0) > f_o(3.155) \tag{25}$$

$$f_o(3.205) > f_o(0.2134) > f_o(2.9863) \tag{26}$$

or as an example

$$1.494\ f_2(0.537) - 1.623\ f_2(2.363) \tag{27}$$

$$< f_o(0.537) - f_o(2.363) <$$

$$1.510\ f_2(0.537) - 1.622\ f_2(2.363)$$

(here the method is slightly different).

Incidentally, it seems that the decimals indicated are not a luxury because these inequalities are very tight and Drs. Levers and Schwela are complaining that the figures are not enough accurate!

One can get many such inequalities. An improvement possible is that it is sufficient to have $R_L(s,\cos\theta_s) - R_L(t,\cos\theta_t)$ of a definite sign. This leads to tighter inequalities. For instance (25) is replaced by

$$f_o(0) > f_o(3.189) \tag{28}$$

This is being studied for more general cases by Dr. Brander who is with us. Before leaving the $\pi_o\pi_o \to \pi_o\pi_o$ let me indicate the important inequality [19]

$$\frac{df_o}{ds} > 0 \quad \text{for } s > 1.7 \tag{29}$$

$$\frac{df_o}{ds} < 0 \quad \text{for } s < 1.12 \tag{30}$$

(This last result has been obtained recently by Auberson [20]) and

$$\frac{d^2 f_o}{ds^2} > 0 \quad \text{in between (by Common [21])} \tag{31}$$

which guarantees the uniqueness of the minimum .

All these considerations can be extended to the cases with isospin [22]. All you have to do is to play with the crossing matrix, but things are not very easy. I shall just give one example of such a relation which is

$$\frac{7}{9}\ f_o^o(0.2937) - \frac{4}{9}\ f_o^2(0.2937) - 0.6146\ f_1^1(0.2937) >$$

$$\frac{1}{3}\ f_o^o(2.4226) + \frac{2}{3}\ f_o^2(2.4226) + 2.510\ \ f_1^1(2.4226)$$

I choose this one because Dr. Levers [23] helped us in finding a mistake of signs in it.

Now are these inequalities useful? Honestly I do not know. It seems that they restrict very severely unitary amplitudes. This is what has been experienced by Wanders [24] and his collaborators and by Bonnier [25], and by LeGuillou and Morel and Navelet [26].

Finally I would like to give an answer to those who complain that all this takes place in the unphysical region. One can in fact write physical sum rules which are consequences of crossing symmetry and involve only physical phase shifts. Such a sum rule has been proposed by Wanders [27] . But one can obtain a large number of them. Let M be a crossing symmetric quantity invariant under permutations of s,t,u. Then it is easy to prove the relation

$$\int \frac{dM_s(s,t)}{dt} \left[\frac{1}{s-t} + \frac{1}{s-4+2t}\right] ds = \int M_s(s,t) \frac{1}{(s-t)} ds$$

if we do not need subtractions in dispersion relations for this specific value of t. If we apply this to the forward amplitude we can take

$$M = \frac{F^{\pi_o \pi_o}}{[(4-s)(4-t)(4-u)]^{\alpha}} \qquad \frac{1}{2} < \alpha < 1$$

in this way we eliminate subtractions near t=0.
For instance with $\alpha = \frac{3}{4}$ we get

$$\int_4^{\infty} \frac{s-2}{s^{7/4}(s-4)^{11/4}} \left[\frac{d}{d\cos\theta} (\mathrm{Re}\,F - \mathrm{Im}\,F)\right] ds =$$

$$= \int_4^{\infty} \frac{(14-3s)}{32} \frac{(\mathrm{Re}\,F - \mathrm{Im}\,F)}{s^{11/4}(s-4)^{3/4}} ds \quad .$$

REFERENCES

1. G. F. Chew and S. Mandelstam, Phys. Rev. 119, 467 (1960).
2. H. Lehmann, Nuovo Cimento 10, 579 (1958).
3. V. Glaser, in Problems of Theoretical Physics, p.68 (Essays dedicated to N.N.Bogoliubov on his 60th birthday), publ. Nauka, Moscow (1969);
   see also: Comptes Rendus de la RCP No. 25, Vol. 7,

Départment de Mathématiques de l'Université de Strasbourg (1969).

4. H. Epstein, V. Glaser, and A. Martin, Communications in Mathematical Physics 13, 257 (1969); in the framework of Wightman axioms see: K. Hepp, Helv. Phys. Acta 32, 639 (1964); there is also an unpublished proof by Jaffé, for Jaffé fields.
5. A. Martin, Nuovo Cimento 42, 930 (1966); 44, 1219 (1966); A. Martin, Proceedings of the 1967,Conference on Particles and Fields, Interscience Publ., New York (1967), p. 244.
6. M. Froissart, Phys. Rev. 123, 1053 (1961); O. W. Greenberg and F. E. Low, Phys. Rev. 124, 2047 (1961); A. Martin, Phys. Rev. 129, 1432 (1963).
7. F. Yndurain, CERN preprint TH. 1122 (1970).
8. A. Martin, High Energy Physics and Elementary Particles, IAEA, Vienna (1965), p. 155.
9. V. F. Müller, Nuovo Cimento 42, 158 (1966).
10. B. Bonnier and R. Vinh Mau, Phys. Rev. 165, 1923 (1968); A. K. Common, Nuovo Cimento 63A, 451 (1969).
11. A. K. Common, unpublished.
12. A. Donnachie, Nuovo Cimento 53A, 933 (1967).
13. A. P. Balachandran and J. Nuyts, Phys. Rev. 172, 1821 (1968).
14. R. Roskies, Phys. Letters 30B, 42 (1969) ; Nuovo Cim. 65A, 467 (1970); and to appear in Journal of Mathematical Physics.
15. J. L. Basdevant, G. Cohen-Tannoudji and A. Morel, Nuovo Cimento 64, 585 (1969).
16. R. Roskies, Yale University preprint (1969).
17. O. Piguet and G. Wanders, University of Lausanne preprint (1969).
18. A. Martin, Nuovo Cimento 63A, 167 (1969).

19. A. Martin, Nuovo Cimento 58A, 303 (1968).

20. G. Auberson, Service de Physique théorique Saclay preprint (1970).

21. A. K. Common, Nuovo Cimento 53A, 946 (1968).

22. G. Auberson, O. Brander, G. Mahoux, and A. Martin, Nuovo Cimento 65A, 743 (1970).

23. R. G. Levers and D. Schwela, University of Bonn preprint 2-73 (1970).

24. O. Piguet and G. Wanders, Nuovo Cimento 57A, 417 (1968); G. Auberson, O. Piguet and G. Wanders, Phys. Letters 28B, 41 (1968).

25. B. Bonnier, Nuclear Phys. B10, 467 (1969); and thesis to be published.

26. J. C. Le Guillou, A. Morel, and H. Navelet, in preparation.

27. G. Wanders, Helv. Phys. Acta 39, 228 (1966).

Acta Physica Austriaca, Suppl. VII, 91–144 (1970)

ASPECTS OF CHIRAL SYMMETRY*

BY

B. RENNER

D.A.M.T.P.Cambridge

Dedicated to Prelate Dr. B. Hanssler

1. IDEA OF CHIRAL SYMMETRY

Most discussions of chiral symmetry start with rather sophisticated theoretical concepts, like current algebra, non-linear group realizations, an unsymmetric vacuum state, Goldstone bosons etc., and only at the end is the formalism put to use to obtain verifiable tests. This approach allows the display of mathematical elegance, but it may not be too helpful in motivating in what way these concepts are actually needed to describe the empirical facts.

In these lectures I will try to go the opposite way, in some sense, starting with some empirical facts and then, inductively, trying to understand their significance. My starting point will be the Goldberger-Treiman relation

$$(2m_N)(g_A/g_V) \approx -(\sqrt{2}\, F_\pi)(\sqrt{2}\, G_{NN\pi}) \qquad (1.1)$$

which connects the axial coupling constant (g_A/g_V) in nucleon β-decay with the pion decay constant F_π and the

---

*Lecture given at IX. Internationale Universitätswochen für Kernphysik, Schladming, February 23 - March 7, 197o.

pion-nucleon Yukawa coupling constant. Sign factors and normalization are subject to wide variation among authors. The most popular, though perhaps not most considerate derivation starts by assuming an unsubtracted dispersion relation in momentum transfer Δ^2 for the form factor of the weak axial divergence

$$\sqrt{2}\langle p|\partial A|n\rangle = i(2m_N)\bar{u}_p\gamma_5 G(\Delta^2)u_n$$

$$\text{with } G(0) = (g_A/g_V) \qquad (1.2)$$

$$2m_N G(\Delta^2) = \frac{-(\sqrt{2}F_\pi m_\pi^2)(\sqrt{2}G_{NN\pi})}{m_\pi^2 - \Delta^2} +$$

$$+ \frac{1}{2\pi i}\int_{9m_\pi^2}^{\infty} \frac{\text{disc } G(s)}{s - \Delta^2}\, ds \qquad (1.3)$$

and then retains only the pion pole contribution for $\Delta^2 \gtrsim 0$, disregarding the three-pion and higher cuts, which start at $\Delta^2=9m_\pi^2$ and above. The lo % disagreement of the Goldberger-Treiman relation is then sometimes quoted as an illustration of the principle that the influence of singularities can be estimated by the inverse of their distance from the point of comparison. Such a principle however would entirely disregard the possibility of singularities having different strengths, i. e. different sizes of pole residues and cut discontinuities. This is particularly relevant in the present case, as the pion pole carries the factor (m_π^2) in its residue which is to be considered small to the same extent as the pole denominator is considered small. For illustration we discuss two simple sources of corrections to the Goldberger-Treiman relation.

(a) As it is suggested by contemporary duality models, there should be recurrences of pion-like states at higher masses, to be termed π^*. The contribution of such a state π^* at mass m^* to the Goldberger-Treiman relation is strictly speaking part of the cut correction, since π^* is expected to be unstable. In the narrow-resonance approximation, however, it would amount to $(-\sqrt{2}\ F_{\pi^*})(\sqrt{2}\ G_{NN\pi^*})$ which is independent of the size of m^*, and some other effect has to operate to suppress either F_{π^*} or $G_{NN\pi^*}$, preferably the former, to maintain approximate validity of the Goldberger-Treiman relation.

(b) From among the states contributing to the cut correction, we pick out $(\rho\pi)$ states in the three-pion channel. Apart from normalization factors, disc $G(s)$ contains the product $\langle O|\partial A|N\rangle\langle N|n\bar{p}\rangle$ with $(\rho\pi)$ states contributing to the intermediate state spectrum $|N\rangle$ at an invariant mass $s = (p^\rho+p^\pi)^2$. To compare these contributions with the ones from the pion pole, let us to a rough approximation regard the dimensionless coupling $(\langle\rho\pi|n\bar{p}\rangle/\sqrt{s})$ as being of the same order of magnitude as the pion-nucleon coupling and concentrate on $(\langle O|\partial A|\rho\pi\rangle\cdot\sqrt{s})$. Just for illustration, let us make a simple model for $\langle O|\partial A|\rho\pi\rangle=(-i)(p^\rho+p^\pi)_\mu\langle O|A^\mu|\rho\pi\rangle$ in terms of two simple Feynman graphs:

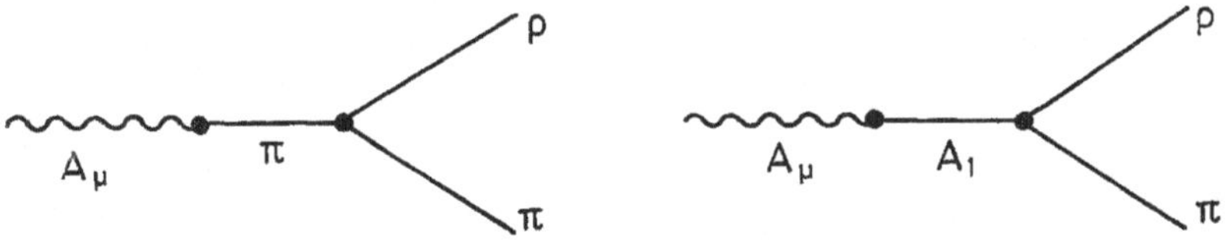

Figure 1: Feynman graph models for $\langle O|A^\mu|\rho\pi\rangle$.

$$\langle 0|A_\mu|\rho\pi\rangle = \frac{(iF_\pi)(p^\rho+p^\pi)_\mu \cdot \{g_{\rho\pi\pi}\varepsilon_\nu^\rho(2p_\pi^\nu)\}}{s-m_\pi^2} +$$

$$+ \frac{f_A\left(-g_{\mu\nu} + \frac{(p^\rho+p^\pi)_\mu(p^\rho+p^\pi)_\nu}{m_A^2}\right) i\{\varepsilon_\nu^{(\rho)} g_{A\rho\pi} + p_\nu^{(\pi)}(\varepsilon_\lambda^{(\rho)} p_\lambda^{(\pi)}) h_{A\rho\pi}\}}{s-m_A^2} \tag{1.4}$$

and

$$\langle 0|\partial A|\rho\pi\rangle \approx \{F_\pi 2g_{\rho\pi\pi} + \frac{f_A}{m_A^2}(g_{A\rho\pi} + \frac{1}{2}(s-m_\rho^2)h_{A\rho\pi})\}(\varepsilon_\nu^\rho p_\pi^\nu \tag{1.5}$$

If on average we regard the quantity $(\varepsilon_\nu^\rho p_\pi^\nu)$ to be of order $\sqrt{s}$, we see that pion-pole dominance in the Goldberger-Treiman relation requires substantial cancellations in the curly bracket to make its contribution negligible as compared to the pion pole contribution F_π.

To provide a mechanism for such a cancellation, one may postulate yet another model for $\langle 0|\partial A|\rho\pi\rangle$, as given by the Feynman graph below

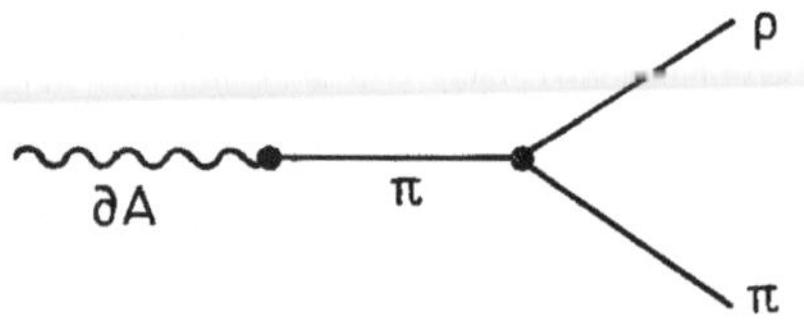

Figure 2: Feynman graph model for $\langle 0|\partial A|\rho\pi\rangle$.

$$\langle 0|\partial A|\rho\pi\rangle = \frac{F_\pi m_\pi^2}{s-m_\pi^2}\{g_{\rho\pi\pi}\varepsilon_\nu^\rho(2p_\pi^\nu)\} \tag{1.6}$$

However this model is, by itself, not very plausible for s being in the domain of the cut. We see that we arrive at a circular situation:[1]

(1) Since the residue of the pion pole is suppressed by a factor m_π^2, it is no longer plausible to neglect the higher singularities merely because of their distance, unless their strength is similarly suppressed down to the order of m_π^2. Nontrivial cancellations would be required to achieve this suppression.

(2) Such cancellations would be generated, if pion pole dominance is postulated on some other grounds than relative distance of singularities. Unless we want to take the success of the Goldberger-Treiman relation as accidental, we are led to seek some mechanism which would allow us to regard all matrix elements $\langle O|\partial A|N\rangle$ as being of order (m_π^2), then we may again use distance of singularities.

In this framework the limit $m_\pi \to o$ suggests itself for examination. Having been led to assume $\langle O|\partial A|N\rangle = O(m_\pi^2)$ for all states $|N\rangle$, we would in the limit $m_\pi \to o$ find the axial vector current conserved, and the Goldberger-Treiman relation would become exact. This is best illustrated in an argument due to Nambu who applies the conservation condition directly to the axial current, exploiting the fact that the pion pole occurs only in a longitudinal form factor and that the transversal form factor is finite at zero momentum transfer:

$$O = i\Delta^\mu \sqrt{2}\langle n|A_\mu|p\rangle = i\Delta^\mu \bar{u}_n \Big(\frac{g_A}{g_V}\gamma_\mu\gamma_5 + \\ + \frac{(\sqrt{2}\,F_\pi)\Delta_\mu \sqrt{2}\,G_{NN\pi}}{\underbrace{\Delta^2-m_\pi^2}_{=0}}\Big)\,u_p \qquad (1.7)$$

Occasionally one meets the argument that the limit $m_\pi \to o$ does not add to the plausibility of the Goldberger-Treiman relation, since in that limit pole and cut would coincide. This argument is only partly true in so far as the limit $m_\pi \to o$ does not by itself adequately describe our hypothesis. What we are led to assume is rather a near-conservation of the weak axial current with $m_\pi \to o$ as a necessary consequence of a symmetry limit in which the cut-discontinuity would vanish. At worst, one expects contributions of the order $m_\pi^2 \log(m_\pi^2/M^2)$ with M being some scaling mass.

A symmetry with conserved axial charges and massless pions is of a different nature than the usual symmetries like for instance isospin. Starting with $[Q, H] = 0$ and applying Q to some single-particle state $|\psi\rangle$ to get $|\psi'\rangle = Q|\psi\rangle$ one expects $|\psi'\rangle$ to be degenerate with $|\psi\rangle$ in energy-momentum, unless $Q|\psi\rangle = 0$. If there is no zero-mass particle in the theory, we find the need for particle multiplets corresponding to the representations of the symmetry group in question. In the present symmetry limit, however, with Q the weak axial charge and with massless pions, the state $|\psi'\rangle$ may also consist of the state $|\psi\rangle$ and a zero-energy pion or in general a coherent admixture of an odd number of zero energy pions.[(2)] There is no longer a need for parity doublets in the symmetry limit, though their existence can of course not be excluded.

The occurrence of zero-mass spinless bosons is a general feature of field theories which contain conserved currents whose charges do not annihilate the vacuum state $Q|0\rangle \neq 0$. Other candidates for such Goldstone-bosons would be the K-meson, associated with the strangeness changing weak axial current and possibly the disputed κ-meson, associated with the strangeness changing weak vector current. The last case is unlikely to lead to realistic results, since SU_3 symmetry seems to be realized rather in terms of particle multiplets.

2. LOW ENERGY THEOREMS

Before proceeding, we should concern ourselves with possible tests of our present hypothesis. This can be done most easily by exploiting pion pole dominance in all possible ways. Since we have argued that the Goldberger-Treiman relation is not plausible by itself, nor is any other application of pion pole dominance, every successful application of pion pole dominance provides non-trivial support to our working hypothesis that hadron dynamics is near to conservation of the weak axial current. In the following we will call it the PCAC hypothesis (partial conservation of axial current).

However in applying pion-pole dominance in four-point and higher functions, there is a nontrivial difficulty. Consider, for instance, the matrix element $\langle\pi|\partial A|\pi\pi\rangle$ in terms of a simple Feynman graph model:

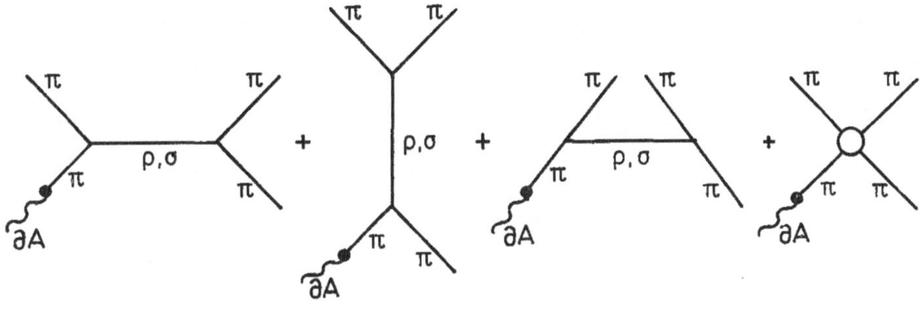

Figure 3: Feynman graph model for $\langle\pi|\partial A|\pi\pi\rangle$ with (ρ, σ) exchange at low energies, the contact graph denoting any higher structure.

In extrapolating in virtual mass q^2, carried by the axial vector divergence, one cannot avoid affecting other channel variables, due to the Mandelstam relation

$$3m_\pi^2 + q^2 = s+t+u$$

There are no reliable grounds for making any particular choice of extrapolation, unless one knows something about off-shell pion interactions beforehand. Only if $(\pi\pi)$ interactions are not too strong at low energies, will this effect be unimportant. In that way, the absence of a low mass σ-resonance is practically a condition for the applicability of PCAC and it does not appear an accident that trouble is found in the $(\Sigma\pi)$ system with the prominent $Y_0^*(1405)$ and $Y_1^*(1385)$ resonances near threshold.[3]

Keeping these precautions in mind we turn to the actual applications.

2.1. Adler Zeroes[4]

A matrix element of a current divergence between states on mass shell can always be written as

$$\langle\alpha|\partial A|\beta\rangle = i(p^{\alpha}-p^{\beta})_{\mu}\langle\alpha|A^{\mu}|\beta\rangle \tag{2.1}$$

It vanishes wherever $(p^{\alpha}-p^{\beta}) = 0$ and $\langle\alpha|A^{\mu}|\beta\rangle$ is finite. The point $(p^{\alpha}-p^{\beta}) = 0$ is often unphysical, if it is near to a physical region and if PCAC is valid and applicable:

$$\langle\alpha|\partial A|\beta\rangle \approx \frac{(F_{\pi}m_{\pi}^2)\langle\alpha|\pi|\beta\rangle}{\underbrace{\Delta^2}_{=0} - m_{\pi}^2} \tag{2.2}$$

then the pion amplitude $\langle\alpha|\pi|\beta\rangle$ should have a zero nearby. This has quite far-reaching consequences, at present we give only a few examples.

(1) In $\pi\pi$-scattering, the point $p^{\pi} = 0$ corresponds to $s=t=u=m_{\pi}^2$. Unless there are strong variations due to low-lying resonances we expect only a relatively weak $(\pi\pi)$ interaction near threshold.

(2) As we did with pions, so we expect Adler-type zeroes with K-mesons, although with a considerably larger margin of error. Accepting now this application of K-meson pole dominance for the axial divergence, we cannot simultaneously use K-meson pole dominance for matrix elements of the parity-violating part of the nonleptonic weak Hamiltonian, since this would make all decay amplitudes vanish.

(3) Adler originally tested his suggestion in (πN) scattering.[4] Here the situation is slightly more complicated, since $\langle \pi N | A_\mu | N \rangle$ is infinite at zero energy-momentum carried by the axial current, due the Born pole singularities from terms like the one given by the Feynman graphs below

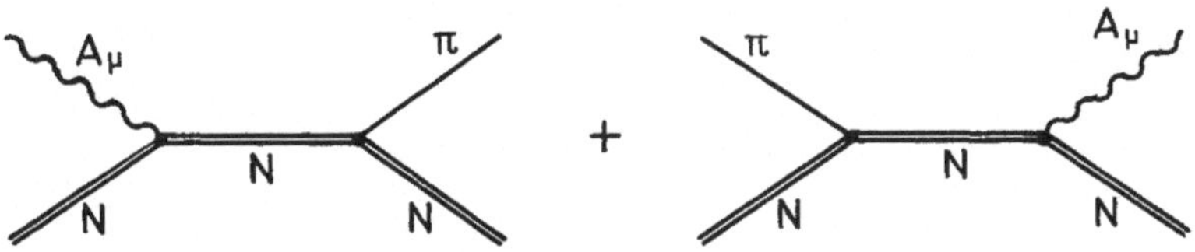

Figure 4: Born graphs in πN scattering .

In such cases it is advisable, not to apply pion pole dominance to the full pion-nucleon scattering amplitude, but to attempt a smooth extrapolation at a reduced amplitude from which all singular terms have been separated which introduce strong variations. In the present case we proceed in two steps

(1) Extract the pion pole terms both from $\langle \pi N | \partial A | N \rangle$ and $\langle \pi N | A_\mu | N \rangle$

$$\langle \pi N | \partial A | N \rangle = \frac{(F_\pi m_\pi^2) \langle \pi N | \pi | N \rangle}{\Delta^2 - m_\pi^2}$$

$$<\pi N|A_\mu|N> \;=\; \frac{(-iF_\pi\Delta_\mu)<\pi N|\pi|N>}{\Delta^2 - m_\pi^2} + <\pi N|A_\mu^T|N> \qquad (2.3)$$

In equ. (2.3), the expression $<\pi N|\pi|N>$ defines an off-mass shell pion-nucleon amplitude which is assumed to be regular about $(\Delta^2 \approx m_\pi^2)$ and $<\pi N|A_\mu^T|N>$ denotes a matrix element of the axial current with the pion pole removed. Equ. (2.1) comes to read

$$i\Delta^\mu <\pi N|A_\mu^T|N> \;=\; F_\pi <\pi N|\pi|N> \qquad (2.4)$$

(2) We extract the nucleon Born poles and make the assumption of smooth extrapolation only for the remaining part of the amplitude which is free from singularities in the low energy domain:

$$(<\pi N|\pi|N> - <\pi N|\pi|N>_{Born})\Big|_{\Delta^2 = m_\pi^2} \approx$$

$$\approx (<\pi N|\pi|N> - <\pi N|\pi|N>_{Born})\Big|_{\Delta_\mu \approx 0} \qquad (2.5)$$

Corrections to equ. (2.5) are of order $O(m_\pi^2)$ and come from two sources

(a) effects from higher pseudoscalar states, as discussed in section 1, being of order $O(m_\pi^2)$ by the hypothesis of near-conservation of the axial current,

(b) effects from the shift in channel variables, as discussed in the beginning of this section, to the same order $O(m_\pi^2)$. Both corrections would vanish in the symmetry limit. We now insert equ. (2.4) into the left-hand side of equ. (2.5) and rearrange:

$$<\pi N|\pi|N>^{Phys} \approx <\pi N|\pi|N>^{Phys}_{Born} + \frac{1}{F_\pi}\,(i\Delta^\mu <\pi N|A_\mu^T|N> -$$

$$-\langle \pi N | \pi | N \rangle_{Born}) \Big|_{\Delta_\mu \approx 0} \tag{2.6}$$

Born terms are not uniquely defined, since only their residue is determined by dispersion relations. Therefore one has the option of choosing a definition of the Born term such that equ. (2.6) takes a simpler form. Wherever spin (1/2) baryons are involved one chooses therefore the Born terms according to the pseudovector coupling model; this makes the bracket on the right-hand side of equ. (2.6) vanish. Any other choice would only have added to the corrections in order $O(m_\pi^2)$ since the difference of any two Born terms is a regular function.

So we finally get the estimate (to zero order in the pion momentum)

$$\langle \bar{\pi}_3 p_4 | \bar{\pi}_1 | p_2 \rangle \approx \left(\frac{2G^2_{NN\pi}}{4m_N^2}\right) \bar{u}_4 (\gamma q_3) \gamma_5 \frac{\gamma(p_2+q_1)+M}{(p_2+q_1)^2-M^2} (\gamma q_1) \gamma_5 u_2 =$$

$$= -(2G^2_{NN\pi}) \bar{u}_4 \gamma_5 \frac{\gamma(p_2+q_1)+M}{(p_2+q_1)^2-M^2} \gamma_5 u_2 - \left(\frac{G^2_{NN\pi}}{m_N}\right) \bar{u}_4 u_2 + O(q_1) \tag{2.7}$$

The first term is the Born pole in a dispersion relation, the second term is the one predicted and confirmed by Adler.[4]

2.2. Low Energy Theorems for Several Pions

As it is generally known, the most interesting low energy theorems are derived with two pions taken to zero energy. An extended concept of pion pole dominance is used here, with the axial divergence as the pion interpolating field

$$\langle f\pi'(q')|i\pi(q)\rangle = (-i)\,\frac{(m_\pi^2-q^2)(m_\pi^2-q'^2)}{F_\pi^2\; m_\pi^4}\times$$
$$\times \int e^{iq'x}\langle f|T^*\{\partial A'(x),\ \partial A(0)\}|i\rangle (dx)^4 \tag{2.8}$$

Equ. (2.8) as it stands, is a definition rather than an assumption. On the pion mass shell it represents an LSZ reduction formula with a specially chosen pion interpolating field $\phi_\pi(x) = \partial A(x)/F_\pi m_\pi{}^2$, and off the pion mass shell the right-hand side defines the left hand side. The star on the T-product (T^*) takes recognition of the fact that we do not insist on any strict time-ordering, if this should turn out to be awkward, in particular we demand that the off-shell expression should have the same covariance properties as the on-shell expression.
We now make the standard partial integrations

$$\langle f\pi^\alpha(q')|i\pi^\beta(q)\rangle = (-i)\frac{(m_\pi^2-q^2)(m_\pi^2-q'^2)}{F_\pi^2\; m_\pi^4}\{q'^\mu q^\nu \int e^{iq'x}\times$$
$$\times\langle f'|T^*\{A_\mu^\alpha(x),A_\nu^\beta(0)\}|i\rangle(dx)^4+\frac{(-i)}{2}\,q'^\mu\int e^{-iqx}\delta(x_0)\times$$
$$\times\langle f|[A_\mu^\alpha(0),A_0^\beta(x)]|i\rangle(dx)^4+\frac{(-i)}{2}\,q^\mu\int e^{iq'x}\delta(x_0)\times$$
$$\times\langle f|[A_0^\alpha(x),A_\mu^\beta(0)]|i\rangle(dx)^4+\frac{1}{2}\int e^{-iqx}\delta(x_0)\times$$
$$\times\langle f|[\partial A^\alpha(0),A_0^\beta(x)]|i\rangle(dx)^4-\frac{1}{2}\int e^{iq'x}\delta(x_0)\times$$
$$\times\langle f|[A_0^\alpha(x),\partial A^\beta(0)]|i\rangle(dx)^4\} \tag{2.9}$$

and, having assumed covariance, we can always choose frames of reference such that the space components of q and q' vanish and we need only commutators of generalized charges with local operators

$$[\int A_o^\alpha(\bar{x},o)(d\bar{x})^3, A_\mu^\beta(\bar{y},o)] \overset{\text{def}}{=} iV_\mu^{\alpha\beta}(\bar{y},o) \tag{2.10a}$$

$$[\int A_o^\alpha(\bar{x},o)(d\bar{x})^3, \partial A^\beta(\bar{y},o)] \overset{\text{def}}{=} i\sigma^{\alpha\beta}(\bar{y},o) \tag{2.10b}$$

That $V_\mu^{\alpha\beta}$ should be a vector and $\sigma^{\alpha\beta}$ a scalar, follows simply from covariance; furthermore it is clear, from integrating equ. (2.10a) over y-space, that, up to a chargeless component which does not contribute to (2.9), we have

$$V_\mu^{\alpha\beta}(y) = e^{\alpha\beta\gamma}V_\mu^\gamma(y) \tag{2.11}$$

In equ. (2.9) we have averaged over two possible orders of partial integration. Demanding that the two orders produce the same result, we derive one more relation

$$\sigma^{\alpha\beta}(y) - \sigma^{\beta\alpha}(y) = e^{\alpha\beta\gamma}\partial V^\gamma(y) \tag{2.12}$$

In the chiral symmetry limit where $\partial A \to o$ we obviously have $\sigma \to o$ and by (2.12) we reach the trivial statement $\partial V \to o$, i. e. the commutator (2.10a) of a conserved charge and a conserved current produces another conserved current. The non-trivial aspect of this statement however is the following: until we are able to master those corrections to pion-pole dominance which are specifically associated with the non-conservation of the axial current, like the cut contributions in the Goldberger-Treiman relation, we will not be able to put into question any theoretical statement on the conservation of V_μ. Putting the same message in different words: theories which want to make pion-pole dominance plausible to some degree of accuracy, simply have to accept conservation of the vector current to the same degree of accuracy.[5]

In a realistic theory there is only one candidate for a conserved vector current with isospin one: the isospin current, of which V_μ may be a yet undetermined multiple.

To find this coefficient one has to try and identify the commutator in (2.9) with extrapolated pion-nucleon scattering data. We proceed as we did in section 2.1. We first extract the pion-pole graphs from the off-shell amplitudes both in equs. (2.8) and (2.9) and retain an equation for off-shell scattering amplitudes

$$F_\pi^2 \langle f\pi^\alpha(q')|i\pi^\beta(q)\rangle = \tag{2.11}$$

$$= (-i)\{q'^\mu q^\nu \int e^{iq'x} \langle f|T^*\{A_\mu^{T\alpha}(x), A_\nu^{T\beta}(o)\}|i\rangle (dx)^4 +$$

$$+ e^{\alpha\beta\gamma}(q^\mu + q'^\mu)/2 \langle f|V_\mu^\gamma|i\rangle - \langle f|(\sigma^{\alpha\beta} + \sigma^{\beta\alpha})/2|i\rangle\} +$$

$$+ O(m_\pi^2)(\text{three-pion cut corrections})$$

As the next step we extract the Born terms, if there are any in the problem, again we find it convenient to use the pseudovector coupling form. As it is well known, a model-independent result requires elimination of the first term in (2.11). This can only be done by discarding quantities of second order in the pion momenta, and we finally obtain, by specializing to the forward direction and to the isospin-antisymmetric pion channel the well known result

$$F_\pi^2 (\langle f\pi^+(q)|i\pi^+(q)\rangle - \langle f\pi^-(q)|i\pi^-(q)\rangle) =$$

$$= (\text{Born terms}) + 2q^\mu \langle f|V_\mu^{(3)}|i\rangle + O(m_\pi^2) + O(q^2) \tag{2.12}$$

Since with pseudoscalar mesons the elastic Born terms occur always in p-wave, it can be checked explicitly that they cancel out at the scattering threshold up to order $O(m_\pi^2)$. This is why the tests are most conveniently given at the scattering threshold.

In fixing the crucial scale factor α in the commutator of axial currents

$$[\int A_o^\alpha(\bar{x},o)(d\bar{x})^3, A_\mu^\beta(o)] = ie^{\alpha\beta\gamma}V_\mu^\gamma(o) = ie^{\alpha\beta\gamma}Z\ I_\mu^\gamma(o) \quad (2.13),$$

we can use strong interaction physics to conclude Z>o; since we can express the low-energy theorem (2.12) also in terms of unsubtracted dispersion relation and saturate with known resonances, for instance in the $(\pi\pi)$ system:

$$Z = F_\pi^2 \int_{(2m_\pi)^2}^{\infty} \frac{ds}{(s-m_\pi^2)^2} (\sigma^{tot}_{\pi^+\pi^-}(s) - \sigma^{tot}_{\pi^+\pi^+}(s)) \frac{2p_\pi^{Lab} m_\pi}{\pi} \quad (2.14)$$

using the empirical fact that for exotic channels

$$\sigma_{\pi^+\pi^+}(s) < \sigma_{\pi^+\pi^-}(s)$$

However we have no rational grounds on which we could constrain the size of Z, i. e. the scale of the weak axial current (which reappears in terms of F_π^2 on the right-hand side of (2.14)), as long as we understand weak interactions only to first order. It is well known from the Adler-Weisberger relation that Z = 1, as originally suggested by Gell-Mann(6) when proposing an $SU_2 \times SU_2$ algebra of weak currents

$$[\int V_o^\alpha(\bar{x},o)(d\bar{x})^3, V_\mu^\beta(o)] = ie^{\alpha\beta\gamma}V_\mu^\gamma(o) \quad (2.15a)$$

$$[\int V_o^\alpha(\bar{x},o)(d\bar{x})^3, A_\mu^\beta(o)] = ie^{\alpha\beta\gamma}A_\mu^\gamma(o) \quad (2.15b)$$

$$[\int A_o^\alpha(\bar{x},o)(d\bar{x})^3, A_\mu^\beta(o)] = ie^{\alpha\beta\gamma}V_\mu^\gamma(o) \quad (2.15c)$$

so to give a precise meaning to weak interaction-universality through the (at that time conjectured) non-linear relations (2.15). A negative factor Z in equ. (2.15c) would have produced the non-compact algebra SO(3,1) whose unitary representations are infinite-dimensional, which in physical terms would have attributed great weight to the exotic channels.

An immense literature exists on the art of extrapolating from the soft-meson points to the physical region. We performed this extrapolation by explicitly separating terms with strong dynamical variations, such as pion pole terms or Born terms, and applying a smoothness assumption to the remaining amplitude. This is being done in some equivalent form by more or less every author. The proposals begin to differ when it comes to the point which variables to keep constant when performing the extrapolation onto the pion mass-shell. At present, the most popular proposal is one made by Fubini and Furlan,[7] of extrapolating so as to keep the virtual pion at rest relative to the target and thereby to achieve a spin-parity selection rule among the possible correction terms. Since these technical points are not in the main line of our argument, we cannot go further here.

As soon as we come to three and more pions, the extraction of singularities becomes prohibitively complicated, and the choice of variables to keep constant in the extrapolation becomes increasingly treacherous, considering configurations like the one below,

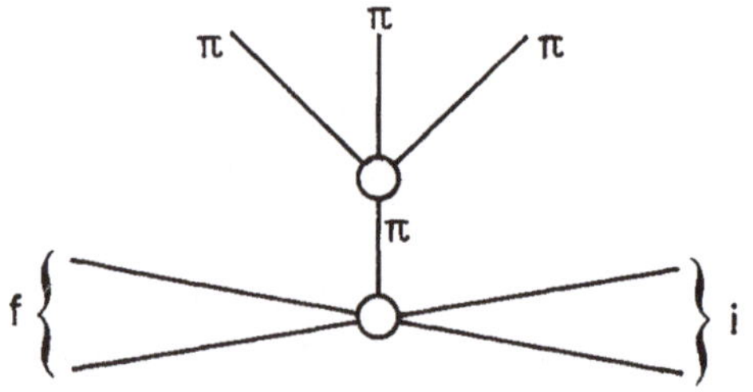

Figure 5: Contribution to a three-pion amplitude .

which are very sensitive to extrapolation errors, since in the propagator, we have divided by a small quantity.

3. GENERALIZED WARD IDENTITIES AND CHIRAL LAGRANGIANS

The next step forward was provided in a paper by Schnitzer and Weinberg.[8] These authors made the important suggestion to represent matrix elements of currents or current divergences by set of simple pole graphs

$$W_{\mu\nu\lambda}(p,q) = -\iint (dx)^4 (dy)^4 e^{ipx}\, e^{iqy} \times$$

$$\times \langle o|T^*\{A_\mu^\alpha(x), A_\nu^\beta(y), V_\lambda^\gamma(o)\}|o\rangle \qquad (3.1)$$

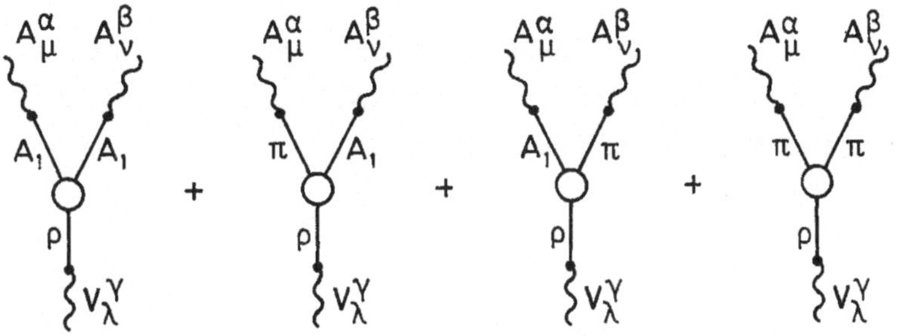

Figure 6: Model for $W_{\mu\nu\lambda}$ in equ. (3.1) .

$$W_{\mu\lambda}(p,q) = -\iint (dx)^4 (dy)^4 e^{ipx}\, e^{iqy} \langle o|T^*\{A_\mu^\alpha(x), \partial A^\beta(y), V_\lambda^\gamma(o)\}|o\rangle \qquad (3.2)$$

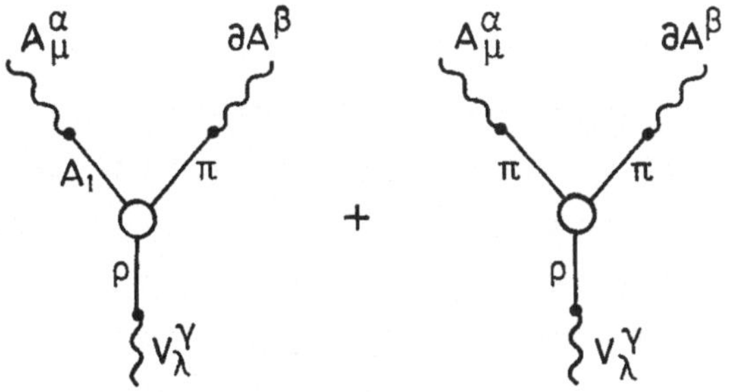

Figure 7: Model for $W_{\mu\lambda}$ in equ. (3.2) .

$$W_\lambda(p,q) = -\iint (dx)^4 (dy)^4 e^{ipx} e^{iqy} \langle o | T^* \{ \partial A^\alpha(x), \partial A^\beta(y), V_\lambda^\gamma(o) \} | o \rangle \tag{3.3}$$

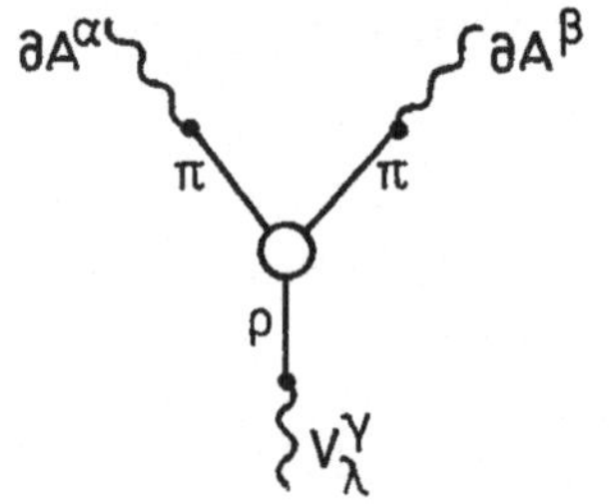

Figure 8: Model for W_λ in equ. (3.3) .

with yet undetermined effective vertex functions, to express the content of current algebra in terms of generalized Ward identities like:

$$o = iq^\nu W_{\mu\nu\lambda}(p,q) + W_{\mu\lambda}(p,q) + ie^{\alpha\beta\gamma}\{ \int (dz)^4 e^{i(p+q)z} \times$$

$$\times \langle o | T^* \{ V_\mu(z), V_\lambda(o) \} | o \rangle - \int (dz)^4 e^{ipz} \langle o | T^* \{ A_\mu(z), A_\lambda(o) \} | o \rangle \} \tag{3.4}$$

$$o = ip^\mu W_{\mu\lambda}(p,q) + W_\lambda(p,q) + ie^{\alpha\beta\gamma} \int (dz)^4 e^{iqz} \times$$

$$\times \langle o | T^* \{ \partial A(z), A_\lambda(o) \} | o \rangle \tag{3.5}$$

and to study their implications on the vertices. This method has the great virtue of providing a simultaneous solution to all current algebra constraints obtainable from an n-current amplitude once one has decided which resonances one wants to take account of explicitly in the extrapolation procedure; one would introduce these in the pole approximation and let the influence of the others generate a momentum-dependence in the effective vertices. The solutions for these vertices are obviously never unique, and one may then choose the one with the least momentum dependence or the one with the best asymptotic properties etc.

The method has been extended straightforwardly to amplitudes involving four currents or divergences, including graphs like in figure 9

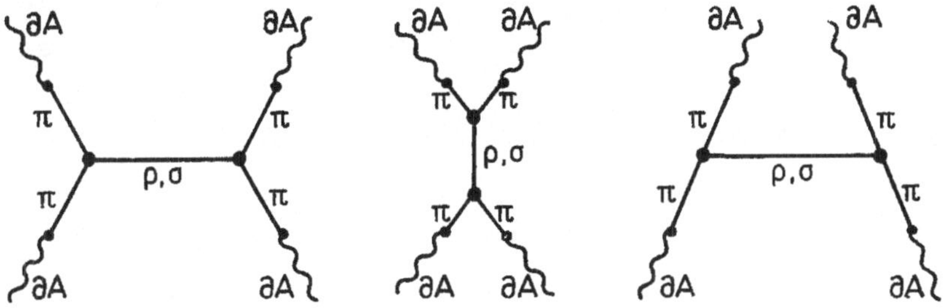

Figure 9: Model for pion four point function .

in representing an expression like

$$\int (dx)^4 (dy)^4 (dz)^4 e^{ip_1 x} e^{ip_2 y} e^{-ip_3 z} \times$$

$$\times \langle o | T^* \{ \partial A(x), \partial A(y), \partial A(z), \partial A(o) \} | o \rangle$$

It has been most interesting to observe that the graphs in figure 9 by themselves are usually not sufficient to produce a solution to the generalized Ward identities, one requires a kind of contact term to be added

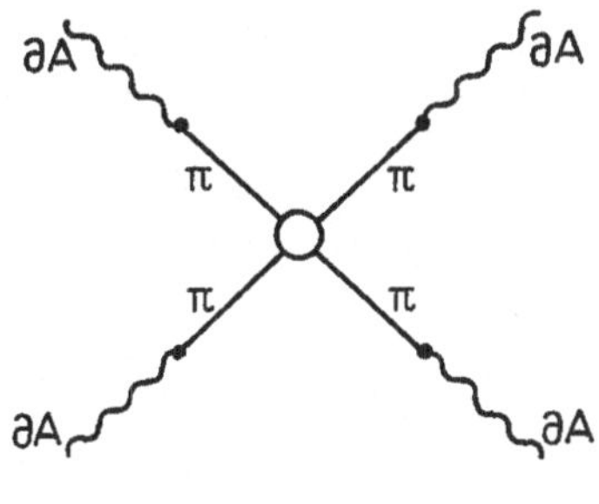

Figure 1o: Contact term in pion four-point function.

This contact term is welcome on what might appear to be totally different grounds. To provide a realistic model for low energy ($\pi\pi$) scattering one will want to see the ($\rho\pi\pi$) and ($\sigma\pi\pi$) vertices enter the calculation with their correct strengths, which is determined by fitting the pole model to the corresponding contribution in an unsubtracted dispersion relation. Such a dispersion relation, however, always operates with one channel-variable fixed: a pole approximation to a ($\pi\pi$) forward scattering dispersion relation will not contain the t-channel exchange graphs from figure 9. The contact term in figure lo offers a possibility to have them cancelled wherever the dispersion relation is unsubtracted or balanced wherever it is subtracted. To realize that this mechanism does in fact operate let us imagine that we add exchange graphs corresponding to higher resonances into our model. On certain low-energy points, however, off mass shell, the amplitude is uniquely fixed already through the target isospin or by an Adler zero, and any newly added exchange graphs are so constrained by the Ward identities that at these special points they either cancel among themselves or with a modified contact term. This property, which by the way is shared by the tree graphs of chiral Lagrangians, has been explicitly checked in a number of instances and has termed a duality-like property(9) of the current-algebra constrained pole graphs. It is not in general shared by any other set of tree graphs constructed by following naively the rules of perturbation theory.

The analysis of generalized Ward identities, as discussed here, continues to suffer from some disadvantages:

(1) Unitarity is taken account of at most approximately, by replacing the most prominent cut

contributions by resonance poles. Effects due to finite widths and proper continua are entirely disregarded. Work is in progress to overcome this difficulty: instead of pole models, functions with proper cut structure (at least in one channel) have been employed in generalized Ward identities, with the discontinuity determined by unitarity. Satisfactory models have been constructed for the $(\pi\pi)$ s-wave (isospin zero) and p-wave phase shifts. Instead of linear equations for effective vertex functions one has now integral equations to solve.[10] This branch of work is still in development.

(2) In any practical calculation, one starts by making some smoothness assumptions at the amplitude with the highest number of axial and vector mesons involved - a starting point where there is little intuition available. One would rather prefer a formalism where the input assumptions are to be made at an amplitude with a small number of pions and then exploited to derive information on amplitudes involving large numbers of mesons. Such a formalism is provided by chiral Lagrangians.

The basic purpose of chiral Lagrangians is quite simple: to find a solution to all generalized Ward identities simultaneously, involving a given set of particles, in the pole approximation. At present, this is the only purpose and the only achievement of chiral Lagrangians, but as is often the case in physics, different formulations of the same object may differ in their potential to suggest the appropriate next step. Let us approach the essential ideas of chiral Lagrangians in several steps.

(1) A simultaneous solution of all generalized Ward identities in a given problem is provided in any theory where the current algebra commutators are satisfied by virtue of canonical commutation relations, which within

their scope of application are not invalidated through higher order renormalization effects. If the theory contains yet unconstrained coupling constants, then the Ward identities are correct to every order in these constants, particularly in the lowest non-trivial order, i. e. the one given by tree graphs, where the renormalization effects do not yet occur. At this point one ought to emphasize very strongly that although tree graphs are formally equivalent to lowest-order perturbation theory, they are not interpreted as such, because the coupling constants are nothing like bare coupling constants and the fields are nothing like fundamental fields; the tree graphs should rather be seen as a simultaneous pole approximation to all possible dispersion relations in the problem, because this is where the coupling constants have been taken from.

(2) If on the level of tree graphs one wishes to retain the pion-pole dominance approximation to matrix elements of the axial divergence and the ρ and (π, A_1) dominance approximation to matrix elements of the vector and axial currents, one has to define the canonical fields so that the following equations hold

$$\partial A(x) = F_\pi \, m_\pi^2 \, \pi(x) \tag{3.6}$$

$$V_\mu(x) = f_\rho \, \rho_\mu(x) \tag{3.7}$$

$$A_\mu(x) = f_A \, A_\mu(x) - F_\pi \, \partial_\mu \, \pi(x) \tag{3.8}$$

A_μ ... axial current, A_μ ... A-meson field

Equ. (3.6) is the PCAC-field equation of Gell-Mann and Levy, equs. (3.7) and (3.8) are the current-field identities. We will now discuss the presently accepted method to achieve these objectives, and in section 4 we will comment on the question to what extent this is the only possible way of proceeding.

Postponing for the moment the realization of (3.7) and (3.8) one starts by constructing a Lagrangian L_o which is invariant under $SU_2 \times SU_2$ transformations. Detailed constructions of this kind have been presented by Weinberg,[11] the most comprehensive discussion (on an accessible mathematical level) has been given by Coleman, Wess and Zumino[12], we will later construct an example following methods of Chang and Gürsey[13] and of Cronin.[14] One constructs the currents by application of Noether's theorem.

The first result to be specified is the infinitesimal transformation of the fields under the generators Q of the $SU_2 \times SU_2$ group

$$\phi_\alpha(x) \to \phi_\alpha(x) + i\varepsilon[\phi_\alpha(x), Q_\beta] \overset{\text{def}}{=} \phi_\alpha(x) + i\varepsilon f_{\alpha\beta}[\phi(x)] \qquad (3.9)$$

With respect to isospin transformations, $f_{\alpha\beta}[\phi(x)]$ will usually be taken linear in ϕ, since the particles occur in isospin multiplets. With respect to transformations generated by axial charges, $f_{\alpha\beta}[\phi(x)]$ will in general not be taken linear in ϕ, since we do not observe parity doubling of particle states. We define

$$j_\mu^\beta(x) = (-i) \sum_\alpha \frac{\delta L_o}{\delta(\partial_\mu \phi_\alpha(x))} f_{\alpha\beta}[\phi(x)] \qquad (3.1o)$$

and find

$$[\phi_\gamma(o), \int j_o^\beta(\bar{x},o)(d\bar{x})^3] = f_{\gamma\beta}[\phi(x)] \qquad (3.11)$$

by virtue of canonical commutators. So we identify

$$Q_\beta = \int j_o^\beta(\bar{x},o)(d\bar{x})^3 \qquad (3.12)$$

which guarantees the group algebra. On this stage of the construction, one has still $\partial j = o$ for all currents in the problem, since L_o was constructed to be invariant.

One may immediately notice that one can add to L_o any term L_1 which depends on the fields only - and not their derivatives - without changing the definition of the current or the commutation relations, i. e. one can break the symmetry without upsetting current algebra. This idea is exploited to realize equ. (3.6)

$$\partial A^{\alpha} = i[\int A_o^{\alpha}(x)\,(d\bar{x})^3, L_1] = F_{\pi}\, m_{\pi}^2\, \pi^{\alpha} \qquad (3.13)$$

by choosing L_1 to depend suitably on $\pi(x)$ only. We see that in this class of models, the limit $m_{\pi}^2 \to o$ eliminates the entire chiral symmetry breaking. We will give an example later.

The only known realization of the current-field identities, equs. (3.7) and (3.8), has to adopt a generalized Yang-Mills formalism as described by Lee, Weinberg and Zumino,(15) with vector and axial vector mesons described by generalized gauge fields. Since the subject of vector and axial vector mesons in chiral Lagrangians has been very amply discussed by S. Gasiorowicz(16) in this school last year, we need not go into details here.

If one restricts one's attention to predicting matrix elements of hadrons on mass shell, one need in fact not explicitly realize equs. (3.6)-(3.8) in one's model, one merely has to convince oneself that one's model is equivalent to some other model which realizes (3.6)-(3.8), in a way to be discussed below.

It has been known for long already that the predictions of field theories for S-matrix elements on the mass-shell do not depend on any specific choice of interpolating fields, in particular they are invariant against canonical field transformations of the type

$$\phi = \chi F(\chi) \quad \text{with} \quad F(o) = 1 \tag{3.14}$$

Note that such transformations do not alter the form of the bilinear terms in L: the particle content of the theory, as specified by the asymptotically free part of the Lagrangian remains the same.

Coleman, Wess and Zumino(12) made the further observation that this invariance with respect to transformations of the type (3.14) does not only apply to the matrix elements of the full S-matrix, but separately also to its representation in terms of tree graphs alone. We omit reproducing the proof here, but we rather turn to the construction of a specific example which will be of use to us in our further discussions.

3.1. Construction of a Chiral Lagrangian

For later use we will present Cronin's construction(14) of a chiral Lagrangian model for pseudoscalar mesons, the essential ideas of which go back to early work of Gürsey. Though at present we are only considering the $SU_2 \times SU_2$ chiral algebra, we wish to draw attention to the fact that the present construction would be equally possible in the case of $SU_3 \times SU_3$ with a nonet of pseudoscalar mesons, using λ-matrices instead of τ-matrices below. We begin with a matrix of hermitean meson fields

$$\phi = \sum_{i=1}^{3} \tau_i \, \pi_i \qquad \left(\text{later } \phi = \sum_{i=o}^{8} \lambda_i \, \phi_i\right) \tag{3.15}$$

and define a nonlinear matrix function

$$M(f\phi) = \sum a_n (if\phi)^n \tag{3.16}$$

To keep the argument of M dimensionless, the constant f is given dimension $[mass]^{-1}$; the coefficients a_n have been chosen real in order to achieve a simple parity transformation of M

$$PM(f\phi)P^{-1} = M^{+}(f\phi) \tag{3.16a}$$

We now have to introduce the $SU_2 \times SU_2$ transformations for M. We have to realize two commuting SU_2 subgroups through transformations on M, the one generated by (V+A), the other by (V-A) charges and the two subgroups connected to each other by parity. We prescribe
(1) transformation generated by (V+A), (parameters α_k) $\phi \to \phi'$ such that

$$M(f\phi') = \exp(i\alpha_k \tau_k/2) M(f\phi) \tag{3.17a}$$

(2) transformation generated by (V-A), (parameters β_k). $\phi \to \phi''$ such that

$$M(f\phi'') = M(f\phi) \exp(-i\beta_k \tau_k/2) \tag{3.17b}$$

Obviously, these two transformation laws are converted into each other by the parity transformation defined in equ. (3.16a).

Equations (3.17) have many solutions. The correct behaviour of M under isovector transformations ($\alpha_k = \beta_k$)

$$M(f\phi + f\delta_V\phi) = M(f\phi) + i\alpha_k [\tfrac{1}{2} \tau_k, M(f\phi)] \tag{3.18a}$$

is guaranteed by linear transformations of the pion fields in isospin

$$\delta_V \phi = i\alpha_k [\tfrac{1}{2} \tau_k, \phi] \tag{3.18b}$$

A transformation generated by axial charges ($\alpha_k = -\beta_k$)

$$M(f\phi + f\delta_A\phi) = M(f\phi) + i\alpha_k \{\tfrac{1}{2} \tau_k, M(f\phi)\}_{+} \tag{3.19}$$

can in general be realized only by a nonlinear transformation of the pion fields. As an explicit form of M, we quote one of the examples given by Cronin(14)

$$M(f\phi) = \frac{1+if\phi}{1-if\phi} \qquad (3.20a)$$

$$\delta_A\phi = \frac{\alpha_k}{2f}\,(\tau_k + f^2\phi\tau_k\phi) \qquad (3.20)$$

This solution also holds in the SU_3 case. It is not the only possible solution, yet $M(f\phi)$ is not entirely arbitrary. From equ. (3.17) follows

$$M^+(f\phi')M(f\phi') = M^+(f\phi)M(f\phi) \qquad (3.21a)$$

$$M(f\phi')M^+(f\phi') = \exp(i\alpha_k\tau_k/2)M(f\phi)M^+(f\phi)\exp(-i\alpha_k\tau_k/2) \qquad (3.21b)$$

Since both $[M(f\phi'),M^+(f\phi')] = 0$ and $[M(f\phi),M^+(f\phi)] = 0$ it follows from equs. (3.21) that $(M(f\phi)\cdot M^+(f\phi))$ is a multiple of the unit matrix. The least redundant form of the theory is obtained by putting

$$M(f\phi)\;M^+(f\phi) = 1 \qquad (3.22)$$

Obviously, equ. (3.22) implies relations between the expansion coefficients a_n. These relations can be so arranged as to leave the odd coefficients a_{2n+1} unconstrained and to have the even ones determined. For normalization we fix $a_1=1$.

A chiral invariant kinetic Lagrangian is now proposed

$$L_o = \frac{1}{4f^2}\,\mathrm{Tr}\,(\partial_\mu M^+\partial_\mu M) = \frac{1}{4}\,\mathrm{Tr}\,(\partial_\mu\phi\partial_\mu\phi) + O[\phi^4] \qquad (3.23)$$

and the axial current is constructed according to Noether's theorem

$$A_\mu^\alpha = \frac{(-i)}{2f^2} \mathrm{Tr}\{\partial_\mu M^+\{\frac{1}{2}\tau^\alpha, M\}_+\} = \frac{(-1)}{f}\partial_\mu\phi^\alpha + O[\phi^3] \qquad (3.24)$$

This identifies $1/f=F_\pi$ with the pion decay constant. It is not possible to construct a chiral invariant meson mass term from $M(\phi)$; this should be obvious, since if there is no parity doubling in the theory the mesons have to be massless in the symmetry limit.

Any mass term has to break chiral symmetry. To keep isospin invariance, it must be of the form

$$L_m = \mathrm{Tr}\{g(M)\} + \text{h. c.} \qquad (3.25)$$

For the axial divergence we get

$$\partial A^\alpha = i[\int A_0^\alpha(x)(dx)^3, L_m = (-i)\mathrm{Tr}\{g'(M)\{\frac{1}{2}\tau^\alpha, M\}_+\} + \text{h. c.} \qquad (3.26)$$

To realize the PCAC identity $\partial A^\alpha = \frac{1}{f} m_\pi^2 \phi^\alpha$, we have to satisfy

$$dg(M(\phi))/dM = i\frac{m_\pi^2}{4f}\frac{\phi}{M(\phi)} \qquad (3.27)$$

One may use this relation to determine either $g(M)$ or $M(\phi)$, once the other quantity is known. A specification of $M(\phi)$ does not have any general significance, since this is just a choice of a particular interpolating field, which can be changed by a canonical field transformation (see equ.(3.14)). The dependence of $g(M)$ on $M(\phi)$, however, has an invariant significance in terms of the transformation properties of L_m under the chiral algebra. Such a characterization is of the same nature as for instance the statement that SU_3-symmetry is broken by an octet Hamiltonian, and obviously such a statement has a meaning independently of any particular choice of fields. Any individual term L_N

$$L_m^N = \mathrm{Tr}\ \{[M(\phi)]^N\} + \text{h. c.} \tag{3.28}$$

is the isoscalar member of a finite dimensional operator multiplet with respect to the chiral algebra. It consists of operators

$$A_N(\alpha_1 \ldots \alpha_N) = \mathrm{Tr}\ \{ \prod_{i=1}^{N} (\tau_{\alpha_i} M(\phi))\} + \text{h. c.}$$

$$\alpha_i = (o,1,2,3) \tag{3.29}$$

with $A_N(o,o, \ldots o) = L_m^N$. Since all factors in (3.29) commute, only symmetrized products enter. The highest isospin realized in this multiplet is N and there are all isospins from 0 to N contained in the multiplet A_N (isospin J, if $(N-J)$ coefficients α_i=o and J coefficients $\alpha_i \neq o$ in a symmetrized and traceless combination). Algebraically such multiplets are described in terms of their dimensionality under the left-handed and right-handed SU_2 factor groups in $SU_2 \times SU_2$ as $(N+1,N+1)$ or in spinor analogy $(N/2,N/2)$. The operators in the set A_N are in fact a reducible representation with $(N+1,N+1)$ being the highest multiplet contained in the set, this is however not relevant for the discussion to follow).

Let us illustrate all these points in a calculation of $(\pi\pi)$ scattering in the tree graph approximation. We take

$$M(\phi) = 1+if\phi + \frac{1}{2}(if\phi)^2 + a_3(if\phi)^3 + (a_3 - \frac{1}{8})(if\phi)^4 + \ldots$$

$$g(M) = \frac{m_\pi^2}{4f^2}(\alpha M + \frac{(1-\alpha)}{4} M^2) \tag{3.3o}$$

where we have allowed for two different forms of chiral symmetry breaking with a parameter α to express their relative weight. We find

$$\langle \pi_a(q_1)\pi_b(q_2)|\pi_c(-q_3)\pi_d(-q_4)\rangle =$$

$$= \sum_{perm} (-)\mathrm{Tr}\{\tau_{i1}\cdot\tau_{i2}\cdot\tau_{i3}\cdot\tau_{i4}\}(if)^4\{\frac{1}{4f^2}(\frac{(-1)}{4}(q_{i1}+q_{i2})(q_{i3}+q_{i4})+$$

$$+ 2a_3 q_{i1}(q_{i2}+q_{i3}+q_{i4})) + \frac{m_\pi^2}{4f^2}(2\alpha(a_3 - \frac{1}{8}) + \frac{(1-\alpha)}{2} \times$$

$$\times (\frac{1}{4} + 2a_3 + 2(a_3 - \frac{1}{8}))\} =$$

$$= \sum_{perm} (-)\mathrm{Tr}\{\tau_{i1}\cdot\tau_{i2}\cdot\tau_{i3}\cdot\tau_{i4}\}f^2\{\frac{1}{16}(q_{i1}+q_{i2})^2+\frac{a_3}{2}(m_\pi^2-q_{i1}^2)-\frac{1}{16}\alpha m_\pi^2\}$$

$$= \frac{(-)}{F_\pi^2}\{\delta_{ab}\delta_{cd}(s-\alpha m_\pi^2)+\delta_{ac}\delta_{bd}(t-\alpha m_\pi^2)+\delta_{ad}\delta_{bc} \times$$

$$\times (u-\alpha m_\pi^2)+(\delta_{ab}\delta_{cd}+\delta_{ac}\delta_{bd} +$$

$$+ \delta_{ad}\delta_{bc})2a_3(4m_\pi^2-\sum q_i{}^2)\} \qquad (3.32)$$

We see now explicitly that the value of the parameter a_3 does not affect the predictions for reactions on mass shell, since it describes only a choice of the pion field, however the parameter α does clearly affect the predictions.

To reproduce the PCAC equation $\partial A = F_\pi\, m_\pi^2\, \phi_\pi$ and so to guarantee Adler zeroes, one has to correlate a_3 with α through $2a_3=(\alpha-1)$ using equ. (3.27). With this choice we rewrite equ. (3.32) in a form which will be useful to us later

$$\langle \pi_a(q_1)\pi_b(q_2)|\pi_c(-q_3)\pi_d(-q_4)\rangle =$$

$$= \frac{1}{F_\pi^2}\{\delta_{ab}\delta_{cd}(m_\pi^2-s)+\delta_{ac}\delta_{bd}(m_\pi^2-t) +$$

$$+ \delta_{ad}\delta_{bc}(m_\pi^2-u)+(\alpha-1)(\delta_{ab}\delta_{cd}+\delta_{ac}\delta_{bd} +$$

$$+ \delta_{ad}\delta_{bc})(s+t+u-3m_\pi^2)\} \qquad (3.33)$$

The Weinberg scattering lengths come out for $\alpha=1$, which is the case of simplest chiral symmetry breaking (isoscalar σ-terms), as adopted by Weinberg.

4. RECONSTRUCTION OF CHIRAL LAGRANGIANS FROM ADLER ZEROES

We have extensively presented the formalism of chiral Lagrangians which provides at present the best model of extrapolating from low energy theorems of current algebra. In developing this formalism we made a number of assumptions such as the $SU_2 \times SU_2$ current algebra (equs. (2.15)), which so far has been properly justified only in the chiral symmetry limit with massless pions. Our construction of chiral Lagrangians made even heavier use of chiral symmetry concepts. Though it has been argued at the very beginning that even the Goldberger-Treiman relation could not be made plausible without advocating the chiral symmetry limit in some form, we cannot yet be certain that our way of developing the idea of chiral symmetry is in fact the only reasonable one.

For this purpose we go back a step, discard current algebra again and assume only pion-pole dominance and in particular the existence of Adler zeroes. We want to understand how much one can deduce merely from the presence of these zeroes. This question has a certain relevance in conjunction with recent claims by Lovelace (17) that there should be connections between chiral symmetry and duality, the latter being realized through the Veneziano model. We will see that Adler zeroes play a very important part in understanding the origin of this claim.

It is not yet possible to present a complete answer since there exists no complete and soluble model to provide the grounds for comparison. A tree-graph model is the nearest to realistic structure available on the side of chiral symmetry and its validity is confined to the low-energy domain, it may be held that the Veneziano model is on a similar level at least for its low-energy properties, since it has poles and no cuts, and its Regge asymptotic behaviour need not be directly relevant to its low-energy properties. It should however be noted that in adopting this point of view one is setting up a framework which is too narrow to even attempt to explain the emergence of Adler zeroes through $\alpha_\rho(m_\pi^2)=1/2$ in the Veneziano amplitude; we will impose Adler zeroes as an input.

As a first step we will construct a tree-graph model involving pions only (both as external and as exchanged particles) and we will allow the irreducible parts of the tree graphs to depend only in second order on the pion momenta. This is the same structure as provided by the chiral Lagrangian model in section 3.2. We will make Adler zeroes our main requirement: every n-pion amplitude is required to vanish whenever any one pion is taken to zero energy-momentum, and the others are on the mass-shell.

For the four-pions amplitude this implies one constraint. Its most general form, according to Weinberg, is given by

$$\langle\pi_a,\pi_b|\pi_c,\pi_d\rangle =$$

$$= \delta_{ab}\delta_{cd}(A+Bs+C(t+u)) +$$

$$+ \delta_{ac}\delta_{bd}(A+Bt+C(s+u)) +$$

$$+ \delta_{ad}\delta_{bc}(A+Bu+C(s+t)) \qquad (4.1)$$

Adler zeroes demand

$$A + Bm_\pi^2 + 2Cm_\pi^2 = o \tag{4.2}$$

and we rewrite

$$\begin{aligned}<\pi_a,\pi_b|\pi_c,\pi_d> &= (B-C)(\delta_{ab}\delta_{cd}(s-m_\pi^2) + \\ &+ \delta_{ac}\delta_{bd}(t-m_\pi^2)+\delta_{ad}\delta_{bc}(u-m_\pi^2)) + \\ &+ C\ (\delta_{ab}\delta_{cd}+\delta_{ac}\delta_{bd}+\delta_{ad}\delta_{bc})(s+t+u-3m_\pi^2)\end{aligned} \tag{4.3}$$

We see that this is of the same form as the result from the chiral Lagrangian (equ. 3.33) if we identify

$$C = (\alpha-1)/F_\pi^2 \tag{4.4a}$$

$$C-B = 1/F_\pi^2 \tag{4.4b}$$

Since the chiral symmetry breaking parameter α is not specified by any general principles we can always adjust it so as to satisfy equ. (4.4a). Equ. (4.4b), however, raises problems, because even if we discard the identification of axial current generated by a chiral transformation with the axial current observed in weak interactions and consequently regard F_π as a free parameter we will still have no guarantee that $(C-B)>o$. In fact this ambiguity corresponds precisely to the undetermined factor Z in the commutator of two axial currents (compare equ. (2.13)) the sign of which distinguishes between the compact algebra $SU_2\times SU_2$ (for $Z>o$) and the unplausible non-compact algebras E_3 (for $Z=o$) and SO(3,1) (for $Z<o$). The distinction between these three alternatives and the fixation of the scale of the axial current are two elements not determined by Adler zeroes or, to our present knowledge, any other consequence of pion-pole dominance alone. For the following we choose the $SU_2\times SU_2$ case, although the arguments for the other

cases would proceed in precisely the same way.

Given any tree graph model with Adler zeroes, we have found that a suitable chiral Lagrangian can be constructed to reproduce the four-pion amplitudes of the model (up to the reservations discussed above). We turn now to the six-pion amplitudes. Since the four-pion amplitudes of both models are identical, so are the reducible contributions to the six-pion amplitudes, and any discrepancy can only occur through the six-pion contact graph. However since both theories have Adler zeroes, so has their difference, i. e. the (possible) difference between the two contact terms has to satisfy all Adler zeroes by itself. For n≥6 pions there is only one possible bilinear form of this kind (up to a multiplicative factor γ)

$$f(p_i) = \gamma\{(\sum_{i=1}^{n} p_i{}^2)-(n-1)m_\pi^2\} \tag{4.5}$$

This can be seen as follows. Any scalar bilinear term, constructed from the momenta of n pions, can be written as

$$f(p^{(i)}) = \sum_{i,j=1}^{n-1} a_{ij}(p^{(i)}\cdot p^{(j)}) \tag{4.6}$$

using energy-momentum conservation. For $f(p^{(i)})\to o$ as $p_\mu^{(1)}\to o$ and $(p^{(i)})^2=m_\pi^2$, $f(p^{(i)})$ must have the form

$$f(p^{(i)}) = \sum_{i=2}^{n-1} b_i(p^{(1)}\cdot p^{(i)}) + \sum_{i=2}^{n} c_i((p^{(i)})^2-m_\pi^2)+d(p^{(1)})^2 \tag{4.7}$$

and a similar form for $p_\mu^{(2)}\to o$

$$f(p^{(i)}) = \sum_{\substack{i=1\\i\neq 2}}^{n-1} b_i'(p^{(2)}\cdot p^{(i)}) + \sum_{\substack{i=1\\i\neq 2}}^{n} c_i'((p^{(i)})^2-m_\pi^2)+d'(p^{(2)})^2 \tag{4.8}$$

For $n\geq 6$ pions, the quantities appearing in equs. (4.7) and (4.8) are not related among each other by energy-momentum conservation, and they are linearly independent since the relations imposed by space-time dimensionality are highly nonlinear ones. Identifying (4.7) and (4.8) implies relations between coefficients:

$$b_2 = b_1' \tag{4.9a}$$

$$b_i = o, b_i' = o \quad (i\geq 3) \tag{4.9b}$$

$$d = c_1', d' = c_2 \tag{4.9c}$$

$$c_i = c_i' \quad (i\geq 3) \tag{4.9d}$$

Further relations are deduced by proceeding to study the limits $p_\mu^{(i)}\to o$ for $i\geq 3$; the final result is the form given in equ. (4.5). Note finally that in the case of mass splitting one would replace equ. (4.6) by

$$f(p^{(i)}) = \gamma\{(\sum_{i=1}^{n}(p_i{}^2/m_i{}^2)-(n-1)\} \tag{4.1o}$$

Turning back to the original problem, we will now show that we can eliminate the discrepancy on the six-pion level without destroying the agreement on the four-pion level by a suitable modification of the chiral symmetry breaking term. Rather than explicitly performing this construction for six pions, let us start on an induction argument.

Given again a tree graph model with Adler zeroes, assume that we have constructed a chiral Lagrangian L_N such that its tree-graph amplitudes agree with those of the model for amplitudes involving up to 2N pions. The amplitudes for (2N+2) pions differ by a contact term of the form (4.5). We now proceed to construct a chiral

Lagrangian L_{N+1} so as to extend agreement to amplitudes with (2N+2) pions. In fact we only change the chiral symmetry breaking term g(M) in equ. (3.25) but since we want to retain the field-theoretic PCAC equation $\partial A^{\alpha}=F_{\pi}m_{\pi}^{2}\phi^{\alpha}$ as a guarantee for Adler zeroes, we also have to change the definition of the pion field and it appears most convenient to start by modifying $M(\phi)$ in order ϕ^{2n+1}

$$M(\phi)\to M'(\phi') = M(\phi')+i\gamma'(\phi')^{2n+1} \tag{4.11}$$

Obviously this change effects only amplitudes for $\geq(2n+2)$ pions. In the kinetic term of the Lagrangian one obtains an additional contribution

$$L_o(\phi)\to L_o'(\phi') =$$

$$= L_o(\phi') + \frac{f\gamma'}{2f^2}\,\mathrm{Tr}\{(\partial_\mu\phi')\partial^\mu(\phi'^{2n+1})\} + O(\phi'^{2n+4}) =$$

$$= L_o(\phi') - \frac{\gamma'}{2f}\,\mathrm{Tr}\{(\partial^2\phi')(\phi')^{2n+1}\} + O(\phi'^{2n+4}) \tag{4.12}$$

We also have a change in the generalized mass term $g(M)\to g'(M')$ which is determined by retaining equ. (3.27) to guarantee Adler zeroes.

We rewrite equ. (3.27)

$$\frac{dg}{d\phi} = \frac{dg}{dM}\cdot\frac{dM}{d\phi} = i\,\frac{m_\pi^2}{4f}\cdot\frac{\phi}{M(\phi)}\left(\frac{dM}{d\phi}\right) \tag{4.13}$$

The right-hand side changes as

$$\frac{dg}{d\phi}\to\frac{dg'}{d\phi'} = \frac{dg}{d\phi'} + i\,\frac{m_\pi^2}{4f}\phi'\frac{d}{d\phi'}\log\left\{\frac{M(\phi')+i\gamma'(\phi')^{2n+1}}{M(\phi')}\right\} =$$

$$= \frac{dg}{d\phi'} - \frac{m_\pi^2}{4f}\gamma'(2n+1)(\phi')^{2n+1} + O(\phi'^{2n+3}) \tag{4.14}$$

which further leads to

$$g(\phi)\rightarrow g(\phi') - \frac{m_\pi^2}{4f}\gamma'\frac{(2n+1)}{(2n+2)}(\phi')^{2n+2} + O(\phi'^{2n+4}) \qquad (4.15)$$

Collecting now the total change in the contact term for (2n+2) pions we get

$$\delta L = -\frac{\gamma'}{2f}(\mathrm{Tr}\{(\partial^2\phi')(\phi')^{2n+1}\} +$$

$$+ m_\pi^2 \frac{(2n+1)}{(2n+2)}\mathrm{Tr}\{(\phi')^{2n+2}\}) + O((\phi')^{2n+4}) \qquad (4.16)$$

which, by adjusting γ', can be used to eliminate any possible discrepancy for the (2n+2)-pion amplitudes and so we can by stepwise construction reproduce any pion-tree graph model with Adler zeroes by a suitably chosen chiral Lagrangian of $SU_2 \times SU_2$ (or E_3 or SO(3.1)) type. This result has first been proved by Ellis[18] in a form very similar to the one presented here.

We see once again that pion pole dominance does in fact, at least in this simple model, force us to adopt the structures of chiral symmetry (up to a crucial scale factor in the commutator of two axial currents): the existence of Adler zeroes alone implies many results of chiral symmetry by itself.

We return now to the Veneziano model. Lovelace[17] has advocated spectacular agreement of the pion-pion scattering lengths as derived from the Veneziano model with the values derived by Weinberg from current algebra. The suggestion was made that this agreement indicates the presence of deeper relations between the two theories. We will try to understand it on a less profound level. As seen in equ. (4.1), the pion-pion scattering amplitude depends on three constants, called there A, B, C; so we need three pieces of information

$$<\pi_a,\pi_b|\pi_c,\pi_d> =$$

$$= \delta_{ab}\delta_{cd}(A+Bs+C(t+u)) +$$

$$+ \delta_{ac}\delta_{bd}(A+Bt+C(s+u)) +$$

$$+ \delta_{ad}\delta_{bc}(A+Bu+C(s+t)) \qquad [\text{of } (4.1)]$$

(1) Adler zeroes: In the current algebra description, Adler zeroes have to be postulated and lead to specify $A+Bm_\pi^2+2Cm_\pi^2=o$. In the Veneziano model, zeroes emerge at the correct point if the ρ-trajectory satisfies $\alpha_\rho(m_\pi^2)=1/2$. We are not able to present any deeper explanation of this relation.

(2) Overall scale: In current algebra, the overall scale is determined through the scale of the weak axial current; in the Veneziano model, this scale is not determined, since no non-linear unitarity relation has been imposed. One may of course arrive at a scale by fitting the residue of the ρ-pole; the result is acceptable and it reflects the fact that the KSFR relation, though neither justified nor accurate, is still approximately in agreement with experiment.

(3) A third input is necessary. Weinberg derives it by assuming the simplest possible chiral symmetry breaking, with isoscalar σ-terms, equivalent to (α=1) in equ. (3.3o) and C=o in equ. (4.4a). In the Veneziano model this third assumption is replaced by a characteristic structure element: the absence of energy-variables associated with exotic channels. In the Veneziano model this is obvious, since there all variables are intimately associated with the pole structure of the model. With the present form of the amplitude

$$<\pi_a,\pi_b|\pi_c,\pi_d> = (B-C)(\delta_{ab}\delta_{cd}(s-m_\pi^2) +$$

$$+ \delta_{ac}\delta_{bd}(t-m_\pi^2)+\delta_{ad}\delta_{bc}(u-m_\pi^2)) +$$

$$+ C(\delta_{ab}\delta_{cd}+\delta_{ac}\delta_{bd}+\delta_{ad}\delta_{bc})(s+t+u-3m_\pi^2) \qquad \text{[of (4.3)]}$$

this implies again C=o, and so the requirement of absence of exotic variables in the four-pion amplitude gives a constraint on chiral symmetry breaking: it has the effect of excluding exotic σ-terms from the four-pion amplitude.

Since there is an infinite number of modes of chiral symmetry breaking in pion reactions, we have to look for further constraints in amplitudes involving six or more pions. However exclusion of exotic channel variables does not produce any further constraints, since it has been shown by Ellis that in amplitudes involving six or more pions any bilinear form in the momenta can be rewritten in terms of channel energies for non-exotic two-pion and three-pion channels, as long as generalized Bose statistics is assumed.

One may proceed to tighten the constraints by admitting only a dependence on energy-variables associated with non-exotic two pion channels,(19) having in mind a situation where the dominant dynamical structure in multipion amplitudes at low energies is meant to approximately reflect exchanges of ρ and σ meson. This constraint does then in fact uniquely determine chiral symmetry breaking: the non-linear σ-model (with only iso-scalar σ-terms) emerges as the only solution, perhaps not unexpectedly.

To see how this comes about recall the structure of the kinetic term in equ. (3.23)

$$L_o = \frac{1}{4f^2} \mathrm{Tr}(\partial_\mu M^+ \partial_\mu M) = - \frac{1}{4f^2} \mathrm{Tr}\{(\partial^2 M^+)(M)\} \qquad (4.17)$$

To make sure that in (4.17) one obtains only channel variables associated with two pion channels one has to require that apart from the linear term, $M(f\phi)$ is even in ϕ. Then we can, by repeated partial integrations[18] rewrite (4.17) into a sum of terms proportional to

$$\mathrm{Tr}\{(\partial^2(\phi^2))((\phi^2)^n)\} = \frac{1}{2}\partial^2(\mathrm{Tr}\{\phi^2\})\cdot\mathrm{Tr}\{(\phi^2)^n\} + \\ + \frac{1}{2}\sum_i \partial^2(\mathrm{Tr}\{\tau_i\phi^2\})\cdot\mathrm{Tr}\{\tau_i(\phi^2)^n\} \tag{4.18}$$

The second term in (4.18) vanishes by Bose-statistics, and we see that only energy-variables of ($J^P=o^+$) channels occur. To guarantee this structure we need

$$M(f\phi) = if\phi+f\sigma[(f\phi)^2] \tag{4.19}$$

$$M^+(f\phi)M(f\phi) = 1 \Rightarrow f\sigma((f\phi)^2) = \sqrt{1-(f\phi)^2} \tag{4.2o}$$

which produces a kinetic term

$$L_o = \frac{1}{4f^2}\,\mathrm{Tr}(\partial_\mu M^+\partial_\mu M) = \frac{1}{4}\mathrm{Tr}(\partial_\mu\phi\partial_\mu\phi+\partial_\mu\sigma\partial_\mu\sigma) \tag{4.21}$$

and a chiral symmetry breaking term $g(M)$ can be constructed in accordance with equ. (3.27)

$$L_m = \Big(\frac{m_\pi^2}{4f^2}\,\mathrm{Tr}\{M(f\phi)-if\phi\} + \text{h. c.}\Big) = \frac{m_\pi^2}{f}\,\sigma \tag{4.22}$$

To summarize: we claim that these are certain structural similarities between the simplest models of broken chiral symmetry and of duality. It appears premature, however, to claim deep relations on the present evidence.

5. EXTENSION TO $SU_3 \times SU_3$

It is widely known that if current algebra is due to some (yet unknown) property of weak interactions, possibly some property of the W-bosons, then it would be very unnatural not to see the strangeness changing currents participate in the algebra. The simplest extension of the $SU_2 \times SU_2$ algebra discussed so far, is an $SU_3 \times SU_3$ structure

$$[Q_V^i, V_\mu^j] = i f^{ijk} V_\mu^k \tag{5.1a}$$

$$[Q_V^i, A_\mu^j] = i f^{ijk} A_\mu^k \tag{5.1b}$$

$$[Q_A^i, V_\mu^j] = i f^{ijk} A_\mu^k \tag{5.1c}$$

$$[Q_A^i, A_\mu^j] = i f^{ijk} V_\mu^k \tag{5.1d}$$

with $\quad Q_V^i = \int V_o^i(x)\,(dx)^3 \quad$ and $\quad Q_A^i = \int A_o^i(x)\,(dx)^3$

and

$$J_\mu^{el} = V_\mu^3 + \frac{1}{\sqrt{3}} V_\mu^8 \tag{5.2}$$

$$J_\mu^W = \cos\theta\,(V+A)_\mu^{(1+i2)} + \sin\theta\,(V+A)_\mu^{(4+i5)} \tag{5.3}$$

where θ is the Cabibbo angle which describes the orientation of the weak current in SU_3-space.

If current algebra is approached from the point of strong interaction symmetries one is confronted with the problem of SU_3 breaking and with the question whether a generalized Goldberger-Treiman relation is valid for the divergence of the strangeness changing weak axial current, for instance in $\Lambda(\beta)$-decay

$$(m_N + m_\Lambda)\,(g_A^{(\Lambda)}) \approx -(\sqrt{2}\,F_K)\cdot(G_{NK\Lambda}) \tag{5.4}$$

Experimentally, little is known about the status of this Goldberger-Treiman relation. The weak-interaction constants have been measured accurately enough; since the entire relation can be multiplied by (sin θ), possible inaccuracies in the estimation of the Cabibbo angle can be bypassed. However the Yukawa coupling constant has so far not been available from a dispersion relation estimate. This is because in ($\bar{K}N$) scattering, there is a strong sub-threshold contribution due to ($\Lambda\pi$) and ($\Sigma\pi$) intermediate states which contains the Y_0^*(14o5) and Y_1^*(1385) resonances. Possible errors in estimating this subthreshold cut strongly influence the estimate of the Λ (and Σ) pole contributions. According to some analysis the relation (5.4) is about as accurate as the pion Goldberger-Treiman relation, according to others it disagrees by a factor of about two.[2o]

As regards purely theoretical arguments, the K-meson Goldberger-Treiman relation is less favoured since the ratio of distances of pole and cut from the origin is $m_K^2/(m_K+2m_\pi)^2 \approx 1/2$, rather than 1/9 as in the pion case. But we have extensively argued that this ratio of distances is not a reliable guide to the accuracy of the relation in any case, and the real question should be to what extent a model of chiral $SU_3 \times SU_3$ symmetry is an acceptable approximation to hadron physics.

In the $SU_3 \times SU_3$ symmetry limit, we would have at least eight massless mesons, and the other particles would occur in mass-degenerate SU_3 multiplets. There should be two chains of symmetry breaking to reach hadron physics from this limit

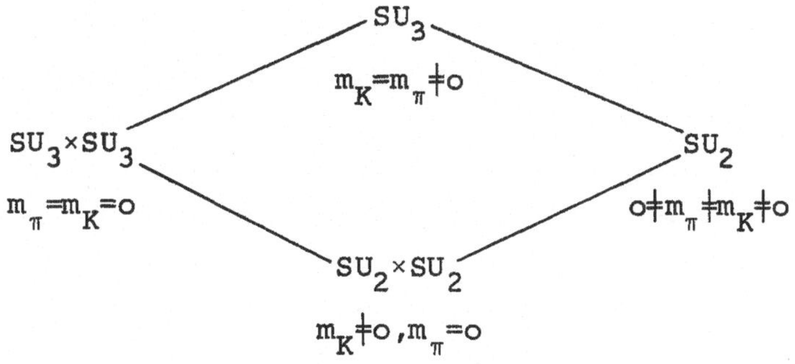

We see immediately that for the pseudoscalar meson system, violations of SU_3 and of chiral symmetries should be strongly connected with each other. In fact the successes of SU_3 for the pseudoscalar meson system are not very plausible unless the pseudoscalar masses are regarded an insignificant perturbation upon a symmetry structure with massless meson. We will present consistent models of this situation, but we wish to recall that it is quite conceivable that our entire interpretation may be wrong, that SU_3 violations are of an entirely different origin and K-meson PCAC is simply not correct beyond the SU_3 symmetry limit. It is fair to say that of the presently known tests of current algebra none are accurate enough to establish the validity of these commutators which are specific to $SU_3 \times SU_3$, on a level of accuracy including SU_3 violations. This is very simple to understand: commutators of strangeness changing axial currents (or divergences) have always been exploited only in the K-meson pole dominance approximation the accuracy of which may well be put into doubt; commutators of strangeness changing vector currents have essentially only been tested in the approximation of single-particle saturation; this approximation is known to be affected by SU_3 violations.

Keeping these precautions in mind, we now turn to the details of the theoretical structure based on the $SU_3 \times SU_3$ current algebra which is only plausible if the chiral $SU_3 \times SU_3$ symmetry limit is in fact relevant to hadron physics. We can construct low energy theorems for K-mesons as we did for π-mesons, we can extend the formalism of generalized Ward identities and we can construct chiral Lagrangians. Again we will illustrate most of our important points using only a pseudoscalar meson model, because it is (for lecturing purposes) the simplest to construct. We will continue to adopt the Gürsey-Cronin construction from section 3.2 because it is most easily generalized to $SU_3 \times SU_3$. This is done without losing generality, since Coleman, Wess and Zumino[12] have shown that, within certain classes, chiral Lagrangians are in fact unique and different methods of construction produce results which are equivalent up to canonical field transformations.

For the present case, the meson matrix is three dimensional

$$\phi = \sum_i \lambda^i \phi^i \qquad (i = o, \ldots, 8) \tag{5.5}$$

We have now the option of either regarding ϕ_o an independent field of a ninth pseudoscalar meson (η' or X) or use a group invariant

$$\det M(\phi) = 1 \Rightarrow \phi_o = f(\phi^i) = O[\phi^3] \tag{5.6}$$

to eliminate ϕ_o and make it a function of the other fields. [In two dimensions, a constraint of the form (5.6) leads to an identity without restrictive content for unitary matrices $M(\phi)$ constructed according to the prescriptions - for further discussion see Cronin.[14]]

If we do not eliminate ϕ_o according to equ. (5.6), we deal with nine mesons and we can extend the $SU_3\times SU_3$ algebra to $U_3\times U_3$, which contains two more SU_3-singlet generators, connected with yet another vector and axial vector current. The vector current may possibly be associated with baryon number, however there is no positive evidence available on the axial current.

The construction of the chiral symmetric kinetic Lagrangian proceeds as in the $SU_2\times SU_2$ case, however some further discussion has to be given to the chiral symmetry breaking term, since now we have SU_3 mass splitting to describe at the same time. Let us study the simplest possible form of chiral symmetry breaking and take

$$L_m = \frac{m^2}{4f^2}\,\mathrm{Tr}\{\sqrt{3/2}(\lambda_o+c\lambda_8)(M(f\phi)+M^+(f\phi))\} \tag{5.7}$$

Algebraically, this type of chiral symmetry breaking operator is described as being a member of an 18-dimensional multiplet of 9 scalars u_i and 9 pseudoscalars v_i transforming under the chiral $SU_3\times SU_3$ algebra as follows

$$[Q_V^i,u^j] = if^{ijk}\,u^k \tag{5.8a}$$

$$[Q_V^i,v^j] = if^{ijk}\,v^k \tag{5.8b}$$

$$[Q_A^i,u^j] =-id^{ijk}\,v^k \tag{5.8c}$$

$$[Q_A^i,v^j] = id^{ijk}\,u^k \tag{5.8d}$$

with

$$L_m = -(u_o+cu_8) \tag{5.9}$$

Given a suitable realization of $M(f\phi)$

$$M(f\phi) = if\phi+\sqrt{1-(f\phi)^2} \tag{5.1o}$$

which is familiar from the σ model, we find

$$\partial A_i = \frac{m^2}{f}\sqrt{3/2}(d_{io\ell}+cd_{i8\ell})\phi_\ell \qquad (5.11)$$

The interpretation of equs. (5.7) and (5.11) depends on whether we regard ϕ_o as an independent or as a dependent field. To obtain the pseudoscalar meson masses, we keep in the generalized mass term in equ. (5.7) the terms of second order in the fields and, if necessary, diagonalize it. In the case of nine independent fields we find from equ. (5.7) ideal mixing among the pseudoscalar mesons with

$$m_\eta^2 = m_\pi^2 \quad \text{and} \quad m_{\eta'}^2 = 2m_K^2 - m_\pi^2 \qquad (5.12)$$

and we find from equ. (5.11) a field theoretic PCAC equation

$$\partial A^i = \frac{m_i{}^2}{f}\phi_i \qquad (5.13)$$

with the guarantee of Adler zeroes for each of the nine mesons. In the case of eight independent fields we find a Gell-Mann Okubo mass formula

$$4m_K^2 = 3m_\eta^2 + m_\pi^2 \qquad (5.14)$$

and Adler zeroes only for π- and K-mesons. Specifically we have

$$\partial A^8 = \frac{m^2}{f}\sqrt{3/2}((\sqrt{2/3} - \frac{c}{\sqrt{3}})\phi_8 + c\sqrt{2/3}\,\phi_o) \qquad (5.15)$$

the latter term being of third and higher order in the meson fields, (see equ. (5.6)).

In this theory, the parameter c describes the relative strength of $SU_2 \times SU_2$ vs SU_3 violation; it is determined by

$$m_\pi^2/m_K^2 = (\sqrt{2}+c)/(\sqrt{2}-c/2) \qquad (5.16)$$

and it comes out as $c \approx -1.25$. In the same theory we find the meson decay constants in $K_{\mu 2}$ and $\pi_{\mu 2}$ decay equal

$$F_K = F_\pi = f \tag{5.17}$$

and we find the $K_{\ell 3}$-decay parameter $F_+(o)=1$. Since we know from experiment that $F_K/F_\pi F_+(o) \approx 1.23$ this implies a 2o% inaccuracy at least.

One may abstract from this model a more general description of chiral symmetry breaking, by demanding that the chiral symmetry breaking energy operator has the same algebraic transformation properties as L_m in equ. (5.9) : being a member of an 18-dimensional operator multiplet which realizes the representation $((3,3^*)+(3^*,3))$ of chiral $SU_3 \times SU_3$. The notation $(3,3^*)$ means that the multiplet consists of operators which under the right-handed SU_3 factor group in $SU_3 \times SU_3$ transform like a triplet and under the left-handed SU_3 factor group in $SU_3 \times SU_3$ like an antitriplet. Since parity interchanges the two factor groups (which are generated by (Q_V+Q_A) and (Q_V-Q_A) respectively, any parity invariant theory must contain multiplets only in the form (a,b)+(b,a). The parameter c would have a significance similar to that of the Cabibbo angle, and its value may be very hard to deduce from first principles. The eighteen operators of the multiplet are partly interpretable in terms of current divergences. We have

$$H_B = -L_B = u_o + cu_8 \tag{5.18}$$

$$\partial V^i = cf_{i8k}\, u_k \tag{5.19}$$

$$\partial A^i = -d_{iok}\, v_k - cd_{i8k}\, v_k \tag{5.2o}$$

in particular

$$\partial A^\pi = -(\sqrt{2} + c)/\sqrt{3}\; v^\pi \tag{5.21a}$$

$$\partial A^K = -(\sqrt{2} - c/2)/\sqrt{3}\; v^K \tag{5.21b}$$

$$\partial A^8 = -(\sqrt{2} - c)/\sqrt{3}\; v^8 - \sqrt{2/3}\; cv^o \tag{5.21c}$$

A test of the theory can be given by identifying the so-called σ-terms which describe the low-energy values of baryon-meson scattering amplitudes in the crossing-even meson forward scattering amplitudes (compare equ. (2.9))

$$\lim_{q\to o} \langle f,M^\alpha(q)|i,M^\beta(q)\rangle = \frac{(i)}{2F_M^2}\{\langle f|[\int A_o^\alpha(x)(dx)^3,\partial A^\beta(o)]|i\rangle +$$

$$+ \langle f|[\int A_o^\beta(x)(dx)^3,\partial A^\alpha(o)]|i\rangle =$$

$$= \frac{1}{2F_M^2}((d_{\beta o\gamma}+cd_{\beta 8\gamma})d_{\alpha\gamma\delta}+(d_{\alpha o\gamma}+cd_{\alpha 8\gamma}) \times$$

$$\times d_{\beta\gamma\delta})\langle f|u_\delta|i\rangle \qquad (5.22)$$

If we consider c as predicted from the meson mass spectrum, we have only one unknown in the problem: $\langle f|u_o|i\rangle$, because the value on $\langle f|u_j|i\rangle$ with (j=1...8) can be computed from the baryon SU_3 mass splitting. Kim and von Hippel have undertaken a systematic comparison of the predictions of equ. (5.22) with experimental knowledge on low-energy baryon-meson scattering and they report good agreement with experiment after fitting the parameter

$$\langle f|u_o|i\rangle \approx [215 \text{ MeV}]\delta_{fi} \qquad (5.23)$$

This value can be quite easily understood. The baryon mass comes from three sources corresponding to the structure of the energy density

$$\theta_{oo} = \tilde{\theta}_{oo} + (u_o+cu_8) \qquad (5.24)$$

(1) an SU_3-symmetric contribution from the $SU_3\times SU_3$ symmetry limit, described by $\tilde{\theta}_{oo}$

(2) an SU_3-symmetric contribution from the chiral symmetry breaking term u_o, which has been estimated in

equ. (5.23)

(3) an SU_3-octet contribution which is observed in Gell-Mann Okubo mass splitting.

For the nucleon we have

$$\langle N|cu_8|N\rangle \approx -210 \text{ MeV} \tag{5.25}$$

and in the chiral symmetry limit

$$\langle N|u_o-\sqrt{2}\, u_8|N\rangle \approx -20 \text{ MeV} \tag{5.26}$$

which within the limited accuracy of the analysis is consistent with zero. Since in the same limit $m_\pi=o$ we may remember an analogy with the free quark model: in the limit of chiral $SU_2\times SU_2$ symmetry free nonstrange quarks would be massless, and it appears that the simultaneous breaking of SU_3 and chiral symmetry primarily affects the masses of systems containing strange quarks.

Despite the consistency of this simplest ansatz of chiral symmetry breaking with experiment, one may of course wish to investigate alternatives or, alternatively, seek ways of presenting the present simple scheme as distinguished or even unique. There are different considerations, the strongest may be minimality: Given the observed quantum numbers of the weak currents, the $SU_3\times SU_3$ current algebra is the minimal algebra, which contains all observed currents and the minimum number of unobserved ones. In a similar way, there are two possible minimal assignments for the current divergences

(a) the present 18-dimensional multiplet: $(3,3^*)+(3^*,3)$

(b) a 16-dimensional multiplet $(8,1)+(1,8)$

This second possibility is certainly not favoured, since it would leave $m_\pi=o$ and lead to gross violations of K-meson pole dominance.

All other multiplets would, apart from the observed current divergences include many other operators, also of exotic quantum numbers. We finally want to report on the problem of reconstructing the $SU_3 \times SU_3$ chiral Lagrangian from Adler zeroes and on its relation to duality models. As in the $(\pi\pi)$ system, so one also observes in the (πK) system a coincidence between the predictions of scattering lengths from broken chiral symmetry (with $(3,3^*)+(3^*,3)$ breaking) and from the Veneziano model. The main element in this correlation are again Adler zeroes for soft π and K-mesons, which in the Veneziano model are due to parallel trajectories, satisfying

$$m_{K^*}^2 - m_\rho^2 = m_K^2 - m_\pi^2 \tag{5.27}$$

A second parameter in low energy $(K\pi)$ scattering is fixed by excluding any explicit dependence on energy-variables associated with exotic channels, which again turns out to be a consequence of the nonexotic σ-terms in $(3,3^*)+(3^*,3)$ chiral symmetry breaking; for the Veneziano model it is an obvious ingredient. Finally there is a scale parameter to be matched.

A complication arises with the η-meson. The most tightly constrained duality-models, like the Veneziano model, appear to require $m_\eta = m_\pi$, as can be seen for instance by postulating Adler zeroes in the $(\eta\pi)$ amplitude. On the chiral symmetry side, the realization of Adler zeroes for η-mesons is uncontroversial for the nonet model (with $m_\pi = m_\eta$ and ideal mixing; compare equ. (5.11)); however they are not a natural outcome of the octet model. For illustration consider the amplitude (no channel is exotic)

$$\langle \eta, K | \pi, K \rangle = A + B(s+u) + Ct \tag{5.28}$$

Imposition of Adler zeroes implies

$$\text{soft } \pi: \quad A + 2B\, m_K^2 + C\, m_\eta^2 = o \tag{5.29a}$$

$$\text{soft K}: \quad A + B(m_\pi^2 + m_\eta^2) + C\, m_K^2 = o \tag{5.29b}$$

soft η: $A+2B\, m_K^2 + C\, m_\pi^2 = o$ (5.29c)

which lead to the solutions $m_\pi^2 = m_\eta^2$ or $m_\pi^2+m_\eta^2 = 2m_K^2$ none of which is satisfied. One can avoid this difficulty by allowing in (5.28) further terms proportional to the virtual meson masses p_i^2, yet this is very much against the spirit of meson-pole dominance, and we prefer to abandon the requirement of Adler zeroes for η-mesons.

Reconstruction of tree graph models for pseudo-scalar-meson amplitudes, assuming only Adler zeroes for π and K-mesons is obviously not sufficiently constrained to lead to unique solutions. This question is still under study. Ellis[18] has shown that one can, up to the scale of the axial current, uniquely obtain the known octet chiral Lagrangian with $(3,3^*)+(3^*,3)$ symmetry breaking if one imposes a list of additional requirements such as:

(1) SU_3 symmetry in the momentum-dependent part of the irreducible meson amplitudes, with respect to linear SU_3 transformations among the octet states, and at most octet symmetry breaking in the momentum independent part (corresponding to symmetry breaking only in a generalized Lagrangian mass term).

(2) No explicit dependence on exotic channel invariants.

(3) Adler zeroes for π and K-mesons.

Whether this is the most natural set of requirements remains to be seen. In any case, it does not explain on what (theoretical) grounds one should reject the solution with ideal nonet mixing among the pseudoscalar mesons, which in many ways appears easier to incorporate into the framework of duality models.

6. SUMMARY

In these lectures it has been argued that the success of the Goldberger-Treiman relation for pions is not plausible by itself and it requires some particular mechanism to justify it. The only such mechanism known so far is the proximity of hadron physics to a world of chiral $SU_2 \times SU_2$ symmetry with massless pions. Chiral symmetry contains current algebra, at least up to [charge, current] commutators, and we have seen that many of their consequences are in fact already implied in a consistent exploitation of the pion-pole dominance principle. The only qualitatively new element in the postulate of current algebra is the choice of a compact algebra ($SU_2 \times SU_2$ over E_3 or SO(3.1)) and of a fixed scale for the weak axial current.

The Adler zeroes in pion amplitudes, which are a consequence of pion pole dominance already, play a crucial part in constraining possible theories and in explaining the observed coincidences between chiral and dual models.

On the level of SU_3 we have a choice between two types of theories, the decision being crucially dependent on the disputed experimental validity of a generalized Goldberger-Treiman relation for K-mesons:

(1) SU_3 symmetry breaking being part of the breaking of a chiral $SU_3 \times SU_3$ symmetric structure and thereby intimately correlated with $SU_2 \times SU_2$ symmetry breaking.

(2) SU_3 symmetry breaking being of totally different origin.

Again we observe correlations between broken chiral symmetry and duality models, but there is a major puzzle why the pseudoscalar mesons appear as an octet and not as an ideally mixed nonet, what to present knowledge would be most natural solution of duality-type constraints.[21]

FOOTNOTES AND REFERENCES

1. For a similar discussion see
   R. F. Dashen and M. Weinstein: Phys. Rev. 183, 1261 (1969)
2. For explicit realizations, see for instance
   G. Kramer and W. F. Palmer: Phys. Rev. 182, 1492 (1969)
3. Compare for instance
   G. Shaw: Phys. Rev. Lett. 18, 1o25 (1967)
4. S. L. Adler: Phys. Rev. 137, B1o22 (1965)
5. For a related argument see
   S. Mandelstam: Phys. Rev. 168, 1884 (1968)
6. M. Gell-Mann: Phys. Rev. 125, 1o67 (1962)
7. S. Fubini and G. Furlan: Ann. Phys. (N.Y.) 48, 322 (1968)
8. H. Schnitzer and S. Weinberg: Phys. Rev. 164, 1828 (1967)
9. B. Renner: Rapporteur's Review at the Lund Conference on Elementary Particles (1969), p. 223
1o. J. J. Brehm, E. Golowich and S. C. Prasad: Phys. Rev. Lett. 23, 666 (1969)
   J. J. Brehm and E. Golowich: University of Massachusetts (Amherst) preprint: Analytic hard-pion methods: the $A_1\rho\pi$ system (1969)
   P. V. Collins: Cambridge preprint DAMTP 7o/3 Current algebra calculation of ($\pi\pi$) s-wave phase shifts (197o)
11. S. Weinberg: Phys. Rev. 166, 1568 (1968)
12. S. Coleman, J. Wess and B. Zumino: Phys. Rev. 177, 2239 (1969)
13. P. Chang and F. Gürsey: Phys. Rev. 164, 1752 (1967)
14. J. Cronin: Phys. Rev. 161, 1483 (1967)
15. T. D. Lee, S. Weinberg and B. Zumino: Phys. Rev. Lett. 18, 1o29 (1967)

16. S. Gasiorowicz: Acta Physica Austriaca, Suppl. 6 (1969), p. 19
17. C. Lovelace: Phys. Lett. 28 B, 264 (1968)
18. J. Ellis: Cambridge preprint DAMTP 7o/7
The Adler zero condition and current algebra
19. H. Osborn: Nuovo Cimento Lett. 2, 717 (1969)
2o. Compare: R. Levi-Setti: Rapporteur's review at the Lund International Conference on Elementary Particles, p. 37o
21. Compare: J. Ellis and B. Renner: Cambridge preprint DAMTP 69/38:
On the relationship between chiral and dual models.

Acta Physica Austriaca, Suppl. VII, 145—165 (1970)

SOME RECENT WORK IN REGGE THEORY: REGGE CUTS, THE ABSORPTION MODEL AND GLAUBER THEORY*

BY

P. V. LANDSHOFF
Department of Applied Mathematics and
Theoretical Physics, University of Cambridge
England

ABSTRACT

It is assumed that scattering amplitudes possess Regge poles, and an analysis is given of the Regge cuts that are then required by unitarity. This is done using perturbation theory. A derivation is given of the cut structure that arises from the absorption model and from the Glauber model of deuteron scattering. The validity of this derivation, and of these models, is then examined.

1. INTRODUCTION

It is well known that perturbation theory is of no use for making numerical calculations in strong interaction physics. But it does, if it is used intelligently, provide [1] a powerful tool for determining the analyticity properties of the scattering amplitude A(s,t). This

---

* Lecture given at IX. Internationale Universitätswochen für Kernphysik, Schladming, February 23 - March 7,1970.

is because the position and nature of the singularities in the Mandelstam variables s,t is controlled by unitarity, and perturbation theory incorporates unitarity exactly, at least in a formal sense. Many of the perturbation theory predictions can be derived using unitarity directly instead [1].

It is also of interest to evaluate the effects of unitarity on the amplitude when it is expressed as a function of the complex variables t and j, where j is the angular momentum in the t-channel. For example, if one assumes the existence of Regge poles, unitarity is thought to generate Regge cuts, corresponding to the simultaneous exchange of two or more Regge poles. Since one's interest in the variable j is for the information it provides on the behaviour of A(s,t) for large s, it is not surprising that it turns out to be important to include multiparticle intermediate states in the s-channel. At the moment the only way we have of studying the effects of incorporating a large number of particles into unitarity is to use perturbation theory. It would be good if we could again use unitarity directly, without bringing in perturbation theory, but at the moment we do not know how to do this. Several attempts have been made to generate Regge cuts using simple models, such as Glauber theory or the absorption model, but none of these models incorporates multiparticle unitarity properly and each yields a cut structure in disagreement with that of perturbation theory, so their detailed predictions should probably be regarded as being of doubtful validity. I shall discuss this further later on.

It is well known that, with a basic ϕ^3 interaction, a Regge pole can be generated by summing an infinite set of ladder diagrams [1]. One might hope that a more realistic interaction, involving spinning particles, will produce much the same results, though this assumption might be dangerous [2]. Certainly the incorporation of spin makes things very much more difficult to work out, so that comparatively little has been done in this direction. I shall ignore spin altogether. Thus I shall suppose that the basic input into the analysis is the Regge pole, and that the only Regge cuts present are those that are then generated by unitarity.

Early work on Regge cuts considered each of the exchanged Regge poles as being explicitly generated by an infinite sum of ladders. For example, a two-Reggeon cut arises from taking a doubly infinite sum of diagrams like the Mandelstam diagram in figure 1 . This analysis is highly complicated [1]. The progress that has been made recently arises from a technique proposed by Gribov [3], which makes it unnecessary to think of an explicit mechanism for generating the Regge poles. This has two advantages:

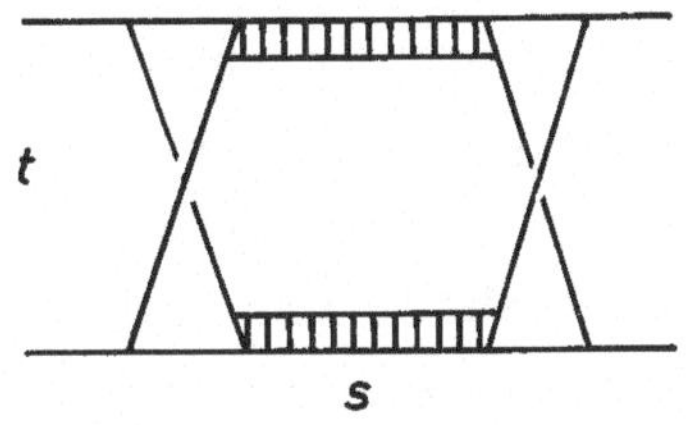

Fig. 1

(i) much of the complication is removed, and (ii) one can insert realistic Regge-pole trajectories instead of making do with those that perturbation theory chooses to generate (I have already said that one should not place too much trust in the numerical predictions of perturbation theory). Thus, for example, instead of taking the doubly infinite sum of diagrams in figure 1 as a model for the two-Reggeon cut, one considers the single diagram of fig.2 in which the wavy lines represent Reggeons with realistic trajectories. (The straight internal lines are ordinary Feynman propagators). A more precise way of thinking about this is to say that what one is doing is to consider the sum of all diagrams with the structure of fig. 3, so that

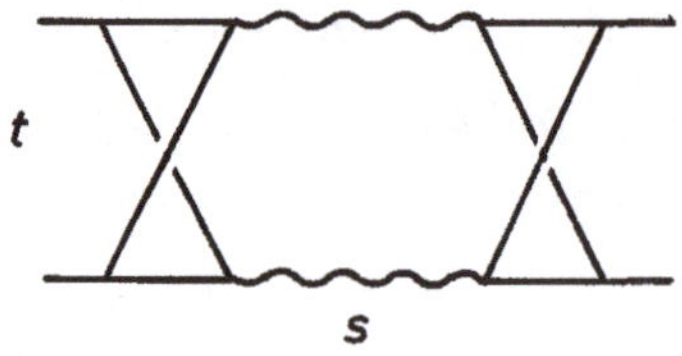

Fig. 2

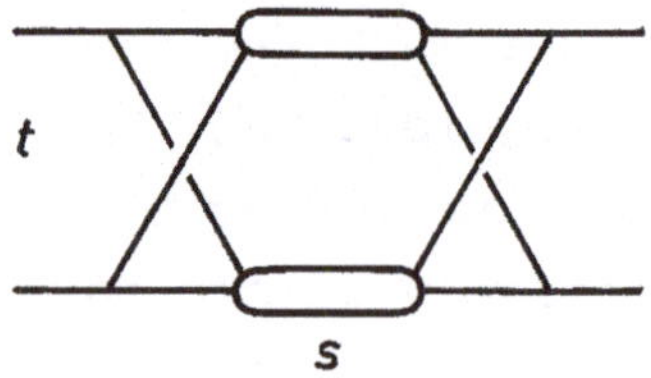

Fig. 3

each of the two "bubbles" represents a complete off-mass-shell scattering amplitude. At large s, the dominant contribution to each of these amplitudes comes from a Regge pole, and fig. 2 is just the result of replacing each of the internal amplitudes by its dominant contribution.

The Amati-Fubini-Stanghellini (AFS) diagram of fig.4 has a similar interpretation. I shall consider this diagram first to get practice. Rather than using the Gribov technique, I shall use a method of analysis due to Rothe [5]. This method is much less powerful than Gribov's but it is simpler when applied to this elementary diagram and so is more suitable for a lecture course.

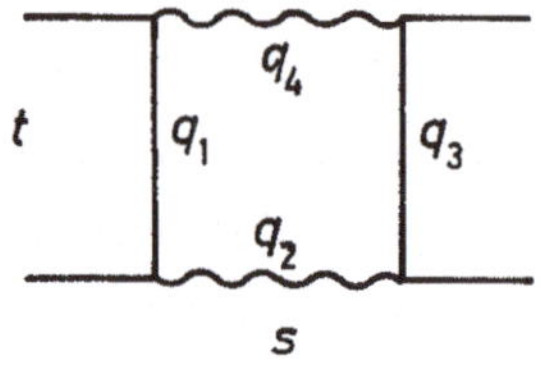

Fig. 4

As is well known, the two-particle discontinuity of the AFS diagram, obtained by putting the lines q_1 and q_3 on the mass shell, has a two-Reggeon cut. However this two-particle discontinuity is only a part of the imaginary part of the diagram; contributions from higher intermediate states cancel the cut, which is not present on the physical sheet of the amplitude for this diagram. This we shall now see; the result serves to emphasize what I said about the importance of incorporating multiparticle unitarity in the analysis.

2. THE AFS DIAGRAM

The graph of figure 4 involves an integration over the four-momentum k that runs round the loop. Make a change of variables

$$d^4k = dq_1^2\, dq_2^2\, dq_3^2\, dq_4^2\, J \tag{2.1}$$

The Jacobian is of the form

$$J = \frac{\theta(D)}{\sqrt{D}} \tag{2.2}$$

where D is a certain function, which for large s behaves like

$$D \sim s^2 f(q_2^2, q_4^2, t) \tag{2.3}$$

This asymptotic form for D is only correct for finite values of q_1^2, q_2^2, q_3^2, q_4^2 . We assume that the dominant contribution to the integral arises from finite values of these variables, so that it is all right to use the limit (2.3) inside the integral. That is, we assume that the two internal amplitudes represented in figure 4 by Reggeons each have asymptotic behaviour of Regge form only when their external masses and momentum transfer are finite; otherwise they go to zero. This property is valid for simple perturbation theory models, such as the infinite sum of ladders; whether it is valid in real life is unknown.

For finite s, D depends on q_1^2, and q_3^2, and so the θ-function in (2.1) constrains the range of integration over these variables. According to (2.3) this constraint disappears for large s, and these variables are then integrated from $-\infty$ to ∞ .

Thus the integral for the graph is, asymptotically,

$$\int_{-\infty}^{\infty} \frac{dq_1^2\, dq_3^2}{(q_1^2-m^2+i\varepsilon)(q_3^2-m^2+i\varepsilon)} \left\{ \int dq_2^2\, dq_4^2\, \frac{\theta(f)}{\sqrt{f}}\, s^{-1+\alpha(q_2^2)+\alpha(q_4^2)} \times \right.$$

$$\left. \times \text{ coupling functions} \right\} \tag{2.4}$$

The coupling functions arise from the four vertices of the graph; for example, the one corresponding to the top left-hand corner depends on q_1^2 and q_4^2 .

All the s-dependence in (2.4) is explicitly displayed, and we seem to have an asymptotic behaviour corresponding to a superposition of Regge poles, that is to a Regge cut extending up to

$$\alpha_c(t) = \max\{\alpha(q_2^2) + \alpha(q_4^2) - 1\} \tag{2.5a}$$

If one works this function out explicitly, it turns out that when the two Regge trajectories involved are the same one obtains

$$\alpha_c(t) = 2\alpha(\tfrac{t}{4}) - 1 \tag{2.5b}$$

Otherwise, if the two trajectories are different and non-linear, $\alpha_c(t)$ turns out to be a horrible function; it usually has cusps [6].

We now show that the discontinuity across the cut is zero, so that the cut does not actually exist on the physical sheet. Consider the q_1^2 integration. The pole from the propagator $(q_1^2-m^2+i\varepsilon)^{-1}$ appears below the integration contour. There are also branch points, from the two coupling functions that depend on q_1^2. On the basis of our knowledge of perturbation-theory amplitudes, we make two assumptions about these coupling functions:

(a) They have only a right-hand cut in q_1^2, and this also appears below the integration contour (fig. 5)

(b) They go to zero for large q_1^2 .

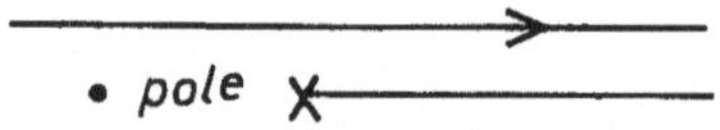

Fig. 5

These properties allow us to complete the contour of integration with an infinite semicircle in the upper half plane, and so result in the integral vanishing. (If, as AFS did, we had put the q_1 line on the mass shell, we should have just picked up the contribution from $q_1^2=m^2$; it is the infinite integration over q_1^2 that gives the zero result, the pole contribution being cancelled by that from the cuts in the coupling functions). Notice that exactly similar arguments apply to q_3^2; that integration also gives zero, independently of what happens with q_1^2.

3. MANDELSTAM'S DOUBLE-CROSS GRAPH

The remedy, if we want to get a cut with non-zero discontinuity, is to replace the left side of the AFS graph by a function that has both left- and right-hand cuts in q_1^2, and the right side by a function having both left and right hand cuts in q_3^2. The simplest such function is the simple cross graph, and with this replacement we arrive at the Mandelstam graph of figure 2.

To analyse this graph, we can still use the squares of q_1 and q_3, the total momenta flowing through the crosses, as two of the variables. One also needs variables internal to the crosses, but we need not consider these explicitly. The variable q_1^2 is the "s" variable for the cross graph; because this graph has an su spectral function it has both right-hand and left-hand cuts in q_1^2 at fixed t. The integration contour in q_1^2 now passes between these cuts, as in figure 6a. Now we cannot close the contour, neither in the upper nor in the lower half plane. The situation for q_3^2 is exactly similar, and the cut discontinuity is now not zero. That is, we do now have a cut attached to a branch point at $j = \alpha_c(t)$, where $\alpha_c(t)$ is given in (2.5).

(a) Fig. 6 (b)

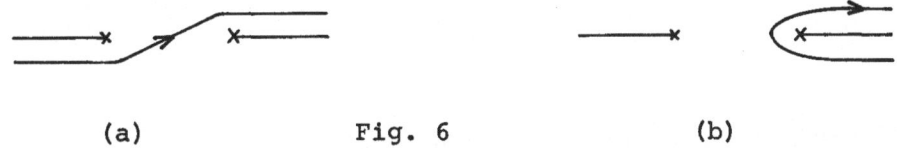

The result can be written in the form [3],[4]

$$\int d^2K\, X^2(K,k)\; s^{-1+\alpha[-(K+\frac{1}{2}k)^2]+\alpha[-(K-\frac{1}{2}k)^2]} \qquad (3.1)$$

where $k^2 = -t$. Here we have converted the integrations over q_1^2 and q_3^2 into one over a two-dimensional momentum K running round the two-Reggeon loop. Both K and k are Euclidean vectors orthogonal to the external momenta, which for large s span a space of two dimensions instead of the usual three. The function X in (3.1) is an integral of the cross function in figure 7 with respect to its energy variable q^2 along the contour of figure 6a. Because the cross function goes to zero faster [1] than $(q^2)^{-1}$ for large q^2, we can deform the integration contour into the shape drawn in fig. (6b), so as to obtain an integration over the imaginary part of the cross graph.

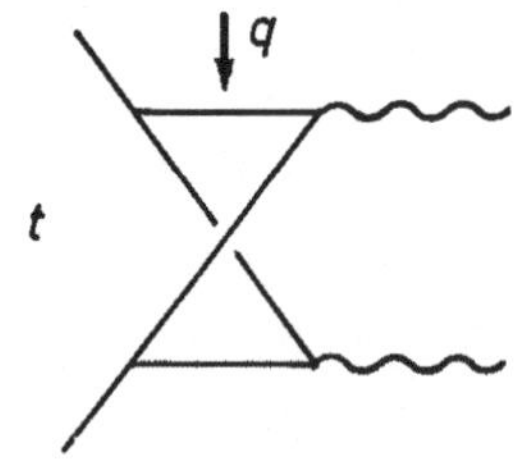

Fig. 7

4. FURTHER DIAGRAMS - THE ABSORPTION MODEL

The cross is merely the simplest thing we can put in the ends of the graph in order to get a Regge cut. Suppose we imagine making all possible insertions, and adding them together, so that we obtain a result that we may represent pictorially as in figure 8 . If we wish, we can include insertions that do not have an su spectral function, and so

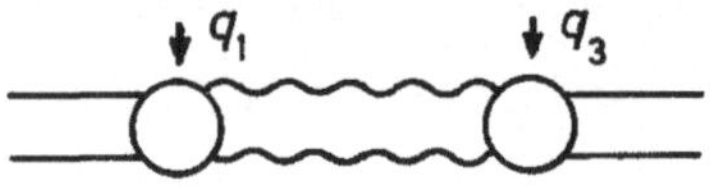

Fig. 8

do not yield a cut, because we know that doing the necessary q_1^2 and q_3^2 integrations will just remove them again.

Suppose we now make the guess (which is in fact wrong, as I shall discuss later) that everything goes through as before. That is, we again obtain (3.1), but now X is replaced by an integral of the imaginary part of the complete amplitude of figure 9. This amplitude represents a complete sum of graphs, and the simple cross is its most elementary non-trivial part.

Because we have taken the precaution of including all graphs in the amplitude of figure 9, including those like the pole graph of figure 10 that do not yield the Regge cut, this amplitude presumably obeys a unitary-like equation. In particular, at least, if the two Reggeons are the same, at t = 0 we expect its imaginary part to be positive. Hence if we integrate the imaginary part over only a portion of the right-hand cut, instead of the whole cut, we obtain a contribution that would not be cancelled if we were to do more of the integration.

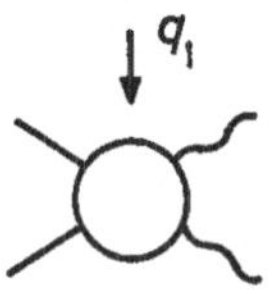

Fig. 9

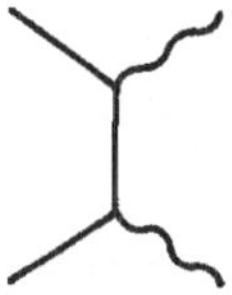

Fig. 10

In particular, if we pick up just the contribution from the pole, figure 10, this is at least part of the correct answer. Taking the pole contribution alone is just what is obtained by calculating the Regge cut term in the absorption model.

Thus, although the set of graphs that have the pole does not contribute directly to the Regge cut, it seems that, if we just take the residues at the poles in q_1^2 and q_3^2 instead of doing infinite integrations over these variables, we get some sort of estimate of the Regge cut.

This estimate could presumably be improved by adding in [8] contributions from some of the resonances in q_1^2 and q_3^2. Later I shall explain why the arguments I have just given must be treated with some suspicion.

5. GLAUBER THEORY

Let us see how the foregoing applies [9] to the Glauber theory of the scattering of a projectile particle on a deuteron. At t=0 the Glauber amplitude is

$$T_d(s,t=0) = T_p(\tfrac{1}{2}s,0) + T_n(\tfrac{1}{2}s,0) +$$

$$+ i\int d^2K\ G(K^2)T_p(\tfrac{1}{2}s,-K^2)T_n(\tfrac{1}{2}s,-K^2) \qquad (5.1)$$

Here T_p and T_n are the on-mass-shell amplitudes for the scattering of the projectile on a proton and on a neutron, respectively, while G is related to the dnp vertex function.

The expression (5.1) can be derived using Feynman graphs [10] and making certain approximations. The first two terms on the right correspond to a single scattering of the projectile on one of the constituent nucleons of the deuteron and arise from two Feynman graphs, each of the structure of figure 11a. The last term represents

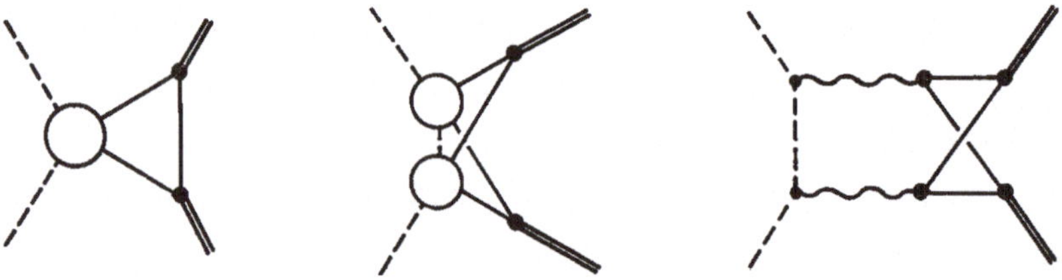

(a) (b) (c)

Fig. 11

double scattering of the projectile, once on each of the nucleons, and arises from Feynman graphs of the structure of figure 11b.

If one inserts in (5.1) Regge-pole asymptotic behaviour for the amplitudes T_p and T_n, one obtains from the double scattering term a large-s behaviour just like that corresponding to a cut (figure 11c) attached to a branch point at $j=\alpha_c(t)$, with $\alpha_c(t)$ given by (2.5). However, we already know that the graph of figure 11b should not result in such a cut. That is, it seems that the cut in (5.1) has been introduced as a result of the approximations made in evaluating the Feynman graph of figure 11b. Because of the important part played by Glauber theory in the analysis of the Serpukhov data, it is an urgent problem to try and understand whether these considerations make the theory invalid at large energy. I cannot give a final answer to this, but it will be useful to go into the matter in a little detail.

Consider first the evaluation of the Feynman graph of figure 11a. This graph contains a projectile/nucleon scattering amplitude with the nucleons off the mass shell, and two dnp vertex functions with the nucleons again off the mass shell. On the abstract theoretical level, there is no unique definition of such off mass shell functions, because there is no unique definition for the nucleon field. If we change our definition of the nucleon field, the graph of figure 11a will correspond to a set of graphs in which various amounts of structure replace the internal nucleon lines. To fix our definition of the nucleon field, and hence of the off mass shell functions, we must somehow impose the condition that, if possible, the graph of figure 11a by itself shall represent the bulk of the single-scattering contribution. This we try to do by heuristic arguments, which rely heavily on the smallness of the deuteron binding energy δ and the fact that we are dealing with forward scattering. Because δ is small, we can regard the deuteron as being composed of two nucleons that are almost at rest in the rest frame of the deuteron, and because t=0 they remain so after the scattering. Thus we can think of the deuteron non-relativistically. The asymptotic part $\exp\{-r\sqrt{M\delta}\}/r$ of the deuteron wave function corresponds to a momentum space wave function peaked near $\underline{P} = 0$ with a width $\Delta P \sim \sqrt{M\delta}$, so the masses of the two nucleons differ from their physical mass M by an amount of order δ . In terms of the dnp vertex function, the statement is that when it is multiplied by the attached nucleon propagators, it is very strongly peaked about values of the nucleon masses near the mass shell, so that the dominant part of the Feynman integral for figure 11a arises from values of the internal nucleon masses differing from M by an amount at most of order δ. This is why the first two terms on the right hand side of (5.1) are on mass shell amplitudes. Similarly, the dnp vertex functions in fig. 11b will constrain the internal nucleons there to be near the mass shell. A simple calculation shows that, provided the laboratory energy E of the incident projectile particle X is not too large, the kinematics will then also constrain the internal virtual X particle to have a mass near its mass shell. In the laboratory frame the four-momentum of the incident projectile is $(E,\underline{P}_X)$, where $E^2-\underline{P}_X^2 = m^2$, the square of the mass of the X-particle. The four-momentum of the internal nucleon on which it first scatters is $(M+\varepsilon_1, \underline{\varepsilon}_2)$ where, according to what I have said, $\varepsilon_1 = O(\delta)$ and $|\underline{\varepsilon}_2|=O(\sqrt{M\delta})$. The square of the mass of the internal X particle is the square of the sum of these two four-vectors:

$$m^2 + O(E\sqrt{M\delta}) \tag{5.2}$$

Because δ is small, this is close to m^2 if E is not too large. Hence for such values of E it is a reasonable approximation to use the on mass shell X-nucleon scattering amplitudes to evaluate the graph, and this results in the Glauber answer in (5.1).

But as E is increased, the mass of the internal X will depart further and further from its physical value m. According to what I said in section 2, the X-nucleon amplitude will then become numerically smaller than the on mass shell amplitude, so that it appears that (5.1) is no longer correct. We can make an extremely rough estimate of the energy E_o where this begins to happen, if we make a guess as to how rapidly the X-nucleon amplitude varies as we vary the mass of the X-particle. The characteristic scale in strong interaction physics seems to be about one nucleon mass (for example, Regge trajectories have a slope of about M^{-1}), so my guess is that it is reasonably good to use the on mass shell amplitudes T_p and T_n until (5.2) becomes of the order of $(m+M)^2$. If this is correct, the formula (5.1) represents a good approximation to the graph of figure 11b up to incident energies of the order of

$$E_o = m \sqrt{M/\delta} \quad ; \qquad (5.3)$$

when the projectile is a proton, $E_o \approx 20$ GeV.

An interesting theoretical point emerges from this. Although the Feynman integral for figure 11b does not have a cut in the complex j-plane, for quite large values of the energy its numerical properties are similar to those corresponding to such a cut. We shall encounter a similar situation again in section 6, where the analytic structure in the j-plane is all-important at <u>infinite</u> energy, though not directly relevant at large but finite energy. In that context, the corresponding value of E_o will turn out to be about the mass of the sun!

To return to the Glauber theory, the obvious thing to consider next is replacing the simple pole structure on the left of figure 11c by a complete two-particle/two-Reggeon amplitude, as we did in section 4. If we then went through an argument exactly similar to that in section 4, we should conclude that figure 11c with the internal X-particle on the mass shell, does after all give some sort of estimate of the double scattering term [11]. In fact this argument predicts that the Glauber estimate of the double-scattering term is too small rather than too big! It also suggests that the estimate can be improved by adding in terms where the intermediate X-particle is replaced by resonances. However, as I suggested at the end of section 4, there is reason to be wary of this argument.

6. ITERATIONS OF CUT DIAGRAMS

In section 4 I put forward the argument that the two Reggeon cut involved an integral of the imaginary part of the complete two-Reggeon/two-particle amplitude of figure 9 with respect to q_1^2 and that this integral might perhaps be approximated by the pole contribution, figure 10. If this is right, we obtain a verification of the absorption model and of Glauber theory, provided corrections from resonance contributions are taken into account.

The difficulty with the argument is that the integral of the imaginary part of the two-Reggeon/two-particle amplitude diverges. If the quantum numbers are appropriate, the amplitude will be dominated at large q_1^2 by a Regge pole, figure 12a . Otherwise, the amplitude will at least contain the very two-Reggeon cut we are trying to investigate, figure 12b. In either case, the amplitude does not go to zero sufficiently rapidly at large q^2 to give a convergent integral.

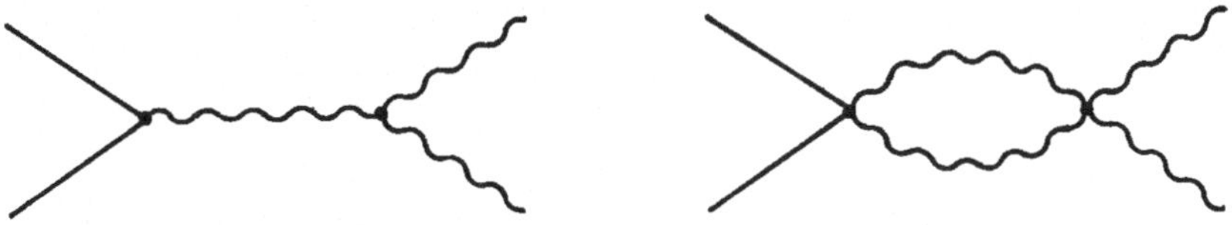

(a) (b)

Fig. 12

This means that the guess made in section 4 is incorrect. That is, the form of the complete diagram of figure 8 is not a straight-forward generalization of the result (3.1) for the simple Mandelstam diagram.

The basic trouble is that the dominant contribution to the integral for figure 8 does not arise solely from finite values of q_1^2 and q_3^2, so that one cannot take limits under the integral as suggested, for example, by (2.3).

An associate feature is that the form (3.1), being a straight-forward superposition of Regge poles, corresponds to a j-plane singularity of the type $\log(j-\alpha_c)$, and it can be shown [12] that such an infinite singularity is not allowed by unitarity. That is, the asymptotic behaviour $s^{\alpha_c(t)}/\log s$ corresponding to (3.1) is fiercer than is allowed. The simplest model that does not have this shortcoming [13] is where the Mandelstam diagram is augmented by an infinite sum of its iterations, as in figure 13. One can show formally that the discontinuity corresponding to the two-Reggeon cut generated by this infinite series takes the rather simple form [4]

$$\text{disc}\, A(j,k^2) = \int d^2K\, F\tilde{F}\,\delta[j-\alpha[-(K+\tfrac{1}{2}k)^2] - \alpha[-(K-\tfrac{1}{2}k)^2]+1] \qquad (6.1)$$

The function F is obtained from the sum of diagrams in

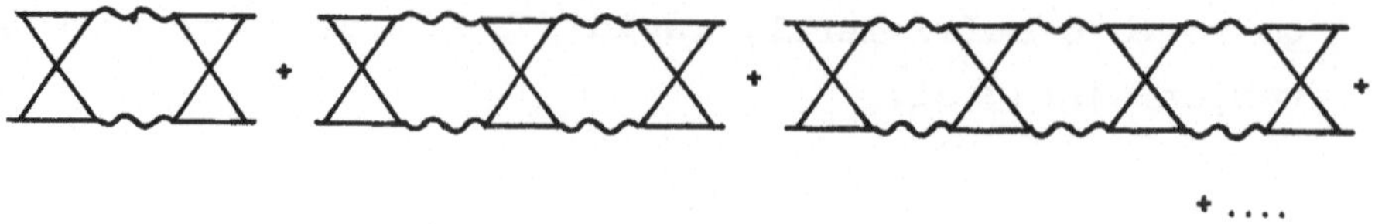

Fig. 13

figure 14, and so itself has a two-Reggeon cut; $\tilde{F}$ is the continuation of F round this cut. It will be recognized that the form (6.1) is reminiscent of a discontinuity

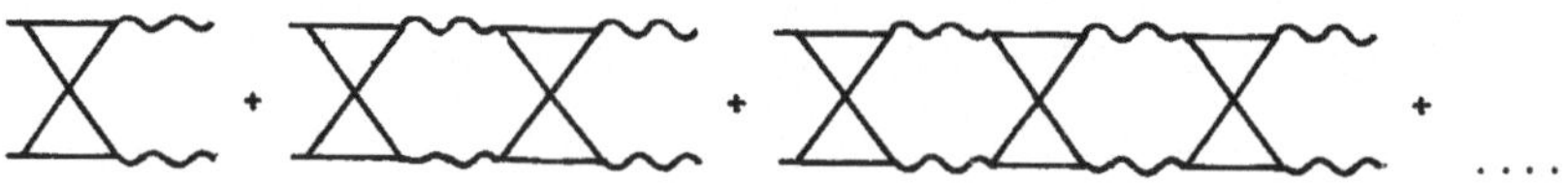

Fig. 14

associated with an ordinary unitary integral.

We are not able to calculate the asymptotic behaviour associated with (6.1) except in the case where the cut discontinuity is too small to be interesting for practical purposes, but it is nevertheless of some theoretical interest to consider this case. Suppose the contribution from the first term in the series of figure 13 is written

$$\frac{\Delta^2 (s/s_o)^{\alpha_c(t)}}{\log s/s_o} \qquad (6.2)$$

Here I have introduced the usual scale parameter s_o of Regge analysis, taken from phenomenology to be about $(1 \text{ GeV})^2$. The expression (6.2) then defines a dimensionless function $\Delta(t)$, which is related to X in (3.1). Then for $\Delta \ll 1$ a crude calculation gives for the sum of all the graphs in figure 13 something like

$$\frac{\Delta^2 (s/s_o)^{\alpha_c(t)}}{\log s/s_o} \left(\frac{1}{1-\Delta \log \log s/s_o}\right)^2 \qquad (6.3)$$

We see that at infinite s the effect of adding up all the iterations is to multiply the form (6.2) obtained from the simplest term by a factor proportional to $(\log \log s/s_0)^{-2}$. This in fact means that the infinite singularity $\log(j-\alpha_c)$ arising from the simplest term has been converted into something more gentle, as required by unitarity.

However, in this context the notion of infinite s is totally irrelevant to experiment, since even when s is chosen to be the square of the mass of the sun, $\log \log s/s_0 \approx 5$. At present-day laboratory energies $\log\log s/s_0 \approx 1$ and so for small Δ the factor in brackets in (6.3) is about 1. That is, at laboratory energies the infinite sum of graphs is well represented by its first term. But the answer (6.3) only applies when $\Delta \ll 1$, which is not very interesting if cuts are to be of any experimental importance. Presumably for larger values of Δ and at laboratory energies the contribution from the simplest graph should again be multiplied by some number that is effectively a constant, but we do not know how to calculate this number.

If one makes the optimistic, though not very plausible, guess that this number is near to one even when Δ is not small, the conclusion is that for practical purposes one can represent the function F in (6.1) by the first term in figure 14. That is, one can ignore the Regge cuts in F and so perhaps believe the derivation of Glauber theory that I gave in section 5; for the absorption model, as discussed in section 4, there is still the problem of the divergence of the integral of the imaginary part of the two-Reggeon/two-particle amplitude arising from a possible t-channel Regge pole.

7. MISCELLANEOUS

(i) Suppose the t-channel quantum numbers allow a Regge pole as well as a Regge cut. Then one has to sum an infinite sum of diagrams like those in figure 15 . This is

Fig. 15

very hard to do, but one might expect the result to contribute to the Regge amplitude something like [14]

$$\frac{\beta(t)}{j-\alpha_o(t)-\Sigma(j,t)} \tag{7.1}$$

Here $\alpha_o(t)$ is the input Regge pole trajectory and $\beta(t)$ its residue function. Σ is the sum of the iterated cut diagrams without the pole; as we have seen, we do not know how to calculate it.

One thing we can see is that the iteration of the Regge pole has renormalized its position, just as in conventional renormalization theory. The experimentally observed trajectory is given by

$$j-\alpha_o(t)-\Sigma(j,t) = 0 \tag{7.2}$$

If the cut discontinuity is at all appreciable, so that Σ is not numerically small, this renormalization may be expected to be quite significant. Thus it is something of a mystery why observed trajectories are so very linear.

It seems difficult to use an expression like (7.1) to derive any realistic estimate of the asymptotic behaviour of the amplitude when the Regge pole is close to the cut.

(ii) I have not discussed multi-Reggeon cuts, fig.16. It is easy to compute [3] the effects of these if one ignores the iteration complication. One finds that, as one increases the number n of Reggeons exchanged, the contributions alternate in sign. The n-Reggeon exchange produces a branch point which, for equal trajectories, is at

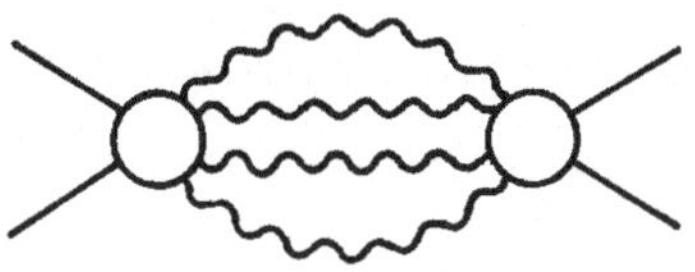

Fig. 16

$$j = n\,\alpha(t/n^2) - n + 1 \qquad (7.3)$$

If all the exchanged Reggeons are Pomerons, then at t=O all expressions (7.3) reduce to 1, for any n. Hence if forward scattering at high energy is controlled by the Pomeron, it may be necessary to take account of the fact that it is not just a simple pole that is being exchanged.

(iii) My analysis of Regge cuts has been based on perturbation theory. If we can cast the results of the analysis in a form that may be expressed independently of perturbation theory language, there is some hope that they then have general validity, that is that one day it will be possible to get the same results directly from unitarity, without bringing in perturbation theory at all. As we have seen, the structure of Regge cuts is highly complicated, so it is not surprising that so far only a start has been made on this program [15].

What has been done is to ignore the complications that arise from the necessity for iteration, and obtain a definition of the two-Reggeon/two-particle amplitude that then appears in the cut discontinuity, in terms of the two-Reggeon/two-particle amplitude that arises in a multiperipheral approximation to the six-point function. The latter is depicted in figure 17. The connection between the two sorts of two-Reggeon/two-particle amplitudes arises because, if one bends round the external lines in figure 17 and joins them up in a suitable way, one arrives

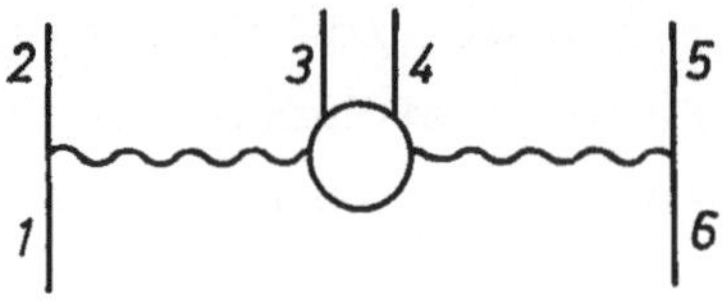

Fig. 17

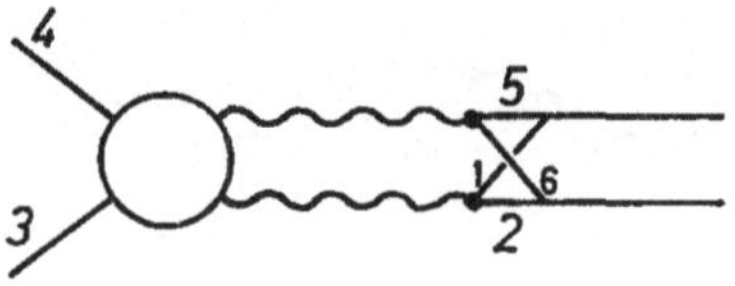

Fig. 18

at figure 18. This is a cut diagram, with a simple cross on one side and a complete two-Reggeon/two-particle amplitude on the other.

REFERENCES

1. R. J. Eden, P. V. Landshoff, D. I. Olive, and J. C. Polkinghorne, The Analytic S-Matrix, Cambridge Univ. Press 1966.
2. H. Cheng and T. T. Wu, Phys. Rev. Letters 22, 666 (1969).
3. V. N. Gribov, Soviet Phys. JETP 26, 414 (1968). A concise account of the Gribov technique is to be found in section 2 of reference 4 .
4. P. V. Landshoff and J. C. Polkinghorne, Phys. Rev. 181, 1989 (1969).
5. H. Rothe, Phys. Rev. 159, 1471 (1967).
6. S. M. Negrine, Nuovo Cim. 43A, 1101 (1969); C. I. Tan and J. M. Wang, Phys. Rev. Letters 22, 1152 (1969).
7. V. N. Gribov and A. A. Migdal, Soviet Journal of Nuclear Physics 8, 581 (1969).
8. A. B. Kaidalov and B. M. Karnakov, Phys. Lett. 29B, 372 and 376 (1969).
9. P. V. Landshoff, Phys. Rev. 177, 2531 (1969); V. N. Gribov, Zh. Eksp. Teor. Fiz. 56, 892 (1969).

10. E. S. Abers, H. Burkhardt, V. L. Teplitz, and C. Wilkin, Nuovo Cim. 42A, 365 (1966): L. Bertocchi and A. Capella, Nuovo Cim. 51A, 369 (1967).
11. Related conclusions have been reached in potential theory by D. R. Harrington, Phys. Rev. 184, 1745 (1969).
12. J. Bronzan and C. Jones, Phys. Rev. 160, 1494 (1967).
13. J. C. Polkinghorne, Nuclear Physics B6, 441 (1969). This model is also important for an understanding of the so-called Gribov-Pomeranchuk phenomenon; see D. I. Olive and J. C. Polkinghorne, Phys. Rev. 171, 1475 (1968).
14. An expression of this form has been verified in a simple model by A. R. Swift, J. Math. Phys. 6, 1472 (1965); see also V.N.Gribov, ref. 3 .
15. P. V. Landshoff, Nuclear Physics B15, 284 (1970).

Acta Physica Austriaca, Suppl. VII, 166—179 (1970)

VENEZIANO - LIKE MODEL FOR A TWO-CURRENT AMPLITUDE*

BY

P. V. LANDSHOFF
Department of Applied Mathematics and Theoretical Physics, University of Cambridge England

ABSTRACT

Some properties of the usual Veneziano beta-function are reviewed. A model is then constructed for an amplitude involving two charged isovector currents. It is found that when one incorporates certain desirable features, in particular the correct Fubini-Gell-Mann fixed pole at j=1, the model automatically satisfies the Bjorken scaling law. It is shown how the model may be extended to describe deep inelastic electron-proton scattering.

1. THE VENEZIANO BETA-FUNCTION

It will be useful for me to begin by going through some features of the familiar Veneziano Beta-Function, for which I choose to write the integral representation

$$B(-\beta(s),-\alpha(t)) = \int_0^\infty \frac{dx}{x} \left(1+\frac{1}{x}\right)^{\beta(s)} (1+x)^{\alpha(t)} \tag{1.1}$$

* Lecture given at IX. Internationale Universitätswochen für Kernphysik, Schladming, February 23 - March 7,1970.

This representation converges for $\beta(s)$, $\alpha(t)$ having negative real parts, though the B-function can be continued into the right half planes of these variables.

From near x=0 one obtains the contribution

$$\int_0 dx\, x^{-\beta(s)-1}(1+x)^{\alpha(t)+\beta(s)} \approx \sum_{J=0}^{\infty}\binom{\alpha(t)+J}{J}\frac{1}{J-\beta(s)} \quad (1.2)$$

The binomial coefficient $\binom{\alpha(t)+J}{J}$ is a polynomial of degree J in $\alpha(t)$. If we take $\alpha(t)$ to be a linear function of t

$$\alpha(t) = \alpha_o + t\alpha' \quad (1.3)$$

we can express $\binom{\alpha(t)+J}{J}$ as a linear combination of Legendre polynomials $P_\ell(\cos\theta)$, where $\ell=0,1,2,\ldots J$ and θ is the angle of scattering in the s-channel. Thus the pole at $\beta(s)=J$ corresponds to a "leading" particle of spin J, together with daughter particles of spins J-1, J-2,...1,0 of the same mass.

Similarly, if we take $\beta(s)$ to be linear, we obtain from near $z=\infty$ a similar structure for the t-channel poles. Note that the poles in the two channels come from different parts of the integration region and the residue of one does not contain the other: the two channels are dual. Consider now the asymptotic behaviour, $|\alpha(t)|\to\infty$ with s fixed. Because of the convergence properties of the representation (1.1), it is easiest to take the limit with $\mathrm{Re}\,\alpha(t)\to-\infty$. The dominant contribution then evidently comes from where the factor $(1+x)^{\alpha(t)}$ is largest, that is near x=0. This suggests that we should make the change of variable

$$x = \frac{x'}{-\alpha(t)} \quad (1.4)$$

and assume that the dominant contribution arises from finite x'. Then

$$(1+x)^{\alpha(t)} \to e^{-x'} \quad (1.5)$$

and (1.1) becomes

$$[-\alpha(t)]^{\beta(s)} \int_0^\infty dx' \, x'^{-1-\beta(s)} \, e^{-x'} = \qquad (1.6a)$$

$$= \Gamma(-\beta(s))[-\alpha(t)]^{\beta(s)} \qquad (1.6b)$$

Because of (1.3) this is of Regge-pole form. Notice that it arises from the same region of integration, $x \gtrsim 0$, as the s-channel poles (1.2), and the poles of the gamma function in (1.6b) are just the leading poles of (1.2). Similarly, Regge-pole behaviour in the crossed channel arises from near $x=\infty$.

Of course my procedure of changing integration variable and then taking a limit inside the integral is not mathematically rigorous as I have described it, but it can be made so. An indication, though not a proof, that all is well is the fact that the resulting integral (1.6a) converges.

It requires rather more work to show that (1.6) is still correct as $\alpha(t) \to \infty$ in the right-half of the complex plane, and I shall not go through this. Let me just say that (1.6) applies in all directions except near the positive real axis. This exception is important; it explains why the result (1.6) appears to have a cut along the positive real $\alpha(t)$ axis, even though we know that in fact the beta-function has only poles, and no cuts. The reason that (1.6) breaks down on the positive real axis is that this is where all the poles (1.2) are situated. It seems reasonable to suppose that, if we took account of unitarity and so introduced a cut, these poles would be depressed below the real axis on to the unphysical sheet, and then (1.6) would after all apply along the positive real axis.

2. THE FUBINI - AMPLITUDE

Consider the amplitude for the "scattering" of the negatively charged isovector current on a proton (fig. 1)

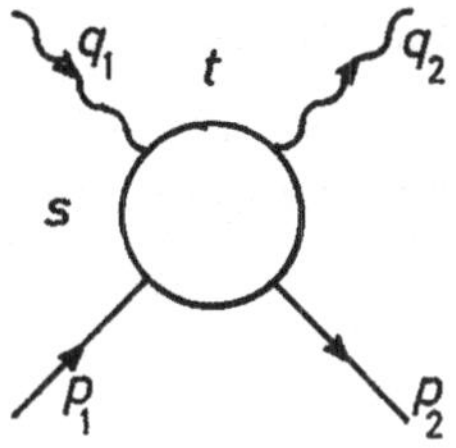

Fig. 1

$$i\int d^4x\, e^{iq_2x} \langle p_2|[j_+^\mu(x), j_-^\nu(o)]|p_1\rangle \theta(x^o) \qquad (2.1)$$

To begin with, I shall suppose that the proton has zero spin, though I shall give it the correct isospin. My metric is (+ ---).

Associated with the amplitude (2.1) there is a tensor $T^{\mu\nu}$. Expand this tensor in terms of $P=(p_1+p_2)$, q_1 and q_2:

$$T^{\mu\nu} = P^\mu P^\nu A(s,t,q_1^2,q_2^2) + P^\mu q_1^\nu B + P^\mu q_2^\nu C + \ldots \qquad (2.2)$$

Of all the invariant amplitudes A,B,C,... in (2.2), the Fubini amplitude A is of the greatest interest, both theoretically and experimentally, and I shall concentrate on this entirely. This is the amplitude that obeys the famous Fubini sum rule [1]

$$\frac{1}{\pi}\int_{-\infty}^{\infty} ds\ \mathrm{Im}\ A(s,t,q_1^2,q_2^2) = F(t) \qquad (2.3)$$

where the elastic form factor F(t) is defined by

$$\langle p_2|j_-^\mu(o)|p_1\rangle = P^\mu F(t) \qquad (2.4)$$

In order that the sum rule (2.3) may be satisfied, it is

necessary that, as $|s|\to\infty$ with t, q_1^2, q_2^2 fixed,

$$A(s,t,q_1^2,q_2^2) \sim -\frac{F(t)}{s} \tag{2.5}$$

The form factor F(t) has poles at $\alpha(t)=1,2,3,\ldots$, where $\alpha(t)$ is the exchange-degenerate ρ-meson trajectory. Because the current j_-^μ is a spin-one operator, for $J>2$ the pole at $\alpha(t)=J$ is to be regarded as being due to the spin-one daughter of the corresponding leading particle. The amplitude A has a similar pole structure in each of its variables q_1^2, q_2^2.

Being a double-flip amplitude, A has poles at $\alpha(t)=2,3,\ldots$. This is apparently inconsistent with the fact that the asymptotic form (2.5) has a pole at $\alpha(t)=1$. The explanation [2] is that (2.5), and the Fubini sum rule, is only valid for $\alpha(t) < 1$. For larger values of t the fixed pole at j=1, which gives (2.5), no longer dominates, and instead the leading behaviour comes from the ρ-meson trajectory:

$$A(s,t,q_1^2,q_2^2) \sim f(t,q_1^2,q_2^2)\, s^{\alpha(t)-2} \tag{2.6}$$

At $\alpha(t)=1$ the function f has a pole, which just cancels that arising from (2.5).

Let me just mention the effects of current conservation, for example

$$T^{\mu\nu} q_{1\nu} = -P^\mu F(t) \tag{2.7}$$

For general values of the variables this imposes a linear constraint among the invariant amplitudes:

$$\tfrac{1}{2}(s-u)A+q_1^2B+q_1\cdot q_2C = -F(t) \tag{2.8}$$

Since I am not going to construct models for B and C this does not give me a constraint, except at the special values $q_1^2=0=q_1\cdot q_2$ for which $q_2^2=t$. Then

$$A = \frac{F(t)}{M^2-s} \tag{2.9}$$

where M is the nucleon mass.

The model for A that I shall describe is given in a recent preprint by Landshoff and Polkinghorne [3]. In that paper it is shown how one can take at least partial account of (2.9), but so far we have not learnt very much from doing this. So here I shall ignore (2.9).

3. CONSTRUCTION OF A(s,t)

To get poles simultaneously at $\alpha(q_1^2)=1,2,3,\ldots$ and $\alpha(q_2^2)=1,2,3,\ldots$ introduce two integration variables u_1 and u_2 and include in the amplitude the factor

$$\boxed{\int_0^\infty \frac{du_1 du_2}{u_1 u_2}\left[1+\frac{1}{u_1}\right]^{\alpha(q_1^2)-1}\left[1+\frac{1}{u_2}\right]^{\alpha(q_2^2)-1}} \tag{3.1}$$

At the same time, we want poles at $\alpha(t)=2,3,4,\ldots$, so introduce another integration variable z and put in

$$\boxed{\int_0^\infty \frac{dz}{z}\left[1+\frac{1}{z}\right]^{\alpha(t)-2}} \tag{3.2}$$

The s-channel is dual to the t-channel,so the poles in s must come from the other end $z=\infty$ of the z-integration. If we let the nucleon trajectory be $\beta(s)$ and remember that at the moment we are taking the nucleon to have zero spin, we want poles at $\beta(s)=0,1,2,\ldots$. So include the factor

$$\boxed{[1+z]^{\beta(s)}} \tag{3.3}$$

Consider now the leading pole at $\beta(s)=J$. As in section 1, we get this by expanding the integrand near the part of the integration that produces the pole, that is near $z=\infty$, and then picking out the term of degree J in $\alpha(t)$ from the binomial coefficient. This gives

$$\frac{[\alpha(t)]^J}{J!}\,\frac{1}{J-\beta(s)}\int_0^\infty \frac{du_1}{u_1}\,[1+\frac{1}{u_1}]^{\alpha(q_1^2)-1}\int_0^\infty \frac{du_2}{u_2}[1+\frac{1}{u_2}]^{\alpha(q_2^2)-1} \tag{3.4}$$

The residue of the pole factorizes into a function of q_1^2 times the same function of q_2^2 , as it should (figure 2).

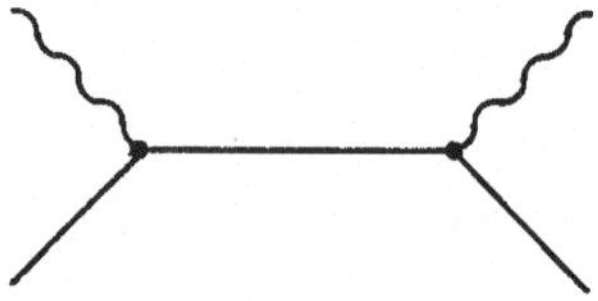

Fig. 2

If we put J=0 we obtain the elastic form factor for the nucleon (α' is the common slope of the α and β trajectories):

$$\frac{1}{\alpha'}\int_0^\infty \frac{du_1}{u_1}\,[1+\frac{1}{u_1}]^{\alpha(q^2)-1} \tag{3.5a}$$

An obvious fault is that this integral diverges at the upper end. The simplest way to put this right is to change (3.5a) by including in it an extra factor $[1+u_1]^{-m}$, where m is any positive constant. (This is merely the simplest choice; there is considerable freedom here). We also want a normalizing constant, to make the form factor equal to one when t=0. Thus the form factor becomes

$$F(q^2) = \frac{1}{B(1-\alpha_o,m)}\int_0^\infty \frac{du_1}{u_1}\,[1+\frac{1}{u_1}]^{\alpha(q^2)-1}\quad [1+u_1]^{-m} \tag{3.5b}$$

(The integral is, of course, a beta function). For large q^2,

$$F(q^2) \sim \text{constant} \times (q^2)^{-m} \tag{3.6}$$

as can be seen by making the change of variable,

$$u_1 = -\,\alpha(q^2)\,u_1'$$

and taking the limit under the integral.

In order to make the extra factors we have just put

in arise from (3.4), we must include in A(s,t) the additional factor

$$\alpha' N^2 \times \left[\begin{array}{c} \text{a term that becomes equal to } (1+u_1)^{-m}(1+u_2)^{-m} \\ \text{when } z \to \infty \end{array} \right] \tag{3.7}$$

where $$N^{-1} = B(1-\alpha_o, m) \tag{3.8}$$

so that F(O) = 1 .

Notice that there is no constraint arising from a factorization requirement for the pole residues in the t-channel; these residues are just a constant times the two-current/particle vertex (figure 3). Also, I do not attempt to impose factorization on the daughter poles in the s-channel.

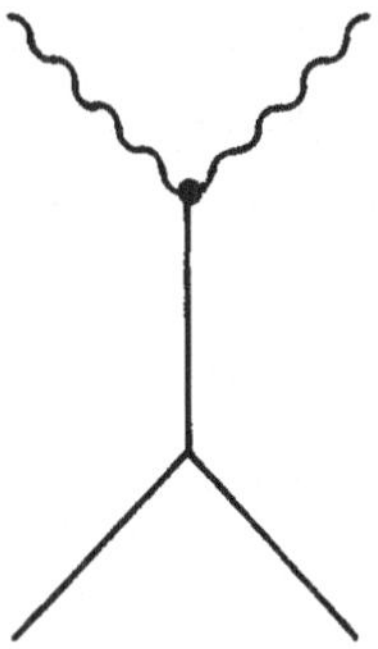

Fig. 3

The required Regge-pole asymptotic behaviour in the s and t channels arises respectively from z=O and z=∞ and is linked with the crossed channel resonance poles, as in section 1. But we also want the j=1 fixed pole (2.5) arising for large s. Because this term is independent of q_1^2 and q_2^2, but involves t, it must come from the region

$$u_1 = u_2 = \infty$$
$$z \text{ finite} \tag{3.9}$$

Thus we must change (3.3), in such a way that the factor whose exponent is $\beta(s)$ shall be near to one not only for z=O (which gave the ρ-trajectory), but also in the region (3.8). The change must not affect the factorization at the s-channel poles, so the new expression must reduce to

(3.3) when $z \to \infty$. I take the new factor to be

$$\boxed{\left[1 + \frac{z}{1+ \dfrac{u_1\, u_2}{(1+u_1+u_2)\,(\lambda+\mu z)}}\right]^{\beta(s)}} \tag{3.10}$$

where λ and μ are constants. Notice that I have arranged that both u_1 and u_2 must become large in order to get the bracket near to one; otherwise there would be extra, unwanted terms in the asymptotic behaviour. There is some arbitrariness in the factor $(\lambda+\mu z)$; I would have used a more complicated function $\phi(z)$ such that $\phi(\infty) = \infty$ (so as not to upset the factorization of the s-channel poles) and $\phi(0)$ is finite (so as not to upset the $s^{\alpha(t)-2}$ contribution to the asymptotic behaviour).

In order to get the residue of the j=1 pole to be F(t), I must arrange that the factor (3.7) becomes equal to $(1+z)^{-m}$ when u_1 and u_2 are large. So I choose it to be

$$\boxed{\alpha' N^2 \left[1+ \frac{z(u_1+u_2+u_1 u_2)}{1+z+u_1 u_2}\right]^{-m}} \tag{3.11}$$

I also need one further factor to get the j=1 pole correct

$$\boxed{\left[1+ \frac{N(u_1+u_2)}{(1+z)\,(\lambda+\mu z)}\right]^{-1}} \tag{3.12}$$

The result (2.5) then comes from making the changes of variable

$$u_1 = -\beta(s)u_1' \quad , \quad u_2 = -\beta(s)u_2' \tag{3.13}$$

and taking the limit under the integral.

Altogether I have

$$A(s,t,q_1^2,q_2^2) = \alpha' N^2 \int_0^\infty \frac{du_1 du_2 dz}{u_1 u_2 z} \, [1+ \frac{1}{u_1}]^{\alpha(q_1^2)-1} \quad \times \tag{3.14}$$

$$\times \left[1+\frac{1}{u_2}\right]^{\alpha(q_2^2)-1}\left[1+\frac{1}{z}\right]^{\alpha(t)-2}\left[1+\frac{z}{1+\frac{u_1\,u_2}{(1+u_1+u_2)(\lambda+\mu z)}}\right]^{\beta(s)}\times$$

$$\times\quad\left[1+\frac{z(u_1+u_2+u_1u_2)}{1+z+u_1u_2}\right]^{-m}\left[1+\frac{N(u_1+u_2)}{(1+z)(\lambda+\mu z)}\right]^{-1}$$

Consider now the Bjorken limit [4]

$$s\to\infty,\quad q_1^2 = q_2^2 = q^2$$

$$\frac{s}{q^2} = 1-\omega\,,\qquad \omega\ \ \text{finite} \tag{3.15}$$

The three brackets whose exponents are infinite in this limit all take values near to one in the region (3.9), and the changes of variable in (3.13) give the answer

$$-\frac{\phi(t,\omega)}{s} \tag{3.16}$$

where

$$\phi(t,\omega) = N\int_0^\infty dz\,\frac{(\lambda+\mu z)(1+z)^{-m}}{z(\lambda+\mu z)+\frac{1}{1-\omega}}\left[1+\frac{1}{z}\right]^{\alpha(t)-1} \tag{3.17}$$

The result (3.16) is the "scaling law" conjectured by Bjorken; in the present model it has appeared automatically as a result of the other inputs, in particular the Fubini-Gell-Mann fixed pole at j=1. Note that when $\omega > 1$ the denominator in (3.17) vanishes in the region of integration, and so ϕ acquires an imaginary part. The presence of this cut is explained in a way similar to the cut that apparently appears in the asymptotic behaviour of the beta-function, as I discussed in section 1.

I should say that in all my discussion of asymptotic behaviour I have only considered the variables going to infinity in the left half plane. I am unable to discuss the right half plane. Notice also that the meson dominance model, where the current amplitude is approximated by the hadron amplitudes obtained from the residues of the poles in q_1^2 and q_2^2, does not have the j=1 fixed pole. Nor does it have the scaling limit (3.16); it is purely Regge-pole-like in its asymptotic behaviour. In fact the residue at $\alpha(q_1^2) = \alpha(q_2^2) = 1$ is an ordinary Veneziano beta-function.

4. DEEP INELASTIC ELECTRON SCATTERING

Of course my function A(s,t) is unrealistic, if only because it does not contain the u-channel Δ-trajectory. But let us see what we can do with it.

In the u-channel, the process (2.1) has pure isospin $\frac{3}{2}$ so by crossing A(u,t) is the s-channel $I=\frac{3}{2}$ amplitude. The amplitude A(s,t) itself is a known mixture of $I=\frac{1}{2}$ and $I=\frac{3}{2}$ in the s-channel. So we know both of the two independent amplitudes $I=\frac{1}{2}$ and $I=\frac{3}{2}$ separately. From this we can calculate the amplitude for the scattering of the neutral isovector current on a proton:

$$A(s,t) + A(u,t) \tag{4.1}$$

The unit normalization is determined by looking at the nucleon pole residue in the s-channel. Notice that (4.1) has even isospin in the t-channel, so the trajectory $\alpha(t)$ is now to be regarded as the exchange-degenerate ω trajectory*. Thus we have made use of the known isospin degeneracy of the meson trajectories.

Now, a photon is made up of a mixture of isovector and isoscalar parts. The isospin degeneracy of the meson trajectories makes it plausible to assume that these two parts scatter equally, so that the scattering of a virtual photon on a proton is also given by (4.1). Again the normalization is determined from the proton pole.

The j=1 fixed pole is correctly absent from (4.1); it cancels between the two terms:

---

* Remember that in the double-flip amplitude the first pole is at $\alpha(t)=2$.

$$\frac{F(t)}{-s} + \frac{F(t)}{-u} \tag{4.2}$$

But the Bjorken scaling limit is present:

$$\frac{\phi(t,\omega)}{-s} + \frac{\phi(t,-\omega)}{-u} \tag{4.3}$$

This can be compared with experiment. The experiment is [5] the inelastic high-energy scattering of an electron on a proton, which takes place by virtual photon exchange. When summed over all final-state hadron configurations, the differential cross section involves the imaginary part of the t=0 proton-spin-averaged photon/proton elastic amplitude in just the kinematic configuration (3.15) appropriate to the Bjorken limit, with $\omega>1$. It is found that the experimental results correspond entirely to the double-flip amplitude, and the scaling law holds remarkably well.

At t=0, I can regard (4.1) as the proton-spin-averaged amplitude, instead of pretending that the proton has zero spin. We saw that, for $\omega>1$, the first term in (4.3) has an imaginary part, which is related to a dimensionless function νW_2 that is plotted by the experimentalists.

To compare the explicit form of νW_2 obtained from my $\phi(t,\omega)$, I have to choose the parameters α_0, m, λ,μ . It turns out that the gross features of νW_2 are surprisingly insensitive to the parameters. The parameter m is determined from the asymptotic behaviour (3.6) of the elastic form factor (this is the form factor that is usually called F_1); a value of 2.5 seems reasonable. For the intercept of the meson trajectory, take $\alpha_0 = \frac{1}{2}$.

If μ is chosen to be less than about 0.1, it turns out that, except for very near $\omega=1$, νW_2 takes a very simple approximate form. For the choice $\lambda=1$, this is

$$\nu W_2 = \frac{8}{3\pi} \frac{(\omega-1)^{3/2}}{\omega^2} \tag{4.4}$$

I have plotted this roughly in figure 4, in the continuous curve. This figure also contains the experimental data. The broken curve is the "theoretical" form of νW_2 for the choice $\lambda = \mu = 1$.

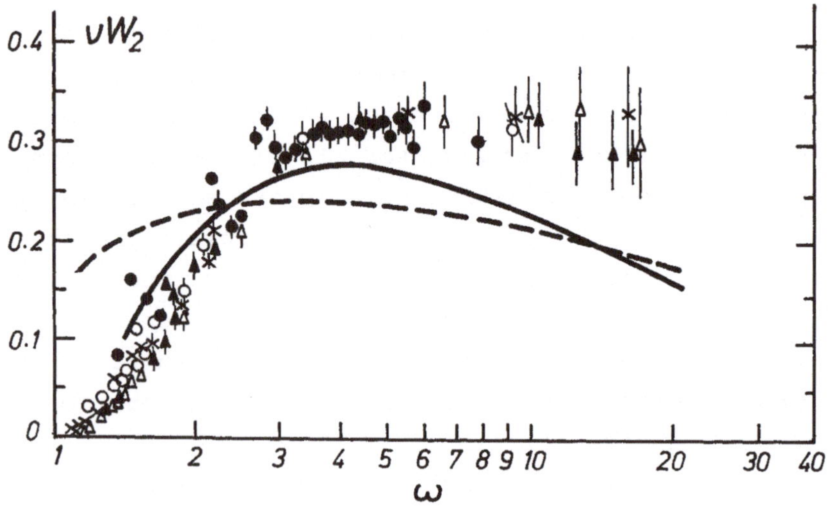

Fig. 4

I leave you to draw your own conclusions. A possible point of view is that, if the model is refined, it will be able to explain the data exactly. Certainly the overall scale of νW_2, which is basically determined by the normalization condition $F(0)=1$ for the form factor, is quite close to what is needed. The alternative point of view [6] is that the data cannot be explained completely unless the Pomeron is included. On this point I merely remark that, in this case, the Pomeron coupling function has to behave in a very special way for large photon mass:

$$g(q^2) \propto (q^2)^{-1/2} \tag{4.5}$$

If the coupling function goes to zero more rapidly than this for large q^2, the Pomeron will not contribute in the Bjorken limit, even if it is important for explaining the data in the scattering of physical ($q^2=0$) photons.

REFERENCES

1. S. Fubini, Nuovo Cimento 43, 475 (1966).
2. J. B. Bronzan, I. S. Gerstein, B. W. Lee, and F. E. Low, Phys. Rev. Lett. 18, 32 (1967).
3. P. V. Landshoff and J. C. Polkinghorne, Cambridge preprint, DAMTP 70/5 .
4. J. D. Bjorken, Phys. Rev. 179, 1547 (1969).
5. M. Briedenbach et al., Phys. Rev. Lett. 23, 935 (1969).
6. H. Abarbanel, M. L. Goldberger, and S. B. Treiman, Phys. Rev. Lett. 22, 500 (1969); H. Harari, Phys. Rev. Lett. 22, 1078 (1969).

Acta Physica Austriaca, Suppl. VII, 180—213 (1970)

ABSORPTION CORRECTIONS AND CUTS IN THE REGGE POLE MODEL*

BY

G. E. HITE
University of Karlsruhe, Germany

The concept of absorptive corrections to amplitudes and the possible existence of cuts in the complex angular momentum plane have been known to particle physicists since 1962. They have been reintroduced to allow the Regge pole model to describe features of experimental data that the simple pole model could not explain. For example, the narrow ($\Delta t \sim \mu^2 = m_\pi^2$) forward peaks seen in reactions like $pn \to np$, $\gamma N \to \pi^\pm N$, and $\pi^+ p \to \rho^0 \Delta^{++}$ which are apparently dominated by π-exchange, cannot be explained by a simple Reggeized pion. For a time, explanations of these reactions were attempted using several Regge poles, both conspiring and evasive (simple). Realizing that a simple Reggeized pion is essentially, for small values of t, the same as an elementary pion, and remembering the success of the one-pion-exchange-model with absorption corrections in describing resonance production in meson-baryon scattering, people soon started to consider absorptive corrections to simple Regge pion exchange to explain such reactions.

Whereas the concept of conspiring trajectories is not at all physically motivated, those of diffractive

* Lecture given at IX. Internationale Universitätswochen für Kernphysik, Schladming, February 23 - March 7,1970.

scattering and multiple rescattering are intuitive to all of us. The real problem is not in convincing ourselves of the need to include absorptive and rescattering effects but how to do so correctly. It seems to be clear that this question is not answered and probably will not be answered in the near future. The consensus of opinion of those using absorptive corrections is that they might not be completely correct but it's a step in the right direction. They argue that their absorptive corrections result in cut-like contributions which have the known properties of the Mandelstam cuts. Unfortunately their diagrams are more closely related to those studied by D. Amati, S. Fubini and A. Stanghellini which do not lead to a cut in the j-plane.

In these lectures I will discuss the methods presently used to introduce absorption corrections, the ways in which they correspond and differ from the cut contributions suggested by S. Mandelstam and the success they are having in describing the structure of experimental differential cross sections. Throughout the lectures I will avoid complications associated with spin and signature factors whenever possible.

ABSORPTIVE CORRECTIONS, EIKONAL APPROXIMATIONS AND REGGE POLES [1]

The need to correct single particle exchange amplitudes can easily be understood from unitarity restrictions on inelastic partial wave amplitudes. Writing the elastic amplitude as

$$f_{el} \equiv \frac{2}{\sqrt{5}} F = ƛ \sum_{\ell} (2\ell+1)\, P_{\ell}(z_s)\,(S^{\ell}-1)/2i \ ,$$

$$(ƛp = 1 \ , \ S^{\ell} = \exp\{2i\delta^{\ell}\})$$

$$\sigma_{el} = \int |f_{el}|^2 d\Omega = \pi\lambda\!\!\!^{-2} \sum_{\ell} (2\ell+1)|S_\ell - 1|^2$$

with unitarity (Optical Theorem)

$$\sigma_{tot} = 4\pi \ \mathrm{Im}\ f_{el}(z_s=1) = 2\pi\lambda\!\!\!^{-2} \sum_{\ell} (2\ell+1)(1-\mathrm{Re}\ S^{\ell})$$

one obtains

$$\sigma_{in} = \sigma_{tot} - \sigma_{el} = \pi\lambda\!\!\!^{-2} \sum_{\ell} (2\ell+1)(1-|S^{\ell}|^2)$$

Thus it is apparent that the partial wave cross sections must share a common amount $\pi\lambda\!\!\!^{-2}$. In most un-unitarized peripheral models (i.e. OPE and Regge) there is no problem for large values of ℓ but for small values the partial wave amplitudes are too large and violate unitarity. Classically this means that while the simple particle exchange amplitude correctly describes the long range part of the interaction it fails to describe scattering at small separations or impact parameters. A good example of this is the pion exchange contribution to $np \to pn$. Because of the pions spin parity 0^-, it must decouple in the forward direction and thus the amplitude has the form

$$\frac{t}{t-\mu^2} = 1 + \frac{\mu^2}{t-\mu^2} = P_o(z_s) - \frac{\mu^2}{2p^2} \sum_{\ell} (2\ell+1) P_\ell(z_s) \times$$
$$\times Q_\ell\left(1+\frac{\mu^2}{2p^2}\right)$$

Consequently this large s-wave contribution violates the unitarity bound. (In the case of Regge poles there is a multiplicative factor of $\exp\{(\alpha' \log s/s_o)t\}$ which automatically reduces the s-wave contribution (built in absorption?)) . This suggests that one must correct such amplitudes to account for the existence of other competing channels, both elastic and inelastic, which must share the low partial wave cross section.

Originally, absorption corrections were taken into account through relativistic generalizations of the

W.K.B. method of potential theory. New derivations have been given of this method which avoided assumptions previously thought necessary. In particular, the assumption that the primary interaction was short ranged was shown to be unnecessary.

Taking into account just the effects of the elastic channel one has

$$A^{\ell Abs}_{ab} = (S^{\ell}_{aa})^{1/2} A^{\ell}_{ab} (S^{\ell}_{bb})^{1/2} = e^{i\delta^{\ell}_{a}} A_{ab}\, e^{i\delta^{\ell}_{b}} \tag{1}$$

where S^{ℓ}_{aa} and S^{ℓ}_{bb} are the S-matrix elements for elastic scattering for states a and b respectively. This expression implies that the effect of the elastic channels is just to impart a phase shift to the primary scattering amplitude. It should be realized that the elastic phase shift δ^{ℓ} is essentially imaginary due to the presence of the strong inelastic channels. Thus in a sense, the presence of inelastic channels besides the one considered is included in the correction.

Since absorptive effects are assumed not to modify the high partial waves the equation

$$A^{\ell Abs}_{ab} = A^{\ell}_{ab} - [1-(S^{\ell}_{aa})^{1/2}(S^{\ell}_{bb})^{1/2}]\, A^{\ell}_{ab}$$

is more useful for calculations since it gives the full amplitude as an unabsorbed amplitude minus a rapidly converging partial wave series.

If one makes the usual assumption that $S^{\ell}_{aa} = S^{\ell}_{bb}$ where

$$S^{\ell} = 1+2i\left(\frac{2p}{W}\right) A^{\ell}_{el} \xrightarrow[S\to\infty]{} 1+2i\, A^{\ell}_{el}$$

one has

$$F^{Abs}_{ab}(s,t) = F_{ab}+2i\left(\frac{2p}{W}\right)\sum_{\ell}(2\ell+1)P_{\ell}(x)A^{\ell}_{el}A^{\ell}_{ab} =$$

$$= F_{ab}+2i\left(\frac{2p}{W}\right)\sum(2\ell+1)P_{\ell}\frac{1}{2}\int dy\, F_{el}(s,t(y))P_{\ell}(y)\frac{1}{2}\int dz F_{ab}(s,t(z)) \times$$

$$\times\, P_{\ell}(z) \quad =$$

$$= F_{ab}+2i\left(\frac{2p^3}{\pi W}\right)\int\int_{-1}^{1} dydz\, \frac{\Theta(K)}{\sqrt{K}}\, F_{el}(s,t(y))\, F_{ab}(s,t(z)) = \tag{2a}$$

$$= F_{ab}+ \frac{i}{2pp'}\left(\frac{2p}{\pi W}\right)\int\int_{-\infty}^{0} dt_1 dt_2 \frac{\Theta(K)}{\sqrt{K}}\, F_{el}(s,t_1)F_{ab}(s,t_2) \tag{2b}$$

where

$$K \equiv 4p^4(1-x^2-y^2-z^2+2xyz) \xrightarrow{s\to\infty}$$

$$-t_1^2-t^2-t_2^2+2(t_1t_2+t_1t+t_2t) \equiv \tau$$

Changing variables from z to ϕ where $z=xy+\sqrt{(1-x^2)(1-y^2)}\times$ $\times\cos\phi$ one obtains

$$F_{ab}^{Abs}(s,t(\Theta)) = F_{ab}(s,t(\Theta))+ \frac{i}{2\pi}\left(\frac{2p}{W}\right)\int d\Omega_y\, F_{el}(s,t(\Theta_1))\cdot$$

$$\cdot\, F_{ab}(s,t(\Theta_2)) \tag{2c}$$

where

$$\cos\phi = \frac{\cos\Theta_2-\cos\Theta\cdot\cos\Theta_1}{\sin\Theta\cdot\sin\Theta_1}$$

Since we are describing high energy peripheral reactions the amplitude should receive contributions from all partial waves and thus the sum over partial waves can be approximated by an integral. If in addition we restrict ourselves to small angles we can use

$$P_\ell(\Theta) \sim J_o((2\ell+1)\sin\tfrac{\Theta}{2}) \xrightarrow[s\to\infty]{} J_o(\Delta b)$$

where

$$\Delta^2 = -t \quad \text{and} \quad b = \hbar(\ell+ \tfrac{1}{2})$$

[In general $d^j_{\lambda\mu} \simeq (-1)^{m-n}\left(\frac{1+\cos\Theta}{2}\right)^{(m+n)/2} J_{m-n}(\Delta b)$, with

$$2m =|\lambda+\mu|+|\lambda-\mu|; \quad n = |m-|\lambda-\mu||\,]$$

and Eq. (4) to obtain an impact representation.

$$F_{ab}(s,t) = 2p^2\int_0^\infty b\, db\, J_o(\Delta b)S_{aa}^{1/2}(b^2)\; A(b^2,s)\; S_{bb}^{1/2}(b^2) \tag{3}$$

which is just the inelastic analog of the Glauber impact representation for the elastic amplitude,

$$f_{el}(s,t) = (\tfrac{2}{W})F_{el} = ip\int_0^\infty b\,db\,J_o(\Delta b)(1-S(b^2)) \tag{4}$$

These expressions, or Fourier Bessel transforms, are easily inverted to give

$$ip(1-S(b)) = \int_0^\infty \Delta d\,\Delta J_o(\Delta b)\;f_{el}(s,-\Delta^2)$$

At high energies the elastic peaks are very diffractive and can be easily fitted by

$$f_{el}(s,-\Delta^2) = \frac{i\sigma_{tot}}{4\pi\bar{\lambda}}(1+i\varepsilon)\,\exp\{-\tfrac{1}{2}a\Delta^2\}$$

where a is the slope of the forward peak and varies slowly with energy and the real quantity ε is about 0.1 for πN scattering. This gives

$$S(b^2) = 1 - C(1+i\varepsilon)\,\exp\{-\frac{b^2}{2a}\} \tag{5}$$

where $C = \sigma_{tot}/4\pi a$, must be less than unity for consistency.

So far we have only considered corrections to inelastic amplitudes due to elastic scattering in the initial or final channel, corresponding to the diagrams in Fig. 1a. It seems reasonable that one should equally include other intermediate states as shown in Fig. 1b .

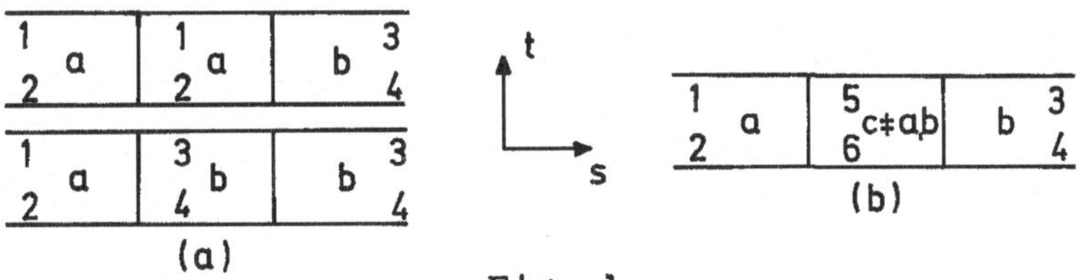

Fig. 1

Generalizing one would write

$$A_{ab}^{Abs} \cong A_{ab} + iA_{aa}A_{ab} + iA_{ab}A_{bb} + i\sum_{c\neq a,b} A_{ac}A_{cb}$$

It is difficult to estimate the contribution of the new

term. It has been suggested that this could be roughly accounted for by multiplying the elastic correction terms by a factor λ which would be of order 2 for most reactions.

Up to now we have not mentioned what the unabsorbed amplitude is. In the O.P.E. model with absorption corrections it was assumed to be the Born term for elementary particle exchanges. This of course led to an incorrect energy dependency for exchanges other than those of spin zero particles. This difficulty is easily overcome in the Regge pole model where the exchanged Regge poles easily give good energy dependence.

Denoting Regge pole exchanges by R and P in the inelastic and elastic cases respectively we symbolically can write the inelastic amplitude as

$$F_{inel} \simeq R + 2i\, P \times R + \ldots \tag{6}$$

$\times \equiv \int$ over all intermediate variables.

Using Arnolds suggestion of $S = e^{2iP}$, the elastic amplitude can be written as

$$F_{el} = \frac{S-1}{2i} = P + iP \times P + \ldots \tag{7}$$

CUTS IN THE COMPLEX J-PLANE [2]

Before discussing the more detailed properties of cuts, let us recall how their existence in the j-plane affects the scattering amplitude. Consequently consider a Sommerfeld-Watson transformation for the t-channel partial wave decomposition of the amplitude,

$$F(z_t, t) = \sum_j (2j+1) A^j(t) P_j(z_t) = \oint_C \frac{A(j,t) P_j(-z_t)}{\sin \pi j}\, dj \tag{8}$$

where the contour C encloses the integers along the real axis of the j-plane. Then assuming we have only poles and cuts in the right hand side of the j-plane, we can deform the contour, see Figure 2

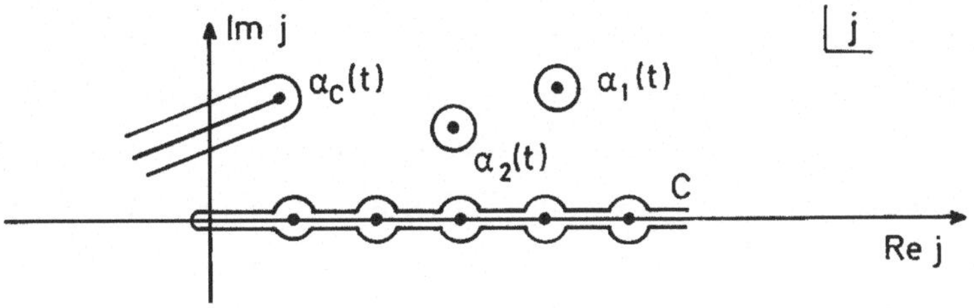

Fig. 2

to obtain

$$F(z_t,t) = \sum_\alpha \frac{\beta(\alpha,t)P_\alpha(-z_t)}{\sin\pi\alpha} + \sum_{\text{cuts}} \oint^{\text{cuts}} \frac{P_\alpha(-z_t)}{\sin\pi\alpha} A(\alpha,t)\,d\alpha +$$

$$+ \text{ L.H. contour integral}$$

Taking the limit of large s and using

$$P_\alpha(-z_t(s)) \underset{s\to\infty}{\longrightarrow} (-z_t)^\alpha$$

we obtain

$$F(z,t) \underset{s\to\infty}{\longrightarrow} \sum_\alpha \frac{\beta(\alpha,t)\left(\frac{s}{so}\, e^{-i\pi}\right)^\alpha}{\sin\pi\alpha} + \sum_{\text{cuts}} \int^{\alpha_c(t)} \frac{\left(\frac{s}{so} e^{-i\pi}\right)^\alpha}{\sin\pi\alpha} \text{Disc}_\alpha A(\alpha,t)\,d\alpha$$

where we have absorbed kinematic and s_o factors into the residue function β and the discontinuity function. [Note, to include signature we only need to replace $e^{-i\pi\alpha}$ by $(1+ \tau e^{-i\pi\alpha})$ where τ is the signature factor.]

The existence of cuts in the complex angular momentum plane was first proposed by D. Amati, S. Fubini and A. Stanghellini from a study of the Feynman diagram shown in Fig. 3a. S. Mandelstam subsequently showed that if one considers all the dispersion diagrams corresponding to the Feynman diagram, he would find the cut was canceled. Mandelstam then considered the diagram shown in Fig. 3b and concluded the resulting cut need not be cancelled. The essential difference between the two diagrams is that while there is no third double spectral function for the A.F.S. diagram the two sets of crossed lines in the Mandelstam diagram each result in third double-spectral functions, ρ_{su} . This is a general property in that if both crosses are replaced by "blobs", there can be a cut only if each "blob" gives a nonzero ρ_{su} . (The way in which the existence of a nonzero ρ_{su} prevented the cancellation of the cut, was demonstrated by P. Landshoff during this conference.)

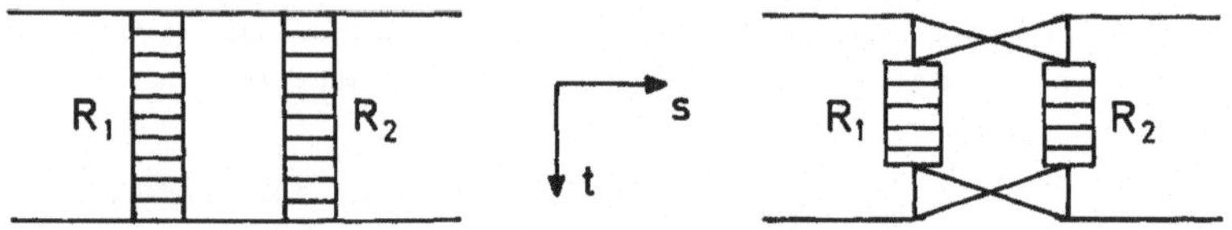

(a) (b)

Fig. 3

The Mandelstam diagram actually leads to two types of moving cuts, depending on whether or not one of the ladders (Regge poles) is replaced by an elementary particle. The branch point of the first type of cut resulting from the exchange of two ladders (Regge poles) is given by

$\alpha^{(1)}_{R_1R_2}(z) = \alpha_{R_1R_2}[x,y(x)]$ where $z=t^{1/2}$, y is a solution of $\partial\alpha_{R_1R_2}/\partial y = 0$ and $\alpha_{R_1R_2}(x,y) = \alpha_{R_1}(x-y) +$

$$+\alpha_{R_2}(y) - 1 \ . \qquad (10a)$$

For simple linear trajectories this gives the familiar result

$$\alpha^{(o)}_{R_1 R_2}(t) = \alpha^{o}_{R_1} + \alpha^{o}_{R_2} - 1 + t\,\alpha'_{R_1}\,\alpha'_{R_2}/(\alpha'_{R_1} + \alpha'_{R_2}) + \ldots$$

The branch point of the second type of cut, due to the exchange of a ladder and an elementary particle, is given by

$$\alpha^{(2)}_{RE}(x) = \alpha_R(x-m_E) + S_E - 1 = \alpha_{RE}(x,m_E) \qquad (10b)$$

Although the second type of cut provides a mechanism to allow fixed poles at wrong signature points to avoid a conflict with t-channel unitarity, unlike the first type of cut it does not contribute to the asymptotic energy behaviour of amplitudes. Consequently we will consider only type one cuts and will leave off the superscript.

It is beyond the scope of these lectures to consider the mathematics of Mandelstam cuts, but it is useful to summarize some of their properties:

a) If n Regge poles, $\alpha_i(t)$, i=1,...,n , are exchanged the leading cut has a branch point given for $t \leq 0$, by

$$\alpha^c_n(t) = \mathrm{Max}[\sum_{i=1}^{n} \alpha_i(t_i)] - n + 1$$

where the quantities $\sqrt{-t}$ and $\sqrt{-t_i}$ form a closed polygon.

b) While the cuts contribute to both parity states the signature of a cut is given by the product of the signatures of the individual Regge poles times minus one raised to an integer given by $[1-(-1)^B-2B]/4$ when B is the sum of their baryon numbers.

c) The discontinuity across the cut in the j-plane near $j= \alpha^c_n$ is given by

$$\mathrm{Disc}_j\, A_n(j,t) \sim \Gamma_n(t)[\alpha^c_n(t) - j]^{n-2}$$

which gives an asymptotic contribution of the form (see Eq. (2)),

$$F^c_n(s,t) \sim \frac{\Gamma_n(t)\; s^{\alpha^c_n(t)}}{(\log s + \frac{i\pi}{2})^{n-1}}$$

d) When all n poles are Pomerons the function $\Gamma_n(t)$ contains a factor $(-1)^n$.

e) Cut contributions are non-factorizable.

AFS cut contributions have essentially the same branch points, discontinuity functions, and other properties similar to those of Mandelstam cuts.

ABSORPTION CORRECTIONS AND MANDELSTAM CUTS [3]

It is interesting to see how the absorption corrections lead to cut-like contributions and whether they can be identified with the Mandelstam cuts. The first question is easily answered by taking the high energy limit of the integral in eq. (2b),

$$F^{Abs}(s,t) \overset{\to}{=} R(s,t) + \frac{2i}{\pi s} \iint_{-\infty}^{0} dt_1 dt_2 \, \frac{\Theta(\tau)}{\sqrt{\tau}} \, R(s,t_1) \, R'(s,t_2)$$

where $\tau = -t^2-t_1^2-t_2^2+2(tt_1+tt_2+t_1t_2)$ is four times the square of the area of a triangle whose sides are $\sqrt{-t}$, $\sqrt{-t_1}$ and $\sqrt{-t_2}$, i.e. $|\vec{p}_i-\vec{p}_f|$, $|\vec{p}_i-\vec{p}\,'|$ and $|\vec{p}\,'-\vec{p}_f|$ respectively. The Θ-function insures that the triangle is closed. With $R_i \sim s^{\alpha_i(t)}$ one easily sees that integral gives a cut with the desired branch point (see eq. (10a)). If one assumes the Pomeron to be imaginary it is apparent that the terms in the expansion of exp{-2 Im P} oscillate in sign as required for cuts produced by multiple Pomeron exchange.

The answer to whether the cuts correspond to the AFS or Mandelstam cuts is not as easily answered. The diagrams of Fig. 1 are very similar to those of Fig. 3a which lead to the suggestion of AFS cuts. The difficulty is that in the intermediate state the particles are considered to be elementary. For the Mandelstam cut the intermediate state consists of two crosses each of which corresponds to two particles. If the intermediate particles in absorption diagrams were composite (i.e. Regge poles) it may be possible to have the crosses. For example consider the diagrams in Fig. 4

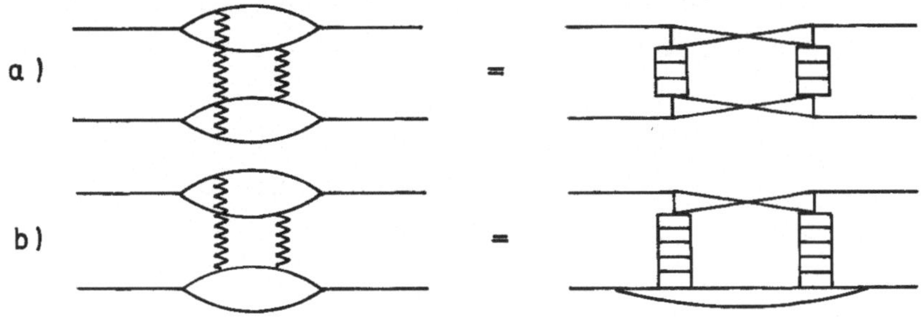

Fig. 4

The first is a Mandelstam type diagram, while the second is not a Mandelstam type diagram since it has only one cross. This suggests that one should consider the intermediate particles as Reggeons with signature and on and off mass shell contributions and hope that off-mass shell contribution can be written in the same form as the on-mass shell contribution. Of course the prescription for continuing off mass shell is not unique and certainly not for the case of a Reggeon.

If one views the Mandelstam diagram as giving the scattering particles composite structure and assumes that absorption corrections are equivalent to such diagrams one comes to the interesting conclusion that the only successful picture of high energy particles is one in which all intermediate particles are considered composite, e.g. as Regge poles. With this in mind it is easy to understand why a model that involves exchanges of elementary particles cannot be successful and that the exchange of Regge poles is the correct description.

ABSORPTIVE AND UNITARITY CORRECTIONS [4]

It has been thought that absorption corrections were of opposite sign to those due to unitarity. Symbolically the unitarity equation for inelastic amplitudes is

$$\mathrm{Im}\,F_{ab} = \mathrm{Re}(F^*_{aa}\times F_{ab}+F^*_{ab}\times F_{bb}) + \mathrm{Re} \sum_{n\neq a,b} F^*_{an}\times F_{bn} \qquad (11a)$$

and for elastic amplitudes is

$$\mathrm{Im}\,F_{aa} = F^*_{aa} \times F_{aa} + \sum_{n\neq a} F^*_{an} \times F_{an} \qquad (11b)$$

If one identifies F_{an} as the amplitude M_{an} occurring in Multipheral Models, which produces

$$\sum_{n\neq a,b} M^*_{an}\, M_{bn} = \mathrm{Im}\, R_{ab} + \text{a low-lying cut}$$

and

$$\sum_{n\neq a} M^*_{an}\, M_{an} = \mathrm{Im}\, P + \text{a low-lying cut}$$

one obtains

$$\mathrm{Im}\,F_{ab} = \mathrm{Im}\,R_{ab} + 2\mathrm{Re}(P^* \times R_{ab}) + \ldots$$

$$\mathrm{Im}\,F_{aa} = \mathrm{Im}\,P + 2\mathrm{Re}(P^* P) + \ldots$$

With P imaginary, the correction suggested by unitarity is just equal and opposite to the correction due to absorption (see Eqs. (6), (7)). The resolution of this problem was given by L. Coneschi who pointed out that the physical production amplitude F_{an} is really M_{an} with absorption corrections i.e. $F_{an} = S^{1/2}_{aa}\, M_{an}\, S^{1/2}_{nn}$. If one assumes absorption in production channels is negligible and writes

$$F^*_{an} \times F_{an} = (1+iP)^*(1+iP)M^*_{an} \times M_{an} + \ldots$$

and

$$F^*_{an} \times F_{bn} = [(1+iP)M_{an}+iR_{ab}M_{bn}]^*[(1+iP)M_{bn}+iR_{ab}M_{an}]+\ldots \quad (12a)$$

one finds

$$\mathrm{Im}\, T_{aa} = \mathrm{Im}\, P + P^* \times P - 2\, \mathrm{Im}\, P \times \mathrm{Im}\, P + \ldots =$$
$$= \mathrm{Im}\, P + \mathrm{Re}(P \times P) \quad + \ldots$$

and

$$\mathrm{Im}\, T_{ab} = \mathrm{Im}\, R_{ab} + 2\mathrm{Re}(P^* \times R_{ab}) - 4\, \mathrm{Im}\, P \times \mathrm{Im}\, R_{ab} + \ldots$$
$$= \mathrm{Im}\, R_{ab} + 2\mathrm{Re}(P \times R_{ab}) + \ldots \quad (12b)$$

This result is thus in agreement with the absorptive correction, independent of P being imaginary. For the elastic case with P imaginary, one is essentially identifying the P with the overlap function $\sum_{a \neq n}{}' M^*_{an} \cdot M_{an}$ plus twice $T^*_{aa} \times T_{aa}$, which is one of the identifications suggested by P. J. O'Donovan.

VARIOUS ABSORPTIVE REGGE POLE AND CUT MODELS [5]

So far we have mentioned three slightly different ways to include absorption corrections. The first, a sum over corrected partial wave amplitudes, which was originally used in the OPE model, is used by the group at Imperial College (Collins et al.) The second, a convolution integral of the amplitudes, is used by the groups at Saclay (Cohen-Tannoudji et al.) and Michigan (Henyey et al.). The third method, an eikonal representation, is used by the group at Argonne (Arnold et al.), people describing pp elastic scattering and some of the people describing pion photo-production.

In addition to these methods, other groups use cut contributions obtained from studying Feynman diagrams (Gribov et al.) or simply introduce contributions that are parametrized to simulate those due to cuts.

Since the three different methods to introduce absorptive corrections are all equivalent at high energies and small angles the essential difference comes from other assumptions made. The assumptions concern:

a) the nature of the input trajectories, e.g. whether their residues contain α-factors, are otherwise constants etc.

b) whether the Pomeron is considered a fixed pole or has a trajectory with a finite slope,

c) whether they apply their model to elastic scattering and

d) whether one includes cut contributions coming from trajectories, none of which are the Pomeron.

In particular, with the acceptance of cuts and thus nonzero third double spectral functions, one can have the Mandelstam-Wang fixed poles at wrong signature points which effectively remove the α-factors. If this is assumed the dips in differential cross sections are then explained as regions where cut and pole contributions are both large and are partially cancelling each other. Alternatively if one assumes the residue function have zeroes due to α-factors, then the convolution integral would give smaller cut contributions and the dip would result from the vanishing of the otherwise dominate Regge pole contribution. Of course these two methods do not predict the same energy behaviour of the dips. The first method would suggest that dips move to small t values asymptotically since the cut would tend to dominate more at large s. In the second method the dip would be stationary, but it would asymptotically vanish as the cut begins to dominate.

Concerning the Pomeron, Arnold, Cohen-Tannoudji et al. and those working with the Hybrid Model assume the Pomeron is a fixed pole. In particular, in the Hybrid Model for elastic proton scattering, the residue of the Pomeron is assumed to be proportional to the square of the proton form factor and, in elastic pion proton scattering, to the product of the pion and proton form factors. (It should be noted that a fixed Pomeron necessitates the existence of many additional undesirable singularities in the j-plane). Others, utilizing the flatness of the effective α resulting from multiple Pomeron exchanges,have been able to take the slope of the Pomeron comparable to that of other trajectories.

Only the people working with the eikonal representation and the Saclay group address themselves to the problem of elastic scattering. The Saclay group, in attempting to incorporate multiparticle unitarity, finds

$$F_{ab} = A\ \delta_{ab} + R_{ab} + 2i\int d\Omega\ A\ R_{ab}$$

where A corresponds to a fixed-pole Pomeron contribution. The eikonal representation describes elastic scattering by Eq. (4), with the eikonal function, $-i \log S(b^2)$, given by the Pomeron and other trajectories that contribute to elastic scattering.

None of the absorption calculations properly treat cut contributions coming from the exchange of two Reggeons, neither of which are the Pomeron. These should be important in double charge exchange reactions like $\pi^- p \to K^+\Sigma^-$, $K^- p \to K^+\Xi^-$, $\pi^+\Sigma^- \to \pi^-\Sigma^+$, which should be dominated by ρK^*, KK^* and $\rho\rho$ cuts respectively. Considering the smallness of the experimental cross sections for three reactions it might be that only cuts involving Pomeron exchange are important. The factor λ used by Michigan group is a rather crude attempt to consider these inelastic contributions.

The Michigan group, in addition to the multiplication factor λ , uses residues without α-factor zeroes and consequently is very much in favour of explaining all dips and cross-over effects as due to interference between cuts and Regge poles. Dips in their model are due to vanishing of helicity flip amplitudes in the region of t approximately -0.6, whereas cross-over effects which occur at small values of t,i.e.-0.2 , are due to vanishing of helicity non-flip amplitudes. The cut contribution is smaller for helicity flip amplitudes because they do not have the lower partial waves which require large absorption corrections. This model predicts that both cross-over points and dips should move to small values of t at high energies. This mechanism must be given credit for being one of the first to explain cross-over effects in a natural way. The difficulty of this model is that the Regge poles tend to only be the "quarks" that produce the cuts which in turn explain the major part of the data.

ABSORPTIVE REGGE POLE MODEL AND EXPERIMENTAL DATA [6]

Before we discuss particular reactions, it is useful to consider some arguments which depend on general properties of reactions. It has been assumed for some time that cut effects, in general, are not important at small values of t where the major contribution to total cross sections is found. Strong support for this is seen in reactions such as double charge exchange where the branch points of possible cut contribution are positive at t=0, but the cross sections are still down by a factor of 1/100 from other reactions. Also there are many relations, that are experimentally verified which would not be expected to hold if cuts were present to destroy the factorizability of the amplitudes.

Another demonstration of the smallness of cut contributions can be obtained by evaluating the cut contribution in the forward direction, i.e. when y=z,

$$f_{el}^{\substack{Abs\\cut}} = \frac{2}{w} F_{el}^{\substack{Abs\\cut}} \cong \frac{p}{4\pi} \int (\frac{2}{w}p)^2 \, d\Omega + .. \cong - \frac{p}{4\pi} \int |f_{el}|^2 d\Omega =$$

$$= - \frac{p}{4\pi} \sigma_{el} \quad .$$

Then

$$\frac{f_{el}^{\substack{Abs\\cut}}}{\mathrm{Im}\, f_{el}} = \frac{f_{el}^{\substack{Abs\\cut}}}{\mathrm{Im}\, f_{el}^{\substack{Abs\\cut}} + \mathrm{Im}\, f_{el}^{P}} = - \frac{\sigma_{el}}{\sigma_{tot}}$$

or

$$\left| \frac{f_{el}^{\substack{Abs\\cut}}}{f_{el}} \right| = \left| \frac{f_{el}^{\substack{Abs\\cut}}}{f_{el}^{\substack{Abs\\cut}} + f_{el}^{P}} \right| \le \frac{\sigma_{el}}{\sigma_{tot}}$$

Since this ratio is experimentally of the order of 15 to 30 % for $\pi^{\pm}$ p, $K^{\pm}$p, pp and $\bar{p}$p data, one would conclude that even in elastic scattering the cut contributes less than 30 % to the full amplitude in the forward direction.

1. π Dominated Reactions [7]

As is well known an elementary pion decouples from reactions in the forward directions due to its unnatural parity. Consequently to produce a pion at an equal mass vertex there must be a small momentum transfer. If you consider absorption as rescattering it is easy to envision the pion being produced and then the rescattering processes supplying the compensating momentum transfer. By considering the convolution integral for t=0, it is apparent that the rescattering process can contribute momentum transfers on the order of a^{-1}, where a is the slope of the diffraction peak.

It is interesting to consider the problem in terms of partial waves. As mentioned earlier the absorption corrections, to a rough approximation, will absorb the s-wave and change $t/(t-\mu^2)$ to $\mu^2/(t-\mu^2)$. This of course corresponds to evaluating the residue at μ^2 which, being a known coupling constant, allows a simple check of the possible success of the model.

a) $\gamma p \to \pi^+ n$, $\gamma n \to \pi^- p$

Since these two reactions are equal for small t, it is apparent that of the two possible sets of trajectories (ρ, B) and (π, A_1, A_2) one dominates the forward scattering. The reactions show extremely narrow forward peaks of width μ^2 , strongly suggestive of pion dynamics (see Fig. 5a). The data is well approximated for small values of t by the expression for elementary pion exchange with t replaced by μ^2 in the numerator, as suggested above. Consequently the data is easily fit within an absorptive pion model with secondary contributions from other singularities such as A_2, ρ or ρP cut. Some ρ or B contribution is necessary to explain the difference of the two reactions away from the forward direction.

An easy way to understand the success of the model is to consider unabsorbed pion and cut correction separately (see Fig. 5b). In the figure the difference between area between the two curves is just the absorptive pion contribution. In this way it is seen that a rather smooth cut contribution is sufficient to explain the forward peaking. This has prompted several groups to parametrize it as a background contribution.

One difficulty with the absorptive method is that a normal slope for the pion trajectory leads to a dip-bump phenomenon with the bump at about $\sqrt{-t} \sim 0.3$. There is a slight shoulder in the data in this region. In order to predict this shoulder it has been found that the slope of the π trajectory has to be rather small. This is under-

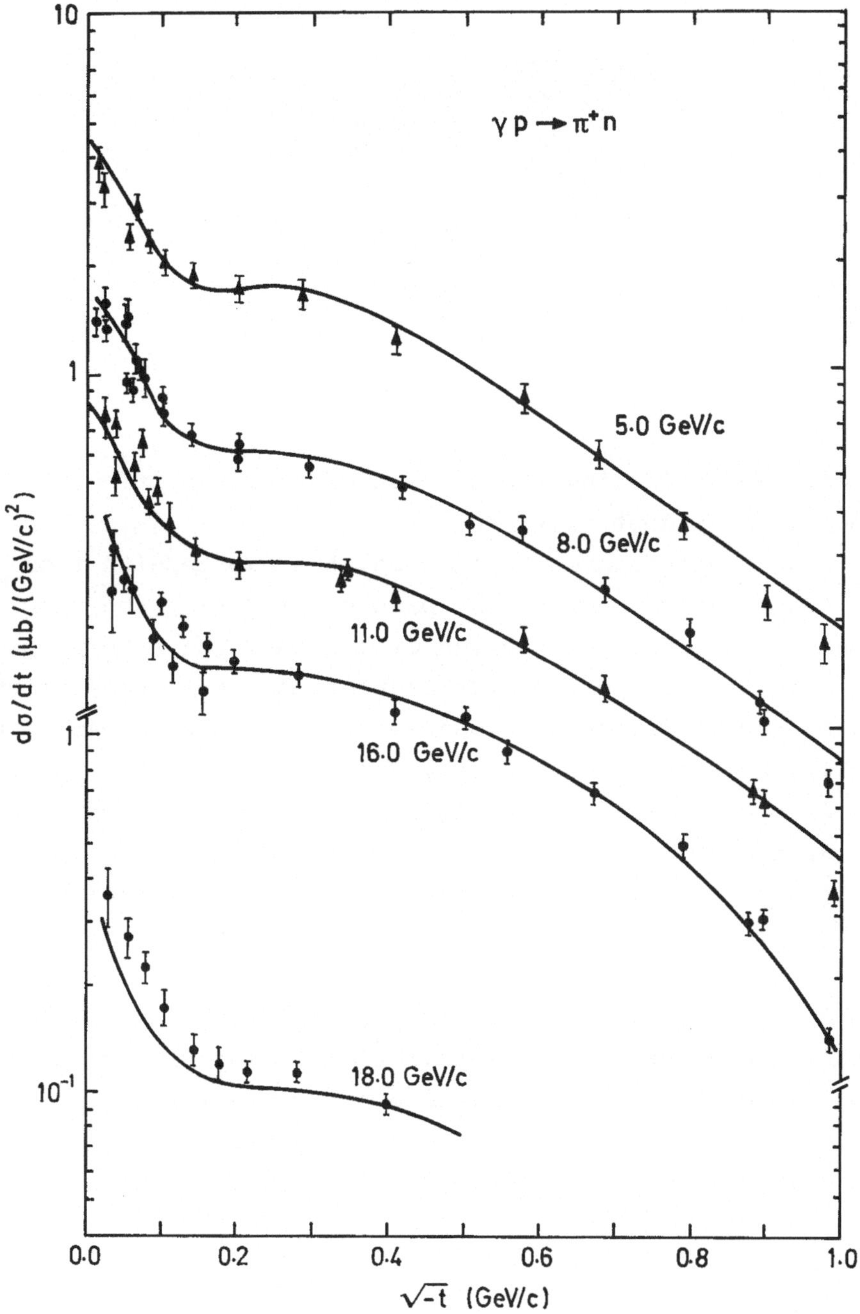
γp → π⁺n
5.0 GeV/c
8.0 GeV/c
11.0 GeV/c
16.0 GeV/c
18.0 GeV/c
dσ/dt (μb/(GeV/c)²)
√-t (GeV/c)
10
1
1
10⁻¹
0.0
0.2
0.4
0.6
0.8
1.0

Fig. 5a

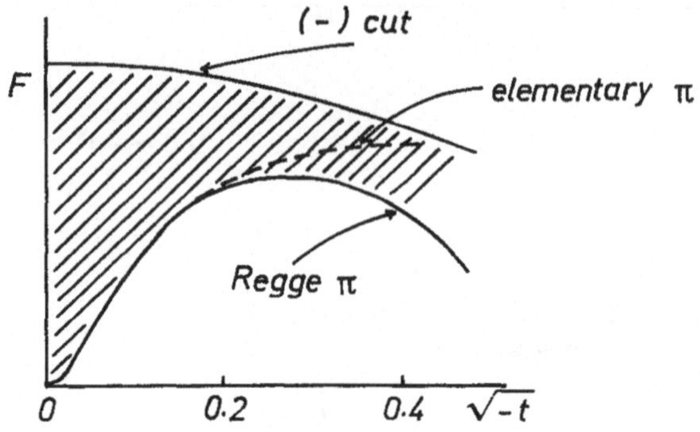

Fig. 5b

standable considering the success of the absorptive OPE model where essentially $\alpha=0$.

In the t-channel amplitudes, the dip results from the cancellation between the pion contribution and the unnatural parity contribution of the cut. The natural parity contribution of the cut reduces this zero to a slight dip in the differential cross section. This zero was found earlier by using Finite Energy Sum Rules and was interpreted in pion conspiracy models as a zero of the pion residue at $t \cong -\mu^2$.

Some people have applied the Vector-Dominance theory to couple the γ through a ρ meson. Consequently one would then include absorption corrections in both the initial and final states. It has been suggested that one should consider a direct photon interaction in addition to a mediated one. For pure ρ dominance one would write $d\sigma(\gamma N\rightarrow\pi N)=(1-\rho_{00})d\sigma(\rho N\rightarrow\pi N)$ where the density matrix ρ_{00} would subtract out the longitudinal ρ contribution.

b) $pn\rightarrow np$, $\bar{p}p\rightarrow\bar{n}n$ and $\pi^+p\rightarrow\rho^0\Delta^{++}$

These reactions with available data tend to display sharp forward peaks with widths of the order of μ^2. The data have been fit using an absorptive pion model. It should be realized that these reactions are just the ones that caused a great deal of controversy when attempts were made to fit them using π, A_1 and A_2 trajectories with several different conspiracy combinations. One objection to such fits was, that somehow the pion played only a minor role, contrary to intuition.

In an absorption model the problem with factorization in the conspiring pion model, as suggested by LeBellac, is easily avoided since cuts do not factorize. A cut contributes to both parity states in the t-channel and consequently is self-conspiring.

2. Elastic Scattering and the Pomeron [8]

Most models for elastic scattering are based on the eikonal representative with the assumption that the eikonal function $\chi_q(= -i\log S(b^2))$ is given by the single-scattering amplitude. The Regge poles, thus, are taken to be the "optical potential" that drives the rescattering processes. In these descriptions, the Pomeron is a rather phenomenological object to which the normal Regge poles are added.

As mentioned before diffractive scattering is thought to be associated with the composite structure of the scattering particles. Chou and Yang suggested a novel way to treat the scattering of two composite particles. Their descripture of the scattering of two diffuse matter clouds is essentially an extension of that of Glauber's model when a wave is diffracted as it passes through a composite system, namely the deuteron. The model is rather simple and it is interesting to recall it.

Consider two composite particles, described by densities ρ_1 and ρ_2 respectively that scatter with impact parameter $\bar{b}$. Then one has

$$\chi(\bar{b}) \propto \int D_1(\bar{b}' - \bar{b})\, D_2(b')\, d^2b'$$

where

$$D_i(\bar{b}) = \int_{-\infty}^{\infty} \rho_i(\bar{b},z)\, dz \quad .$$

Using the relation between the composite particle's density and its form factor,

$$\rho_i(\vec{r}) = \int e^{-ikr}\, F(-k^2)\, d^3k$$

one has

$$D(\bar{b}) = (2\pi) \int d^2q\, F(-q^2)\, e^{-iqb} \qquad (-q^2=t) \quad .$$

Thus

$$\chi(\bar{b}) \propto \int F_1(-q^2)\, F_2(-q^2)\, e^{iqb}\, d^2q$$

$$\propto \int \Delta\, d\Delta\, J_o(\Delta b)\, F_1(-\Delta^2)\, F_2(-\Delta^2) \quad , \quad \Delta=+q^2$$

$$(\text{i.e. } J_o(b\Delta) = \frac{1}{2\pi} \int_0^{2\pi} e^{ib\Delta\cos\phi}\, d\phi) \quad .$$

Since we are talking about the Pomeron single scattering term, we take

$$\chi_\infty(\bar{b}) = i\, K \int \Delta d\Delta\, J_o(\Delta b)\, F_1(-\Delta^2)\, F_2(-\Delta^2)$$

or

$$\chi_\infty(s,t) = i\, K\, F_1(t)\, F_2(t)$$

where the subscript is to remind us that this is the asymptotic eikonal function.

Before we can apply the expression we must decide what we can use for the form factors. Chou and Yang suggested that electromagnetic density of a particle should also be a reasonable descripture of its density for strong interactions and thus identified the form factors with the electromagnetic form factors. These form factors are generally dipoles of the form

$$F_{em} = (1-t/m^2)^{-2} \qquad \text{with } m^2 \sim 1 \quad .$$

Now how does this work for pp elastic scattering? The answer was given by Durand and Lipes and is shown in Fig. 6.

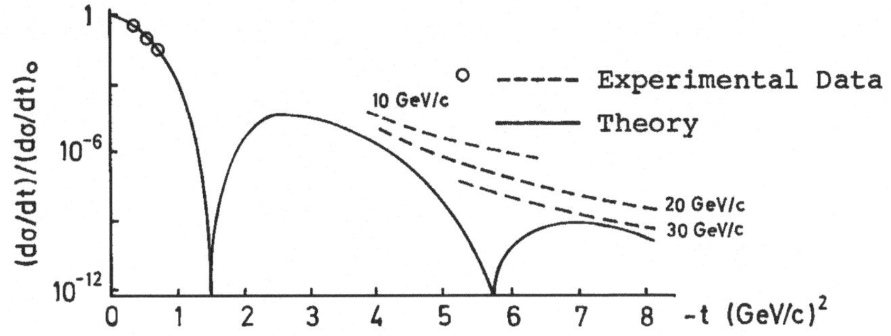

Fig. 6

With K adjusted to fit the total cross section the fit for small $|t|$ values was very good, but for larger values than about 1.0 the data were much higher but seemed to be asymptotically approaching the predicted curves. It was found that the strong zeroes could be smoothed by giving K a small imaginary part consistent with the forward amplitude.

The zeros in the case when K is real are essentially the result of various orders of scattering becoming important. This case be easily seen by considering an expansion in χ,

$$f_{el} = -ip\int_0^\infty b\ db\ J_o(\Delta b)(e^{i\chi}-1) = -ip\int_0^\infty bdbJ_o(\Delta b)\sum_{n=1}^{\infty}\frac{(i\chi)^n}{n!}$$

If we approximate $\chi(s,t)$ by

$$\chi(s,t) = i\ K\ \exp\{t\ R^2/2\}\ ,$$

then

$$\chi(b^2) = i\ \frac{K}{R^2}\ \exp\{\ -b^2/2R^2\}$$

and

$$f_{el} = i\,\frac{pK}{R^2}\sum_{n=1}^{\infty}\frac{(-\frac{K}{R^2})^{n-1}}{n\,n!}\exp\{\frac{tR^2}{2n}\} \quad \text{(note factor of } (-1)^n) .$$

Thus the contribution for one and two Pomeron exchanges are

$$f^{(1)} = i(\frac{pK}{R^2})\exp\{\frac{tR^2}{2}\} , \qquad f^{(2)} = -i(\frac{pK}{R^2})(\frac{K}{4R^2})\exp\{\frac{tR^2}{4}\}$$

These two terms would give a zero at $t= -0.6$ for normal values of the K and R^2 . Because the third term in the sum is positive this zero will be shifted to larger values of $|t|$ and closer to the value of $t_o \approx -1.0$ found with the form factors .

Now let us consider the data and the predicted curve. Obviously what is missing in the model are the nonasymptotic contributions due normal Regge poles. Using essentially the "optical potential" idea, Chiu and Finkelstein suggested adding the contribution due to the exchange degenerate trajectories $P'(f^o)$ and ω. They then wrote

$$\chi(s,t) = \chi_\infty(s,t) + \beta(s/s_o)^{\alpha_\omega}$$

with $\alpha_\omega = \frac{1}{2} + t$.

With the introduction of two new parameters β and s_o, they were able to give a reasonable fit to the data. In particular, they fit the break in the data at $t \sim -1.0$ and predicted a second break at about -5.5 . They explain why the experimental curves approach the asymptotic curves slower for larger values of $|t|$.

The use of a fixed Pomeron pole is not appealing or necessary. In particular Frautschi and Margolis (also Kaplan and Schiff at Orsay) dropped the concept of form factors and considered as a normal Regge pole with nonzero slope. They make three predictions

i) $d\sigma/dt$ continues to fall without reaching an asymptotic form,

ii) σ_{pp}^{tot} will rise from 40 mb at 30 GeV/c to about 50 mb at infinity .

iii) Re F(t=0)/Im F(t=0) will change from minus to plus as $s \to \infty$.

By using $P = \frac{1}{2}\,\beta(s\,e^{-i\pi/2})^{\alpha_P} = i\beta\,e^{tR^2/2}$ where $R^2 = 2\alpha' \log s - i\pi$, the first of the predictions can easily be understood. In particular

$$4\pi \lambda\!\!\!^{-}\, \mathrm{Im}\, f_{el}(t=0) \to \sigma_{tot}^{\infty}\,[1 - \frac{K}{8\alpha' \log s} + O((\log s)^{-2})]\ .$$

(Note such mechanism might be indicated in the Serpukhov data.) The value of α' found by Frautschi and Margolis was about 0.8 , whereas that of Kaplan and Schiff was 0.3 .

The $\bar{p}p$ data are similar to that of pp except the diffraction peak is larger and falls off faster with the dip or break now at t=-0.6.

The Chiu and Finkelstein model easily gives the correct features with the dip at the correct value of t . In $\bar{p}p$, the ω contribution changes sign and effectively $\beta_\omega \to \beta_\omega\ e^{-i\pi\alpha}$. In this way they fit the total cross section for $\bar{p}p$ and also obtain the correct t value for the cross over effect.

Some comment should be made also of the Argonne fits to $\pi^{\pm}p$ elastic data. They again use dipole form factors and obtain the general features. They have difficulties with the cross over zero (t=-0.2) which they like to explain by an α-factor zero in the nonflip amplitude at t=-0.5 being shifted to smaller values of $|t|$ by the cut contribution. Unfortunately the shift is not sufficient to move the zero to smaller values than t=-0.3 .

3. The ρ and A_2 and Charge Exchange Reactions [9]

In the Regge pole model ρ and A_2 give the dominate contribution to $\pi^- p \to \pi^0 n$ and $\pi^- p \to \eta n$ respectively and together dominate to $K^- p \to \bar{K}^0 n$ and $K^+ n \to K^0 p$.

The pronounced dip in the differential cross section for $\pi^- p \to \pi^0 n$ at t=-0.6 and the subsequent bump are well known features. Formerly it was assumed that this dip was due to the vanishing of the nonflip amplitude at the nonsense point where $\alpha_\rho(-0.6)=0$. It was also realized that the nonzero polarization required another contribution in addition to the ρ . Consequently contributions such as a ρ', a ρP cut have been suggested.

With the interest in large cut contributions it has been possible to explain the dip as due to a cancellation between the ρ and the cut contributions. In particular if one parametrizes the input amplitudes by

$$F_\rho = c_\rho \exp\{t\, R_1^2/2\} \qquad R_1^2 = \alpha_\rho' \log(s/s_o)$$

$$F_{el} = R_o^2\, c_o \exp\{t\, R_o^2/2\} \qquad R_o^2 = \text{const.} \quad \text{(i.e. P as a fixed pole)}$$

then

$$A_{CE} = A_\rho(b^2)(1+2iA_{el}) = \frac{c_\rho}{R_1^2} \exp\{-b^2/4\, R_1^2\}(1-c_o\exp\{\frac{-b^2}{4R_o^2}\})$$

which gives

$$F_{CE} = c_\rho[\exp\{t\, R_1^2\} - c_o(R_2^2/R_1^2)\exp\{t\, R_2^2\}] ,$$

$$R_2^{-2} = R_1^{-2} + R_o^{-2}$$

Thus one easily sees how the first term, which is the unabsorbed ρ contribution, can be cancelled by the second term due to absorption. Note for large s that the second term contains the (1/logs) factor and the slope of its effective trajectory is zero as expected for a cut formed from the exchange of two Regge poles one of whose trajectory has zero slope.

If one parametrizes the ρ contribution with a nonsense zero in the residue, then the convolution integral will give a much smaller cut contribution. Since both methods are self-consistent only experiment will be able to determine which is correct. Most absorption people have chosen the first method and have been able to give good predictions for the differential cross section and polarization. Available data seem to show neither the shift of the dip to smaller values of $|t|$ as predicted by the pole-cut cancellation model nor a filling in of the dip as predicted by the α-factor model. If anything the dip seems to be moving slightly to larger values of $|t|$ with increasing energy.

Unfortunately both methods predict that the effective trajectory should bend and become tangent to the trajectory of the branch point of the cut. This is in contradiction to the effective trajectory, found from fitting the data with $d\sigma/dt = C(t)\, s^{2\alpha_{eff}-2}$, which is linear out to t=-1.0 and shows no tendency to bend.

None of the reactions $K^-p\to\bar{K}^on$, $K^+n\to K^op$ and $\pi^-p\to\eta n$ show any dip-bump phenomena near t=-0.6 . The first two reactions have been fit well with an A_2 absorptive model and the latter with an $\rho+A_2$ and absorption. It is apparently rather easy to get a simultaneous fit to fit $\pi^-p\to\eta n$ and $K^-p\to\bar{K}^on$ with an absorptive $\rho+A_2$ model. Different polarizations predictions have been made for $K^-p\to\bar{K}^on$ which could explain either a slow variation or rapid variation.

4. Other Reactions [10]

One of the most important reactions not discussed above is π^o photoproduction. The π^o photoproduction differential cross section shows a rather broad peak with a maximum at about t=-0.1 and with a dip-bump structure starting at t=-0.6 . Fits have been made with a dominate ω contribution with addition contribution coming from cuts, i.e. ωP^n, and the B trajectory or a simple parametrized cut.

It is known that a fit of the energy variation of the data with $d\sigma = C(t) s^{2\alpha_{eff}-2}$ gives a flat α_{eff} in the range of 0 to 0.5 . Consequently it is necessary to use cut contributions whose branch points are independent of t, i.e. ωP^n with large n . Another interpretation of this is that the reaction is dominated by a fixed pole. Without the assumption of vector dominance, unitarity would not forbid the existence of such a pole.

New fits of backward $\pi^{\pm}p$ scattering with absorption have had to retain the older explanation for the dip seen only in π^+p near t = -0.2 . This was the point where α_N passed - 1/2 causing the N contribution to vanish.

Cut contributions have been used to explain the observed non-zero polarization in $K^- n \to \Lambda^o \pi$ which duality diagrams suggest would be zero.

CONCLUDING COMMENTS

Perhaps the best way of concluding this series of lectures is to summarize the reasons a Regge cut enthusiast would give as evidence for the existence of cuts:

a) The sharp forward peaks of many reactions that should be dominated by π exchange are easily understood by taking a moderate amount of absorption into account. Thus our intuitive feeling for the importance of pion in strong interactions is gratified.

b) In this model one only needs to introduce Regge poles corresponding to known particles and then turn the absorption crank to find the additional contributions necessary for non-zero polarizations etc.

c) With cut contributions one can work with simple evasive poles whose residues vary smoothly. This is certainly better than having to accept many unphysical conspiring trajectories and residue functions with rapid variations for small values of t.

d) The apparent flatting of $\alpha_{eff}(t)$ at large values of $|t|$, for such reactions as pp elastic scattering, is essentially accounted for by the dominance of cuts whose branch points are mostly independent of t.

e) The existence of dips, and secondary maxima in differential cross section are essentially diffractive minima indicating the importance of absorption. In particular the breaks in pp elastic scattering are essentially transition regions between different rescattering contribution. An interference between cut and pole contribution gives a natural explanation of cross-overs.

f) The concepts of diffractive scattering and rescattering are very natural and should be included in any realistic scattering model.

g) There is much theoretical evidence that cuts do exist in the complex angular momentum plane.

Of course he would have to admit that the model has some difficulties both experimentally and theoretically. Experimentally the difficulties come from:

a) $\alpha_{eff}(t)$ for $\pi^- p \rightarrow \pi^0 n$ decreasing linearly for $0<-t<1.0$ and showing no curvature or flattening as necessary for model using a ρ and a ρP cut.

b) $\alpha_{eff}(t)$ for π^0 photoproduction remaining constant for $0<-t<1.0$ which is difficult to explain if the ω and

ωP cut are supposed to dominate.

Theoretically, there are even more difficulties coming from:

a) Whether the absorptive cut-like contribution can be associated with the Mandelstam cuts or only with the A.F.S. cuts. The latter are known to be cancelled by multiparticle intermediate state contributions. Even if the identification with the Mandelstam cuts can be made, one must justify the use of $s^{\alpha^c}/\log s$ which implies that $A(j,t)$ (see eq. (9)) has a logarithmic divergence at α^c and thus is in violation of unitarity.

b) Lack of a basic understanding of the Pomeron. Is the Pomeron all of elastic scattering or just the single scattering part of the amplitude? Is it a fixed, moving pole or what?

c) Whether α-factors should be included in residue functions or not.

d) Whether one should include diagrams with intermediate states other than the initial or final, i.e. should one include cuts contributions that do not include the Pomeron and should dominate reactions such as double charge exchange.

e) Whether one can justify the use of Regge pole exchanges as the primary scattering input. If absorption can be identified with a j-plane cut, then this problem of double counting is avoided since Regge poles and cuts are two distinct types of j-plane singularities.

The Regge pole model with absorption has to be given credit for maintaining simplicity by requiring only Regge poles associated with physical particles. But the use of large cut contributions that tend to dominate the pole contributions is certainly unappealing. The explanation of dips of differential cross sections coming from cancellation between cut and pole contributions implies that the correlation of such dips with wrong signature values of trajectories was purely accidental.

In conclusion the addition of absorptive corrections to the Regge pole model is an improvement. It is apparently not entirely correct but does provide us with a relatively simple model that explains many features that could not be explained by a pure pole model.

REFERENCES

1. H. Abarbanel and C. Itzykson, Phys. Rev. Letters 23, 53 (1969); J. S. Ball and W. R. Frazer, Phys. Rev. Letters 14, 747 (1965); L. Durand and V. T. Chiu, Phys. Rev. 139, B 646 (1965); R. J. Glauber, Lectures in Theoretical Physics vol.1, p. 315 (N.Y. 1959); J. Finkelstein and M. Jacob, Nuovo Cim. 56A, 681 (1968) M. Lévy and J. Sucher, Phys. Rev. 186, 1656 (1969); R. Omnes, Phys. Rev. 137, B 64 (1965); R. L. Sugar and R. Blankenbecler, Phys. Rev. 183, 1387 (1969); R. Torgerson, Phys. Rev. 143, 1195 (1966).
2. D. Amati, S. Fubini and A. Stanghellini, Phys.Lett. 1, 29 (1962); D. Amati, S. Fubini and A. Stanghellini, Nuovo Cim. 26, 897 (1962); D. Branson, Phys. Rev. 179, 1608 (1969); C. B. Chiu and J. Finkelstein, Nuovo Cim. 57A, 650 (1968); S. Mandelstam, Nuovo Cim. 30, 1113, 1127, 1148 (1963); H. J. Rothe, Phys. Rev. 159, 1471 (1967); J. C. Polkinghorne, Nuovo Cim. 56A, 755 (1968) J. H. Schwarz, Phys. Rev. 162, 1671 (1967); C. Wilkins, Nuovo Cim. 31, 377 (1964).
3. R. C. Arnold, Phys. Rev. 153, 1523 (1967); M. E. Ebel and R. J. Moore, Phys. Rev. 177, 2470 (1969).
4. J. Caneschi, Phys. Rev. Letters 23, 254 (1969); P. O'Donovan, Phys. Rev. 185, 1902 (1969); J. Finkel-

stein and M. Jacob, Nuovo Cim. 56A, 681 (1968); R. J. Rivers and L. M. Saunders, Nuovo Cim. 58A, 385 (1968).

5. S. B. Chiu and J. Finkelstein, N. C. 57A, 650 (1968); V. N. Gribov, Sov. J. Nucl. Phys. 5,138 (1967); S. Mandelstam and L. Wang, Phys. Rev. 160, 1490(1967); M. Ross, Irvine Conference, Univ. of California, Dec. 1969 (see footnotes 7, 8, 9 and 10 for members of the various groups mentioned in the text)

6. P. Freud and P. O'Donovan, Phys. Rev. Letters 20, 1329 (1968); R. J. N. Phillips, Phys. Rev. Letters 20 , 1329 (1968).

7. D. Amati, G. Cohen-Tanoudji, R. Jengo, Ph. Salin, Phys. Lett. 26B, 510 (1968); M. Blackmon and G. Kramer, "Regge Pole Model with Absorptive Cuts for Photoproduction of Charged Pions", Argonne April 1969; J. Froyland and D. Gordon, Phys. Rev. 177, 2500 (1969); G. Kramer, K. Schilling and F. Stodolsky, Nucl. Phys. B5, 317 (1968) ; D. K. Ross, Phys. Rev.166, 1521 (68); M. Ross, Irvine Conf. (above)

8. R. C. Arnold and M. Blackmon, Phys. Rev. 176, 2082 (1968); M. Blackmon and Goldstein, Phys. Rev. 179, 140 (1969); C. B. Chiu, Rev. of Mod. Phys. 41, 640 (1969); S. A. Dunne, Phys. Rev. Letters 19, 1299 (1967); S. Frautschi and B. Margolis, Nuovo Cim. 56A (1968); J. M. Kaplan and D. Schiff, Orsay 1969/64; H. Yabuki, Phys. Rev. 177, 2209 (1969).

9. R. Arnold, Phys. Rev. 140, B1023 (1965); M. Blackmon, Phys. Rev.178,2385(1969); M.Blackmon and G.Goldstein, Phys. Rev. 179, 1480 (1969); G. Cohen-Tannoudji, G. Morel, H. Navelet, Nuovo Cimento 48, 1075 (1967); P. Collins, B.Hartley, B. Moore, K. Moriarty,I.C.T.P. 69/9; C. B. Chiu, RMP 41, 640 (1969); B. Hartley, B. Moore, K. Moriarty, ICTP/69/35 and ICTP/69/2 (This

paper contains references to 30 other papers on the absorptive Regge pole model); F. Henyey, G. Kane, J. Pumplin and M. Ross, Phys. Rev. 182, 1579 (1969); S. Frautschi and B. Margolis, Nuovo Cim. 56A, 1155 (1968); V. Lany, D. Gross, I. Muzinich, V. Triplity, Phys. Rev. Letters 18, 149 (1967);

10. M. Blackmon, G. Kramer and K. Schilling, "Regge pole Model with Absorptive Cuts for Photoproduction", Argonne February 1969; A. Contogouris, J. P. Lebrun and G. Von Bochman, Nucl. Phys. B13, 246 (1969); M. Ross, Irvine Conference (see above).

Acta Physica Austriaca, Suppl. VII, 214–264 (1970)

LECTURES ON REGGE PHENOMENOLOGY*

BY

R. J. N. PHILLIPS
Rutherford Laboratory, Chilton, Berkshire,
England

ABSTRACT

These lectures are about the comparison of Regge theory (and folklore) with experiment.

The first lecture surveys the properties of Regge-pole exchange amplitudes, and the places where simple pole dominance seems to fail. The second deals with Regge cuts, various prescriptions for calculating them, and what experiments are helpful. The third is about finite energy sum rules, their use to predict high energy scattering, and some problems with asymptopia.

Duality and the Veneziano model are omitted, to avoid double-counting (see Kugler's lectures).

---

*Lecture given at IX. Internationale Universitätswochen für Kernphysik, Schladming, February 23 - March 7, 197o.

1. REGGE POLES[1]

A typical t-channel boson Regge pole contribution, to a high energy s-channel amplitude, has the form

$$f(s,t) = \gamma(1+\tau e^{-i\pi\alpha})/\sin \pi\alpha \left[\frac{s-u}{2s_o}\right]^{\alpha} \tag{1}$$

where $\alpha(t)$, $\gamma(t)$ and $\tau=\pm 1$ are the trajectory, residue function and signature of the pole, respectively. This form applies for instance to a spin-independent invariant amplitude, or to t- or s-channel helicity amplitudes away from kinematical singularities. It is guaranteed to the leading power in s only; terms of relative order 1/s are model-dependent, because of possible daughter-poles and other corrections. We could as well have written s^{α} in place of $(s-u)^{\alpha}$ above. We normally suppose α and γ to be real in the s-channel scattering region, though in principle two trajectories could meet at some t and become complex thereafter[2].

Things are more complicated for baryon Regge poles (in the u-channel, say). According to conventional theory[3], baryon poles meet in pairs at u=o, a kinematical branch point. They correspond to pairs of states that differ just by parity, and MacDowell symmetry relates their trajectories and residues by

$$\alpha^{+}(\sqrt{u}) = \alpha^{-}(-\sqrt{u}), \quad \gamma^{+}(\sqrt{u}) = -\gamma^{-}(-\sqrt{u}) \tag{2}$$

where the suffix $\pm$ indicates the quantum number $\tau P=1$, τ being the signature. Essentially, the correct variable here is not u but $\sqrt{u}$. For u>o, $\alpha^{\pm}$ and $\gamma^{\pm}$ are effectively independent; for u=o they coincide in pairs; for u<o they are complex conjugate pairs.

In πN scattering, for example, we end up with "parity-preserving" u-channel amplitudes $F^{\pm}$,

$$F^{\pm} = \mp A-(\sqrt{u}\pm m_N)B, \tag{3}$$

$$= \gamma^{\pm}(1+\tau e^{-i\pi(\alpha-\frac{1}{2})})/\sin\pi(\alpha-\tfrac{1}{2})\cdot(\frac{s-t}{2s_o})^{\alpha-\frac{1}{2}} \tag{4}$$

from which more familiar amplitudes can be re-assembled.

Here are some physical consequences, for a single Regge pole

(i) Connection with particles. Reggeon-exchange is a workable realisation of the particle-exchange idea. High spin gives no trouble provided $\alpha(t)\leq 1$ for $t\leq o$ (see (iii) below). This was one of the first reasons for using Regge poles; it is still good.

(ii) Crossing. High energy scattering in the s and u channels are symmetrical, for a Regge pole in the t-channel. Essentially, corresponding amplitudes are the same within a sign given by the signature τ. If the s-channel is described by Eq. (1), the crossed amplitude is

$$f(u,t) = \tau\ \gamma(1+\tau e^{-i\pi\alpha})/\sin\pi\ \alpha[\frac{u-s}{2s_o}]^{\alpha}. \tag{5}$$

(iii) Energy-dependence is given by α alone; it offers a way to measure α.

$$\frac{d\sigma}{dt} = X(t)s^{2\alpha(t)-2} \tag{6}$$

$$\frac{d\sigma}{du} = Y(u)s^{2\alpha(u)-2} \tag{7}$$

Elementary-particle exchange would give fixed spin J in place of α: trouble for $J>1$.

(iv) Shrinking of $d\sigma/dt$, as s increases, follows from $d\alpha/dt>o$.

(v) Phase is related to energy-dependence, being given by α alone through the factor $1+\tau\exp(-i\pi\alpha)$, for boson poles. This follows from analyticity and crossing, so is a restatement of points (ii) and (iii).

(vi) Spin-dependence is specific. Particular spin-parity classes of Regge poles allow particular kinds of spin-flip.

(vii) Helicity-flip and non-flip amplitudes have the same asymptotic s-dependence.

(viii) Helicity-flip and non-flip also have the same phase.

(ix) Factorisation. If the t-channel process is $ab\to cd$, the t-channel Regge residues factorize

$$\gamma(t) = \tilde{\gamma}_{ab}(t)\ \tilde{\gamma}_{cd}(t) \tag{8}$$

Thus $\tilde{\gamma}$ refers to a Reggeon vertex. This relates different processes.

(x) Exchange degeneracy. Long ago Arnold argued[4], from the weakness of "exchange" potentials generating boson Regge poles, that they occur in pairs with approximately the same $\alpha(t)$, same $\gamma(t)$, opposite τ. Recently duality gave new arguments for the same pairings, for both bosons and baryons, but the residues γ are only equal for high energy scattering in an "exotic" channel[5]. (If the residues of an exchange-degenerate pair are equal in the s-channel, say, the resulting invariant amplitudes here are purely real. For boson exchange, the crossed u-channel amplitudes are the same, except for an extra phase $\exp(-i\pi\alpha)$. For fermion exchange, the crossed-channel relations is less simple, unless $\alpha^+ = \alpha^-$ (and both are then real); in this case, the t-channel amplitudes are the same,

from this phase.
The Pomeranchuk pole (or Pomeron, P) is not supposed to be exchange-degenerate.

(xi) Wrong-signature nonsense zeros. Poles in physical amplitudes correspond to particles; they must not have "nonsense" properties such as imaginary mass, or spin $J<o$, or coupling to amplitudes with net helicity $|\lambda_1-\lambda_2|>J$. Whenever the factor $1/\sin \pi\alpha$ threatens to produce a "nonsense" pole, it must be killed by a zero in $\gamma(t)$ or in the signature factor $1+\tau \exp(-i\pi\alpha)$. With exchange-degeneracy, the signature factor cannot vanish for both partners at once, so $\gamma(t)$ must have a zero. For the partner with "wrong" signature, however, the signature factor vanishes anyway and the amplitude is zero at the nonsense point. Through factorization, these zeros persist in non-exotic channels where the residues are not equal, and one partner may even be forbidden.

It was once believed, from simple models, that wrong-signature nonsense zeros would occur quite generally; but no general argument exists. An appeal to duality and exchange degeneracy, as empirical facts, seems the best basis.

Points (i) and (iii) agree with experiment. Peaks at small t (small u) occur in high energy scattering when, and only when, there are known particles in the t (or u) channel. These particles can often be correlated on linear trajectories with slopes $\alpha' \approx 1$ GeV^{-2}. The energy-dependence of peaks given by a particular class of exchange agree broadly with each other, and with the extrapolation of α (6).

The crossing property (ii) is used rather than tested. Any amplitude can be divided into crossing-even and crossing-odd parts; each part requires Regge poles of suitable signature. So far there have been enough Regge poles, with suitable α and τ, to fit data.

To check details of one-pole predictions, we take processes where one pole dominates. The best examples are $\pi^- p \to \pi^o n$ and $\pi^- p \to \eta^o n$, believed dominated by ρ and A_2, respectively, with exchange-degenerate $\alpha(t)$. These confirm points (i), (iii), (iv), (v), (vii) and (x)[7-1o].

Factorisation (ix) is hard to check directly. The simplest things it predicts[1o, 11] are inaccessible, like

$$\sigma_T^{\infty}(\pi\pi) = [\sigma_T^{\infty}(\pi N)]^2/\sigma_T^{\infty}(NN).$$

It relates the spin-dependence of πN, KN and NN scattering[11, 12], but to test this the various Regge terms must be separated. We get a rough check on factorisation for the Pomeron, assuming it dominates elastic scattering and diffractive N^* production, from the experimental equality

$$\frac{d\sigma/dt(\pi N\to\pi N^*)}{d\sigma/dt(\pi N\to\pi N)} \approx \frac{d\sigma/dt(NN\to NN^*)}{d\sigma/dt(NN\to NN)} \tag{9}$$

that holds for 1o-2o GeV/c, at small t[13].

Exchange-degeneracy (x) seems well obeyed by the boson mass spectrum, especially by the ρ and A_2 trajectories (see Fig. 1), but these same trajectories seem to diverge for t<o[8, 1o] (see Fig. 2). Also exchange-degeneracy of $\alpha(t)$ would predict equality for the crossed processes (with isospin reflection),

$$d\sigma/dt(K^- p\to\bar{K}^o n) = d\sigma/dt(K^+ n\to K^o p) \tag{1o}$$

assuming $A_2 \pm \rho$ exchanges dominate, since the Regge terms are $\pi/2$ out of phase and cross-terms vanish. This fails by a factor 2 at 2-3 GeV/c, but improves at 5 GeV/c.[14] Exchange-degeneracy of $\gamma(t)$, as required by duality in exotic channels, gives many more predictions[5] - but they belong in Kugler's lectures.

Now for some troubles.

(a) Charge-exchange polarization. The phase-equality (viii) for single-Regge amplitudes forbids polarization, that depends on terms like $\mathrm{Im}(f_1 f_2^*)$. But polarization is found, in the alleged one-pole examples above[15]. For $\pi^- p \to \eta^o n$ one can appeal to a second A_2 trajectory, since the A_2 resonance is split[16]. But for $\pi^- p \to \pi^o n$ there is no known nearby trajectory, to explain the rather slow s-dependence (polarization varies $\sim s^{|\alpha_1 - \alpha_2|}$). Analysis suggests[17] a second pole ρ' would need intercept $\alpha(o) \approx o$, but the lowest mass $J^P = 1^-$ candidate[18] is at 1.55 GeV, suggesting an intercept $\simeq -1$.

A more likely explanation is a Regge cut, resembling a nearby secondary pole.

(b) Crossover mechanism. High energy pp and $\bar{p}p$ scattering are supposed to be dominated by the P, f^o (alias P') and ω Regge poles. A_1, A_2, ρ, π and B are present but small, since they control $np \to pn$ and $\bar{p}p \to \bar{n}n$ charge exchanges, which are small. We can write amplitudes symbolically:

$$A(pp) = P + P' - \omega,$$
$$A(\bar{p}p) = P + P' + \omega \qquad (11)$$

Now $d\sigma/dt(\bar{p}p)$ is bigger and more steeply t-dependent than $d\sigma/dt(pp)$ at t=o; hence the two intersect, near

$t=-0.1=t_o$. See Fig. 3. If we think of black diffracting spheres this is quite natural; $\sigma_T(\bar{p}p)$ is bigger because of annihilation channels, hence the black-sphere radius is bigger, hence both the value and slope of $d\sigma/dt$ at $t=o$ are bigger. In terms of Regge poles, however, the crossover requires the ω non-flip residue to change sign at $t=t_o$ (since non-flip scattering dominates here). Factorizing, this residue is a square, $\gamma=(\tilde{\gamma}_1)^2$; hence $\tilde{\gamma}_1$ must contain the factor $(t-t_o)^{1/2}$. However, other NN scattering amplitudes with helicity-flip at one vertex or both, have residues factoring like $\tilde{\gamma}_1\tilde{\gamma}_2$ and $(\tilde{\gamma}_2)^2$. If all these residues are real and free from singularities at t_o, as normally expected, then γ_2 also must contain the factor $(t-t_o)^{1/2}$. Similarly, by considering chains of inelastic processes with and without a nucleon vertex, with residue functions like $\tilde{\gamma}_1\tilde{\gamma}_j$, $(\tilde{\gamma}_j)^2$ $\tilde{\gamma}_j\tilde{\gamma}_k$, $(\tilde{\gamma}_k)^2$... etc., we see that all ω factor functions in all processes must have at least this factor $(t-t_o)^{1/2}$, and all ω residue functions vanish at $t=t_o$. This is a universal zero[7].

This zero agrees nicely with the observed crossover in $K^{\pm}p$ scattering (again dominated by $P+P'\pm\omega$), but does not show up universally.

In $\pi N\to\rho N$, the I=O exchange is supposedly dominated by ω: this part can be isolated by forming the combination

$$2d\sigma/dt(I_t=o)=d\sigma/dt(\pi^+p\to\rho^+p)+d\sigma/dt(\pi^-p\to\rho^-p)-d\sigma/dt(\pi^-p\to\rho^on), \tag{12}$$

but it shows a maximum rather than a zero near t_o, at 4 and 8 GeV/c[19].

In $\gamma p\to\pi^op$, ω is again supposed to dominate, but $d\sigma/dt$ shows no minimum at t_o. Analysis of high-energy data[2o] and FESR constraints[21] indicate no zero in the

ω residue.

One way out is to have another strong ω-like term, ω'[22]. The crossover is explained if the imaginary part of the $\omega+\omega'$ nonflip amplitude changes sign; this can happen without any special residue zeros. (The previous trouble came through phase and factorisation constraints on a single ω pole). There is no plausible Regge-pole candidate for ω', since ϕ is not supposed to couple to $\bar{N}N$, so presumably ω' is a cut.

(c) Pion Conspiracy. The s-dependence and very sharp t-dependence of the forward peak in $pn \rightarrow np$ charge exchange, with $d\sigma/dt \sim s^{-2} \exp(50t)$, strongly suggest that π exchange is responsible. But normal π-exchange (and any other Regge pole it can interfere with) vanishes at t=o, giving a sharp dip instead. This is because π contributes with equal strength to NN helicity amplitudes with net flip 0 and 2 in the s-channel:

$$f_\pi(++,\ --) = f_\pi(+-,\ -+) \qquad (13)$$

Angular momentum conservation makes the latter vanish at t=o, so the former must vanish too. The only way out, in a pure Regge-pole framework, is to invoke a conspiracy[23] with an opposite-parity Regge pole c, that contributes differently:

$$f_c(++,\ --) = -f_c(+-,\ -+) \qquad (14)$$

By arranging that π and c contributions to $f(+-,\ -+)$ cancel at t=o the angular momentum constraint can be satisfied, $f(++,\ --)$ need not vanish, and a forward peak results.

In $\gamma p \to \pi^+ n$, a similar peak again suggests π exchange, but is forbidden for a non-conspiring π.

Both reactions can be fitted by a π+c conspiracy[24, 25], but the π residue has to fall sharply to zero or a deep minimum near -t=o.o2-o.o5. A linear zero would be universal (like the ω crossover); a quadratic zero or a deep minimum would propagate at least into reactions with an $\bar{N}N$ vertex. Either way, we expect a zero or minimum in the π-exchange in $\pi N \to \rho N$; analysis suggests, on the contrary, that π exchange has a maximum hereabouts[26].

There are other troubles for the simple pion conspiracy. It predicts a forward dip in $\pi^+ p \to \rho^o \Delta^{++}$ at very small t[27], but actually a peak is seen[28].

Much fancier schemes have been proposed, with A_1 playing the major role instead[29], but all ultimately founder on factorisation difficulties[3o]. The available poles alone, even with conspiracies, cannot explain the data.

Introducing Regge cuts is an obvious solution[31]. They can conspire to give nonvanishing terms at t=o; destructive interference with non-conspiring π can then give a sharp peak.

(d) Missing dips. Wrong-signature nonsense zeros apparently give dips in many cross-sections; ρ exchange gives a dip at α=o in $\pi^- p \to \pi^o n$; N_α- exchange gives a dip at $\alpha=-\frac{1}{2}$ in backward $\pi^+ p \to p\pi^+$. The problem is then to explain why dips are not seen in some other cases (Harari gave a survey recently[32]).

The absence of a dip in backward $K^+ p \to pK^+$, from Λ_α at $\alpha=-\frac{1}{2}$, is explained by the presence of an exchange-degenerate Λ_γ pole-as required by duality-filling in the dip[33].

The absence of a dip in backward π^+ and π^o photoproduction, from N_α at $\alpha=-\frac{1}{2}$, can similarly be blamed on an N_γ contribution[34]. But this time there is no duality argument for N_γ; there is no more reason to expect it here than in backward $\pi^+ p\to p\pi^+$ or $\pi^- p\to n\pi^o$ - where dips are seen and N_γ is neglected. It is a bit unsatisfactory to bring in N_γ thus arbitrarily.

Other missing dips can also be argued away, but some arguments seem ad hoc.

(e) Missing baryons. Linear trajectories $\alpha=a_o+a_2u$ seem to fit baryon spectra quite well: see Fig. 1b. If α is indeed even in $\sqrt{u}$ (as also required by duality), the opposite parity doublets on $\alpha^{\pm}$ trajectories should be degenerate in mass. But opposite-parity partners to the well-known baryons are not seen; opposite-parity baryons are seen, but their SU_3 F/D ratios are quite different[35].

One explanation is that the residue mysteriously vanishes for all the missing partners.

Alternatively, Carlitz and Kislinger[36] suggest the missing trajectory is on an unphysical sheet, because of a fixed cut in the J-plane (absent in the usual theory[3]). Consequences for high-energy scattering have not yet been explored.

(f) Serpukhov data. The new $\pi^- p$, $\pi^- n$, $K^- p$, $K^- n$ σ_T data from Serpukhov[37], for 25-65 GeV/c, disagree with Regge pole extrapolations[17, 38]. Theory predicted a steady decrease of σ_T through this range, but the data are essentially flat. (We return to this question in lecture 3).

2. REGGE CUTS

From early days, Regge practitioners worried about cuts. Amati, Fubini and Stanghellini[41] generated a cut by iterating Regge-pole exchanges, to satisfy unitarity: Fig. 4a. Mandelstam[42] showed that this diagram, calculated completely, gives no cuts but double-crossed diagrams do: Fig. 4b.

Nobody yet knows how to calculate all the cuts, but we know some general properties[42]. If α_c is the J-plane branch point of a signatured amplitude, the cut contribution appears as a continuum of poles:

$$f^{cut} = \int^{\alpha_c} d\alpha\rho(\alpha) f^{Regge}(\alpha), \tag{15}$$

$$\sim \beta(t)\{1+\tau e^{-i\pi\alpha_c(t)}\}(\frac{s}{so})^{\alpha_c(t)}/(\ln s)^{\lambda+1} \tag{16}$$

assuming $\rho(\alpha)\sim(\alpha_c-\alpha)^\lambda$ near α_c. Eq. (16) shows the asymptotic form, neglecting terms $x(\ln s)^{-1}$, (but it can be misleading to neglect the latter in practice). Unitarity requires $\rho(\alpha_c)=o$, so $\lambda>o$.[43]

For two linear trajectories $\alpha_i(t)=\alpha_i(o)+\alpha_i' t$, the 2-Reggeon branch point is also linear:

$$\alpha_{12}(t) = \alpha_1(o)+\alpha_2(o)-1+t\alpha_1'\alpha_2'/(\alpha_1'+\alpha_2') \tag{17}$$

This equation also gives the rule for combining α_{12} and α_3 to get the 3-Reggeon branch point α_{123}, and so on. If all Reggeons except 1 are Pomerons with $\alpha_p(o)=1$, the branch points coincide with α_1 at t=o and lie above it for t<o. Thus we expect P-induced cut corrections to dominate asymptotically over poles for t<o; at t=o they are unfavoured, but only logarithmically.

The signature of a 2-boson cut is the product $\tau_1\tau_2$; for a 2-baryon cut it is $-\tau_1\tau_2$ [44].

A Regge pole belongs to a unique spin-parity class. A 2-Reggeon cut does not[45]; it can have contributions of both parities, and can conspire with itself.

Unknown cuts have so many free parameters that people first chose to ignore them hoping they would prove unimportant at lab. energies. Later the deficiencies of pure Regge-polology prompted a more positive attitude. Prescriptions were made for calculating cuts from poles, without introducing many new parameters. Though arbitrary, some prescriptions have been quite successful.

What general properties distinguish cuts from poles? The extra logarithmic factors in the s-dependence are not distinguishable in practice. Almost any sort of exchange-process or spin-dependence can be generated by <u>some</u> Regge pole or poles; but we then know what s-dependence to expect, from knowing the trajectories. If we find an "unexpected" effect, with s-dependence incompatible with known trajectories, it presumably comes from a cut. Here are some examples.

i) <u>Unexpected σ_T.</u> A sum of Regge poles gives the form

$$\sigma_T = \sum_i \sigma_i \, s^{\alpha_i(o)-1} \tag{18}$$

We expect P to have $\alpha_P(o)=1$ and the leading pole corrections to have $\alpha_i(o) \simeq o.5$. If the data suggest one or more additional terms with $\alpha(o) \simeq o.8$, say, they cannot credibly be assigned to poles. This in fact happens with the new Serpukhov results for $\pi^- N$ and $K^- N$ [37].

ii) Unexpected $d\sigma/dt$. Sometimes the leading poles are forbidden but the leading cuts are not. If σ were a meson with $J^{PG}=0^{+}$, then in the process $\pi^{\pm}p\to\sigma^{\pm}p$ parity would forbid P exchange but not the two - P cut. The latter would give slowly varying $d\sigma/dt \sim (\ln s)^{-\beta}$ at $t=0$, whereas the highest Regge pole A_2 would give $d\sigma/dt \sim s^{-1}$ ($\delta(962)$ is a possible candidate for σ).

iii) Unexpected spin-effects. In NN scattering, all the leading poles P, P', ..., A_2 have natural parity and can give only a part of the maximum possible spin dependence. For instance, they make the Wolfenstein depolarization parameter D=1, and at t=o they give no spin-dependence at all. However, the PP, PP' etc. cuts can give more general spin-dependence - with s-dependence quite distinct from the lower-lying unnatural-parity poles(46).

iv) Exotic exchanges. No Regge poles are known for exchanges with "exotic" quantum numbers, not included in the usual $\bar{q}q$ and qqq Quark Model configurations. But multi-Reggeon cuts can easily have exotic quantum numbers. This is the cleanest place to look for them. Any exotic particle will presumably be so massive, and its trajectory so low, that its effects will rapidly vanish with increasing energy. For instance, an exotic baryon with spin 3/2 and mass 2 GeV would give $\alpha(0)\approx-2.5$, $d\sigma/du \sim s^{-7}$.

The simplest exotic cases are double-charge-exchanges(47), such as $\pi^{-}p\to\pi^{+}\Delta^{-}$, $K^{-}p\to\pi^{+}\Sigma^{-}$, $K^{-}p\to pK^{-}$, with leading cut $\rho\rho$, ρK^{*} and $K^{*}\Delta$. Up to a few GeV/c, all these cross sections fall very rapidly, like s^{-8} or s^{-10}; high energy data are lacking, or consist of upper limits(48). The cuts should vary much more slowly, $d\sigma/dt \sim s^{-2}$-s^{-3}, but model estimates explain the discrepancy[49]. The cut contributions are smaller than the low energy cross sections; future measurements at 5 GeV/c or above should show a marked change.

Another approach is to look for interference between exotic and non-exotic exchanges, when both are allowed. Such effects are first-order in the former, which may be an advantage. For example, if t-channel isospin is $I_t \leq 1$, we predict definite ratios

$$\frac{\sigma(\gamma n \to K^+\Sigma^-)}{\sigma(\gamma p \to K^+\Sigma^0)} = 2, \quad \frac{\sigma(\gamma n \to \pi^+\Delta^-)}{\sigma(\gamma p \to \pi^+\Delta^0)} = 3, \quad \frac{\sigma(\gamma p \to \pi^-\Delta^{++})}{\sigma(\gamma n \to \pi^-\Delta^+)} = 3 \qquad (19)$$

that seem to be substantially violated by recent data[32, 50], indicating exotic exchanges of order 10% in the amplitudes.

Most of the time, however, we work with cuts that are not clearly distinguishable from Regge poles. There is then little chance of determining them directly from data, and we need some theory or model. So let us consider some prescriptions for cuts.

The popular approach is to identify the Regge-pole exchange amplitude as the first term in some series; the higher terms are then given by iterations of the first term, generating cuts. The characteristic form of cut amplitude (15) appears explicitly, if these iterations are written as momentum-space convolutions. For the present we avoid explicit convolutions. To fix notation we write the S-matrix, T-matrix and K-matrix formally:

$$S = 1 + iT = (1 + \tfrac{1}{2}iK)/(1 - \tfrac{1}{2}iK) \qquad (20)$$

Our discussion may be taken as referring to a particular partial wave, or to the complete operators (in which case multiplication implies integration over phase space).

The eikonal approximation gives T in terms of the Born amplitude T_B:

$$T \approx -i(e^{iT_B}-1) = T_B + \tfrac{1}{2} i T_B^2 - \tfrac{1}{6} T_B^3 \ldots \tag{21}$$

Arnold[51] proposed to replace T_B by the sum of Regge-pole exchanges T_R; the higher terms in the series then generate cuts without new parameters. Frautschi and Margolis did the same[52]. These authors all used impact-parameter representations rather than partial wave sums, but the differences from our discussion are inessential. In practice, for elastic scattering, all calculations are restricted to elastic intermediate states.

The K-matrix expansion is another starting-point:

$$T = K(1-\tfrac{1}{2} i K)^{-1} = K + \tfrac{1}{2} i K^2 - \tfrac{1}{4} K^3 \ldots \tag{22}$$

Lovelace[53, 54] proposed to replace K by T_R. This agrees to second order with the eikonal approach.

The absorption model[55], with Regge exchange as the Born term, gives the inelastic $b \to a$ matrix element

$$T^{ab} = T_R^{ab}(\tfrac{1}{2} S^{aa} + \tfrac{1}{2} S^{bb}) \tag{23}$$

(Alternative formulations use the geometric rather than arithmetic mean of S^{aa} and S^{bb}; in practice one often takes $S^{aa} \approx S^{bb}$ and there is no difference). The eikonal and K-matrix approaches give, to second order

$$T^{ab} = T_R^{ab}(1 + \tfrac{1}{2} i T_R^{aa} + \tfrac{1}{2} i T_R^{bb}) + \tfrac{1}{2} i \sum_{c \neq a,b} T_R^{ac} T_R^{cb} \tag{24}$$

This agrees with the absorption model through second order, if we ignore effects of intermediate states $c \neq a,b$.

Authors of the Michigan school[56, 57] argue that diffractively produced states c are not negligible but will enhance the cuts; they therefore multiply the absorption-model cuts by an empirical factor λ.

Hybrid models. Chiu and Finkelstein[58, 59] put an empirical non-Regge "diffraction" term in place of P, in the eikonal expansion. This makes little practical difference, except for effects of P shrinking; some authors take $\alpha_P' \approx 0$ anyway. Cohen-Tannoudji et al.[60] tried another hybrid formula: $S=S_{diff}\, S_R$, where S_R excludes P; for inelastic processes this gives the absorption model again.

AFS cut. Assuming that Regge poles need unitarity corrections only from elastic intermediate states gives[41, 64]

$$\mathrm{Im}\, T^{aa} = \mathrm{Im}\, T_R^{aa} + T_R^{aa}\, T_R^{aa*} \tag{25}$$

$$\mathrm{Im}\, T^{ab} = \mathrm{Im}\, T_R^{ab} + \mathrm{Re}(T_R^{aa*}\, T_R^{ab}) \tag{26}$$

to second order, putting $T^{aa} \approx T^{bb}$. These agree with the imaginary parts implied by Eqs. (21) - (24), except for the Hermitian conjugation in the final term. But this difference is crucial; since elastic amplitudes T^{aa} are dominantly imaginary, it reverses the sign of Im T (cut). Experiment shows the AFS prescription is wrong. It gives the wrong sign for polarization in $\pi^- p \to \pi^0 n$, while the absorption model can give the right sign.[61, 62] The AFS sign is also wrong for elastic scattering, assuming the Serpukhov data[37] indicate a negative cut contribution to Im T^{aa} (see lecture 3).

Thus there is broad agreement among the surviving prescriptions, at least to second order: the eikonal or K-matrix expansion for elastic or diffractive processes (with P exchange), the absorption model for inelastic reactions. There are sharp differences, however, on how to parametrize the Regge-pole input.

The Argonne school[63] assumes the Regge poles have exchange-degeneracy and hence have wrong-signature nonsense zeros. When such zeros are present at small t (or u) in an inelastic process, the corresponding sign changes of the Regge amplitude cause internal cancellations in the cut convolutions and the cuts are somewhat suppressed. The cuts being a relatively small correction at small t, at lab. energies, their effect is simply to displace the nonsense dips of $d\sigma/dt$ a little. Since the cuts interfere destructively (T^{aa}, T^{bb} mainly imaginary in Eq. (24)), they generally move the dips to smaller t-values. The essential point is that this approach preserves all the nonsense dips and zeros, not much displaced.

The Michigan school[56], in contrast, sees no reason to assume nonsense zeros. Starting with zero-free Regge poles, the cuts are relatively stronger and their destructive interference can generate dips in $d\sigma/dt$. Since dips were the original evidence for nonsense zeros, they argue, the latter have now no experimental support. In fact, to get the dips at the right places, the cut must be stronger than in the naive absorption model Eq. (23). It is therefore multiplied by an empirical factor $\lambda \approx 2$, depending on the process, argued to be an effect of diffractive contributions in the last term of Eq. (24). In this "Strong Absorption Model" the folklore says cuts tend to generate dips for s-channel non-flip processes near t (or u) = = -o.2, and for helicity-flip-1 processes near t=-o.6, at lab. energies (as $s\to\infty$, the dips generally tend logarithmically toward t=o). Hence we may expect a dip near t=-o.2 or -o.6, if non-flip or helicity-flip are dominant; but if both are important, we expect no dips at all. Actually the dip positions are sensitive to λ and other parameters; in one example, the helicity-flip dip is moved to u=-o.2 by taking $\lambda\approx 3$[73]. The Michigan approach is thus extremely flexible.

In spite of their strong differences, it seems that both the Argonne and Michigan approaches can approximate all the data they have confronted[31, 62-68]. It is therefore important to find tests to distinguish between them.

Test 1. All the places where nonsense dips are missing mean trouble for Argonne, just as for pure Regge polology (see lecture 1). Explanations can usually be found, but some seem pretty ad hoc. On the other hand, Harari[32] argued that there is a discernible systematics among the dips, at least for those near t=-o.6; here dips seem to occur when, and only when, s-channel helicity-flip dominates, agreeing with Michigan folklore.

Test 2. Consider $\pi^- p\to\pi^0 n$ and $\pi^- p\to\eta^0 n$, with ρ and A_2 exchange respectively. In both cases $d\sigma/dt$ has a shoulder at small t indicating dominant helicity-flip. There is a dip in the former case near t=-o.6, no dip in the latter (Fig. 5)[69]. This is straightforward for Argonne; α=o is a wrong-signature nonsense point for ρ but not for A_2, so they expect a dip in the first case only[62, 67]. For Michigan there is no real difference between the two cases (apart from an inessential phase due to signature); with helicity-flip dominant, we might expect a dip near t=-o.6 <u>in both cases</u>. But a Michigan-style fit can still be made, for example by taking $\lambda \approx 1.5$ to get a medium-weak cut. A steep exponential t-dependence for ρ-exchange then produces a dip in one case, while a mild exponential for A_2 de-emphasizes the dip and moves it outward, in the other case.

<u>Test 3</u>. Consider the dip near u=-o.2, in $d\sigma/du$ ($\pi^+ p \to p \pi^+$). According to Argonne, this comes from the $\alpha=-\frac{1}{2}$ nonsense zero of N, in both flip and non-flip amplitudes. According to Michigan, this is simply evidence of non-flip dominance. (Note: in backward scattering, f_{++} and f_{+-} represent net flip and non-flip, not the other way round). Thus we can decide between them if we can determine the flip amplitude(7o). There are not yet enough data for a complete high-energy analysis, but we can look at the upper end of the CERN πN phase shift analyses(71), at 2.o1 and 2.o7 GeV/c. In Figure 5 I have plotted real and imaginary parts of f_{++} and f_{+-} for the i-spin combination $I_u=\frac{1}{2}$, appropriate to N_α-exchange. The "flip" f_{++} results suggest this term is strong and develops a zero near u=-o.1 or -o.2, supporting Argonne. But the "non-flip" f_{+-} results show a zero near u=o or o.1 instead (the second zero of Im f_{+-} seems just a phase effect) - and this does not support anybody. Either 2 GeV/c is too low for our purposes, or else N_γ exchange is obscuring the results.

<u>Test 4</u>. Measurements of $d\sigma/du$ ($K^- n \to \Lambda \pi^-$) show remarkable structure(72): see Figure 7. The implications are discussed in ref. 73. A strong minimum in the backward direction suggests dominant helicity-flip f_{++}; there is another minimum near u=-o.2. This is straightforward for Argonne: N_α-exchange gives a nonsense zero at u=-o.2, just

as in $\pi^+p\to p\pi^+$. Michigan is in trouble if you believe the folklore, that with dominant helicity flip a dip is predicted near u=-o.6. However, by exploiting λ and other parameters the dip can be moved almost anywhere; with λ=3.1 an excellent fit to data can be made[(73)]. This shows that Michigan does not really predict the dip positions, and cannot easily be excluded by this kind of test.

Test 5. Duality between resonances and Regge-pole exchanges is a rather successful concept (see Kugler's lectures). It requires exchange-degeneracy and nonsense zeros, and thus supports Argonne. Now the usual resonance-pole duality ignores cuts completely, whereas cuts are crucial for Michigan; so we may ask if duality is restored by taking the Michigan poles and cuts together, instead of poles alone? The answer is no. If we take the elastic amplitude to be purely imaginary, as commonly done, the absorption-model cuts have this important property[(74)]: if the input "Born" term has a constant phase, the final amplitude has the same constant phase - and vice versa. Duality requires the high energy amplitude in exotic channels to be real; we now see that even in the presence of absorptive cuts this requires the Regge pole terms to be real - and this in turn implies exchange degeneracy and nonsense zeros.

None of these tests are yet conclusive, perhaps, but they show how a choice between Argonne and Michigan may be made eventually.

Finally, we discuss some general physical effects.

Elastic scattering. The eikonal and K-matrix prescriptions give power series in (iT_R), with positive coefficients. Since T_R for elastic scattering is mainly positive imaginary (and we ignore inelastic intermediate states), successive cut terms are also approximately imaginary and alternate in sign. Successive terms also have a weaker t-dependence than the input (a property of the convolution), if $T_R \sim e^{At}$, then $(T_R)^n \sim e^{At/n}$. Thus a series of diffraction minima are generated, shown schematically in Fig. 8. This is just like non-Regge multiple scattering models[75]. As $s \to \infty$ we expect the amplitude to become more nearly imaginary, and this structure in $d\sigma/dt$ to become more marked; this contrasts with structure induced by secondary Regge poles, that fades away asymptotically.

Anselm and Dyatlov[76] considered how to sum Mandelstam cuts; at large t they found a smooth behaviour $d\sigma/dt \sim \exp(-A\sqrt{-t})$ with slow oscillations superposed. Frautschi and Margolis[52] discussed the structure in pp scattering, using the eikonal method and simplifying to P exchange alone. Their analysis, extended to new Serpukhov data on shrinking at small t[77], favours a Pomeron slope $\alpha_P' \approx 1$[78].

Crossover. The crossovers in pp, $\bar{p}p$ and K^+p, K^-p scattering are attributed to ω-cut interference, but in practice there is difficulty getting them in the right place, especially for Argonne. In the Argonne approach, a nonsense zero of all ω residues at $\alpha_\omega = 0$ $(t \approx -0.4)$ provides a natural crossover, but at the wrong place. The relatively weak cuts cannot move the crossover to $t=-0.15$; it usually gets no further than $t=-0.3$[66].

A similar crossover is seen in π^+p, π^-p scattering, again near $t=-0.15$ though the exact position is hard to fix.

Here a nonsense zero of ρ residues provides a natural crossover at $\alpha_\rho=0(t\approx-0.6)$, and again the cuts have trouble moving it far enough, despite what the authors say[62].

Effective alphas. We can describe the energy-dependence of a cross-section in some energy interval by an effective trajectory $\alpha_{eff}(t)$, found by fitting Eq. (6) or (7). Asymptotically, the multi-P cuts dominate to make α_{eff} flat. At lab. energies α_{eff} is determined by poles and cuts together. Near a dip, the relative importance of poles and cuts fluctuates rapidly and the predicted α_{eff} shows bumps or kinks[79, 80]. Figure 9 illustrates α_{eff} predictions for $\pi^-p\to\pi^0n$, for the range 5-20 GeV/c, for three ρ + cut models: one Michigan, one Argonne, one intermediate[81]. They are to be contrasted with the smooth $\alpha_{eff}=0.57+0.91t$ found experimentally[8]. Both the slope and the kinkiness of α_{eff} are problems for the cut model.

The energy-dependence may well prove a difficulty in elastic scattering too. For $t<1$, α_{eff} is negative for both πp and pp[82], whereas the multi-P branch points condense near + 1. It will require clever cancellations to reconcile the two.

Cut signature. The absorption model gives a $\rho\rho$ cut in $\pi^-p\to\pi^0n$, although this cut should have positive signature[44] and be forbidden. Something is wrong with the prescription. One solution is to apply the eikonal method to $\pi^{\pm}p$ states instead of i-spin eigenstates, that gives two series in $(T_P+T_P\pm T_\rho)$; charge-exchange is the difference, automatically odd in ρ.

3. ANALYTICITY AND ASYMPTOTICS

Analyticity relates low- and high-energy amplitudes.

Suppose $f(\nu)$ is a scattering amplitude, real analytic in the ν-plane, with cuts from ν_o to ∞ and $-\nu_o'$ to $-\infty$ (and maybe some isolated poles) on the real axis. If we are working at fixed t, ν could be the crossing-antisymmetric variable $\nu=(s-u)/(4m_N)$. Physics is on the real axis: s-channel physics for $\nu>\nu_o$ and u-channel physics for $\nu<-\nu_o'$. We can get useful relations between physical things by integrating $f(\nu)$ - or some function of f - around a closed contour C, as in Figure lo.

For simplicity, let us suppose $f(\nu)$ is anti-symmetric: $f(\nu)=-f(-\nu)$. The left and right-hand cuts are then trivially related. There is little loss in generality, since any function is a sum of symmetric and antisymmetric parts. Then applying Cauchy's formula to the integral of $f(\nu)/(\nu^2-\nu'^2)$, and letting the semi-circles of contour C tend to infinity, we get a familiar dispersion relation for real ν',

$$\mathrm{Re}\, f(\nu') = \frac{2\nu'}{\pi}\, P\int_{\nu_o}^{\infty} \frac{\mathrm{Im}\, f(\nu)d\nu}{\nu^2 - \nu'^2} \tag{27}$$

provided $|f(\nu)|<|\nu|^{1-\varepsilon}$ as $|\nu|\to\infty$.

Such relations are constraints on physics. Because the dispersion integrals cover a wide range of energies, these constraints correlate low and high energy behaviour. Because the dispersion integrals go to infinity, they appear to offer us a handle on asymptotic behaviour.

Finite Energy Sum Rules (FESR) offer a very practical way of relating low and high energy scattering(83-86). They may be derived from Cauchy's theorem

$$\int_C f(\nu)d\nu = 0$$

evaluating the high-energy part of the integral $|\nu|\geq N$ by assuming a Regge pole expansion of f:

$$f(\nu) = \sum_j \gamma_j \frac{1-\exp(-i\pi\alpha_j)}{\sin \pi\alpha_j} \nu^{\alpha_j} \quad \text{for} \quad |\nu|\geq N, \tag{28}$$

Cauchy's theorem for the contour C then gives the FESR,

$$\int_{\nu_0}^{N} \text{Im } f(\nu)d\nu = \sum_j \gamma_j \frac{N^{\alpha_j+1}}{\alpha_j+1} \tag{29}$$

The right-hand side above comes simply from integrating Eq. (28) around the part of the contour with $|\nu|\geq N$. Since Eq. (28) has no singularities out here, we can distort the contour in many ways, but it is most convenient just to reduce this part to upper and lower semicircles with $|\nu|=N$. It must be emphasized[86] that we only need a finite contour C, avoiding all questions of convergence.

The FESR Eq. (29) directly relates low-energy scattering amplitudes to high-energy parameters. It offers a way to determine the latter. So far we have only one relation, against two parameters for each Regge pole, but many more relations can be found as follows.

Moment sum rules come from taking $\nu^m f(\nu)$ instead of $f(\nu)$. Since we have assumed f antisymmetric, the odd-moment contour integrals vanish by symmetry and only even moments give sum rules. (Schwarz[87] showed the missing odd-moment sum rules can be recovered, by assuming a Mandelstam representation with no third double-spectral function, but Mandelstam cuts[42] violate this assumption).

Inverse sum rules. Eq. (29) exploits the low-energy integral of Im f only. To bring Re f into the game we replace $f(\nu)$ by $-iqf(\nu)$, where $q=(\nu^2-\nu_o^2)^{1/2}$. This also is a real analytic function with the same cuts, and the resulting sum rule has $-\int \mathrm{Re}\, f(\nu) q\, d\nu$ on the left.

Continuous moment sum rules (CMSR) come from considering $(-iq)^{\beta} f(\nu)$ instead of $f(\nu)$, where β is a continuous real parameter(85). The resulting sum rule includes all those above as special cases:

$$\int_{\nu_o}^{N} \{\cos(\tfrac{1}{2}\pi\beta)\ \mathrm{Im}\ f - \sin(\tfrac{1}{2}\pi\beta)\ \mathrm{Re}\ f\} q^{\beta} d\nu = \sum_j \gamma_j \frac{N^{\alpha_j+\beta+1}}{\alpha_j+\beta+1} \frac{\cos\frac{\pi}{2}(\alpha_j+\beta)}{\cos\frac{\pi}{2}\alpha_j} \tag{30}$$

omitting terms of order ν_o^2/N^2.

Given a low-energy analysis for $|\nu|<N$, we can evaluate sum rules and try to determine the high-energy parameters α_j, γ_j. Here are some practical considerations however.

(i) Though we have an infinite number of sum rules, they do not give an infinite number of constraints in practice. The β-dependence of CMSR is smooth, and its finer details are lost in experimental error.

(ii) It is therefore essential to truncate the Regge series (28) after a few terms. This imposes an inevitable coarseness on the solution.

(iii) Regge cuts can in principle be included, as continua of poles, but must in practice be represented by few parameters. The pole terms are often assumed to represent cuts also.

(iv) Low-energy analyses are mostly restricted to $\nu \lesssim 2$ GeV, far short of the point ($\sim$5 GeV) where smooth Regge behaviour is normally supposed to begin. Working with $N \approx 2$ GeV is a bold step. The fact that it seems to work suggests that non-Regge fluctuations are integrating to zero in the sum rules.

(v) For very high moments β, the sum rules reduce simply to the Regge formula Eq. (28) evaluated at $\nu=N$. Since non-Regge fluctuations then have no chance to average out, high moments are unsafe.

(vi) Apart from the oscillating factor $\cos\frac{\pi}{2}(\alpha+\beta)/(\alpha+\beta+1)$, the series of Regge terms on the right-hand side of the sum rules (29) and (3o) converges at the same rate as the Regge expansion for $f(\nu)$ itself, Eq. (28), evaluated at $\nu=N$. A Regge pole that is important (negligible) in $f(N)$ is equally important (negligible) in the sum rules.

(vii) The sum rules give amplitudes directly - unlike scattering data, that are quadratic functions of amplitudes and have to be unscrambled. The sum rules exploit the unscrambling previously done at low energies. In particular, they give spin dependence directly.

(viii) The sum rules need not be taken alone; they can be included with high-energy experimental data in Regge analyses.

Within the practical limitations, there have been many interesting and successful applications of FESR. Here are a few examples.

(a) Dolen, Horn and Schmid[84] were the first to show that FESR are a practical predictive tool. They studied $\pi^- p \to \pi^0 n$, with Regge ρ-exchange, over a range of t-values. Now the invariant amplitudes A' and B for this process (corresponding to t-channel non-flip and helicity-flip[88]) are found from Regge-pole fitting to high-energy data to change sign near t=-o.15 and t=-o.6, respectively. Dolen, Horn and Schmid successfully predicted the signs, the sign changes and the approximate magnitudes of the ρ residues in A' and B, although their cut-off was as low as N=1.1 GeV. This generated confidence in the method.

(b) Ferro-Fontan et al.[89] set out to predict high-energy πN elastic and charge exchange amplitudes, using CMSR and a low-energy input with N<2 GeV. Their predictions up to 2o GeV seem to agree with data, within 2o% in the amplitude.

(c) The term B^+, meaning the contribution from even signature exchanges P, P' etc., is ill determined by high energy πN data. Various arguments, plus least-squares data fitting, suggested νB^+ was large, negative and steeply t-dependent compared to A'^+[7]. Sum rules indicated, on the contrary, an approximate relation $\nu B^+ \approx A'^+$[9o]. Recent measurements of spin-rotation parameters, that are sensitive to B^+, show the sum rules are right[91, 17].

The method seems to work quite well up to 2o GeV. How far can we take it, and how reliably? Remark (vi) above shows we cannot expect unlimited success. Any term that is unimportant near $|\nu|$=N will be ill-determined; we can imagine many such terms that can be very important for $|\nu|$>>N. For example, if we try to approximate a segment of a cut by a pole, we are involved in approximation like the following:

$$\int_{\alpha-\Delta}^{\alpha+\Delta} d\alpha' \nu^{\alpha'} = 2\Delta\nu^{\alpha}\{\frac{\sinh(\Delta \ln\nu)}{\Delta \ln\nu}\}, \tag{31}$$

$$= 2\Delta\nu^{\alpha}\{1 + \frac{1}{3!}(\Delta \ln\nu)^2 + \dots\} \tag{32}$$

The pole approximation - the first term in (32) - is fine for $\nu=1$. As ν increases the approximation gradually gets worse; it can be improved by reducing Δ (taking a smaller segment), but not indefinitely. The logarithmic factors in a cut cannot be faked asymptotically by a few power terms.

We see therefore that FESR cannot predict reliably to very high energy. This is reasonable enough. We are trying to make an analytic continuation from the low-energy region $|\nu|<N$, but our low-energy input is not exact; the further we extrapolate, the more uncertain our results will be. Another way to see this is to remember the derivation of FESR, on a finite contour. As long as our truncated Regge series (28) is a good approximation on the appropriate part of the contour (which may be just semicircles of radius N), the FESR will be satisfied. The approximation may fail completely for $\nu \gg N$, but the sum rules do not care.

What about the ordinary dispersion relations, then? They are derived on an infinite contour. The usual reason for pushing the contour to infinity is just to get rid of the unphysical integration on the semicircles (at infinity it is negligible, given enough convergence). However, when evaluating a dispersion integral to infinity, one always parametrizes the high energy part in a simple form - such as Regge poles. Given such a form, that can be continued analytically into the complex ν plane, the "unphysical" semicircle integrals can be evaluated in terms of physical parameters and there is no need to avoid them any more. We can write Finite Contour Dispersion Relations(92). With our antisymmetric amplitude f, for instance, Eq. (27) is replaced by

$$\mathrm{Re}\, f(\nu') = \frac{2\nu'}{\pi}\int_{\nu_0}^{N} \frac{\mathrm{Im}\, f(\nu)}{\nu^2-\nu'^2}d\nu + \frac{\nu'}{\pi i}\int_{C'} \frac{f(\nu)d\nu}{\nu^2-\nu'^2} \qquad (33)$$

where C' is the upper semicircle from $N+i\varepsilon$ to $-N+i\varepsilon$, and $N>|\nu'|$. A parametric form is needed for the C' integral. If there is enough convergence we can let $N\to\infty$ and get the conventional result (27), but the point is that we do not need to take this limit. The high energy part of the integral is the same, for any distortion of the contour that crosses no singularities. The finite contour form has no problems of convergence at infinity, making subtractions, etc. It also demonstrates that a dispersion relation will be satisfied if the high energy parametrization is a good approximation on the closing contour. There is no direct test of asymptotics.

This explains why FESR and dispersion relations(93) failed to predict the Serpukhov total cross sections.

The Serpukhov σ_T data(37) for π^-p, π^-n, K^-p and K^-n were a big surprise. Previous results(94, 95) showed a steady fall with increasing momentum, up to 25 GeV/c. Regge pole extrapolations, with and without FESR constraints, predicted a continued fall(17, 38). But the Serpukhov results indicate all these cross sections are essentially constant, from 25 to 65 GeV/c. Figure 11 shows σ_T plotted against $\ln p_L$ $(\nu\approx p_L)$. We have equated $\sigma_T(\pi^-n)=\sigma_T(\pi^+p)$ by charge symmetry; unfortunately the present Serpukhov results are only for negative beams.

Figure 12 shows the same data more dramatically, plotted against $p_L^{-1/2}$, (following Horn[96]). Here infinity is in sight, at the right-hand side of the graph. The usual leading Regge poles, with $\alpha_P(o)=1$ and $\alpha_i(o)\approx o.5$ $(i\neq P)$, give $\sigma_T\approx A+B\nu^{-1/2}$, that is a straight line on this plot. The old data are roughly compatible with such straight lines, with common asymptotic limits for $\pi^{\pm}p$ and $K^{\pm}p$, satisfying the Pomeranchuk theorem. The new data abruptly spoil this tidy picture.

The new data raise two separate questions:

(i) Why the sudden change of slope in σ_T?

(ii) Will $\pi^{\pm}p$ and $K^{\pm}p$ satisfy the Pomeranchuk theorem?

To discuss these questions, we need to separate the amplitudes into even- and odd-signature parts: $A'^{\pm}(\pi p) = \frac{1}{2}A'(\pi^-p)\pm\frac{1}{2}A'(\pi^+p)$, and similarly for $A'^{\pm}(Kp)$. The optical theorem gives $\sigma_T=\text{Im } A'(t=o)/p_L$ and the Regge pole approximations can be written (simplifying the signature factors),

$$A'^{+} = \sum_j \gamma_j \, \nu^{\alpha_j} e^{-i\pi\alpha_j/2} , \tag{34}$$

$$A'^{-} = \sum_j \gamma_j \, \nu^{\alpha_j} i e^{-i\pi\alpha_j/2} . \tag{35}$$

<u>Pomeranchuk limit</u>. Since there are no K^+p data above 2o GeV/c, and the $\pi^+p(=\pi^-n)$ data have big errors (because of the Glauber correction for a deuteron target), it is quite possible that $\pi^{\pm}p$ and $K^{\pm}p$ cross sections continue to approach each other as smoothly as before. In other words, the anomalous behaviour may be in A'^{+} alone. If so, we predict that the K^+p (and K^+n) cross sections, that stay remarkably constant from 5 to 2o GeV/c, begin to

rise thereafter. This is an important point for future experiments.

Alternatively, maybe the Pomeranchuk theorem fails. It is based on fallible assumptions, as recently emphasized by Martin(97) and Eden(98). What Regge mechanism can cause such a failure, by giving $\mathrm{Im}\, A'^{-}/\nu \to \sigma^{(-)} \neq o$? Regge poles cannot; to get the right ν-dependence they need $\alpha=1$, but then the amplitude is purely real and cannot contribute (eq. 35). Regge cuts, giving pole continua, also fail. But a Regge dipole(92, 99), giving essentially the derivative of a Regge pole term, has the right properties:

$$A'^{-}(\text{dipole}) = (\ln\nu - i\pi/2)\gamma\nu^{\alpha} i e^{-i\pi\alpha/2} \quad . \tag{36}$$

With $\alpha=1$, $\mathrm{Im}\, A'^{-}$ violates the Pomeranchuk theorem. However, $\mathrm{Re}\, A'^{-}$ then dominates asymptotically over $\mathrm{Im}\, A'^{-}$ (this is a general consequence, not limited to dipoles(97, 98), with remarkable consequences at t=o.

(a) $\dfrac{\mathrm{Re}\, A'(\pi^{+}p)}{\mathrm{Im}\, A'(\pi^{+}p)} \approx - \dfrac{\mathrm{Re}\, A'(\pi^{-}p)}{\mathrm{Im}\, A'(\pi^{-}p)} \to +\infty$, the sign $+\infty$ being taken from the sign of the experimental Pomeranchuk violation; and similarly for $K^{+}p$, $K^{-}p$.

(b) $\dfrac{d\sigma}{dt}$ (elastic or charge-exchange) $\sim (\ln\nu)^2$.

(c) $d\sigma/dt$ must exhibit $(\ln\nu)^2$ shrinking - stronger than the logarithmic shrinking of a normal Regge pole contribution - to keep σ (elastic) finite. Finkelstein(100) showed that $(\log)^2$ shrinking is hard to get. Regge dipoles cannot achieve it. He constructed a fearsome example that gives cuts for $t \neq o$, reduces to a dipole at $t=o$, and shrinks as $(\log)^2$.

The processes $\pi^-p\to\pi^on$ and $K_2^op\to K_1^op$ offer the cleanest experimental approach, since they depend on $A'^-(\pi p)$ and $A'^-(K^op)$ alone. We may expect the asymptotic constancy of Im A'^-/ν and increase of Re A'^-/ν to give a more observable effect here than in elastic scattering(1o1).

Kink in σ_T. Accepting this change of slope as established, several explanations are offered.

(i) Regge cuts. With the eikonal or K-matrix prescription, the two - P cut contributes negatively $\sim -(\ln s)^{-1}$ to σ_T. Secondary Regge pole contributions are positive but decrease more rapidly, $\sim s^{-1/2}$, so eventually the cut dominates, σ_T levels out and approaches the asymptotic limit from below - as sketched in Figure 13a. Frautschi and Margolis(52) made exactly this prediction, in an analysis of pp scattering. Dean(1o2) found similar cut effects, in a multiple-scattering quark model. Barger and Phillips(1o3) used this idea to make quantitative fits to the new data. Though the cross-sections above 3o GeV/c diverge from previous expectations, the dispersion-relation calculations of Re A' are not much affected below 2o GeV/c - confirming our arguments above.

A physical consequence, when Regge cuts are strong, is that asymptopia is very far off. With a power law for σ_T, the asymptotic limit is approached for large s; with cuts, the limit is not approached until ln s is large.

(ii) Ionization point. Horn(96) suggested the kink in σ_T may be at an "ionization point", where s-channel resonances cease to exist (e. g. near $s=9m_Q^2$ in a quark model.) Through duality, Regge poles could be a reasonable description up to this point but not beyond it. He suggested σ_T values are constant from here to infinity, violating the Pomeranchuk theorem. This is not inconsistent with dispersion relation constraints.

(iii) Regge dipole[92]. If we are ready to consider a dipole in A'^-, why not in A'^+? The change of signature brings some difference.

$$A'^+(\text{dipole}) = -(\ln\nu - \tfrac{1}{2}i\pi\alpha)\gamma\nu^\alpha \, e^{-i\pi\alpha/2} \tag{37}$$

For $\alpha=1$ the striking effect is a logarithmic increase in $\sigma_T^{(+)} \sim \gamma \ln\nu$ asymptotically. (This is consistent with the Froissart bound $\sigma_T < c(\ln s)^2$). As with the cut explanation, this slowly rising term eventually dominates over falling Regge pole terms, causing σ_T to flatten out and finally to rise; but now the asymptotic rise is not to a finite limit. An example is shown in Figure 13(b).

One consequence is that σ(elastic) can now increase logarithmically, so that a dipole term in A'^- need not have $(\log)^2$ shrinking; ordinary logarithmic shrinking is enough.

The Regge dipoles suggested for $A'^{\pm}$ may be approximations for more complicated terms[100], and there is no immediate need to identify them with physical particles.

(iv) Complex alphas. Chew and Snider[104] proposed complex trajectories, suggested by the multi-peripheral model. If $\gamma=\gamma_0 \exp(i\phi)$ and $\alpha=\alpha_R+i\alpha_I$ are the complex residue and trajectory of a pole, then a pair of conjugate poles contributes

$$\sigma_T^{(+)} = \text{Im } A'^{+}/\nu$$

$$= \gamma_o \nu^{\alpha_R - 1} \{ e^{\frac{1}{2}\pi\alpha_I} \sin(\alpha_I \ln\nu - \tfrac{\pi}{2}\alpha_R + \phi) - e^{-\frac{1}{2}\pi\alpha_I} \sin(\alpha_I \ln\nu + \tfrac{\pi}{2}\alpha_R + \phi) \} \quad (38)$$

This introduces oscillations versus $\ln\nu$ with period $2\pi/\alpha_I$. There is also a damping factor $\nu^{\alpha_R - 1}$, so unless α_I is substantial the oscillations do not appear clearly. The Pomeranchuk trajectory must be real, or we get $\sigma_T < 0$. Figure 13(c) sketches a possible example. Such oscillations give kinks in σ_T.

If α_ρ were complex, $\sigma_T^{(-)} = \frac{1}{2}\sigma_T(\pi^- p) - \frac{1}{2}\sigma_T(\pi^+ p)$ would oscillate in sign, but there is no experimental evidence for this.

(v) Size resonances. Lovelace[105], recently emphasizing analogies between the Veneziano model and nuclear physics, conjectured that high energy physics might contain analogies to the "size resonances" of the nuclear optical model. These could then give anomalies in σ_T. Since size resonances are roughly speaking related to fitting integral numbers of wave lengths into a box, we may expect them to be equally spaced in $\sqrt{s}$, as sketched in Fig. 13(d).

A final surprise from Serpukhov are the results for $\sigma_T(\bar{p}p)$. These show no kink, and when plotted against $P_L^{-\frac{1}{2}}$ seem quite consistent with a straight-line extrapolation to a common pp, $\bar{p}$p Pomeranchuk limit. See Figure 14.

I leave the explanation as an exercise for the reader.

REFERENCES

1. See any text-book on the subject, for example,
   P. D. B. Collins and E. J. Squires, "Regge poles in
   particle physics", Springer Tracts in Modern Physics
   Vol. 45 (Springer-Verlag, 1968);
   V. Barger and D. Cline, "Phenomenological theories
   of high energy scattering" (W. A. Benjamin,
   New York, 1968)

2. N. Bali et al, Phys. Rev. $\underline{161}$, 1450 (1967);
   J. S. Ball and F. Zachariasen, Phys. Rev. $\underline{179}$,
   346 (1969);
   A. E. A. Warburton, Nuovo Cim. $\underline{32}$, 122 (1964)

3. V. N. Gribov, Soviet Physics - JETP $\underline{16}$, 1080 (1963)

4. R. C. Arnold, Phys. Rev. Letters $\underline{14}$, 657 (1965)

5. H. Harari, Phys. Rev. Letters $\underline{20}$, 1395 (1968)

6. D. R. O. Morrison, Phys. Letters $\underline{22}$, 528 (1966)

7. R. J. N. Phillips and W. Rarita, Phys. Rev. $\underline{139}$,
   B 1336 (1965)

8. G. Hohler et al, Phys. Letters $\underline{20}$, 79 (1966)

9. F. Arbab and C. B. Chiu, Phys. Rev. $\underline{147}$, 1045 (1966)

10. R. J. N. Phillips and W. Rarita, Phys. Letters
    $\underline{19}$, 598 (1965)

11. M. Gell-Mann, Phys. Rev. Letters $\underline{8}$, 263 (1962), and
    proceedings of 1962 CERN conference

12. V. N. Gribov and I. Ya. Pomeranchuk, Phys. Rev.
    Letters $\underline{8}$, 343, 412 (1962)

13. P. G. O. Freund, Phys. Rev. Letters $\underline{21}$, 1375 (1968)

14. D. Cline et al, Phys. Rev. Letters $\underline{23}$, 1318 (1969)

15. P. Bonamy et al, Phys. Letters $\underline{23}$, 501 (1966) and
    proceedings of Heidelberg conference (1967);
    D. Drobnis et al, Phys. Rev. Letters $\underline{20}$, 274 (1968)

16. D. M. Austin et al, Phys. Rev. 173, 1573 (1968);
T. J. Gajdicar and J. W. Moffat, Phys. Rev. 181, 1875 (1969)
17. V. Barger and R. J. N. Phillips, Phys. Rev. Letters 21, 865 (1968); Phys. Rev. 187, 221o (1969)
18. M. Davier et al, preprint, SLAC-PUB-666 (1969)
19. A. P. Contogouris et al, Phys. Rev. Letters 19, 1352 (1967)
2o. M. P. Locher and H. Rollnik, Phys. Letters 22, 696 (1966);
P. di Vecchia and F. Drago, Phys. Letters 24B, 4o5 (1967);
J. P. Ader et al, Nucl. Phys. B3, 4o7 (1967)
21. P. di Vecchia et al, Nuovo Cim. 55A, 8o9 (1968)
22. V. Barger and L. Durand, Phys. Rev. Letters 19, 1295 (1967)
23. D. V. Volkov and V. N. Gribov, Soviet Phys. - JETP 17, 72o (1963);
E. Leader, Phys. Rev. 166, 1599 (1968)
24. R. J. N. Phillips, Nucl. Phys. B2, 394 (1967)
25. J. S. Ball et al, Phys. Rev. Letters 2o, 518 (1968);
P. di Vecchia et al, Phys. Letters 27B, 296, 521 (1968);
D. P. Roy and S. Y. Chu, Phys. Rev. 171, 1762 (1968)
26. G. V. Dass and C. D. Froggatt, Nucl, Phys. B8, 661 (1968)
27. M. Le Bellac, Phys. Letters 25B, 524 (1967);
F. Arbab and J. D. Jackson, Phys. Rev. 176, 1796 (1968)
28. M. Aderholz et al, Phys. Letters 27B, 174 (1968)
29. F. Arbab and R. C. Brower, Phys. Rev. 175, 1991 (1968)
3o. G. Fox, G. Ringland and P. Thews, unpublished analyses
31. J. Froyland and D. Gordon, Phys. Rev. 177, 25oo (1969);
M. L. Blackmon et al, Nucl. Phys. B12, 495 (1969);

F. Henyey et al, Phys. Rev. 182, 1579 (1969);
J. D. Jackson and C. Quigg, Phys. Letters 29B, 236 (1969)

32. H. Harari, Liverpool Conference on Electron and Photon Interactions (1969)
33. V. Barger, Phys. Rev. 179, 1371 (1969)
34. V. Barger and P. Weiler, Phys. Letters 3oB, 1o5 (1969)
35. Remark due to C. Lovelace
36. R. Carlitz and M. Kislinger, Phys. Rev. Letters 24, 186 (197o)
37. J. V. Allaby et al, Phys. Letters 3oB, 498 (1969)
38. V. Barger et al, Nucl. Phys. B5, 411 (1968)
39. Rosenfeld compilation, N. Barash-Schmidt et al, Rev. Mod. Phys. 41, 1o9 (1969);
V. Barger and M. Olsson, Phys. Rev. 151, 1123 (1966)
4o. K. Foley et al, Phys. Rev. Letters 11, 425, 5o3 (1963)
41. D. Amati et al, Phys. Letters 1, 29 (1962)
42. S. Mandelstam, Nuovo Cim. 3o, 1127, 1148 (1963);
V. N. Gribov et al, Phys. Rev. 139, B185 (1965);
V. N. Gribov, Soviet Phys. - JETP 26, 414 (1968)
43. J. B. Bronzan and C. E. Jones, Phys. Rev. 16o, 1494 (1967)
44. D. Branson, Phys. Rev. 179, 16o8 (1969)
45. V. N. Gribov, Leningrad preprint, report to 1966 Berkeley conference
46. D. Branson et al, Phys. Letters 25B, 141 (1967);
R. J. N. Phillips, Phys. Letters 25B, 517 (1967)
47. R. J. N. Phillips, Phys. Letters 24B, 342 (1967)
48. O. I. Dahl, Phys. Rev. 163, 143o (1967);
G. Belletini, proceedings of Vienna conference (1968);
D. M. Dauber et al, Phys. Letters 29B, 6o9 (1969)

49. C. Michael, Phys. Letters 29B, 23o (1969) and proceedings of the 1969 Regge Cut Conference at Madison
5o. A. M. Boyarski et al, reports submitted to Liverpool Conference (1969)
51. R. C. Arnold, Phys. Rev. 153, 1523 (1967)
52. S. Frautschi and B. Margolis, Nuovo Cim. 56A, 1155; 57A, 427 (1968)
53. C. Lovelace, Nucl. Phys. B12, 253 (1969)
54. G. Cohen-Tannoudji et al, CERN report TH-1oo3 (1969) make a modified K-matrix assumption
55. See J. D. Jackson, Rev. Mod. Phys. 37, 484 (1965) for early references;
H. D. D. Watson, Phys. Letters 17, 72 (1965);
R. J. Rivers, Nuovo Cim. 63A, 697 (1969)
56. Ross, Kane, Henyey and co-workers
57. For a counter-argument, see D. G. Ravenhall and H. W. Wyld, Phys. Rev. Letters 21, 177o (1968)
58. C. B. Chiu and J. Finkelstein, Nuovo Cim. 57A, 649 (1968); 59A, 92 (1969)
59. See also H. Abarbanel et al, Phys. Rev. 177, 2458 (1969)
6o. G. Cohen-Tannoudji et al, Nuovo Cim. 48, 1o75 (1967)
61. J. Finkelstein and M. Jacob, Nuovo Cim. 56A, 681 (1968);
R. J. Rivers and L. M. Saunders, Nuovo Cim. 58A, 385 (1968)
62. J. N. J. White, Phys. Letters 27B, 92 (1968);
F. Schrempp, Nucl. Phys. B6, 487 (1968);
R. C. Arnold and M. L. Blackmon, Phys. Rev. 176, 2o82 (1968);
C. Michael, Nucl. Phys. B8, 431 (1968)
63. Arnold, Blackmon and co-workers

64. F. Henyey et al, Phys. Rev. Letters 21, 946 (1968); 21, 1782 (1968); Phys. Rev. 182, 1579 (1969)
65. R. L. Kelly, G. L. Kane and F. Henyey, Michigan preprint (December 1969)
66. M. L. Blackmon and G. R. Goldstein, Phys. Rev. 179, 148o (1969), and Syracuse preprints SU-12o6-211 (1969), SU-12o6-218 (1969)
67. M. L. Blackmon et al, Phys. Rev. 183, 1452 (1969); M. L. Blackmon, Phys. Rev. 178, 2385 (1969)
68. T. Roth and G. H. Renninger, Carnegie-Mellon preprint (1969)
69. A. V. Stirling et al, Phys. Rev. Letters 14, 763 (1965); M. Wahlig and I. Mannelli, Phys. Rev. 168, 1515 (1968);
O. Guisan et al, Phys. Letters 18, 2oo (1965)
7o. Test suggested by M. Ross and by F. Drago
71. Revised results: private communication from C. Lovelace (January, 1969)
72. D. J. Crennell et al, Phys. Rev. Letters 23, 1347 (1969)
73. F. Drago et al, Rutherford Laboratory preprint (in preparation)
74. V. Barger and R. J. N. Phillips, Phys. Letters 29B, 676 (1969)
75. T. T. Chou and C. N. Yang, Phys. Rev. 17o, 1591; 175, 1832 (1968);
L. Durand and R. Lipes, Phys. Rev. Letters 2o, 637 (1968)
76. A. A. Anselm and I. T. Dyatlov, Phys. Letters 24B, 479 (1967)
77. G. G. Beznogikh et al, Phys. Letters 3oB, 274 (1969)
78. K. S. Kölbig and B. Margolis, Phys. Letters 31B, 2o (197o)

79. R. J. Rivers, Nuovo Cim. 58A, 1oo (1968)
80. Emphasized by G. C. Fox, proceedings of Stony Brook Conference (1969)
81. Parameters taken from C. Michael, ref. 62
82. N. E. Booth, Phys. Rev. Letters 21, 465 (1968); K. Huang and S. Pinsky, Phys. Rev. 174, 1915 (1968)
83. A. A. Logunov et al, Phys. Letters 24B, 181 (1967); K. Igi and S. Matsuda, Phys. Rev. Letters 18, 625 (1967)
84. R. Dolen, D. Horn and C. Schmid, Phys. Rev. Letters 19, 4o2 (1967); Phys. Rev. 166, 1768 (1968)
85. Y. C. Liu and S. Okubo, Phys. Rev. Letters 19, 19o (1967)
86. V. A. Meshcheriakov et al, Phys. Letters 25B, 341 (1967)
87. J. H. Schwarz, Phys. Rev. 159, 1269 (1967)
88. V. Singh, Phys. Rev. 129, 1889 (1963)
89. C. Ferro-Fontan et al, Nuovo Cim. 58A, 534 (1968)
90. V. Barger and R. J. N. Phillips, Phys. Letters 26B, 73o (1968)
91. B. Amblard et al, report submitted to the Lund Conference (1969)
92. V. Barger and R. J. N. Phillips, Madison preprint COO-262 (197o)
93. K. J. Foley et al, Phys. Rev. 181, 1775 (1969)
94. W. Galbraith et al, Phys. Rev. 138, B913 (1965)
95. K. J. Foley et al, Phys. Rev. Letters 19, 33o (1967)
96. D. Horn, Phys. Letters 31B, 3o (197o)
97. A. Martin, CERN report TH. 1o75 talk at Stony Brook Conference (1969)
98. R. J. Eden, Riverside preprint UCR-34P1o7-1o5; Phys. Rev. (in press)
99. Regge dipoles have previously been introduced, for

different reasons, by G. Frye, Phys. Rev. 129, 1453 (1963);
R. Kreps and J. Moffat, Phys. Rev. 175, 1942 (1968);
Y. Tomozawa, Michigan preprint (1969);
T. Gajdicar and J. Moffat (ref. 16)

1oo. J. Finkelstein, Phys. Rev. Letters 24, 172 (197o)

1o1. K. Kleinknecht, Phys. Letters 3oB, 514 (1969)

1o2. N. W. Dean, Phys. Rev. 182, 1695 (1969)

1o3. V. Barger and R. J. N. Phillips, Phys. Rev. Letters 24, 291 (197o)

1o4. G. F. Chew and D. R. Snider, Phys. Letters 31B, 75 (197o)

1o5. C. Lovelace, talk at Irvine Conference (1969); CERN report TH. 1123 (197o).

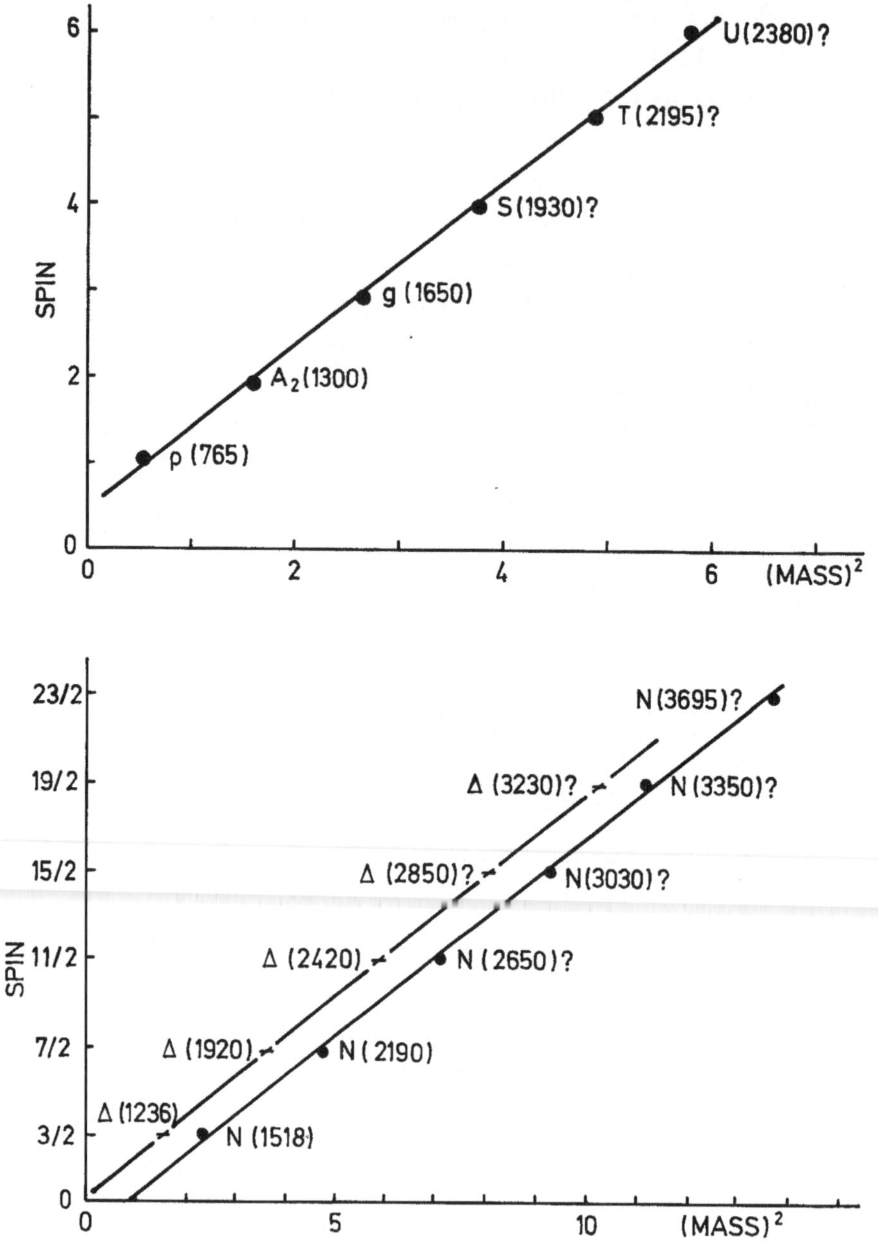

Fig. 1. Empirical trajectories (ref. 39)

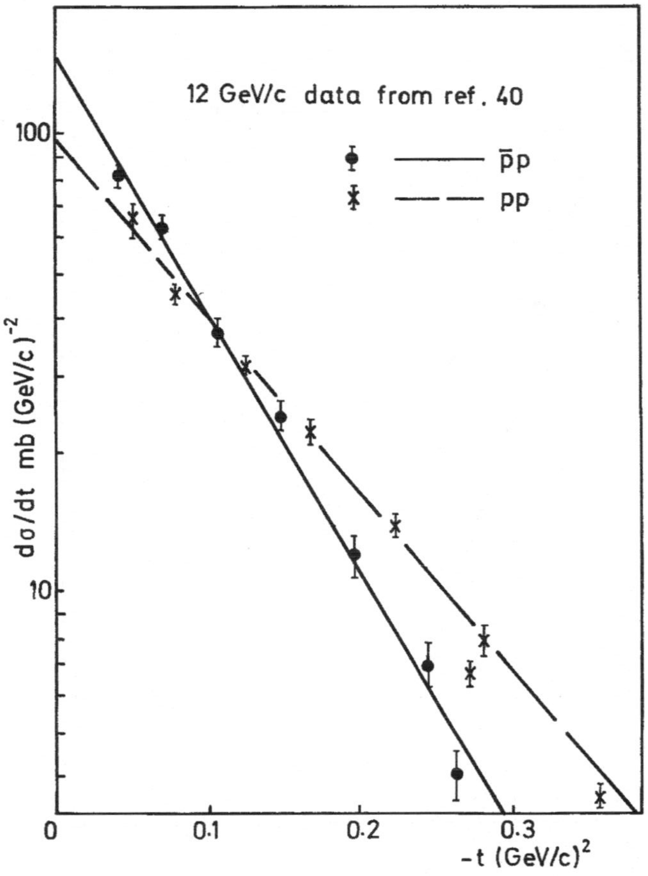

Fig. 2. Crossover effect

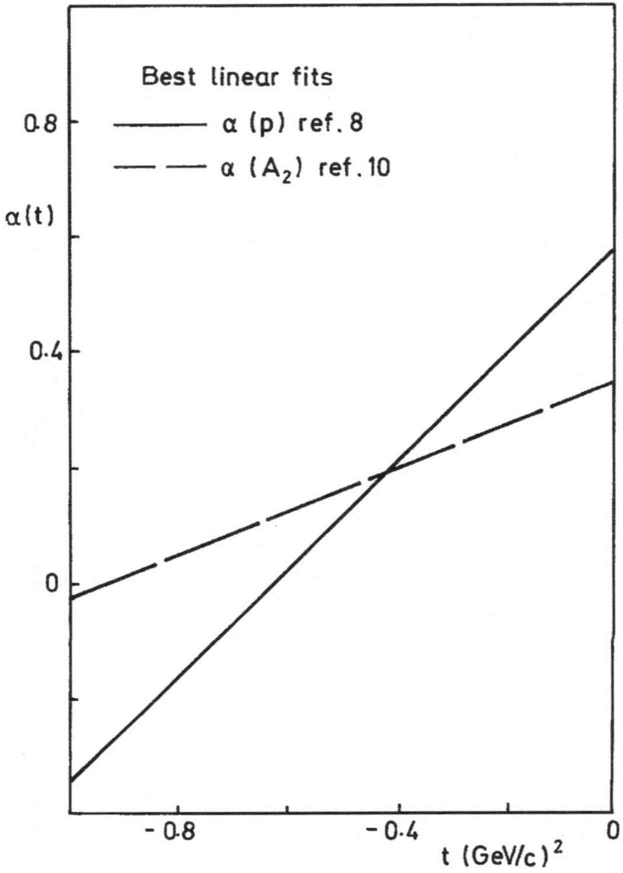

Fig. 3

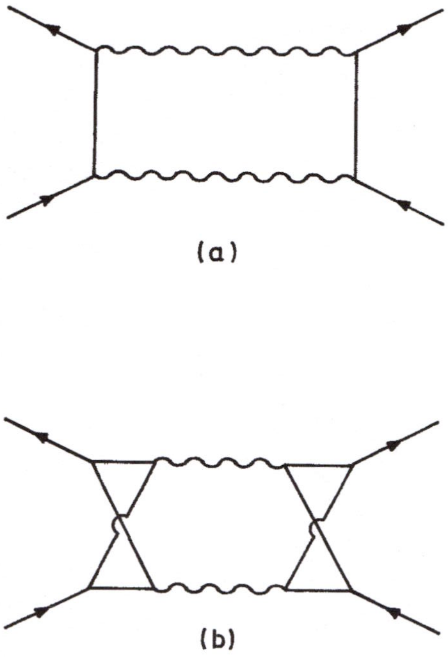

Fig. 4. Double Reggeon-exchange graphs

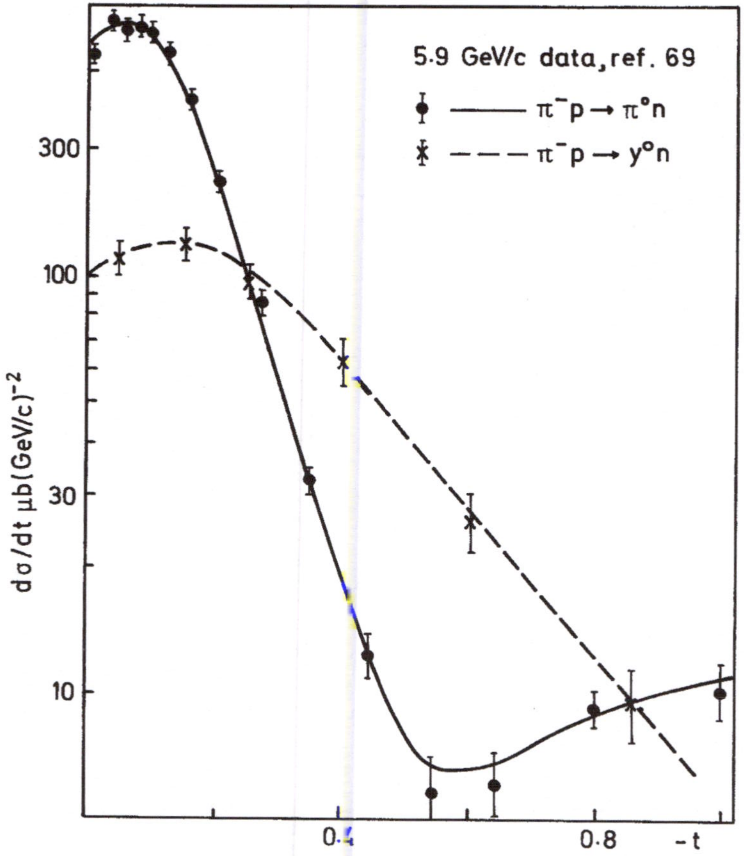

Fig. 5

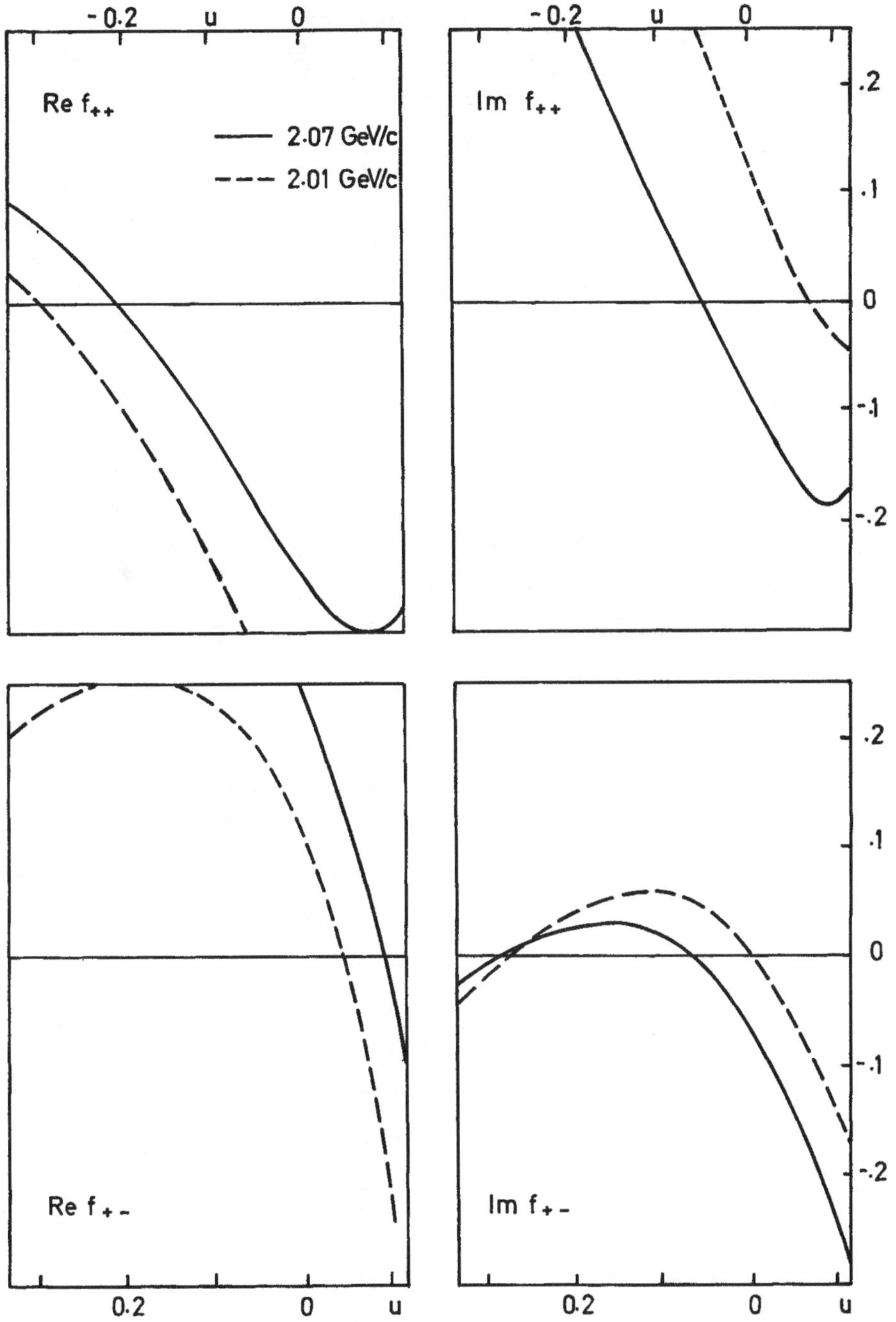

Fig. 6. πN amplitudes from CERN phase shifts

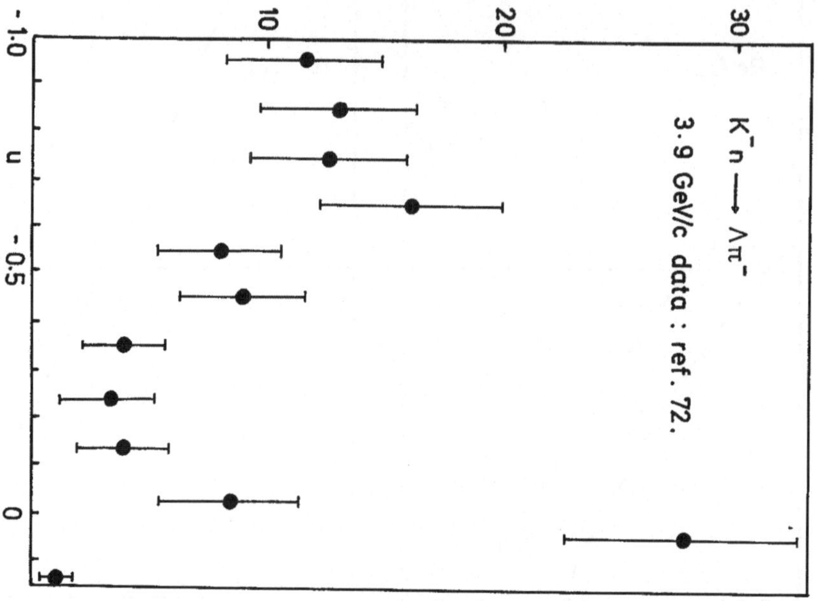
dσ/du μb(GeV/c)$^{-2}$
30
20
10
-1.0
u
-0.5
0
K⁻n ⟶ Λπ⁻
3.9 GeV/c data : ref. 72.

Fig. 7

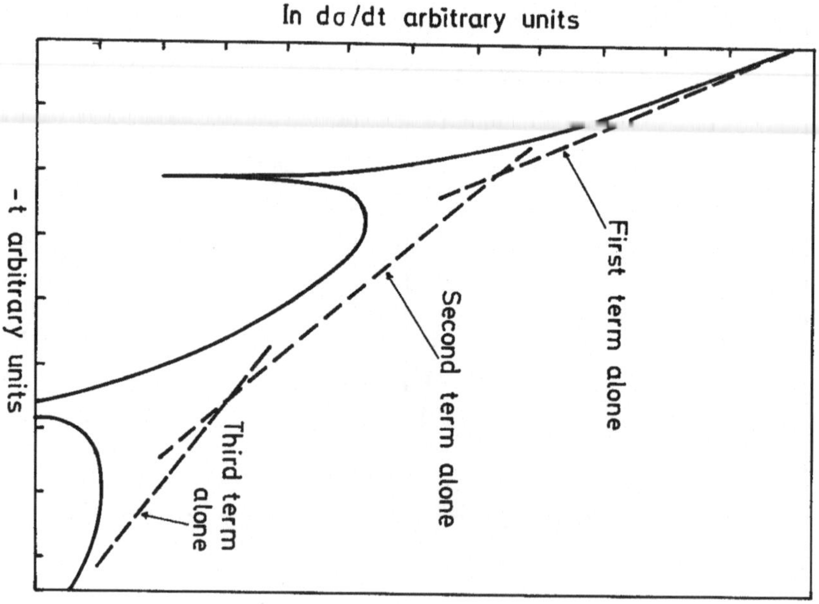
ln dσ/dt arbitrary units
-t arbitrary units
First term alone
Second term alone
Third term alone

Fig. 8

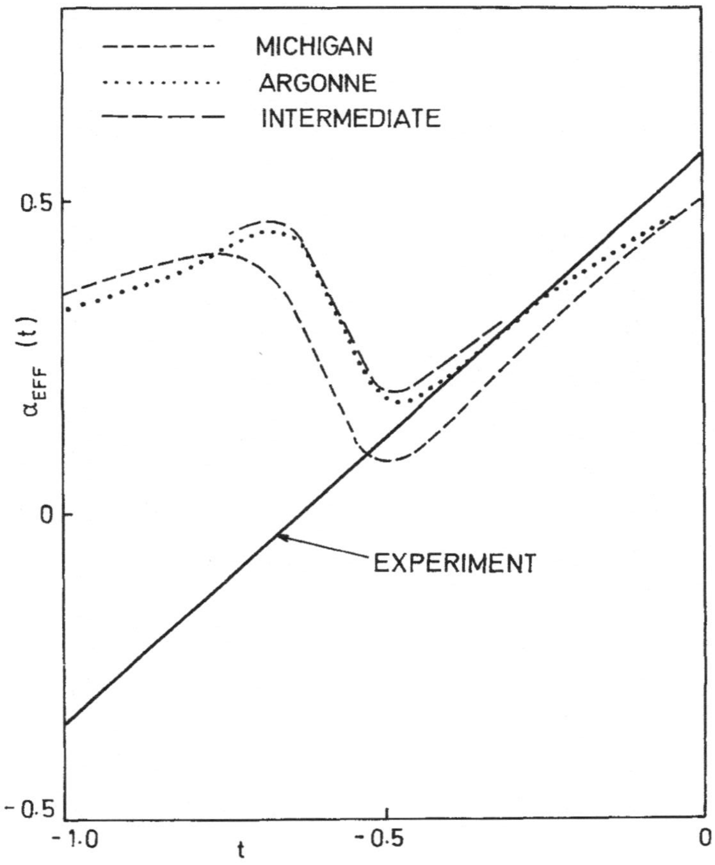

Fig. 9. Effective alphas

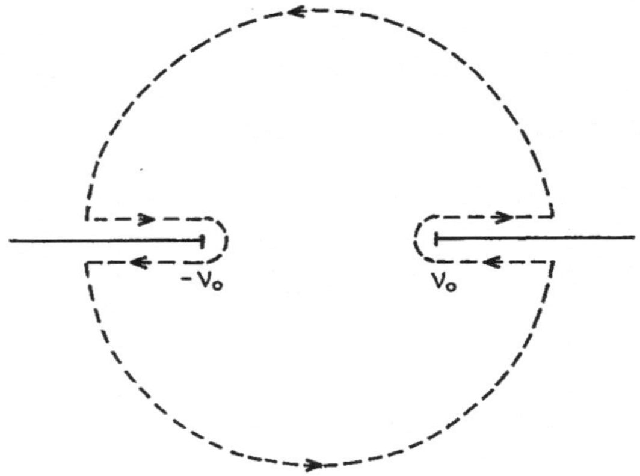

Fig. lo. Integration contour C

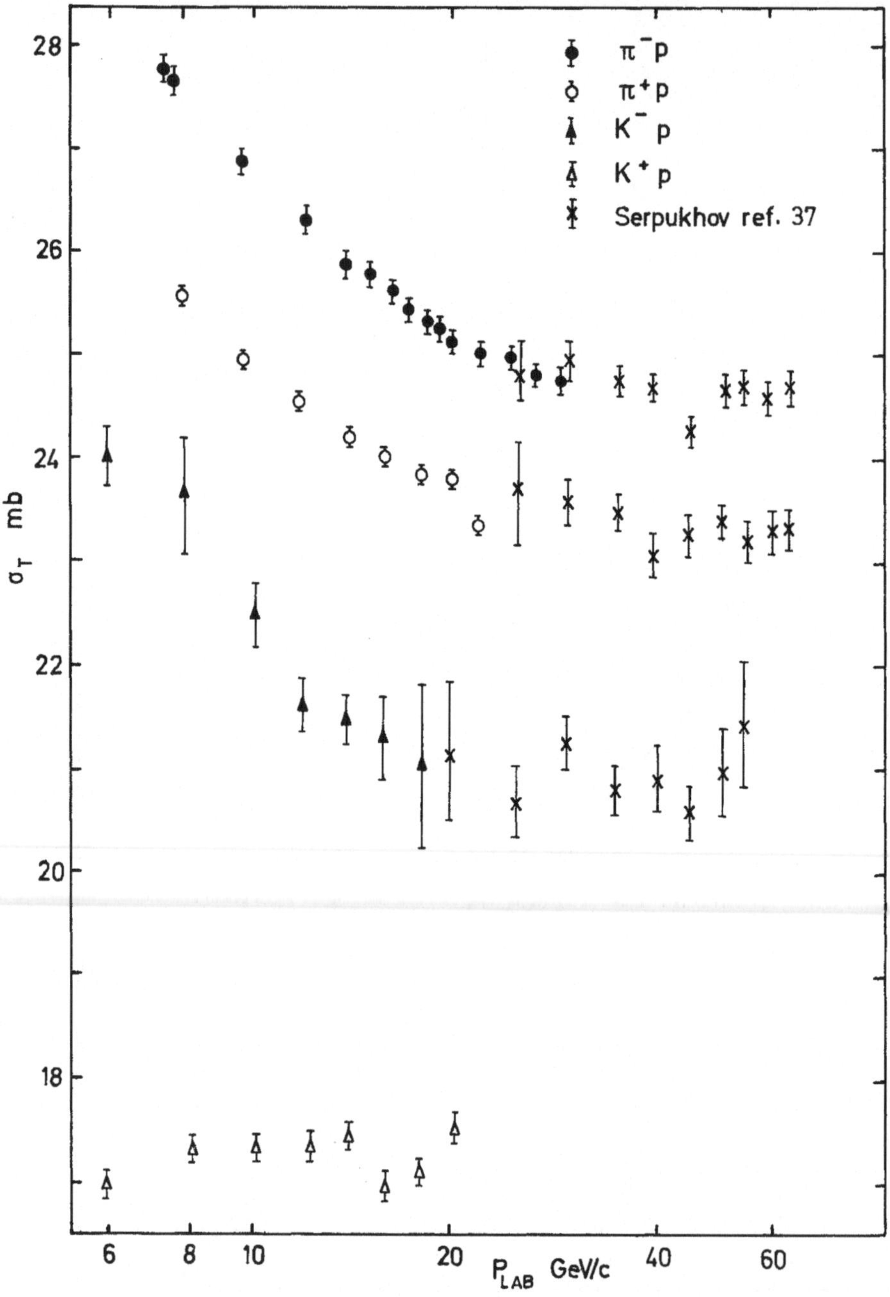
28
26
24
22
20
18
σ_T mb
6
8
10
20
40
60
P_{LAB} GeV/c
$\pi^- p$
$\pi^+ p$
$K^- p$
$K^+ p$
Serpukhov ref. 37

Fig. 11

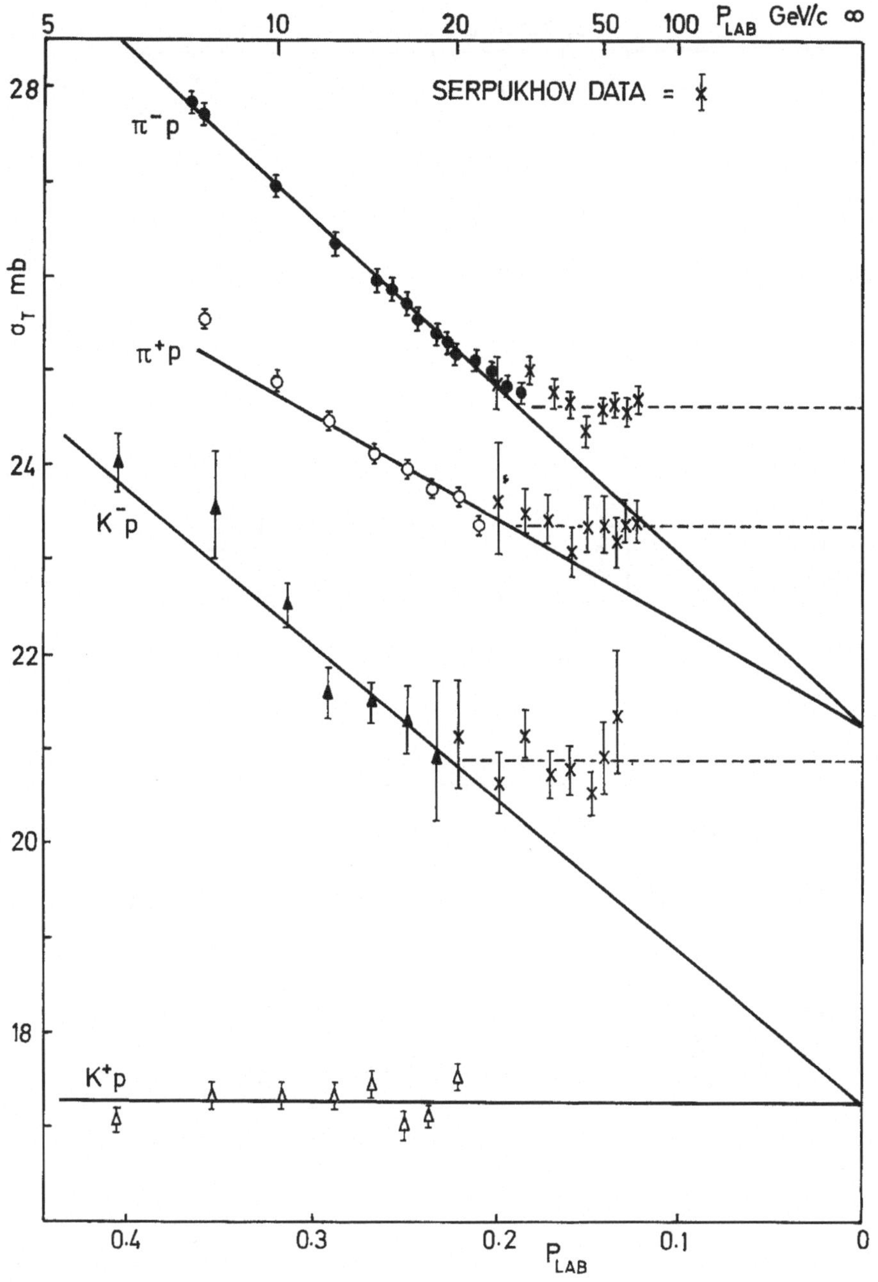

Fig. 12. σ_T data

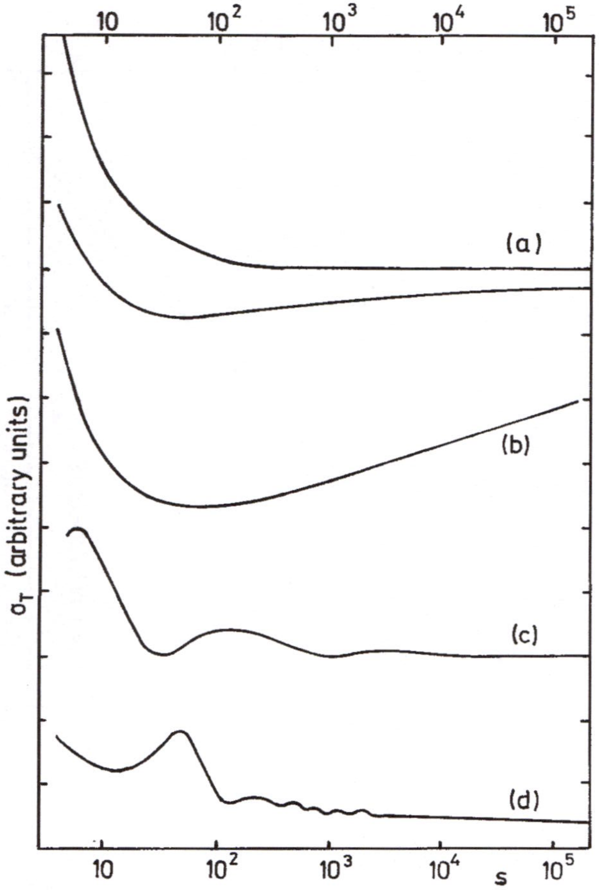
10
10^2
10^3
10^4
10^5
σT (arbitrary units)
(a)
(b)
(c)
(d)
s

Fig. 13

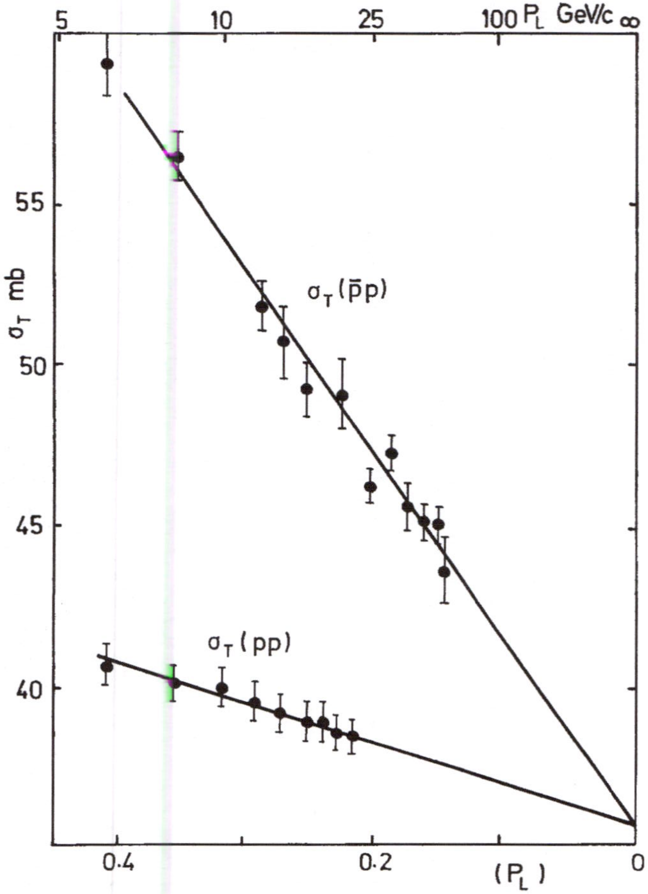
5
10
25
100 PL GeV/c ∞
σT mb
55
50
45
40
σT(p̄p)
σT(pp)
0.4
0.2
0
(PL)

Fig. 14

Acta Physica Austriaca, Suppl. VII, 265–307 (1970)

DIFFERENTIABLE MANIFOLDS*

BY

E. HLAWKA
Mathematisches Institut der Universität
Wien, Austria

1. BASIC CONCEPTS

Let M be a set. An n-dimensional chart c on M is a pair (U,ϕ), where U is a subset of M and ϕ a bijective mapping of U onto an open set of R^n. U is called the domain of c, ϕ the coordinate mapping of c. For each $p \in U$ we can write (see Fig. 1)

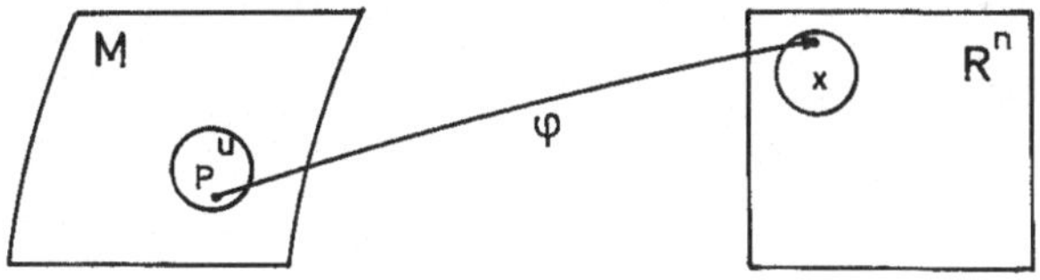

Fig. 1

$$\phi(p) = (\phi_1(p),\ldots,\phi_n(p)) = (x_1,\ldots,x_n)$$

We call $(x_1,\ldots,x_n)$ the local coordinates of p in the chart (U,ϕ). The notation can be made more explicit by writing $x_1,\ldots,x_n$ in the form

$$(x_1(p),\ldots,x_n(p))$$

* Lecture given at IX. Internationale Universitätswochen für Kernphysik, Schladming, February 23 - March 7, 1970.

or, in short, $x(p)$.

If $p \varepsilon U$, then we say $c=(U,\phi)$ is a chart at p. Now let us suppose that we have two charts $c=(U,\phi)$, $c'=(U',\phi')$ on M. We call the charts C^r-admissible or compatible if (see Fig. 2)

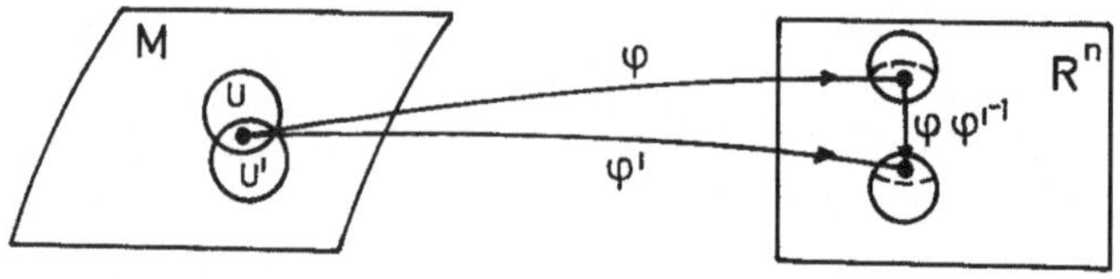

Fig. 2

1) $\phi(U \cap U')$, $\phi'(U \cap U')$ are open in R^n
2) $\phi o \phi'^{-1} : \phi(U \cap U') \rightarrow \phi'(U \cap U')$ and $\phi' o \phi^{-1} : \phi'(U \cap U') \rightarrow$ $\rightarrow \phi(U \cap U')$ are C^r-isomorphisms.

Here r is an integer, ∞ or ω (ω means that the functions are analytic). Unless otherwise stated, we shall assume that $r=\infty$. We define a set $A = (U_i, \phi_i)_{i \varepsilon I}$ of charts on M to be a $\underline{C^r\text{-atlas}}$ (or subatlas) if all charts of A are compatible and the set $\underset{i}{\cup} U_i$ of all domains U_i covers M. Two atlases A_1, A_2 are called equivalent if and only if $A_1 \cup A_2$ is an atlas.

An n-dimensional differentiable manifold is a pair (M,γ), where γ is an equivalence class of atlases on M. The union A^* of the atlases in γ: $A^* = \cup(A | A \varepsilon \gamma)$ is the maximal atlas of (M,γ), and a chart (U,ϕ) of A^* is called a compatible local chart of the manifold. We can alternatively write the manifold (M,γ) as (M, A^*). It is important to remark that for a given set M more than one differential structure can exist. Very often this is disregarded, however, and consequently γ is considered fixed. (M,γ) can then be identified with M.Only if the charts are assumed

assumed to be C^r-compatible is the manifold called a C^r-manifold. For $r=\omega$ the manifold is called analytic.

If (U,ϕ), (U',ϕ') are charts for a manifold, we call the mapping $\sigma_{UU'}=\phi' o \phi^{-1}$ (defined on $\phi'(U \cap U')$, wherever $U \cap U'$ is not empty) a transition mapping (see Fig. 2). The coordinate representation of the transition mapping $\sigma_{UU'}=\sigma$ is given by $\sigma=(\sigma_1,\ldots,\sigma_n)$ and thus $\phi'=\sigma o \phi$ becomes $\eta_i=\sigma_i(\xi_1,\ldots,\xi_n)$ $i=1,\ldots,n$ (ξ_i are the coordinates on R^n; $\xi_i(p)=x_i$) . The determinant of this transformation must be nonvanishing:

$$\frac{\partial(\sigma_1,\ldots,\sigma_n)}{\partial(\xi_1,\ldots,\xi_n)} \neq 0 \quad \text{in} \quad U \cap U' .$$

Examples

1) $M=R^n$; the structure is already defined by the atlas (R^n, Id) (Id = identity mapping)

2) M is an open set in R^n, and (M,Id) is again an atlas

3) S^n (the n-dimensional sphere in R^{n+1}): $x_1^2+\ldots+x_{n+1}^2=1$. Here we need two charts: (U_1,ϕ_1), (U_2,ϕ_2), where $U_1 \cap S^n$ consists of those points for which $x_{n+1}>-1$ and U_2 consists of those points for which $x_{n+1}<1$. Let $\phi_1: U_1 \rightarrow R^n$ be given by $\phi_1(x_1,\ldots,x_{n+1})=x_i/(1+x_{n+1})$ $(i=1,\ldots,n)$, i.e. a stereographic projection from the south pole $(0,\ldots 0,-1)$ to R^n which is regarded as the equatorial plane. Similarly we define ϕ_2 by $\phi_2(x_1,\ldots,x_n)=x_i/1-x_{n+1}$ $(i=1,\ldots,n)$. Then $\phi_1(U_1 \cap U_2)=\phi_2(U_1 \cap U_2)=\{\xi \varepsilon R^n; \xi \neq 0\}$. Thus $\phi_1 o \phi_2$ is defined for all $\xi \neq 0$ and

$$\phi_2 o \phi_1^{-1}(\xi) = \frac{\xi_i}{\xi_1^2+\ldots+\xi_n^2}$$

All conditions are therefore fulfilled.

4) M=R and let $\phi: M \rightarrow R$ be the mapping $\phi(x)=x^3$. Then (M,ϕ) is a chart defining an atlas. But this chart is not compatible with the chart (R,Id) of example 1(n=1) because the mapping $x \rightarrow x^{1/3}$ is not differentiable at 0.

We now define a topology on (M,γ). Let $A=((U_i,\phi_i),$ $i\varepsilon I)$ be an atlas of γ. Then a subset O of M is defined to be open if and only if $\phi_i(O\cap U_i)$ is open in R^n for each U_i. It is easy to show that this definition does not depend on A , i.e. if A' is equivalent to A the two topologies defined with A and A' resp. coincide. Usually we shall assume that the topology defined in this way is Hausdorff and second countable (this holds if an atlas exists with a countable family of charts). Let (M,γ) be a manifold and O an open subset of M, then we can define a differentiable structure on O by defining as atlas A' the intersections $A'(U_i\cap O,\phi_i|(U_i\cap O)),A=(U_i,\phi_i)$. We call (O,A') an open submanifold of (M,A) (A' is the structure induced by A).

If (M,γ) and (N,γ') are two manifolds we can define a manifold structure on $M\times N$. Let $A=(U_i,\phi_i)$ and $A'=(U_i',\psi_i)$ be two atlases on M and N resp. Then a product atlas $A''=\{U_i\times V_j,\phi_i\times\psi_j\}$ can be defined on the product $M\times N$. It is easy to see that the product of compatible atlases gives compatible product atlases.

Now we define the important concept of a differentiable mapping F between two differentiable manifolds (M,γ) and (N,γ'). We say F is a differentiable map (or a C^r-map if the manifolds are C^r-manifolds) if for each $p\varepsilon M$ there exists a chart (U,ϕ) at p and a chart (V,ψ) at F(p) so that $F(U)\subset V$ and $F_{U,V}=\psi o F o\phi^{-1}:\phi(U)\to\psi(V)$ is a C^∞map (or a C^r-map). (Note: if r=0 F is continuous but not differentiable) (see Fig. 3).

If this holds then the same condition holds for any choice of charts (U,ϕ) at p and (V,ψ) at F(p) such that $F(U)\subset V$. We call $F_{U,V}$ a local representation of F with respect to the charts (U,ϕ), (V,ψ). If M and N are m and n dimensional resp. then $\sigma=F_{U,V}$ has the coordinate representation $(\sigma=(\sigma_1,\ldots,\sigma_m))$

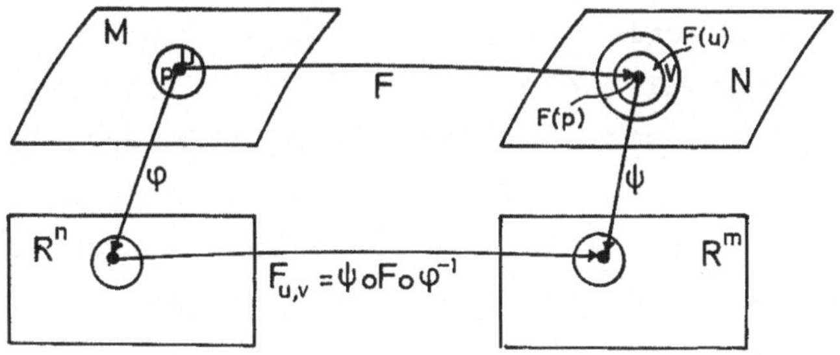

Fig. 3

$$\eta_i = \sigma_i(\xi_1,\ldots,\xi_n) \quad i=1,\ldots,m$$

where σ_i are differentiable functions on $\phi(U)$.

It is obvious that the composition of two differentiable mappings leads again to a differentiable mapping. In the important special case N=R we write $f:M\to R$ and call f a differentiable function on M. More general, let O be open in M, and we consider the function $f:O\to R$ differentiable in O, the class of all these functions we denote by $C^\infty(O)$. Therefore, a function f is differentiable at O if for each $p\in O$ exists a chart (U,ϕ) so that $f_U=f\circ\phi^{-1}$ is differentiable in $\phi(O\cap U_i)$ (see Fig. 4)

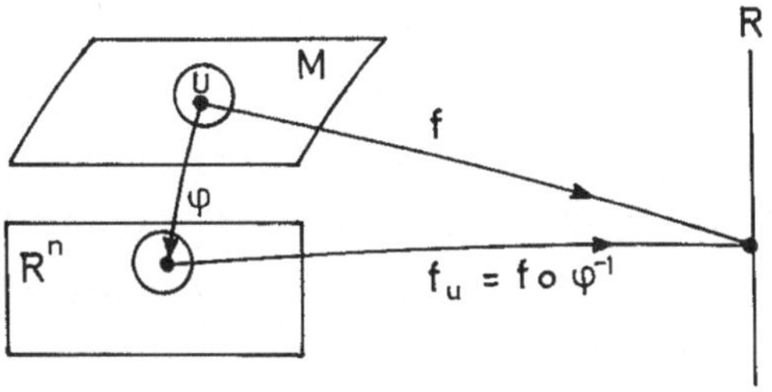

Fig. 4

We say that a function defined on M (or more generally on O) is differentiable at a point $p \in O$ if an open set O' with $p \in O' \subset O$ exists such that f is differentiable in O'. For the class of all functions which are differentiable at p we write $C_p^\infty(O)$. The local representation f_U of f on a chart (U,ϕ) is given in terms of the coordinates $\xi_1,\ldots,\xi_n$ defined on $\phi(O \subset U)$ as $f_U = g(\xi_1,\ldots,\xi_n)$.

The following theorem is very useful: Given an open set $O \subset M$, a function $f \in C^\infty(O)$ and a point $p \in O$, there exists a function $\tilde{f} \in C^\infty(M)$ such that $\tilde{f} = f$ in a neighbourhood of p. The proof makes use of the following lemma:

If C is a compact set on M (M Hausdorff) and V an open set containing C, then there exists a "test function" $\psi \in C^\infty(M)$ which is 1 on C and zero outside of V.

To prove the theorem mentioned above, we assume that f is differentiable in a chart (U,ϕ) with $U \subset O$. Then we can always find a chart (U_1,ϕ) such that $p \in \bar{U}_1 \subset U \subset O$ (as usual $\bar{U}_1$ denotes the closure of U_1, i.e. $\bar{U}_1$ is the smallest closed set containing U_1). We put $c = U_1$, $V = U$ in the lemma given before and construct an appropriate test function ϕ. Putting

$$\tilde{f}(q) = f(q)\phi(q) \quad \text{for all } q \in U,$$
$$= 0 \quad \text{for all } q \notin U,$$

we obtain a function $\tilde{f}$ with the properties postulated before.

Having introduced the concept of differentiable functions we are able to formulate the concept of differentiable mappings $F:(M,\gamma) \rightarrow (N,\gamma')$ differently. We say F is differentiable if for each $p \in M$ there exists (U,ϕ) at p and (V,ψ) at F(p) with $F(U) \subset V$ and for any differentiable function g on V the function $g \circ F$ is differentiable on U. (See Fig. 5).

In short, F is differentiable if for all functions g which are differentiable in a neighbourhood of F(p) on N, $g \circ F$ is differentiable in a neighbourhood of p on M. F

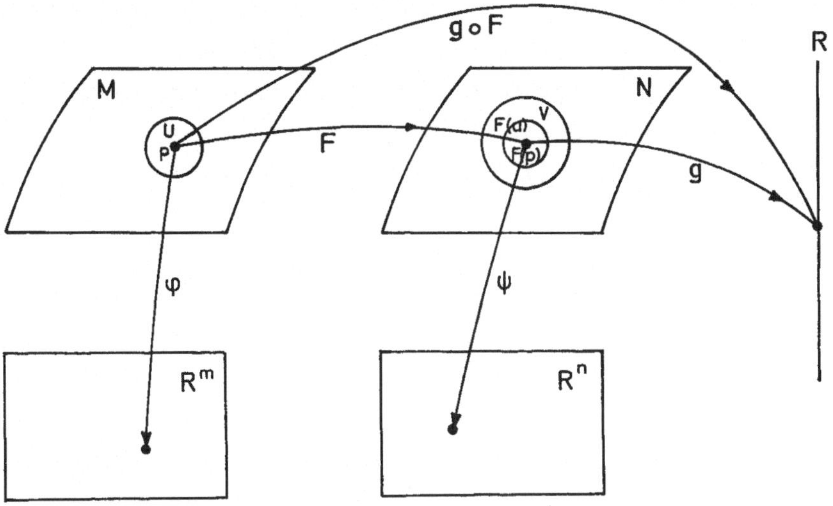

Fig. 5

is called a diffeomorphism if F is bijective and F and F^{-1} are differentiable mappings. F is a local diffeomorphism at a point pϵM if F induces a diffeomorphism of an open neighbourhood of p on an open neighbourhood of F(p) in M.

Now we are able to introduce the rather complicated concept of a submanifold of (M,γ) (first definition). If m is an integer $m \leq n$ we can write $R^n = R^m \times R^q$ $(q+m=n)$. A subset B of M is called m-dimensional if for each pϵB a chart (U,ϕ) exists on M at p such that the mapping $\phi : U \to U_1 \times U_2$ (U_1 open in R^m, U_2 open in R^q) * exists for which $\phi(U \cap B) = U_1 \times \{0\}$. This means that the points of B in (U,ϕ) are precisely the points having coordinates $(x_1,\ldots,x_m, 0,\ldots,0)$. We call such charts admissible for B. B is a submanifold of M if ϕ is a diffeomorphism and the differentiable structure on B is generated by the atlas $\{U \cap B, \phi | U_i \cap B\}$, where (U,ϕ) are those charts of an

* To be more precise $\phi(U) \cap R^m$ is open in R^m and $\phi(U) \cap R^q$ open in R^q .

atlas $\mathcal{A}$ of γ which are admissible for B. For open sets we get the former case of open manifolds.

If B is a submanifold of M, then the inclusion map of B in M is differentiable. If B is closed, then we say B is a closed manifold. An important example: Let O be open in R^n and f a differentiable function, and assume further that, for a point $a\varepsilon O$, $\frac{\partial f}{\partial x_n}(a) = 0$.

Let B be the subset of U consisting of all points p with f(p)=c. Then there exists an open neighbourhood V of a such that $V \subset B$ is a submanifold of R^n.

At the end of these general definitions we are able to define the concept of a Lie group: A finite dimensional manifold G is a Lie group if G is a group and the mappings: $G\times G \to G$: $(x,y) \to x\cdot y$ (group multiplication) and $G\to G: x \to x^{-1}$ are differentiable. The left translation $L_a: G\to G$ with $L_a(x)=a\cdot x$ and the right translation $R_a(x)=x\cdot a$ are diffeomorphisms.

2. TANGENT SPACE

Let M be a C^∞ manifold, p a point of M and $C_p^\infty(M)$ the set of all functions on M which are differentiable at p. A tangent vector V_p at p is defined to be a realvalued function on $C_p^\infty(M)$ having the properties:

1) $V_p(cf) = cV_p(f)$

2) $V_p(f+g) = V_p(f) + V_p(g)$

3) $V_p(fg) = f(p)V_p(g) + g(p)V_p(f)$

for all f,g in $C_p^\infty(M)$. The set of all V_p is a vector space, the tangent space $T_p(M)$ of M at p. Let $c=(U,\phi)$ be a chart at p and $(x_1,\ldots,x_n)=(\phi_1,\ldots,\phi_n)$ coordinate functions. If f is differentiable at p then the function $f_U=f\circ\phi^{-1}$ is differentiable in $\phi(U)\subset R^n$, and we are therefore able to define the derivatives of f to be

$$\frac{\partial f}{\partial x_j}(p) = \frac{\partial f\circ\phi^{-1}}{\partial \xi_j}(\phi(p)) \qquad (j=1,\dots,n)$$

The realvalued function $(\frac{\partial}{\partial x_j})_p$ which associates to f the real number $\frac{\partial f}{\partial x_j}(p)$ is a tangent vector .

It can be shown (see below) that the set of vectors $(\frac{\partial}{\partial x_1})_p,\dots,(\frac{\partial}{\partial x_n})_p$ is a base of $T_p(M)$. The tangent space $T_p(M)$ is thus n-dimensional, i.e. for any $V_p \varepsilon T_p(M)$ there exist constants $c_1,\dots,c_n$ such that

$$V_p = \sum_i c_i (\frac{\partial}{\partial x_i})_p \quad .$$

For all functions $f \varepsilon C_r^\infty(M)$ and for an arbitrary tangent vector V_p, $V_p(f)$ can thus be written in the form

$$V_p(f) = \sum_j c_j (\frac{\partial f}{\partial x_j})_p = \sum_j c_j (\frac{\partial f\circ\phi^{-1}}{\partial \xi_j})$$

If we take the differentiable functions $f=x_i$ we see that $c_i = V_p(x_i)$, because

$$(\frac{\partial x_i \circ \phi^{-1}}{\partial \xi_j}) = \frac{\partial \xi_i}{\partial \xi_j} = \delta_{ij} \quad .$$

Therefore, we have the formula $V_p = \sum_i V_p(x_i)(\frac{\partial}{\partial x_i})_p$. We call this base the natural base of $T_p(M)$ with respect to the chart (U,ϕ) at p. If we take another chart (V,ϕ') at p with coordinate functions $(y_1,\dots,y_n)$, we see immediately that

$$(\frac{\partial}{\partial x_i})_p = \sum_{j=1}^{n} (\frac{\partial y_j}{\partial x_i})_p (\frac{\partial}{\partial y_j})_p$$

where

$$(\frac{\partial y_j}{\partial x_i})_p = \frac{\partial \sigma_j}{\partial \xi_i}(\phi(p))$$

and

$$\sigma = (\sigma_1,\dots,\sigma_n) = \phi' \circ \phi^{-1} \quad .$$

The proof of the theorem concerning $T_p(M)$ is sinplest when M is a ball $B(p,r) \subset R^n$. In this case we use the taylor formula

$$f(x) = f(p) + \sum_j (x_j - p_j) \frac{\partial f(p)}{\partial x_j} + \ldots$$

and apply the functional V_p. Then we have

$$V_p(f) = \sum_j V_p(x_j) \left(\frac{\partial f}{\partial x_j}\right)_p \quad .$$

In the general case we have to work with test functions which have been introduced before.

The connection between the vector concept introduced here and the more elementary one is the following: The elementary vector concept defines the components c_i, $i=1,\ldots,n$, of an n-dimensional vector by means of their transformation properties

$$c_{\bar{i}} = \frac{\partial y_{\bar{i}}}{\partial x_k} c_k$$

If we write

$$V_p = \sum c_i \frac{\partial}{\partial x_i} = \sum c_{\bar{i}} \frac{\partial}{\partial y_{\bar{i}}}$$

we see immediately that the relation between the old and new vector components is exactly the same as given above. The dual space to the tangent space, $T_p(M)$ is the <u>cotangent space</u> $T_p^*(M)$, i.e. the space of all linear functions ω_p on $T_p(M)$. Especially important are the functions $df_p : V_p \to V_p(f)$, the differential of f in the point p. We define a scalar product $\langle\,\rangle$ by

$$\langle df_p, V_p \rangle = df_p(V_p) = V_p(f) = \sum V_p(x^j) \left(\frac{\partial f}{\partial x_j}\right)_p \quad .$$

If (U,ϕ) is a chart at p with the coordinate functions $(x_1,\ldots,x_n)$, the differentials $(dx^1)_p,\ldots,(dx^n)_p$ are a base of $T_p^*(M)$, and we have

$$\langle (dx^i)_p, \left(\frac{\partial}{\partial x_j}\right)_p \rangle = \delta_{ij}$$

This means that $(dx^i)_p$ $(i=1,\ldots,n)$ are the dual base to $\left(\frac{\partial}{\partial x_i}\right)_p$ $(i=1,\ldots,n)$. For any $\omega_p \varepsilon T_p^*(M)$ we have

$$\omega_p = \sum \langle \omega , \frac{\partial}{\partial x_j} \rangle_p (dx^j)_p$$

and thus

$$df_p = \sum_{j=1}^{n} \left(\frac{\partial f}{\partial x_j}\right)_p (dx^j)_p \quad .$$

The Leibniz rule $d(fg)_p = g(p)df_p + f(p).dg_p$ holds as usual. To study the transformation properties of the components of covariant vectors we introduce a second chart (V,ψ) at p with the coordinate functions $(y_1,\ldots,y_n)$. The basis vectors transform according to

$$(dy^j)_p = \sum_i \left(\frac{\partial y^j}{\partial x_i}\right)_p (dx^i)_p \quad .$$

An arbitrary element $\omega_p \varepsilon T_p^*(M)$ can be expressed in the new basis as

$$\omega_p = \sum \alpha_i (dx_i)_p = \sum \beta_i (dy_i)_p$$

with

$$\beta_j = \sum_i \alpha_i \left(\frac{\partial x_i}{\partial y_j}\right)_p$$

This agrees with the usual transformation properties of covariant vector components.

a-covariant and b-contravariant tensors at p are (a+b) linear realvalued functions on $(T_p)^a \times (T_p^*)^b$. The vectorspace of these tensors is denoted by $T^{b,a}(M_p)$.
If (U,ϕ) is a chart at p, any $\beta \varepsilon T^{b,a}(M_p)$ can be represented as

$$\beta = \sum_{i_1 \ldots i_{a+b}} \alpha_{i_1 \ldots i_a}^{i_{a+1} \ldots i_{a+b}} \left(dx^{i_1} \otimes \ldots \otimes dx^{i_a} \frac{\partial}{\partial x_{i_{a+1}}} \otimes \ldots \otimes \frac{\partial}{\partial x_{i_{a+b}}}\right)$$

Especially important is the case b=0 (covariant tensors). In this case an arbitrary tensor $\alpha \varepsilon T^{0,a}(M_p)$ has the form

$$\alpha = \sum_{i_1 \ldots i_a} \alpha_{i_1 \ldots i_a} (dx^{i_1} \otimes \ldots \otimes dx^{i_a}) \quad .$$

If $\omega_1,\ldots,\omega_a \varepsilon T_p^*(M)$, then $\Omega = \omega_1 \otimes \ldots \otimes \omega_a$ is the function of

$T_p^a \to R$: $\Omega(V_p^1 \dots V_p^a) = \omega_1(V_p^1), \dots, \omega_a(V_p^a)$.

In general, if $\omega \varepsilon T^{o,a}, \eta \varepsilon T^{o,b}$ then $\omega \otimes \eta \varepsilon T^{o,a+b}$ is defined by $\omega \otimes \eta(v_i, \dots, v_{a+b}) = \omega(v_1, \dots, v_a)\eta(v_{a+1}, \dots, v_{a+b})$ for all $v_1, \dots, v_{a+b} \varepsilon T_p(M)$. Now we consider the subspace $\Lambda^a_{i_1}(M_p)$ of alternating tensors of the form

$$\omega = \sum_{i_1 < \dots < i_a} \alpha_{i_1 \dots i_a} (dx^{i_1} \wedge \dots \wedge dx^{i_a})_p .$$

If $\omega\ T^{o,a}(M_p)$, then we define

$$\text{Alt}\ \omega = \frac{1}{a!} \sum_{\sigma \varepsilon \gamma_p} \varepsilon_\sigma \omega_\sigma$$

(γ_p is the symmetric group of p elements, ε_σ the signature of σ and $\omega_\sigma(v_1, \dots, v_p) = \omega(v_{\sigma(1)}, \dots, v_{\sigma(a)})$ for all v_1 $v_1, \dots, v_p \varepsilon T_p(M)$. An arbitrary form ω is alternating if and only if $\omega_\sigma = \varepsilon_\sigma \omega$ for all σ. The tensors Altω which are constructed according to the prescription given above are elements of $\Lambda^a(M_p)$ for all $\omega \varepsilon T^{o,a}(M_p)$.

The vectorspace Λ^a has the dimension $\binom{n}{a}$. If $\omega \varepsilon \Lambda^a$, $\eta \varepsilon \Lambda^b$ then the wedge product $\omega \wedge \eta \varepsilon \Lambda^{a+b}$ is defined through the formula

$$\omega \wedge \eta = \frac{(a+b)!}{a!b!} \text{Alt}(\omega \otimes \eta)$$

This product is associative and we have the rule $\omega \wedge \eta = (-1)^{ab} \eta \wedge \omega$. If $\omega^j \varepsilon T_p^*$, j=1,...,n, we have for $V_1, \dots, V_n \varepsilon T_p(M)$,

$$(\omega^1 \wedge \dots \wedge \omega^n)(v_1, \dots, v_n) = \text{Det}(\omega^j(v_j)) .$$

These are the most relevant facts concerning the Graßmann algebra

$$\Lambda(T_p) = \sum_{a=o}^{n} \Lambda^a(T_p) \quad (\text{where } \Lambda^o(T_p) = R) .$$

Symmetric tensors are interesting in the case a=2, b=O, and we have in a chart at p

$$\omega = \sum_{i_1 i_2} g_{i_1 i_2} (dx^{i_1} \otimes dx^{i_2})$$

and $\omega(V_p^1, V_p^2)$ is the "scalar product" $<V_p^1, V_p^2>$ in $T_p(M)$ defined with the help of ω.

We have preferred the natural base, but we see that all remains true if we take any base $(e_1,\ldots,e_n)$ in $T_p(M)$ and the dual base $(e^1,\ldots,e^n)$ in $T_p^*(M)$ instead of $(\frac{\partial}{\partial x_1},\ldots,\frac{\partial}{\partial x_n})$ and $(dx^1,\ldots,dx^n)$ because our definitions were independent of any base main difference to the tradition.

Now let us consider a differentiable mapping F of the manifold M in the manifold N. We know if $g\in C^\infty_{F(p)}(M)$, then $goF\in C^\infty_p(M)$. We define a linear mapping F_{*p} (often called the differential $(dF)_p$ of $T_p(M)$ in $T_{F(p)}(N): X_p \to Y_{F(p)} = F_{*p}V_p$ in the following way: if $g\in C^\infty_{F(p)}$, then $Y_{F(p)}(g) = V_p(goF)$. In short $F_{*p}(V_p(g)) = V_p(goF)$.

Now let us suppose that (U,ϕ) is a chart in p, and (V,ψ) a chart at F(p) with $F(U)\subset V$. Then we have

$$F_{*p}V_p = \sum_{i=1}^{m} V_p(y^i oF)\left(\frac{\partial}{\partial y_i}\right)_{F(p)} = \sum \beta_i \left(\frac{\partial}{\partial y_i}\right)_{F(p)}$$

(x=coordinate function for (U,ϕ), y=coordinate functions of (V,ψ); n=dim M, m=dim N). For

$$V_p = \sum_{j=1}^{n} \alpha_j \left(\frac{\partial}{\partial x_j}\right)_p$$

we have

$$\beta_i = \sum_{j=1}^{n} \alpha_j \left(\frac{\partial(y^i oF)}{\partial x_j}\right)_p$$

or, if we introduce $\sigma = F_{U,V} = \psi oFo\phi^{-1} = (\sigma_1,\ldots,\sigma_n)$ we obtain

$$\beta_i = \sum_{j=1}^{n} \alpha_j \left(\frac{\partial \sigma^i}{\partial \xi_j}\right)(p) \qquad (i=1,\ldots,m)$$

The chainrule holds: if we have three manifolds M,N,L, and mappings $F:M\to N$, $G:N\to L$ then

$$(G\ F)_{*GF(p)} = G_{*F(p)} o F_{*p}$$

Examples: 1) $Id:M\to M$ then $(Id)_*=Id$

2) $F:M\to N$ is a diffeomorphism then $(F^{-1})_*=(F)_*^{-1}$ (at the points resp. p and F(p)). This means that F_{*p} is at each point p a nonsingular mapping of $T_p(M)$ on $T_{F(p)}(N)$. Therefore dim M = dim N.

Definition: Rank of F in point p = Rank of F_{*p}=dim Im F_{*p} (Im = Image of $T_p(M)$ in $T_{F(p)}(N)$ of the linear mapping F_{*p}).

Definition: A mapping F is called an immersion of M in N if for all p the mapping F_{*p} is injective.

We say F is an embedding if F is an immersion and F is a morphism of M in F(M). A subset B of N is called an immersed submanifold of N if there exist a manifold M and an immersion $F:M\to N$ such that F(M)=B. If $F:M\to N$ is an embedding and on B=f(M) we define a differentiable structure such that F becomes a diffeomorphism, then B becomes a submanifold of N.

The following can be shown: Let dim M = dim N and F be an immersion. Then we can find at each $p\varepsilon M$ a neighbourhood U such that $F|U$ is a diffeomorphism of U on a neighbourhood of F(p). Now we can define the concept of a submanifold M of N (second definition) if the inclusion $i:M\to N$ is an immersion. The following theorem is: If M is Hausdorff with a countable base, then we can find an euclidean space R^k, such that there exists an embedding for M in R^k. (According to Whithey we can take k=2.dim M+1). We see that the concept of an abstract manifold is not wider than the concept of manifolds in euclidean spaces).

Now we can also define differentiable curves in M: A curve γ is a mapping (differentiable) from an open interval I of R into M. Therefore, we have the induced mapping $\gamma_{*t}:T_t(I)\to T_{\gamma(t)}(M)$. Now I is one-dimensional and the base of $T_t(I)$ is d/dt, therefore we have only to consider $\gamma_{*t}(d/dt)$ which we denote by $\dot{\gamma}(t)$. We have, therefore, for any differentiable function f on M,

$$\dot{\gamma}(t)(f) = \left(\frac{df\circ\gamma}{dt}\right)_{\gamma(t)}$$

Now let $\left(\frac{\partial}{\partial x_1}\right)_p, \ldots, \left(\frac{\partial}{\partial x_n}\right)_p$ be the natural base and (U,ϕ) a chart of $\gamma(t)$, then

$$\dot{\gamma}(t) = \sum_{i=1}^{n} \frac{dx^i(\gamma(t))}{dt}\left(\frac{\partial}{\partial x_i}\right)_p = \sum \frac{d\phi_i(\gamma)}{dt}\left(\frac{\partial}{\partial x_i}\right)_p ,$$

where $\phi=(\phi_1,\ldots,\phi_n)$ is the coordinate mapping of (U,ϕ). For $M=R^n$ we get, as usual,

$$\dot{\gamma} = \left(\frac{d\gamma_1}{dt},\ldots,\frac{d\gamma_n}{dt}\right) .$$

We call $\dot{\gamma}(t)$ the velocity vector of γ in t. It is easy to show that if p is a point of M then there always exists a curve γ with $\gamma(0)=p$, such that $\dot{\gamma}(0)$ is a given vector $V_p \varepsilon T_p(M)$. (This V_p has the form

$$\sum_{i=1}^{n} \alpha_i\left(\frac{\partial}{\partial x_i}\right)_p) .$$

The proof is very simple: we take a chart (U,ϕ) at p, such that $\phi(U)$ is the ball $B(0,p)$. Then the curve $\gamma:>-p,p<\to M$ defined by $\gamma(t)=\phi^{-1}\circ\sigma(t)$ (note: $\sigma=(\sigma_1,\ldots,\sigma_n)=$ $=(\alpha_1 t,\ldots,\alpha_n t)$) gives immediately the result. Two curves γ_1, γ have contact in a point p on M if for a $t\varepsilon I$ we have $p=\gamma(t)=\gamma_1(t)$ and $\dot{\gamma}(t)=\dot{\gamma}_1(t)$.

Let (U,ϕ) be again a chart at p and consider the curves $\phi\circ\gamma$, $\phi\circ\gamma_1$ in $\phi(0)\subset R^n$. Then these curves have contact in $\phi(p)$ and the converse is also true. This leads to another definition of $T_p(M)$: Consider all curves through p. We say that the curve γ is equivalent to γ_1 if, for any chart (U,ϕ) at p, $\phi\circ\gamma$ and $\phi\circ\gamma_1$ have in $\phi(p)$ the same tangent in the classical sense. It can be shown that this is an equivalence relation in the set of all curves, and the set of equivalence classes is bijective to the vectors of $T_p(M)$. A third definition of $T_p(M)$ is related to this one: We consider triples (U,ϕ,V) where (U,ϕ) chart at p and V is a vector in $\phi(U)\subset R^n$. All (U,ϕ,V) are defined to be equivalent to (U',ψ,W) if the derivative map $\phi\circ\phi^{-1}$ at $\phi(p)$ maps V into W. Then to each class belongs exactly one vector of $T_p(M)$. Let us suppose that F is a differentiable mapping of M in N and γ a curve on M (i.e., a mapping of an interval I in M). Then $F\circ\gamma$ is a curve in N, and we have $(F\circ\gamma)_*=F_*\gamma_*$.

3. TANGENT BUNDLE

Let M be a manifold and $T(M)=\bigcup_{p\in M} T_p(M)$ be the set of all pairs (p,v), where $v\in T_p(M)$. T(M) is called the <u>tangent bundle</u> of M. We have the projection $\pi:T(M)\to M$ defined by $\pi((p,v))=p$. Now we can give T(M) the structure of a differentiable manifold of 2n dimensions, if n is the dimension of M. Let (U,ϕ) be a chart of an atlas $\mathcal{A}$ of M, then we can define a chart $(\bar U,\bar\phi)$ to be the bundle chart associated with (U,ϕ) as follows: $\bar U$ on T(M) is the set of all (p,v), where $p\in U$. $\bar\phi=(\bar\phi_1,\ldots,\bar\phi_{2n})$ is the mapping of $\bar U$ in R^{2n}, such that

$$\bar\phi_i((p,v)) = \begin{cases} \phi_i(p)=x_i(p) & i=1,\ldots,n \\ V(x_i) & i=n+1,\ldots,2n \end{cases}$$

for all (p,v) in $\bar U$. This set of charts $(\bar U,\bar\phi)$ is an atlas of T(M) and the projection π is a differentiable mapping of T(M) in M of rank n. For any point p of M we call $\pi^{-1}(p)$ (i.e. the set of all (p,v) with $\pi(p,v)=p$) the <u>fiber at p</u> [it agrees exactly with $T_p(M)$]. In a similar way we can introduce the cotangent bundle $T^*(M)$ and the tensor bundles $T^{a,b}(M)$ over M.

Furthermore, we want to discuss briefly the concept

of a fiber bundle, which is a quadruple $\xi=(E,B,F,\pi)$, where E,B,F are differentiable manifolds and π is a differentiable mapping from E to B such that $\pi^{-1}(b)$ is diffeomorph to F for each b of B. E is called the totalspace, B the base space, F the fiber and π the bundle projection. Usually one demands that E be locally trivial over B. This means, any $b\in B$ has a neighbourhood U and a diffeomorphism ψ of $\pi^{-1}(U)\to U\times F$, such that $\psi(q)=(\pi(q),\eta(q))$, where $\eta(q)\in F$. If F is a vectorspace then ξ is called a vector bundle over B. T(M) and all tensor bundles over M are vector bundles over M. A further important bundle over M is the bundle of bases over M: Let B(M) be the set of all (n+1)tuples $b=(p,e_1,\dots,e_n)$, where $p\in M$ and $e_1,\dots,e_n$ is an ordered base of $T_p(M)$ and π the projection $\pi(b)=p$. If (U,ϕ) is a chart of an atlas $\mathcal{A}$ of M, let $(\bar{U},\bar{\phi})$ be the following associate chart: $\bar{U}=\pi^{-1}(U)$ and $\bar{\phi}=(\bar{\phi}_1,\dots,\bar{\phi}_n,\bar{\phi}_{11},\dots,\bar{\phi}_{nn})$, where $\bar{\phi}_i(b)=\phi_i(p)$ $(i=1,\dots,n)$ and $\bar{\phi}_{ij}(b)=x_{ij}(p)$, where

$$e_i = \sum_{j=1}^{n} x_{ij}(p)\left(\frac{\partial}{\partial x_j}\right)_p .$$

Again the set of all $(\bar{U},\bar{\phi})$ is an atlas, and π is a differentiable mapping from B(M) in M. Any fiber $\pi^{-1}(p)$ $(p\in M)$ is an n^2-dimensional submanifold and is a diffeomorphism to the group GL(n,R). (General Linear Group).

4. VECTOR FIELDS

A vector field X on M (sometimes called infinitesimal transformation) is a mapping from M to T(M), such that $X(p)\in T_p(M)$ for any $p\in M$, or, if we use the projection $\pi: T(M)\to M$, X is a vector field on M if $\pi\circ X=Id$. A vector field X is differentiable if, for any $f\in C^\infty(M)$, the function $\tilde{X}(f):M\to R$ defined by $p\to\tilde{X}(f)(p)=X(p)(f)$ is differentiable. We also say X is a section in T(M). Let O be an open set in M. Then we consider O as an open submanifold of M. A vector field on O is now the restriction to O of a vector field on M. A differentiable vector field X on M gives a differentiable vector field $X|O$ on any open O of M. If U is especially a domain of a chart (U,ϕ), then any X(p) has the form

$$\sum_{i=1}^{n} \alpha_i(p)\left(\frac{\partial}{\partial x_i}\right)_p \quad \text{for all } p\in U .$$

One easily sees that X is differentiable in U if and only if the $\alpha_i: U\to R$, $p\to\alpha_i(p)$ $(i=1,\dots,n)$ are differentiable functions on U. It can be shown that a vector field X is differentiable on M if, for all charts (U,ϕ) of an atlas A of M, $X|U$ and therefore the corresponding $\alpha_1,\dots,\alpha_n$ are differentiable.

Therefore, for any domain U of a chart (U,ϕ) the vector fields defined by $\frac{\partial}{\partial x_i}: U\to T(M)$, where $p\to\left(\frac{\partial}{\partial x_i}\right)_p$ are differentiable vector fields on U for $i=1,\dots,n$. For a proof of the existence of vector fields on M (except the trivial case $X\equiv 0$) the following theorem is useful: Let Z be a vector field on an open set $B\subset M$ and let $p\in B$. Then there always exists a vector field $\tilde{Z}$ on M and an open neighbourhood A of p, such that $p\in A\subset B$ and $\tilde{Z}|A=Z|A$, i.e. $\tilde{Z}$ is equal to Z in a neighbourhood of A. (Proof: Let C be a compact neighbourhood of p and A the interior of C, and let ψ be a test function with $\psi=1$ on C and $\psi=0$ outside of B. Then $\tilde{Z}=\psi Z$ has the desired properties). If we take any chart (U,ϕ), we get many differentiable vector fields, but as the construction shows, these vector fields have zeros, i.e. there exists a point $p\in M$ where $X(p)=0$. It can be shown that there exist manifolds M - for instance the sphere - such that each differentiable vector field on M has zeros.

Next we characterize vector fields differently. To any vector field X we can associate a mapping $\tilde{X}: C^\infty(M)\to C^\infty(M)$

by means of the definition: $f \in C^{\infty}(M) \to \tilde{X}(f)$. We see immediately the following properties:

1) $\tilde{X}(f+g) = \tilde{X}(f) + \tilde{X}(g)$

2) $\tilde{X}(cf) = c\tilde{X}(f)$

3) $\tilde{X}(f\,g) = f\tilde{X}(g) + g\tilde{X}(f)$

$\tilde{X}$ is thus a derivation on the module $C^{\infty}(M)$ (in general a derivation on any module A is a mapping D of A in itself which is linear and obeys rule 3, (Leibniz rule)). One easily sees that to any derivation D of $C^{\infty}(M)$ there belongs a vector field X: Define $X(p)$ for any $p \in M$ as the mapping of $C^{\infty}(M) \to R$ where $X(p)(f) = (D(f))(p)$. One just has to prove that this definition depends on p only.

Consider the set $\chi(M)$ of all vector fields on M or - equivalently - the set $\tilde{\chi}$ of all derivations on $C^{\infty}(M)$. We define the sum X_1+X_2 of two vector fields X_1, X_2 as the mapping: $M \to T(M)$, where $p \to X_1(p)+X_2(p)$. If $c \in R$, then cX is the vector field $p \to cX(p)$. The product hX, (where $h \in C^{\infty}(M)$), can be defined to be the mapping $p \to h(p)X(p)$. We have, therefore, a mapping $C^{\infty}(M) \times \chi(M) \to \chi(M)$, i.e. $\chi(M)$ is a module over $C^{\infty}(M)$ (not a vector space, because $C^{\infty}(M)$ is not a field). Usually $\chi(M)$ has no base, because if a base $X_1, \ldots, X_L$ existed, then these X_i would have to be everywhere different from zero on M. One can show that if a base exists then $L \leq n$. If $L=n$ then M is called parallelizable. On the domain of a chart (U,ϕ) we have a base of n elements, for instance $\frac{\partial}{\partial x_1}, \ldots, \frac{\partial}{\partial x_n}$. Now if X is a vector field on M, then a curve γ is called an integral curve of X if for all $t \in I$ $\dot{\gamma}(t) = X(\gamma(t))$. On the domain U of the chart (U,ϕ) with

$$X = \sum_{i=1}^{n} \alpha_i \frac{\partial}{\partial x_i}$$

this means

$$\frac{dx_i(\gamma)}{dt} = \alpha_i(\gamma(t)) \qquad (i=1,\ldots,n) .$$

It can be shown that for any point p and for any $\varepsilon>0$ there exists an interval $(a-\varepsilon, a+\varepsilon)$ and a curve γ such that $\gamma(a)=p$ and γ is an integral curve of X in a neighbourhood of p. (Proof: Use the theorem of Picard on differential equations). If $X(p)=0$ then the integral curve is the trivial one: $\gamma=p$.

The set of all integral curves of X is called the flow of X. According to the theorem mentioned above at least a local flow belonging to X exists.

Remark: Sometimes it is useful to generalize the concept of a vector field. Let N,M be manifolds and F a mapping $N\to M$ (all differentiable). Then a mapping $X:N\to T(M)$ is a vector field along F if $X(p)\in M_{F(p)}$ for any p of N. It is differentiable if for any $f\in C^\infty(M)$ and all $p\in N$ the function $p\to X(p)(f)$ is differentiable on N. The set of all these vector fields is called χ_F. An important case is N=I (open interval on R) and F a curve γ on M. Then we speak of a vector field along γ.

The Lie product of vector fields: Let us consider two vector fields X,Y as derivations of $C^\infty(M)$ (we may identify X with $\tilde{X}$). Then XY and YX are also linear mappings of $C^\infty(M)$ in itself, but not derivations. But it can easily be seen that $[XY]=XY-YX$ is a vector field. Locally in a chart (U,ϕ) with the natural base $(\frac{\partial}{\partial x_1},\ldots,\frac{\partial}{\partial x_n})$, if

$$X = \sum \alpha_i \frac{\partial}{\partial x_i} \quad , \qquad Y = \sum \beta_i \frac{\partial}{\partial x_i} \quad ,$$

we get

$$[X,Y] = \sum_{i,j} (\alpha_j \frac{\partial \beta_i}{\partial x_j} - \beta_j \frac{\partial \alpha_i}{\partial x_j}) \frac{\partial}{\partial x_i} \quad ,$$

using $[\frac{\partial}{\partial x_i}, \frac{\partial}{\partial x_j}] = 0$. We have the rules

1) $\quad [X,Y] = - [Y,X]$

2) $\quad [fX,gY] = fg[X,Y]+f(Xg)Y-g(Yf)X$

3) $$[X,[Y,Z]] + [Y,[Z,X]] + [Z,[X,Y]] = 0$$ (Jacobi).

Now we define the Lie-derivative of vector fields to be a mapping $L_X: \chi(M) \to \chi(M)$. If X is a fixed vector field, then $L_X(Y)=[X,Y]$ for all $Y \varepsilon \chi(M)$ (often also called Θ_X). It is a linear mapping: $L_X(Y_1+Y_2)=L_X(Y_1)+L_X(Y_2)$. If we define also a mapping $L_X: C^\infty(M) \to C^\infty(M)$ such that $L_X(f)=Xf$, then we have defined L_X as a mapping of $C^\infty(M) \cup \chi(M)$ in itself, and we have in a chart:

$$Xf = \sum_i X_i \frac{\partial f}{\partial x_i} \quad \text{if} \quad X = \sum_i X_i \frac{\partial}{\partial x_i} \quad ,$$

and the rules

1) $$L_X(gZ) = (L_X g)Z + gL_X(Z) \quad ,$$
$$L_X(fg) = gL_X(f) + fL_X(g)$$ (Leibniz rule)

2) $$L_{X_1+X_2}(Z) = L_{X_1}(Z) + L_{X_2}(Z)$$

3) $$L_{[X,Y]} = [L_X, L_Y]$$

4) $$L_X[Y,Z] = [L_X Y,Z] + [Y,L_X Z] \quad .$$

Let F be a mapping of M in N, $X \varepsilon \chi(M)$, $Y \varepsilon \chi(N)$. Then X,Y are called F-related if for all $p \varepsilon M$: $F_{*p}(X(p))=Y(F(p))$. This means that for all $g \varepsilon C^\infty(N)$: $X(g \circ F)=(Yg) \circ F$.

If X is a vector field on M, then it is apparently possible to define a vector field Y on N through the formula $Y(F(p))=F_{*p}(X(p))$, but this is not always true. For instance, F must be surjective, but this is still not sufficient. If F is a diffeomorphism, then it is true that the set of tangent vectors Y on N defined through the formula

$$Y(q) = F_{*F^{-1}(q)}(X(F^{-1}(q)))$$

is a vector field on N. It can be shown that if X_i, Y_i $(i=1,2)$ are F-related, then $[X_1,X_2],[Y_1,Y_2]$ are F-related.

If F is a diffeomorphism of M in itself, then a vector field X on M is called invariant with respect of F if X is F-related to itself. This means that for all p on M, $F_{*p}(X(p))=X(F(p))$, or that for all $g\varepsilon C^{\infty}(M)$, $X(g\circ F)=(Xg)\circ F$.

An important example: Let G be a Lie group and let F be a left translation L_a $(a\varepsilon G)$. A vector field X on G is called left invariant if it is invariant for all L_a. If X is left invariant (ε the neutral element of G) then it follows from the definition that for all $a\varepsilon G$, $L_{a*}(X(\varepsilon))=$ $=X(a)$. This shows that to any $X(\varepsilon)\varepsilon T_{\varepsilon}(G)$ there is one and only one left invariant vector field. Therefore the module of the left invariant vector fields on G is an n-dimensional vector space over R isomorphic to the vector space $T_{\varepsilon}(G)$. On $T_{\varepsilon}(G)$ we can define a Lie-multiplication: If $X(\varepsilon)$, $Y(\varepsilon)\varepsilon T_{\varepsilon}(G)$, then we define $X(\varepsilon)\circ Y(\varepsilon)=[X,Y](\varepsilon)$, where X,Y are the left invariant vector fields generated by $X(\varepsilon)$,$Y(\varepsilon)$. Therefore T(G) can be given the structure of a Lie algebra: The Lie algebra $\mathcal{G}$ of the Lie-group G. The same results can be obtained with the help of right transformation. More generally, let G be a transformation group on a manifold M. Suppose that G is effective on M and define as translation $L_{\alpha}(x)=\alpha x$ $(\alpha\varepsilon G, x\varepsilon M)$. Again we can consider vector fields on M which are invariant against translations.

5. COTANGENT FIELDS OR PFAFFIAN FORMS

A cotangent field ω is a mapping $M\to T^{*}(M)$, i.e. to any $p\varepsilon M$ there corresponds a $\omega_p\varepsilon T^{*}_p(M)$. ω is called differentiable if for any vector field $X\varepsilon\chi(M)$ the function $\langle\omega,X\rangle=\omega(X):M\to R$ with $p\to\omega(X)(p)=\omega_p(X(p))$ is differentiable for all p . To any cotangent field ω corresponds a mapping $\tilde{\omega}$: $\tilde{\chi}(M)\to C^{\infty}(M)$ with $\tilde{\omega}(\tilde{X})=\omega(X)$ for all X. We have the rules:

$$\tilde{\omega}(\tilde{X}_1+\tilde{X}_2) = \tilde{\omega}(\tilde{X}_1) + \tilde{\omega}(\tilde{X}_2) \quad ,$$

$$\tilde{\omega}(f\tilde{X}) = f\tilde{\omega}(\tilde{X})$$

This means that $\tilde{\omega}$ is a linear functional on the module $\tilde{\chi}(M)$ over the ring $C^\infty(M)$. If we have a linear functional $\tilde{\omega}$, then we can define a cotangent field ω as follows: for any $p \in M$ we define ω_p by $\omega_p(X(p)) = \omega(X)(p)$. This is possible, because it can be shown that $\tilde{\omega}(\tilde{X})(p) = \omega(X)(p)$ depends not on X but only on X(p). The proof is not easy and has two steps. First: if in a neighbourhood U of p: X=Y then $\omega(X) = \omega(Y)$ in U.

Secondly: If $X(p) = Y(p)$ then $\omega(X)(p) = \omega(Y)(p)$. After the first step it is enough to consider the domain U of a chart and here all things are simple, considering the following facts: Let us consider cotangent fields on an open set U on M and especially on the domain U of a chart (U,ϕ): In any point $p \in U$, we have $\omega_p = \sum_i \alpha_i(p)(dx^i)_p$ and it can easily be shown: ω is differentiable in U if the n functions $\alpha_i : U \to R$ defined by $p \to \alpha_i(p)$ are differentiable. Especially if f is a function on U, then the cotangent field on U: $p \to (df)_p$ is differentiable and is called df. Further it is easy to see that the n differentials $dx^1, \ldots, dx^n$ are a base of all ω on U and we have

$$\omega = \sum_{i=1}^{n} \alpha_i \, dx^i \quad \text{with } \alpha_i = \omega\left(\frac{\partial}{\partial x_i}\right) \quad .$$

The base $dx^1, \ldots, dx^n$ is dual to the base $\frac{\partial}{\partial x_1}, \ldots, \frac{\partial}{\partial x_n}$ of the vectorfields in U with $dx^i\left(\frac{\partial}{\partial x_j}\right) = \delta_{ij}$. Therefore the cotangent fields are also called differential forms. More generally, let us consider a differentiable function f on M, then we define the differential df as the cotangent field $p \to (df)_p$ and we have for all $X \in \chi(M)$: $df(X) = X(f)$. Therefore cotangent fields always exist on M and we have the rules: $d(f+g) = df + dg$, $d(fg) = fdg + gdf$. Again we shall identify $\tilde{\omega}$ with ω. The set of all $\tilde{\omega}$ or ω we call $\chi^*(M)$. It is again a module, because $\omega_1 + \omega_2$, $f\omega$ are defined in an obvious way.

We define now a Lie derivation $L_X: \chi^*(M)\to\chi^*(M)$ where X is a fixed vector field: $(L_X\omega)$ is a cotangent vector field such that for all $Y\varepsilon\chi(M)$ $(L_X\omega)(Y)=X(\omega(Y))-\omega([X,Y])$. We have $L_X(Y)=[X,Y]$ and for any function f: $L_Xf=X(f)$, therefore we have $(L_X\omega)(Y)=L_X(\omega(Y))-\omega(L_XY)$ for all ω and Y. If in a chart $\omega = \sum_i \alpha_i dx^i$, $X=\sum X_i\frac{\partial}{\partial x_i}$

$$L_X\omega = \sum_{i,j} (X_i\frac{\partial\alpha_i}{\partial x_j} + \alpha_j \frac{\partial X_j}{\partial x_i})\, dx_i \quad .$$

We have the rules:

$$1)\qquad L_{fX}\omega = \omega(X)df + f(L_X\omega)$$

$$2)\qquad L_{fX}Y = f(L_XY) - df(Y)X$$

$$3)\qquad L_{fX}g = f\, L_X\, g \quad .$$

If F is a mapping of M in N and ω is a cotangent field on N then we can define a cotangent field $F^*\omega$ on M if for any p on M: $F^*\omega(p)\,\varepsilon T_p^*(M)$ is the covector such that for all $V_p\varepsilon T_p(M):F^*\omega(p)(V_p)=\omega_{F(p)}(F_{*p}(V_p))$. It can be shown that $F^*\omega$ is differentiable on M and we say ω is pulled back from N to M. If $\omega=dg$ $(g\varepsilon C^\infty(N))$, then $F^*dg=d(g\circ F)$. For charts (U,ϕ) on M, (V,ψ) on N with $F(U)\,\varepsilon V$ and $\omega=\sum\alpha_i dy^i$,

$$F^*\omega = \sum_{i=1}^{n} (\alpha_i\circ F)F^*(dy^i) = \sum_{i=1}^{n} (\alpha_i\circ F)d(y^i\circ F)$$

If F is a diffeomorphism of M then a form ω is called invariant in respect to F if $F^*\omega=\omega$. If G is a Lie group then a form ω is called a form of Maurer-Cartan if ω is left-invariant for all left translations (resp. right translations). The set of all forms of Maurer-Cartan gives a vector space of n dimensions over R (if G is n-dimensional) and to any $\omega_\varepsilon\,\varepsilon T_\varepsilon^*(M)$ belongs exactly one left invariant form. If X is a left invariant vector field then $\omega_\alpha(X(\alpha))$ is independent of α for all $\alpha\varepsilon G$.

Now we can define arbitrary tensor fields on M. If $T^{a,b}(M)$ is the bundle of a-co- and b-contravariant tensors on M then we define a tensor field $T^{a,b}$ as a mapping $M \to T^{a,b}(M)$ i.e. $p \to T_p^{a,b}(M)$. This mapping is called differentiable if for any set of a-vector fields $X_1,\dots,X_a$ and b-covector fields $\omega^1,\dots,\omega^b$ the functions

$$T^{ab}(X_1,\dots,X_a,\ \omega^1,\dots,\omega^b):M \to R$$

defined by

$$p \to T(X_1,\dots,X_a,\omega^1,\dots,\omega^b)(p) = T_p(X_1(p),\dots,X_a(p),\ \omega^1(p),\dots,\omega^b(p))$$

are differentiable.
In a chart (U,ϕ) at p $T_p^{a,b}$ has the form:

$$T_p^{a,b}(U) = \sum \alpha_{i_1\dots i_a}^{j_1\dots j_b}(p)\,((dx)^{i_1} \otimes \dots \otimes (dx)^{i_a} \otimes (\frac{\partial}{\partial x_{j_1}})_p \otimes \dots \dots \otimes (\frac{\partial}{\partial x_{j_b}})_p)$$

and $T^{a,b}$ is differentiable if and only if all the functions

$$\alpha_{i_1\dots i_a}^{j_1\dots j_b}$$

are differentiable.

With any tensor field T we can associate a linear mapping $\tilde{T}$ with $\tilde{T}:\chi^a \times \chi^{*b} \to C^\infty(M)$ i.e. $(X_1,\dots,X_a,\omega^1,\dots,\omega^b) \to \to \tilde{T}(X_1,\dots,X_a,\omega^1,\dots,\omega^b)$. On the other hand any linear mapping $\tilde{T}$ gives us a tensorfield T. We just have to define for any $p \in M$: $p \to T_p^{a,b} = \tilde{T}(X_1,\dots,X_a,\omega^1,\dots,\omega^b)(p)$ and see immediately that this definition is possible. Therefore we can identify again T with $\tilde{T}$. This means that the set $\mathcal{T}^{a,b}$ of all tensor fields $T^{a,b}$ can be identified with the tensorproduct

$$\underbrace{\chi^* \otimes \dots \otimes \chi^*}_{a} \otimes \underbrace{\chi \otimes \dots \otimes \chi}_{b}$$

as the set of all linear functions, over $\chi^a \times \chi^{*b}$. We define the tensor product $T\otimes T_1$ of two tensor fields T,T_1. For simplicity we suppose covariant fields to be the tensor field $\chi^{o,a+b}\to C^\infty(M)$, i.e. $(X_1,\ldots,X_a,X_{a+1},\ldots,X_{a+b})\to T(X_1,\ldots,X_a)T_1(X_{a+1},\ldots,X_{a+b})$, if T is of degree a and T_1 of degree b.

We can define again a Lie derivative L_X: $\mathcal{T}^{a,b}\to\mathcal{T}^{a,b}$ for a fixed vector field X. If $T\varepsilon\mathcal{T}^{a,b}$ the L_XT is defined by means of the formula:

$$(L_XT)(X_1,\ldots,X_a,\omega^1,\ldots,\omega^b) = L_XT(X_1,\ldots,X_a,\omega^1,\ldots,\omega^b) - $$

$$- \sum_{i=1}^{a} T(\ldots,L_XX_i,\ldots) - \sum_{j=1}^{b} T(\ldots,L_X\omega^j,\ldots)$$

for all $X_1,\ldots,X_1,\ \omega^1,\ldots,\omega^b$.

It can be shown that $L_X(T_1\otimes T_2)=L_XT_1\otimes T_2+T_1\otimes L_XT_2$ (Leibniz rule). The cotensor fields and especially the alternating cotensor fields ω (which are also called differential forms) are of importance. A tensor field ω of degree a is also called a form of degree a. The set of all alternating ω of degree a on M is denoted by $\Lambda^a(M)$. For a=0 let $\Lambda^o(M)$ be $C^\infty(M)$. For a=1 we get the covector field or differential forms of degree one or pfaffians. For a>n $\Lambda^a(M)=0$ (n=dimM), for a=n $\Lambda^a(M)$ is one-dimensional. As before we define $\mathrm{Alt}\omega=\sum_\sigma\varepsilon_\sigma\omega_\sigma$ for tensors in a point p. The wedgeproduct $\omega\wedge\eta\ \varepsilon\Lambda^{a+b}$ of the forms ω,η is defined by $\frac{(a+b)!}{a!b!}\mathrm{Alt}(\omega\otimes\eta)$, $\omega\varepsilon\Lambda^a(M)$, $\eta\varepsilon\Lambda^b(M)$. As before we have $\omega\wedge\eta=(-1)^{a\cdot b}\eta\wedge\omega$. If a=b=1 then $(\omega\wedge\eta)(X_1,X_2)=\omega(X_1)\eta(X_2)= -\omega(X_2)\eta(X_1)$ for all $X_1,X_2\varepsilon\chi(M)$. From the above formula it follows immediately:

1) $$L_X(\omega\wedge\eta) = L_X\omega\wedge\eta + \omega\wedge L_X\eta$$

2) $$L_X\omega(X_1,\ldots,X_a) = (L_X\omega)(X_1,\ldots,X_a) + $$

$$+ \sum_{i=1}^{a}(-1)^{i+1}\omega(L_XX_i,X_1,\ldots,\hat{X}_i,\ldots,X_a)$$

($\hat{}$ means that X_i must be deleted). If $f_1,\ldots,f_a \varepsilon C^\infty(M)$, then $df_1 \wedge \ldots \wedge df_a \varepsilon \Lambda^a(M)$. If U is the domain of a chart any $\omega \varepsilon \Lambda^a(U)$ has the form

$$\sum_{i_1<i_2\ldots<i_a} \alpha_{i_1\ldots i_a} dx_{i_1} \wedge \ldots \wedge dx_{i_a}$$

therefore $\Lambda^a(U)$ is a module of $\binom{n}{a}$ dimension.

Now we introduce linear mappings $d\colon \Lambda^a(M) \to \Lambda^{a+1}(M)$ (exterior derivative of E.Cartan). If $\omega \varepsilon \Lambda^a(M)$, we define $d\omega \varepsilon \Lambda^{a+1}(M)$ in the following way:

$$d\omega(X_1,\ldots,X_{a+1}) = \frac{1}{2} \sum_{i=1}^{a+1} (-1)^{i+1} (L_{X_i} \omega(X_1,\ldots,\hat{X}_i,\ldots,X_{a+1}) +$$

$$+ (L_{X_i}\omega)(X_1,\ldots,\hat{X}_i,\ldots,X_{a+1}) \text{ where } X_1,\ldots,X_{a+1} \varepsilon \chi(M).$$

Without the use of the Lie derivative this can be written as

$$d\omega(X_1,\ldots,X_{a+1}) = \sum_{i=1}^{a+1} (-1)^{i+1} X_i \omega(X_1,\ldots,\hat{X}_i,\ldots,X_{a+1}) +$$

$$+ \sum_{i<j} (-1)^{i+j} \omega([X_i X_j],\ldots,\hat{X}_i,\ldots,\hat{X}_j,\ldots,X_{a+1})$$

For a=0 then ω=f and therefore $d\omega$=df. For a=1 (pfaffian forms) we get:

$$d\omega(X_1,X_2) = X_1(\omega(X_2)) - X_2(\omega(X_1)) - \omega([X_1 X_2]) .$$

In the domain U of a chart (U,ϕ) we have

$$d\omega = \sum d\alpha_{i_1\ldots i_s} \wedge dx^{i_1} \wedge \ldots \wedge dx^{i_a}$$

if

$$\omega = \sum \alpha_{i_1\ldots i_a} dx^{i_1} \wedge \ldots \wedge dx^{i_a}$$

If for instance (this is sufficient) $\omega = f\, dx^{i_1} \wedge \ldots \wedge dx^{i_a}$ then

$$d\omega = df \wedge dx^{i_1} \wedge \ldots \wedge dx^{i_a} = \sum_{j>a} \frac{\partial f}{\partial x_j} dx_j \wedge dx^{i_1} \wedge \ldots \wedge dx^{i_a} .$$

We have the rules

1) $$d(\omega \wedge \eta) = d\omega \wedge \eta + (-1)^{\deg \omega} (\omega \wedge d\eta)$$

2) $d(d\omega)=0$ (Lemma of Poincaré) (more exactly $d_{a+1}(d_a\omega) = 0$ if $\deg\omega=a$)

3) $d(L_X\omega) = L_X(d\omega)$

Some examples: Take the Maxwell equations

1) $\operatorname{rot} E = -\frac{1}{c}\frac{\partial B}{\partial t}$

2) $\operatorname{rot} H = \frac{4\pi}{c} I + \frac{1}{c}\frac{\partial D}{\partial t}$

3) $\operatorname{div} D = 4\pi\rho$

4) $\operatorname{div} B = 0$, $E = (E_1, E_2, E_3)$, $B = (B_1, B_2, B_3)$,

$H = (H_1, H_2, H_3)$, $D = (D_1, D_2, D_3)$,

$I = (I_1, I_2, I_3)$

and introduce

$$\alpha = c \sum_{i=1}^{3} (E_i dx_i \wedge dt) + B_1 dx_2 \wedge dx_3 + B_2 dx_1 \wedge dx_3 + B_3 dx_1 \wedge dx_2$$

$$\beta = - \sum (H_i dx_i \wedge dt) + D_1 dx_2 \wedge dx_3 + D_2 dx_1 \wedge dx_3 + D_3 dx_1 \wedge dx_2$$

$$\gamma = (I_1 dx_2 \wedge dx_3 + I_2 dx_1 \wedge dx_3 + I_3 dx_1 \wedge dx_2) \wedge dt - \rho dx_1 \wedge dx_2 \wedge dx_3$$

Equations 1) and 4) simply become $d\alpha=0$, while 2) and 3) can be expressed as $d\beta+4\pi\gamma=0$. It is convenient to introduce a linear mapping $i_X: \Lambda^{a+1}(M) \to \Lambda^a(M)$ for a fixed vector field X by $i_X\omega(X_1,\dots,X_a)=\omega(X,X_1,\dots,X_a)$ for all $X_1,\dots,X_a \in \chi(M)$ (if $a=-1$, then $i_X f=0$) . i_X is called contraction and $i_X\omega$ the inner product of X and ω . We have the rules

1) $i_X(\omega\wedge\eta) = i_X\omega\wedge\eta + (-1)^{\deg\omega}\omega\wedge(i_X\eta)$

2) $i_{FX}\omega = f\, i_X\omega$

3) $i_X\, df = L_X f$

4) $L_X\omega = i_X(d\omega) + d(i_X\omega)$

5) $L_{fX}\omega = f\, L_X\omega + df \wedge i_X\omega$

6) $i_X i_X = 0$

A form ω is called exact if there exists an η such that $\omega = d\eta$; ω is called closed if $d\omega = 0$. Every exact form is closed; the converse is not true in general, but if M is a convex open set in an euclidean space then every closed form is exact (Lemma of Poincaré).

Let F be a mapping of M in N and T a covariant vector field on N of degree a. Then we define a covariant vector field F^*T on M and call it the pull back of T (the special case a=1 has already been considered).

By the following construction we associate to each $p \in M$ a vector from $T_p^{a,0}(M)$: $p \to (F^*T)_p \in T_p^{a,0}(M)$ such that for all

$$X_{p_1}, \ldots, X_{p_a} \in T_p(M)$$

$$F^*T(X_{p_1}, \ldots, X_{p_a}) = T_{F(p)}(F_{*p}(X_{p_1}), \ldots, F_{*p}(X_{p_a}))$$

It can be shown that F^*T is differentiable. If T is a form $\omega \in \Lambda^a(N)$ then $F^*\omega \in \Lambda^a(M)$ and we have the rules

1) $F^*(\omega \wedge \eta) = F^*\omega \wedge F^*\eta$

2) $F^*(d\omega) = d(F^*\omega)$

3) If $F: M \to N$, $G: N \to P$ then $(GF)^* = F^*G^*$.

If F is a diffeomorphism and $G = F^{-1}$ then $i_{G^*X} G^*\omega = G^* i_X \omega$. Let be (U, ϕ) a chart on M with coordinate functions $(x_1, \ldots, x_n)$; (V, ψ) a chart on N with coordinate functions $(y_1, \ldots, y_m)$ and F a mapping between M and N with $F(U) \subset V$. If $\omega \in \Lambda^a(N)$, i.e.

$$\omega = \sum \alpha_{i_1 \ldots i_a} dy^{i_1} \wedge \ldots \wedge dy^{i_a} ,$$

and if we use the equation: $F^*(dh) = d(h \circ F)$ it follows immediately that

$$F^*\omega = \sum \alpha_{i_1, \ldots, i_a} \circ F d(y^{i_1} \circ F) \wedge \ldots \wedge d(y^{i_a} \circ F) .$$

To get $F^*\omega$ in $dx_1, \ldots, dx_n$ we have to use

$$d(y\circ F) = \sum_{i=1}^{n} \frac{\partial(y\circ F)}{\partial x_i}\, dx^i$$

In the special case where M is an open set in R^n, N is an open set in R^m and F is a mapping from M to N it follows that

$$F^*(d\eta_1\wedge\ldots\wedge d\eta_s) = \sum_{h_1<\ldots<h_s} \frac{\partial(F_1,\ldots,F_s)}{\partial(\xi_{h_1}\ldots\xi_{h_s})}\, d\xi_{h_1}\wedge\ldots\wedge d\xi_{h_s}$$

6. ORIENTATION OF A MANIFOLD

Let M be n-dimensional. M is called orientable if there exists a form ω of degree n (also called volume element) so that $\omega_p\neq 0$ for all p. The pair (M,ω) is called an orientable manifold. If $\Omega\in\Lambda^n(M)$ is another form then there exists a function $f\in C^\infty(M)$ such that $\Omega=f\omega$, because $\Lambda^n(M)$ is of dimension 1 . Let us now suppose that M is path connected; if Ω is always nonvanishing then f is always > 0 or f<0 on M. We say $\Omega\sim\omega$ if f>0. The class $[\omega]$ is called the orientation of M.

M has only two orientations (if M is connected) and we write also for the oriented manifold $(M,[\omega])$. The domain U of a chart (U,ϕ) is always orientable, take for instance $\omega=dx_1\wedge\ldots\wedge dx_n$. If M is orientable then it can be shown that there always exists an atlas $A=(U_i,\phi_i)_{i\in I}$ on M such that $\frac{\partial(\sigma_1,\ldots,\sigma_n)}{\partial(\xi_1,\ldots,\xi_n)} > 0$ holds in $\phi_i(U_i\cap U_j)$ where $\sigma=(\sigma_1,\ldots,\sigma_n)=\phi_j\circ\phi_i^{-1}$. (A is then called positive). It can further be shown: If an atlas (U_i,ϕ_i) exists with these properties, then M is orientable, if the index set I is countable. For the proof we use a <u>partition of unity.</u> A family $(g_j)_{j\in J}$ of functions $\in C^\infty(M)$ is called a partition unity if

1) $g_j \geq 0$ for all j,
2) $Sp(g_j)$ = support of g_j=closure of all $p\in M$ with

$g_j(p)>0$ is compact.

3) For any $p\in M$ there exists a neighbourhood V_p of p such that $V_p\cap S_p(g_j)\neq 0$ only for a finite number of j.
4) $\sum_j g_j(p)=1$ for all $p\in M$. This sum is always finite because of 3).

If $\{U_k\}_{k\in K}$ is an open covering of M then a partition $(g_j)_{j\in J}$ of unity is called subordinate of $\{U_j\}$ if for any j there exists a $k(j)\in K$ such that $Sp(g_j\subset U_{k(j)})$.

It can be shown that to any open covering such a subordinate partition of unity exists if M is e.g. Hausdorff and has a countable atlas. If $A=(U_k,\phi_k)$ is a positive atlas and for any U_k we have $\omega_k=dx_1\wedge\ldots\wedge dx_n$ in U_k continued on M (this is always possible) and a partition (g_k) of unity subordinated to U then $\omega=\sum g_k\omega_k$ is a form which is $\neq 0$ on M .

A Lie group G is always orientable: If $\sigma_1,\ldots,\sigma_n$ is a basis of the vector space of all left invariant forms of the first degree then $\sigma_1\wedge\ldots\wedge\sigma_n$ is a volume element of G. The tangent bundle T(M) of M is for any M (orientable or not) always orientable.

For any M there always exists an orientable manifold M^* which is a covering space of M and which is two-sheeted over M if M is not orientable and one-sheeted if M is orientable.

7. RIEMANN METRIC

Let us suppose that g is a symmetric covariant tensor of degree two on M i.e. for all vector fields X,Y we have $g(X,Y)=g(Y,X)$. If for all $X\neq 0$ we have $g(X,X)>0$ then g defines a positive definite metric, because on any tangent space $T_p(M)$ we have a positive definite scalar product $g(X_p,Y_p)=\langle X_p,Y_p\rangle$. We can thus speak of the length

$|X_p|=\sqrt{g(X_p,X_p)}$ of the vector $X_p \in T_p(M)$. We call (M,g) Riemannian manifold. If the weaker property $g(X,Y)=0$ for all $X \in \chi(M)$ then $Y=0$ holds g is non singular and defines a pseudometric or indefinite metric on M. We call the pair (M,g) a pseudo Riemannian manifold.

If U is a domain of a chart (U,ϕ) we get for g:

$$g = \sum_{i,j} g_{ij}\, dx^i \otimes dx^j \quad .$$

Now g is positive definite if in all charts the quadratic form

$$\sum_{i,j} g_{ij}\, \xi^i\, \xi^j$$

(where g_{ij} are functions in U) is positive definite and g is not singular if in all charts $\det(g_{ij}) \neq 0$. It is a trivial exercise to show that to every chart we can find a Riemannian metric, for example $g=\sum dx^i \otimes dx^i$. This is not true for arbitrary differentiable manifolds; but if these manifolds have a countable base, then there exists a positive definite metric.

Proof: Take an atlas $\mathcal{A}=(U_i,\phi_i)_{i\in I}$; let g_i be a positive definite metric on U_i and (γ_i) a partition of unity subordinate to U_i then

$$g = \sum \gamma_i\, g_{k(i)}$$

is a Riemannian metric on M.

Example:

$$M = R^n,\ g = \sum_{i=1}^{p} dx^i \otimes dx^i - \sum_{i=p+1}^{n} dx^i \otimes dx^i \text{ where } p \text{ is } \leq n$$

If p=n the metric is positive definite, if $p \leq n-1$ then it is only non singular and is called Lorentz metric.

Let be (M,g) a Riemannian manifold or an R-space. The Riemannian metric g now gives a pseudo metric on M by the following definition: Let γ be a curve in M, i.e. $\gamma: I \to M$. We know that there exists a covariant vector field X on M with $X(t)=\gamma(t)$ for all $t \in I$. We now define the length of the curve γ by

$$|\gamma|_{a,b} = \int_a^b \sqrt{g(X(t),X(t))}\, dt$$

for all $a,b\in I$. Then we not only consider differentiable curves, but also broken curves $\gamma=\gamma_1+\dots+\gamma_n$, where γ_j are differentiable, and we define $|\gamma|=\sum_j|\gamma_j|$. Then the distance $d(p,q)$ of two points on M is $\inf|\gamma|$, for all broken γ from p to q.

Now let us suppose that g is only nonsingular. If U is the domain of a chart (U,ϕ), then

$$g = \sum_{i,j} g_{ij}\, dx_i \otimes dx_j \quad .$$

Then we consider the nonsingular quadratic form

$$\gamma = \sum_{i,j} g_{ij}\, \xi^i\, \xi^j \quad ,$$

and we can always find a linear transformation

$$\xi^i = \sum \beta_{ij}\, \eta^j \quad , \text{ such that } \gamma = \sum_{i=1}^n \varepsilon_i (\eta^i)^2 \quad ,$$

where $\varepsilon_i = \pm 1$. Now $\varepsilon\gamma=\varepsilon_1,\dots,\varepsilon_n=(-1)^{n-s}$ $(0\leq s\leq n)$ is independent of the transformation and is called the signature of γ . Now let σ_j $(j=1,\dots,n)$ be the vector fields

$$\sum_{i=1}^n \beta_{ij} \frac{\partial}{\partial x_i} \quad .$$

Then we find that the $\sigma_1,\dots,\sigma_n$ are a basis of $\chi(U)$ with $g(\sigma_i,\sigma_j)=\varepsilon_i\delta_{ij}$. The (σ_i) are, therefore, an orthonormal basis of $\chi(U)$, and if $X=\sum\gamma_i\sigma_i$ and $Y=\sum\gamma_i'\sigma_i$, then

$$g(X,Y) = \sum_{i=1}^n \varepsilon_i\, \gamma_i\, \gamma_i' \quad .$$

Now let $\sigma^1,\dots,\sigma^n$ be the basis of $\chi^*(U)$, i.e. $\sigma^1,\dots,\sigma^n$ is the dual basis to $\sigma_1,\dots,\sigma_n$. Therefore, $\sigma^i(\sigma_j)=\delta_{ij}$. Then we have

$$g = \sum_{i=1}^{n} \varepsilon_i \sigma^i \otimes \sigma^i$$

in U and we define $\varepsilon_g = \varepsilon_1, \dots, \varepsilon_n$ as the <u>signature of g in U</u>. Now ε_g depends on U. If M is connected, then ε_g does not depend on U, we can speak of the signature $\varepsilon_g = (-1)^{n-s}$ of the pseudo Riemannian metric of g. For a positive definite metric $\varepsilon_g = 1$. It is unknown, as far as I know, whether there exists, for any ε, a g on M with $\varepsilon_g = \varepsilon$ (except $\varepsilon = 1$).
On any Riemannian manifold we can define a mapping G: $T(M) \to T^*(M)$ in the following way: If X is any vector field, then G(X) is the form ω, such that for all $Y \in \chi(M)$, $\omega(Y) = g(X,Y)$. If U is the domain of a chart (U,ϕ), $X = \sum \alpha^i \frac{\partial}{\partial x_i}$ and $\omega = \sum \alpha_i dx^i$, then

$$\alpha_j = \sum_{j=1}^{n} \alpha^j g_{ij}$$

(lowering and raising indices). Conversely to any ω belongs exactly one X such that $\omega = G(X)$, and we write $X = G^{-1}(\omega)$. Locally, if $\omega = \sum \alpha_i dx^i$ and $X = \sum \alpha^j \frac{\partial}{\partial x_j}$, then $\alpha^j = \sum g^{ji} \alpha_i$, when (g^{ji}) is the inverse matrix to (g_{ij}). If the $\sigma_1, \dots, \sigma_n$ are an orthonormal basis of $\chi(U)$, and the $\sigma^1, \dots, \sigma^n$ a dual basis, then we see that $\sigma^j = G(\sigma_i)\varepsilon_j$. For natural numbers a,b, i,j with $i \le a$, $j \le b$ we have the mapping $G^{ij}: T^{a,b}(M) \to T^{a+1,b-1}(M)$ defined by

$$G^{ij} T(X_1, \dots, X_{a+1}, \omega_1, \dots, \omega_{b-1}) =$$

$$= T(X_1, \dots, \hat{X}_i, \dots, X_{a+1}, \omega_1, \dots, \omega_{j-1} G(X_j), \omega_{j+1}, \dots, \omega_{b+1}).$$

and $G_{ij}: T^{a,b} \to T^{a-1,b+1}$, defined by

$$G^{ij} T(X_1, \dots, X_{a-1}, \omega_1, \dots, \omega_{b+1}) =$$

$$= T(X_1, \dots, X_{i-1} G^{-1}(\omega_i), X_{i+1}, \dots, X_{a-1}, \omega_1, \dots, \hat{\omega}_j, \dots$$

$$\dots, \omega_{b+1}).$$

If F is a mapping from M to N and (N,g) is a Riemannian space, then $(M, F_* g)$ is an R-space too. Now let us suppose that M is orientable with volume element Ω . Then we call the triple (M,g,Ω) an orientable Riemannian space. We now define on M a real valued function f. For any $p\varepsilon M$ let $f(p)$ be $\frac{1}{\Omega(\sigma_1,\ldots,\sigma_n)_p}$ where $(\sigma_1,\ldots,\sigma_n)$ is any orthonormal basis in a chart at p. Then we define $\sigma=f\omega$ and call it the volume element of (M,g,Ω). It can easily be shown that $\sigma^1\wedge\ldots\wedge\sigma^n=\pm\sigma$. We say $\sigma^1,\ldots,\sigma^n$ is positive if + is true. We now define a mapping $*$ (called star operation) such that

$$*:\Lambda^a(M)\to\Lambda^{n-a}(M) \qquad (n=\dim M) .$$

This maps $(\omega\to*\omega)$ where $*\omega$ is the differential form defined as follows: For all vector fields $X_1,\ldots,X_{n-a}$ $(*\omega)(X_1,\ldots,X_{n-a})=\varepsilon_g\alpha(X_1,\ldots,X_{n-a})$, where $\alpha(X_1,\ldots,X_{n-a})=$ $=\omega\wedge G(X_1)\wedge\ldots\wedge G(X_{n-a})$.
If for instance, $\omega\varepsilon\Lambda^0(M)$, i.e., $\omega=f\varepsilon C^\infty(M)$, then we have $*f=f\sigma$.

Some rules:

1) In a chart we have $*(\varepsilon_1\sigma^1\wedge\ldots\wedge\varepsilon_a\sigma^a)=\varepsilon_{a+1}\sigma^{a+1}\wedge\ldots\ldots\wedge\varepsilon_n\sigma^n$

2) If $\omega\varepsilon\Lambda^a(M)$, $*(*\omega)=\varepsilon_g(-1)^{(n-a)a}\omega$.

3) If $\eta,\omega\varepsilon\Lambda^a(M)$, we have $\omega\wedge*\eta=\eta\wedge*\omega$.

Examples:

1) $g(X,Y) = \varepsilon_g*(G(X))\wedge(*G(Y)))$

2) If $M=R^3$, g is the usual scalar product and $x\wedge y$ is the vector product, then: $x\wedge y=G^{-1}(*(g(X)\wedge g(Y)))$.

If we change the orientation, i.e. $\Omega\to-\Omega$ $(-\Omega=\tilde{\Omega})$, then $\tilde{*}=-*$, because $\sigma\to-\sigma$.
Above we defined the derivative d; now we define the co-derivative $\delta:\Lambda^a(M)\to\Lambda^{a-1}(M)$ (induced by d). If $\omega\varepsilon\Lambda^a(M)$, then $\delta\omega=\varepsilon_g(-1)^{n(a-1)}*d*\omega$ (In the literature we have often only $-\varepsilon_g$ or $\varepsilon_g(-1)^{n(a-1)+1}$) . For even n or odd a $\delta\omega=\varepsilon_g*d*\omega$.

<u>Remark</u>: δ is independent of the orientation because the star operation appears twice.

If $f \in \Lambda^0(M)$, then $\delta f=0$, and $\delta \circ \delta=0$. This means $\delta(\delta\omega)=0$. Forms with $\omega=\delta\eta$ are called coexact. Forms with $\delta\omega=0$ are called coclosed. We can further define the Laplace operator Δ. If $\omega \in \Lambda^a(M)$, then $\Delta\omega=(d\delta+\delta d)\omega$. Therefore, Δ is a mapping from $\Lambda^a(M)$ to $\Lambda^a(M)$. If $f \in \Lambda^0(M)$, then $\Delta f=\delta df=$ $=\varepsilon_g * d * df$.

<u>Examples</u>: If $f \in C^\infty(M)$ (let us suppose $M \subset R^n$), then grad $f =$ $= G^{-1}(df)$. We define div $X=\delta G(X)$, where X is a vector field, $\Delta f=\text{div grad } f=\delta df$, $\Delta X=G^{-1}\Delta G(X)$ and rot $X=G^{-1}(*d\, G(X))$, for $n=3$. Again consider the Maxwell equations for the vacuum $B=-\text{rot } E$, div $B=0$, $1/c^2\, \dot{E} + \mu_0 c\, I = \text{rot } B$ and div $E = \frac{\rho}{\varepsilon_0}$. We take the R^4 with the scalarproduct:

$$g(X,Y) = -x_1y_1-x_2y_2-x_3y_3+x_4y_4 ,$$

where $x_4=ct$ and $X=(x_1,\dots,x_4)$, and $Y=(y_1,\dots,y_4)$.
Let

$$I = I_1dx_1+I_2dx_2+I_3dx_3-c\rho dx_4$$

and

$$F = c(B_3dx_1\wedge dx_2+B_2dx_3\wedge dx_1+B_1dx_2\wedge dx_3)+(E_1dx_1+E_2dx_2+ +E_3dx_3)\, dx_4 .$$

Then the Maxwell equations have the form $dF=0$, $\delta F=\mu_0 cI$.

The integration of these equations: According to the lemma of Poincaré there must exist a form $\tilde{\omega}$ with $d\tilde{\omega}= \frac{1}{c}F$. We consider a form $\omega=\tilde{\omega}\wedge d\phi$, and we have also $d\omega= \frac{1}{c}F$. We now choose ϕ such that $\delta\omega=0$. This means $\Delta\phi=-\delta\tilde{\omega}$ (this is possible). We also have $\Delta\phi=\mu_0 F$.

8. CONNEXIONS

At first it is convenient to generalize the concept of tensor fields. Any linear mapping B of $\chi(M)^r \to (\chi(M))^s$ ($r \geq 1$, $s \geq 0$) over the ring $C^\infty(M)$ is called a tensor field of type (r,s). For s=0 and with the understanding that $(\chi(M))^0 = C^\infty(M)$ we get the co-tensor fields which were considered before. To put it more explicitly:

1) $B(X_1,\ldots,X_{i-1},X_i+X_i',\ldots,X_r) = B(\ldots,X_i,\ldots)+ +B(\ldots,X_i',\ldots)$

2) $B(\ldots,fX_i,\ldots) = f\,B(\ldots,X_i,\ldots)$

It can easily be shown that $B(X_1,\ldots,X_r)_p = B(\tilde{X}_1,\ldots,\tilde{X}_r)_p$ if $X_i(p) = \tilde{X}_i(p)$ for $i=1,\ldots,r$ and therefore only depends on $X_1(p),\ldots,X_r(p)$ for any point $p \in M$. We have, therefore, for each p, a mapping of $T_p{}^r(M) \to T_p{}^s(M)$. A <u>connexion, infinitesimal connexion or covariant differentiation</u> on M is a mapping $D: \chi(M)\times\chi(M) \to \chi(M)$. This means that D(X,Y) for any $X,Y \in \chi(M)$ is a vector field. We write $D_X(Y)$ instead of D(X,Y). D satisfies the following rules:

1) $D_X(Y_1+Y_2) = D_XY_1+D_XY_2$

2) $D_X(f\,Y) = Xf.Y + D_XY$

3) $D_{X_1+X_2}Y = D_{X_1}Y + D_{X_2}Y$

4) $D_{fX}Y = f\,D_XY$

Therefore, for fixed Y, $D(Y):\chi(M) \to \chi(M)$, $X \to D_X(Y)$ is a tensor field of type (1,1).

D_XY is called the <u>covariant derivation</u> of Y in the direction X. $(D_XY)_p$ depends only on X(p), and we write $D_{X(p)}Y = (D_XY)_p$.

Examples: 1) $M = (R^n,\mathrm{Id})$. Then $D: D_XY = \sum_{i=1}^{n} X\,b^i . \frac{\partial}{\partial x_i}$ when

$Y=\sum b^i \frac{\partial}{\partial x_i}$ is a connexion, the so-called natural connexion on R^n. We see that

$$D_{\partial_i} \partial_j = 0 \qquad (\partial_i = \frac{\partial}{\partial x_i})$$

2) M parallelizable. This means $\chi(M)$ has a base $X_1, \ldots, X_n$. Then, generalizing example 1), the mapping

$$D : D_X Y = \sum_{i=1}^{n} (Xb^i) X_i \quad ,$$

with $Y=\sum b^i X_i$ is again a connexion, and again $D_{X_i} X_j = 0$

Now consider the domain U of a chart (U, ϕ). Let $(e_1, \ldots, e_n)$ be a base of $\chi(U)$, $(e^1, \ldots, e^n)$ be the dual base of $\chi^*(U)$, $X=\sum a_i e_i$, $Y=\sum b_i e_i$. Then we get

$$D_X Y = \sum_{i,j} a_i (e_i(b_j) e_j + b_j D_{e_i} e_j) =$$

$$= \sum_j (X(b_j) e_j + b_j D_X e_j) \quad ,$$

and, for a point $p \varepsilon U$,

$$(D_X Y)_p = \sum_j X_p(b_j)(e_j)_p + \sum_{i,j} a_i(p) b_j(p) (D_{e_i} e_j)_p \ .$$

Therefore, a_j, b_j and $X\, b_j$ determine $D_X Y$ completely if the fields $D_{e_i} e_j$ are known.

For $e_i = \frac{\partial i}{\partial x_i}$ $(i=1, \ldots, n)$ we get

$$D_X Y = \sum_{i,j} a_i \left(\frac{\partial b_j}{\partial x_i} \frac{\partial}{\partial x_j} + b_j D_{\partial_i} \partial_j\right) \ .$$

Now set

$$D_X e_j = \sum_{i=1} e_{i,j}(X) e_i \ .$$

It is easy to see that e_{ij} are linear over $C^\infty(M)$. This means the $e_{i,j}$ are forms belonging to $\chi^*(M)$. The we have

$$D_X Y = \sum_i (X\, b_i + \sum_j b_j\, e_{ij}(X)) e_i \ .$$

Now let us take $e_i = \partial_i$ and $e^j = dx_j$ and set $e_{ij} = \sum_k \Gamma^i_{jk} dx_k$, where $\Gamma^i_{jk} = e_{ij}(\frac{\partial}{\partial x_k})$. Then we get

$$D_X Y = \sum_i \left(\sum_L a_L \frac{\partial b_i}{\partial x_L} + \sum_{j,k} b_j a_k \Gamma^i_{jk}\right)\frac{\partial}{\partial x_i} .$$

Now we consider the torsion tensor Tor: $\chi^2 \to \chi$
For any $X,Y \in \chi(M)$, $\mathrm{Tor}(X,Y)=D_X Y - D_Y X - [X,Y]$, and the curvature tensor $R_D: \chi^3 \to \chi$, for any X,Y,Z, is (instead of $R(X,Y,Z)$ we write $R(X,Y)Z$) $R(X,Y)Z=[D_X,D_Y]Z-D_{[XY]}Z$, where $[D_X D_Y] = D_X D_Y - D_Y D_X$. If $T_D=0$, then D is symmetric or torsion free.
If we have again the domain U, we can write T(X,Y)

$$T(X,Y) = \sum_{j=1}^{n} T_j(X,Y)e_j \quad ,$$

$$R(X,Y)e_i = \sum_j R_{ij}(X,Y)e_j \quad ,$$

where the T_i and R_{ij} are alternating tensors. Then we get the two equations of E. Cartan

$$de^i = -\sum_{i=1}^{n} e^{ij} \wedge e^j + T_i$$

and

$$de^{ij} = -\sum_k e^{ik} \wedge e^{kj} + R_{ij}$$

If we introduce the vectors

$$\begin{pmatrix} e_1 \\ \vdots \\ e_n \end{pmatrix} = e \quad \begin{pmatrix} T_1 \\ \vdots \\ T_n \end{pmatrix} = T$$

and the matrices $E=(e^{ij})$, $R=(R_{ij})$ we can write, in short,

$$d\,e = -E \wedge e + T$$

$$d\,E = -E \wedge E + R$$

Let us take $e_j=\partial_j$, $e^j=dx_j$ as before. Then we can write

$$T_i = \sum_{j,k} T_{ijk} dx_j \otimes dx_k \quad , \quad R_{ij} = \sum_{k,L} R^i_{ikL} dx_k \otimes dx_L \ .$$

Then we get the well known formulae

$$T_{ijk} = \Gamma^i_{kj} - \Gamma^i_{jk} = T_i\left(\frac{\partial}{\partial k_j}, \frac{\partial}{\partial x_k}\right)$$

and

$$R^i_{jLk} = \frac{\partial \Gamma^i_{jk}}{\partial x_L} - \frac{\partial \Gamma^i_{jL}}{\partial x_k} + \sum_m (\Gamma^i_{mL}\,\Gamma^j_{jk} - \Gamma^j_{mk}\,\Gamma^m_{jL}) \;.$$

D is symmetric if $\Gamma^i_{kj} = \Gamma^i_{jk}$.

If $T=0$, the Bianchi-identity holds:

$$R(XY)Z + R(ZX)Y + R(YZ)X = 0$$

Now let γ be a curve: $I \to M$((I interval) and V a vector field along γ. This means that for any $t \varepsilon I$, $V_t \varepsilon \Gamma_{\gamma(t)}$, differentiable in t (for all $f \varepsilon C^\infty(M)$ is the function $t \to V_{\gamma(t)} f$ differentiable in t).

Then we define a new vector field $\frac{DV}{dt}$ along γ : for fixed t $\frac{DV}{dt} = (D_{\dot{}}V)_{\gamma(t)}$. We have

$$\frac{D(V+W)}{dt} = \frac{DV}{dt} + \frac{DW}{dt} \text{ and } \frac{d(fV)}{dt} = \left(\frac{df\circ\gamma}{dt}\right)_{\gamma(t)} V_{\gamma(t)} + f\,\frac{DV}{dt} \;.$$

If $V = \sum_{i=1}^{n} V_i(\gamma(t))\frac{\partial}{\partial x_i}$ in a chart, we have

$$\frac{DV}{dt} = \sum_j \left(\frac{dV_j(t)}{dt} + \sum_{i,k} \Gamma^j_{ik}(\gamma(t))\; V_k(\gamma(t))\frac{dx_i\circ\gamma}{dt}\right)\partial_j .$$

A vector field V along γ is parallel if $\frac{DV}{dt}=0$ for all t, and it can be shown that for each tangential vector V_o in $\gamma(t_o)$, there exists one and only one vector field V parallel along γ with $V_{\gamma(t_o)} = V_o$.

A curve is geodesic if $D_{\dot\gamma}/dt = 0$. This means that locally

$$\frac{d^2x_j}{dt^2} + \sum \Gamma^j_{ik}\left(\frac{dx_i}{dt}\right)^2 = 0$$

Now consider, instead of an interval I, the square I^2 and let $s: I^2 \to M$ be a surface in M. Then we can consider vector fields V along s. This means $V(x,y) \to T_{s(x,y)}(M)$ for all $(x,y) \varepsilon I^2$ and for all $f \varepsilon C^\infty(M)$ is $T_{s(x,y)} f$ differentiable in (x,y). For example

$$\frac{\partial s}{\partial x} = s_x \frac{\partial}{\partial x} \; , \quad \frac{\partial s}{\partial y} = s_x \frac{\partial}{\partial y}$$

are vector fields along s, where $\frac{\partial}{\partial x}$, $\frac{\partial}{\partial y}$ is the natural base of $T(I^2)$. For fixed y_o we have the curve $s_{y_o}: I: x \to s(x, y_o)$ in M and for fixed x_o the curves $s_{x_o}: I: y \to (x_o, y)$. This means the parameters are x and y resp.

Therefore, we have two vector fields along s_{y_o}, s_{x_o} and derivatives $\frac{DV}{\partial x}$, $\frac{DV}{\partial y}$. Then the following is true:

$$\frac{D}{\partial x}\left(\frac{\partial s}{\partial y}\right) - \frac{D}{\partial y}\left(\frac{\partial s}{\partial x}\right) = T\left(\frac{\partial s}{\partial x} , \frac{\partial s}{\partial y}\right) ,$$

and

$$\frac{D}{\partial y} \frac{D}{\partial x} V - \frac{D}{\partial x} \frac{D}{\partial y} V = R\left(\frac{\partial s}{\partial x} , \frac{\partial s}{\partial y}\right) V \quad .$$

If we have two connexions D, $\bar{D}$, then we can form $B(X,Y) = \bar{D}_X Y - D_X Y$. B is a tensor: the difference-tensor of D and $\bar{D}$. It can be shown that D,$\bar{D}$ have the same geodesic if and only if B=0. Further, for any $\bar{D}$ there exists one and only one connexion D with the same geodesic and without torsion if for all X,Y $D_X Y = \bar{D}_X Y - \frac{1}{2} \bar{T}(X,Y)$.

Now let (M,g) be a pseudoriemannian manifold. Then a connexion D is compatible with g if, for any curve γ and parallel vector fields V,W along γ, g(V,W) is constant or, what is the same, for any vector fields V,W along γ

$$\frac{d}{dt}\, g(V,W) = g\left(\frac{DV}{dt}, W\right) + g\left(V, \frac{dW}{dt}\right) ,$$

or equivalently for all $X,Y,Z \varepsilon T(M)$

$$Z\, g(XY) = g(D_Z X, Y) + g(X, D_Z Y) \quad .$$

Now, if we further demand that the torsion Tor=0, then there exists one and only one D (of Levi-Civita). For any $X,Y \varepsilon \chi(M)$ we define the mapping $\chi(M) \to C^\infty(M)$.

$$Z \to \frac{1}{2}(Xg(Y,Z) + Yg(Z,X) - Zg(X,Y) + g(Z,[XY]) + g(Y,[ZX]) - g(X,[YZ]) = \omega(Z)$$

$\omega\varepsilon\chi^*(M)$ implies $D_X Y = G^{-1}\omega$. Locally we get the well known formula

$$\Gamma^k_{ij} = \frac{1}{2}\sum_r (g^{-1})_{kr}\left(\frac{\partial g_{rj}}{\partial x_i} + \frac{\partial g_{ri}}{\partial x_j} - \frac{\partial g_{ij}}{\partial x_r}\right) .$$

We define the Riemann-Christoffel tensor K of type 0,4:

$$K(X_1 X_2 X_3 X_4) = g(X_1, R(X_3 X_4)X_2)$$

so that

$$K(X_1 X_2 X_3 X_4) = -K(X_2 X_1 X_3 X_4)$$

$$K(X_1 X_2, X_3 X_4) = -K(X_1 X_2 X_4 X_3)$$

$$K(X_1 X_2 X_3 X_4) = K(X_3 X_4 X_1 X_2) .$$

If we further define $A(X,Y) = g(XX)g(YY) - g^2(X,Y)$ and $\bar{K}(X,Y) = K(X,Y,X,Y)/A(X,Y)$, then $\bar{K}(X_p, Y_p) = \bar{K}(X,Y)_p$ is the Riemannian curvature of the two dimensional subspace σ of T_p spanned by X_p and Y_p. There exists also a Riemann-Christoffel-tensor of type 1,3:

$$K(\omega X\ Y\ Z) = \omega(R(Y\ Z)X) .$$

The Cartan equations are now

$$d\sigma = -\ \Omega \wedge \sigma$$

$$dE = -\ \Omega \wedge \Omega R \qquad \sigma = \begin{pmatrix} \sigma^1 \\ \vdots \\ \sigma_n \end{pmatrix} , \quad \Omega = (\sigma^{ij})$$

dual to $\sigma_1, \ldots, \sigma_n$ with $g(\sigma_i, \sigma_j) = \varepsilon_j \delta_{ij}$, where E is antisymmetric. $\sigma^{ij} = -\sigma^{ji}$ because

$$0 = g(D_X \sigma_i, \sigma_j) + g(\sigma_i, D_X \sigma_j) \quad \text{and} \quad \sigma^{ij} = g(D_X \sigma_i, \sigma_j) .$$

Now we can also define D_X for other quantities: $D_X f = X f$ and for all Y, $D_X \omega(Y) = X\omega(Y) - \omega(D_X Y)$. For a $T \varepsilon \mathcal{T}^{a,b}(M)$ $(D_X T)$ is defined through the formula

$$D_X T(Y_1, \ldots, Y_a, \omega_1, \ldots, \omega_b) =$$

$$= (D_X T)(Y_1, \ldots, Y_a) + \sum_{i=1}^{a} T(\ldots, D_X Y_i, \ldots) + \sum_{j=1}^{b} (\ldots, D_X \omega_j, \ldots) .$$

LITERATURE

R. Abraham: Foundations of Mechanics, W. A. Benjamin 1967.

L. Auslander and R. Mackenzie: Introductions to Differentiable Manifolds, Mc Graw-Hill 1963.

H. Flanders: Differentialforms with Applications to Physical Sciences, Academic Press 1963 .

S. Helgason: Differentialgeometry and Symmetric Spaces, Academic Press 1962.

N. J. Hicks: Differentialgeometry, Van Nostrand, Math. Studies 3 (1965).

W. Klingenberg: Riemannsche Geometrie im Großen, Lecture Notes, Springer (1968).

S. Kobayashi-K.Nomizu: Foundations of Differentialgeometry, J. Wiley a.S. 1963.

S. Lang: Introduction to Differentiable Manifolds, Interscience 1962.

G. W. Mackey; Mathematical Foundations of Quantum Mechanics, W. A. Benjamin, 1963.

S. Sternberg: Lecture on Differentialgeometry, Prentice Hall, 1964.

ACKNOWLEDGMENT

I would like to thank Prof. R. Sexl for his final revision, Dr. Breitenecker, Mr. K. Doppel, Mr. P. Schmitt and Miss Ch. Binder for reading the manuscript and for many useful comments improving the text. I also thank Mrs. F. Prokopp who typed the first draft of manuscript for her patience.

Above all, however, I wish to thank Prof. P. Urban for his kind invitation to Schladming and for his encouragement.

Acta Physica Austriaca, Suppl. VII, 308–354 (1970)

GENERAL RELATIVITY AND GRAVITATIONAL COLLAPSE*

BY

R. U. SEXL
Institute for Theoretical Physics,
University of Vienna, Austria

1. A BRIEF SUMMARY OF GENERAL RELATIVITY

In these lectures we shall assume that the reader is familiar with the basic ideas of general relativity. The following summary will serve mainly as an introduction to the notations and conventions used here.

In this first chapter we shall not use the more modern mathematical concepts presented in the lectures of Prof. Hlawka but restrict ourselves to the classical concepts which are used in the standard text books on general relativity.

The physical idea on which Einstein's theory of gravitation is based is the equivalence principle (see Fig.1). Two satellite laboratories circle the earth in free fall. No effects of the presence of the earth's gravitational field will be detectable within these satellites (assuming that they are sufficiently small), so that these free falling systems are equivalent to inertial systems. A laboratory resting on earth is accelerated with respect to these inertial systems. The effect of the earth's gravi-

---

* Lecture given at IX. Internationale Universitätswochen für Kernphysik, Schladming, February 23 - March 7,1970.

tational field is therefore equivalent to the introduction of an accelerated frame of reference.

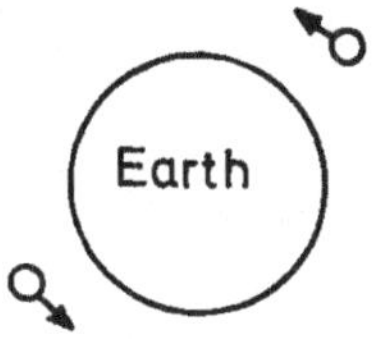

Figure 1 - Earth with two satellites

In the freely falling systems the laws of special relativity are valid. Specifically the distance between two events in space-time will be given by

$$ds^2 = \eta_{ik}\, dx^i\, dx^k \tag{1.1}$$

where $\eta_{ik} = \mathrm{diag}(1,-1,-1,-1)$ (Latin indices have the range $0,\ldots,3$; we use the Einstein convention and put $c=1$, except in those equations which are to be compared directly with experiment). (1.1) is valid, however, only in a special (inertial) coordinate system within the freely falling satellite; if we want to use any other coordinate system or study the space-time geometry in a larger region we have to use the more general Riemannian metric

$$ds^2 = g_{ik}(x)\, dx^i\, dx^k \tag{1.2}$$

The basic task of any theory of gravitation is to determine the metric tensor g_{ik} as a function of the mass-energy distribution and to calculate the equations of motion of given masses in the gravitational field. The latter task is rather simple in the case of structureless (spin 0, uncharged, etc.) test masses. Their motion follows a geodesic which is given by

$$\frac{d^2x^i}{ds^2} + \Gamma_{k\ell}^{\ i} \frac{dx^k}{ds} \frac{dx^\ell}{ds} = 0 \tag{1.3}$$

where $\Gamma_{k\ell}^{\ i}$ are the Christoffel symbols

$$\Gamma_{k\ell m} = \frac{1}{2}(g_{\ell m,k} + g_{km,\ell} - g_{k\ell,m})$$
$$\Gamma_{k\ell}^{\ i} = \Gamma_{k\ell m}\, g^{mi} \quad , \quad g_{\ell m,k} \equiv \frac{\partial g_{\ell m}}{\partial x^k} \tag{1.4}$$

ds is the proper time element measured along the path of the test particles.

The curvature of space-time is determined by the Riemann-tensor

$$R^i_{k\ell m} = \Gamma_{km,\ell}^{\ \ i} - \Gamma_{k\ell,m}^{\ \ i} + \Gamma_{r\ell}^{\ i}\, \Gamma_{mk}^{\ r} - \Gamma_{rm}^{\ i}\, \Gamma_{k\ell}^{\ r} \tag{1.5}$$

By lowering the first index of $R^i_{k\ell m}$ one forms the covariant Riemann tensor

$$R_{ik\ell m} = g_{ri}\, R^r_{k\ell m} \tag{1.6}$$

$R_{ik\ell m}$ has a number of remarkable symmetry properties, listed below

$$R_{ik\ell m} = -R_{ikm\ell}$$
$$R_{ik\ell m} = R_{\ell mik}$$
$$R_{ik\ell m} = -R_{ki\ell m}$$
$$R_{ik\ell m} + R_{imk\ell} + R_{i\ell mk} = 0 \tag{1.7}$$

These symmetry properties reduce the number of independent components of $R^i_{k\ell m}$ to 20.

The Ricci-tensor is defined as

$$R_{ik} = R_{hik\ell}\, g^{h\ell} \tag{1.8}$$

and the curvature scalar as

$$R = R_{ik}\, g^{ik} \tag{1.9}$$

The field equations of general relativity are then given by

$$R_{ik} - \frac{1}{2} g_{ik} R = - \kappa T_{ik} \tag{1.10}$$

where $\kappa = 8\pi G/c^2 = 1.67 \times 10^{-27}$ cm/g and T_{ik} is the mass-energy tensor of the matter distribution considered.

Remark: It is customary to include in (1.10) an additional term, the cosmological constant. This term takes into account energy-momentum density of the vacuum. In this case the source T_{ik} which enters into (1.10) is split into two parts

$$T_{ik} = T_{ik}^{matter} + T_{ik}^{vac} \tag{1.11}$$

Because of invariance reasons T_{ik}^{vac} has to be of the form

$$T_{ik}^{vac} = \lambda/\kappa \cdot g_{ik} \tag{1.12}$$

λ is the cosmological constant, λ/κ is the (unknown) energy momentum density of the vacuum. We shall disregard this term in the sequel, since we are interested mainly in high density situations where $T_{oo}^{matter} \gg \lambda/\kappa$.

The Bianchi identities

$$(R_{ik} - \frac{1}{2} g_{ik} R)^{;k} = 0 \tag{1.13}$$

(we denote covariant derivatives by a semicolon) imply that the field equations (1.10) are consistent only when a conserved source, fulfilling

$$T_{ik;}^{k} = 0 \tag{1.14}$$

is inserted into (1.10). This is completely analogous to the situation in electrodynamics.

The most important solution of the vacuum field equations $R_{ik}=0$ is the Schwarzschild solution with the line element

$$ds^2 = (1-\frac{2GM}{c^2 r})dt^2 - dr^2(1-\frac{2GM}{c^2 r})^{-1} - r^2\, d\Omega^2$$

$$d\Omega^2 = d\Theta^2 + \sin^2\Theta\, d\phi^2 \qquad (1.15)$$

It is the only spherically symmetric vacuum solution of the field equations. It represents, therefore, the gravitational field outside an arbitrary spherically symmetric mass distribution (independent of the internal structure of the mass). Comparing (1.15) with the Newtonian approximation one finds that the integration constant M contained in (1.15) is the mass of the body considered. The length

$$2M = \frac{2GM}{c^2} \qquad (1.16)$$

is the Schwarzschild radius of M . (For the sun M=1.5 km, for the earth M = 0.9 cm .)

For most applications of general relativity it is sufficient to use the linear approximation to the field equations only, which is obtained by putting

$$g_{ik} = \eta_{ik} + 2\psi_{ik} \qquad (1.17)$$

with $\psi_{ik}^2 \sim 0$. In this approximation R_{ik} becomes

$$R_{ik} = \psi_{ik,}{}^{m}{}_{m} + \psi^{m}{}_{m,ik} - \psi^{m}{}_{k,im} - \psi^{m}{}_{i,km} \qquad (1.18)$$

This is still rather complicated. The freedom in the choice of the coordinate system enables one, however, to impose four coordinate conditions on the ψ_{ik}. If these are chosen as

$$\psi_{ik,}{}^{k} = \frac{1}{2}\, \psi^{k}{}_{k,i} \qquad (1.19)$$

(harmonic coordinates), the expression (1.18) for R_{ik} simplifies to

$$R_{ik} = \Box\psi_{ik} \qquad (1.20)$$

The Einstein field equations become in this case

$$\Box\psi_{ik} = -\kappa(T_{ik} - \frac{1}{2}\eta_{ik}T_m{}^m) \qquad (1.21)$$

and are thus decoupled and can be solved with the help of the standard Green's function formalism.

For a slowly moving mass distribution $\Box \approx -\Delta$ and $T_{ik} \approx \rho\delta_i^o \delta_k^o$, where ρ is the mass distribution. Then the equations (1.21) become

$$\Delta\psi_{mm} = 4\pi G\rho \qquad m = 0,1,2,3 \qquad (1.22)$$

The line element becomes therefore in this approximation

$$ds^2 = c^2 dt^2(1+\frac{2U}{c^2}) - d\vec{x}^2(1-\frac{2U}{c^2}) \qquad (1.22)$$

where U is the Newtonian gravitational potential.

2. EXPERIMENTAL TESTS OF GENERAL RELATIVITY

The recent revival of interest in general relativity is largely due to progress in the experimental tests of the theory. We shall therefore discuss the present experimental situation briefly.

The basic test of general relativity is the Dicke-Eötvös (Dicke 1964) experiment. In this experiment the independence of the gravitational acceleration of a test mass on its internal composition is tested with an accuracy of 10^{-11}. This experiment tests the very foundation of Einstein's theory of gravitation, i.e. the principle of equivalence and is therefore of crucial importance.

The red shift of a light ray ascending in a gravitational field has been measured by Pound and Rebka (1960) with the help of the Mößbauer effect with an accuracy of about 1 % . The outcome of the experiment can be predicted, however, from the conservation of energy alone, without any knowledge of the field equations and is therefore no real test of Einstein's theory .

Light deflection is more important. Einstein's theory predicts that a light ray passing close to the edge of the sun will suffer a deflection $\delta = 1.75''$. Present measurements are very inaccurate, $1.5'' \leq \delta \leq 2.2''$.

There is, however, an alternative experiment which contains the same information as light deflection, since both experiments can be calculated from the linear approximation (1.22). This is the Shapiro (1969) experiment (Fig. 2).

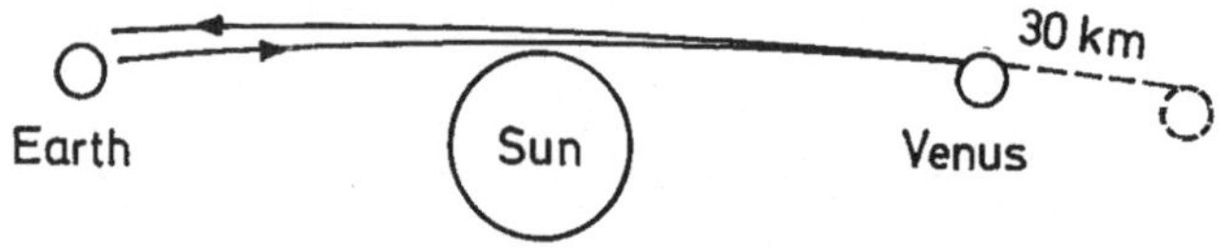

Figure 2 - The Shapiro experiment

A radar beam is passed close to the edge of the sun, is reflected by Venus and its total travel time measured upon its return to earth. General relativity predicts that this beam is not only slightly deflected but also delayed, as compared with Newtonian theory. This delay corresponds to an additional path length of 2×30 km, as shown in Fig. 2 .

This prediction of Einstein's theory has been verified by Shapiro (1969) with an accuracy of about 10 % .

In the course of this experiment Shapiro was able to measure the orbits of Venus and Mercury with great precision and obtained thereby a new determination of the perihelion advance of Mercury confirming the value predicted by general relativity (43.11"/century) to an accuracy of 2 % . This could be the most accurate and important test of general relativity if the question of the solar quadrupole moment were resolved. We shall not enter into a discussion of this complicated question here, but note only that a non-spherical mass distribution in the sun could also contribute to the perihelion advance and no distinction between relativistic and classical effects is possible at present.

3. GRAVITATIONAL WAVES

Equation (1.21) is very similar to the Maxwell equations

$$\Box A^i = -4\pi j^i \tag{3.1}$$

One expects, therefore, that oscillating mass distributions emit gravitational waves, just as oscillating charge distributions emit electromagnetic radiation. The energy loss of an oscillating system of mass can be calculated from (1.21) to be

$$-\frac{dE}{dt} = \frac{G}{5c^5} \sum_{\alpha,\beta=1}^{3} \dddot{Q}_{\alpha\beta} \dddot{Q}_{\alpha\beta} \tag{3.2}$$

where

$$Q_{\alpha\beta} = \int d^3x (x_\alpha x_\beta - \frac{1}{3} \delta_{\alpha\beta} r^2) \rho \tag{3.3}$$

is the quadrupole moment of the mass distribution (dots denote time derivatives as usual).

The calculation leading to (3.2) is not very satisfactory for a number of reasons and many authors have attempted to improve it. Only recently a convincing theory of gravitational radiation has been given by Thorne (1970) and his group at Caltech.

The great difficulty of this work is partially due to the nonlinear nature of gravitation and the fact that for sources which are expected to be strong radiators of gravitational waves, the linear approximation is unapplicable. Because of the quadrupole nature of the radiation it is insufficient, furthermore, to study spherically symmetric mass distributions.

It turns out, however, that the results predicted by the customary formula (3.2) are in good agreement with the more accurate calculations. We shall therefore base our discussion of gravitational waves on (3.2).

Let us consider as an example the radiation emitted by the solar system, considering only earth and sun. Then $Q_{\alpha\beta}$ is of the order of magnitude

$$Q_{\alpha\beta} \approx \mu \cdot R^2 \tag{3.4}$$

where μ is the mass of earth and R the distance earth-sun. Inserting this into (3.2) we obtain

$$-\frac{dE}{dt} = +\frac{G}{5c^5}\mu^2 R^4 \cdot \omega^6 = +\frac{G}{5c^5}\mu^2 R^4 (2\pi/T)^6 \tag{3.5}$$

where $\omega = 2\pi/T$ is the angular frequency of the earth's motion around the sun. (3.5) can be rewritten as

$$-\frac{dE}{dt} = +\frac{G}{5c^5}\frac{\mu^2}{R^2}\left(\frac{2\pi R}{T}\right)^6 = \frac{G}{5c^5}\frac{\mu^2}{R^2} v^6 \tag{3.6}$$

where v is the mean velocity of the earth.

The virial theorem $v^2 \approx GM/R$ (M being the mass of the sun) permits us to simplify (3.6) further

$$-\frac{dE}{dt} = \frac{G}{5c^5}\frac{\mu^2}{R^2}\left(\frac{GM}{R}\right)^3 = \frac{c^5}{5G}\left(\frac{G\mu}{Rc^2}\right)^2\left(\frac{GM}{c^2R}\right)^3 =$$

$$= \frac{c^2}{5G}\left(\frac{m}{R}\right)^2\left(\frac{M}{R}\right)^3 \tag{3.7}$$

The factors in brackets are the dimensionless ratios of (Schwarzschild radius/earth-sun distance); $c^5/5G = 8\cdot 10^{58}$ erg/sec is therefore of the dimension of a power output, its mass equivalent being roughly $40000 M_\odot$/sec ($M_\odot$ = solar mass).

Inserting numerical values into (3.7) it turns out that the power radiated by the earth-sun system is negligible (10 Watt/sec).

Let us apply (3.7) to a more interesting object, e.g. a non-symmetrical neutron star. Then R is the radius of the star, M its Schwarzschild radius and $m \lesssim M$ the Schwarzschild radius of the asymmetric part of the mass distribution.

The energy loss becomes in this case $(m \sim M)$

$$- \frac{dE}{dt} \approx \frac{c^5}{5G} \left(\frac{M}{R}\right)^5 \qquad (3.8)$$

For a neutron star with $M \sim M_\odot$, $M/R \sim 0.1$ this becomes

$$- \frac{dE}{dt} = 10^{54} \text{ erg/sec} \hat{=} 0.4 M_\odot/\text{sec} \qquad (3.9)$$

Obviously the star can radiate at this rate only for a very short time. One expects therefore to observe a short burst of radiation. To estimate the duration and frequency of the radiation we have to understand the physical significance of the ratio M/R. Consider the formation of a star from a highly dispensed mass M_o of gas. During the condensation phase the part of the preassembly mass M_o which corresponds to the gravitational binding energy is radiated away. The final mass of the star is therefore given by

$$M = M_o - GM_o^2/Rc^2 \qquad (3.10)$$

where R is the radius of the star. The relative mass defect $\Delta M/M_o = (M_o - M)/M_o$ becomes

$$\frac{\Delta M}{M_o} = \frac{GM}{Rc^2} = \frac{M}{R} \qquad (3.11)$$

The ratio $\frac{M}{R}$ is therefore (approximately) equal to the relative mass defect of the star and $(\frac{M}{R})$ M is the maximum energy that can be radiated during the formation of a highly collapsed star, i.e. during gravitational collapse.

This energy is much larger than the nuclear energy in the case of highly condensed ($M \sim 0.1\,R$) stars (neutron stars). A neutron star with $M \sim M_{\odot}$ and $M=0.1\,R$ can radiate, therefore, about $0.1\,M_{\odot}$. If we assume for lack of better information that all the energy is emitted in the form of gravitational radiation we expect a burst of radiation lasting for about 1 sec (using (3.9)).

The frequency of the radiation can be estimated as follows: The radiation is due to a mass anisotropy, moving around the neutron star with a velocity v comparable to c . Since the radius R of the star is about 10 km it takes a light ray $c/2\pi R \sim (1/5000)$ sec to propagate around the star. Since v will be somewhat lower than c one expects a frequency of the gravitational wave in the range 500-5000 Hz.

J. Weber (1969) has constructed a gravitational wave detector operating at a frequency of 1660 Hz, which can observe gravitational waves with an intensity $\geq 10^4\,erg/cm^2 sec$. This intensity has to be contained, however, in the very narrow frequency band (width 0.01 Hz) to which the detector is sensitive. The total flux (integrated over the spectrum) can therefore be expected to be about 10^8-$10^9\,erg/cm^2 sec$ during any events observed by Weber. (Compare this to the solar constant $S=10^6\,erg/cm^2 sec$!) Two such detectors have been operating for more than a year at Argonne and Maryland respectively and on the average one burst of gravitational radiation (lasting less than 1 sec) has been observed per day.

Recent directional measurements indicate that this radiation originates in the center of our galaxy (Weber 1970). An energy current of $10^{54}\,erg/sec$ (as estimated in (3.9)) emanating there - at a distance of $2\cdot 10^{22}$cm from the solar system - leads to an energy flux $10^{54}/(4\pi)\,4\cdot 10^{44} \sim$ $\sim 10^8\,erg/cm^2$ in reasonably good agreement with the values deduced from Weber's experiment.

The duration of the observed bursts of gravitational radiation is consistent with the hypothesis that they are due to the collapse of a star leading to the formation of a neutron star. The fact, however, that approximately one event is detected per day is inconsistent with our present ideas on the formation of neutron stars. It is this fact which has led many physicists to doubt the validity of Weber's measurements. However, nobody has come up yet with any valid criticism of the experiment.

The numbers given here (0.1 $M_\odot$ radiated per event, 300 events per year) lead to an annual mass loss of the galaxy of about 30 $M_\odot$ in the form of gravitational radiation. This number is very large compared to the optical mass loss (0.1$M_\odot$ per year) of the galaxy.

The best explanation available presently is that these tremendous energies are due to gravitational collapse. In these lectures we shall, therefore, sutdy the rudiments of the theory of gravitational collapse. It will be impossible, unfortunately, to discuss realistic (asymmetric) models involving the generation of gravitational waves. We hope, however, that the theory of the spherically symmetric collapse presented here can serve as a first introduction to the strange world of high density objects.

4. THE USE OF EXTERIOR FORMS IN RIEMANNIAN GEOMETRY

Before we go into the physics of gravitational collapse we shall show how the methods presented in Professor Hlawka's lectures can be applied to actual calculations in Riemannian geometry. Let us first summarize the most important concepts introduced in these lectures in a very heuristic way, which shows, however, the geometrical concepts quite clearly.

Exterior forms have the general structure

$$\omega = \sum_H a_H(x^1,\ldots,x^n)\,dx^H \tag{4.1}$$

where

$$dx^H = dx^{h_1} \wedge dx^{h_2} \wedge \ldots \wedge dx^{h_r} \tag{4.2}$$

$$h_1 < h_2 \ldots < h_r \tag{4.3}$$

The sum over H is extended over all combinations of the h_r fulfilling (4.3).

The exterior derivative of ω is given by

$$d\omega = \sum_H da_H \wedge dx^H \tag{4.4}$$

$$d(d\omega) \equiv 0 \quad \text{(Poincaré Lemma)} \tag{4.5}$$

The tangent space $T^{(P)}$ in a point P of a manifold is an n-dimensional vector space with basis vectors

$$\vec{e}_i = \left.\frac{\partial}{\partial x^i}\right|_P \tag{4.6}$$

A vector $\vec{v} \in T(P)$ is given by

$$\vec{v} = v^i\,\vec{e}_i \tag{4.7}$$

$\vec{v}$ applied to a function f is (for reasons which will become clear below this is written as $\vec{v}(df)$)

$$\vec{v}(df) = v^i \frac{\partial f}{\partial x^i} \tag{4.8}$$

The components of $\vec{v}$ are given by

$$v^k = \vec{v}(x^k) \tag{4.9}$$

The cotangent space $T^*(P)$ is the dual space to $T(P)$. A basis in $T^*(P)$ is

$$\vec{e}^i = dx^i \tag{4.10}$$

The duality of T^* and T is expressed by

$$\vec{e}_i(\vec{e}^j) = \delta_i{}^j \tag{4.11}$$

Connections define isomorphic mapping between the vector spaces in neighboring points (Fig. 3)

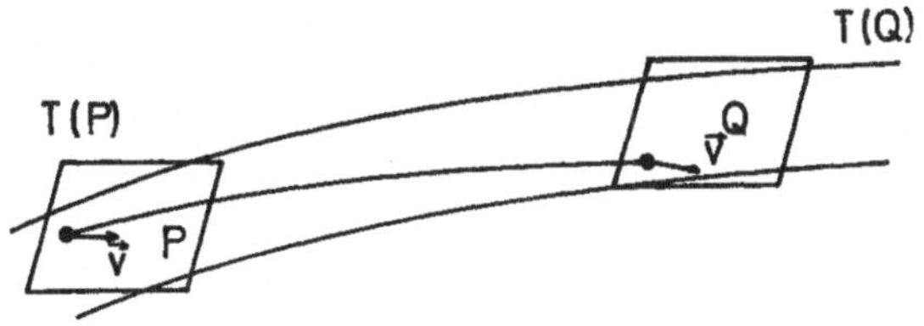

Figure 3 - Connections

When one compares two vectors at different points P,Q of an Euclidean space, one transports one vector parallely from P to Q and compares the vectors there; i.e. one maps the vector space attached to P in some way onto the vector space attached to Q. If one takes two neighboring points P (coordinates x^k) and $Q(x^k+dx^k)$ of a manifold the basis vectors $\vec{e}_i$ of T(P) will in general not be mapped into the basis vectors at Q but differ from these by $d\,\vec{e}_i$ (Fig. 4)

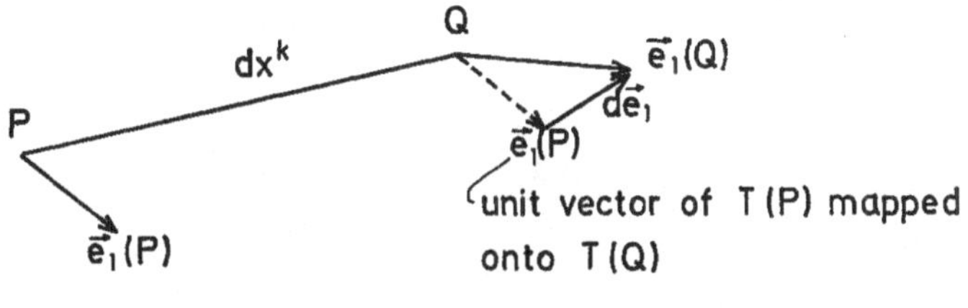

Figure 4 - Mapping of unit vectors P → Q .

Since the $\vec{e}_j$ form a basis of the vector space T(Q), $d\,\vec{e}_i$ can be expressed in the general form

$$d\,\vec{e}_i = \omega^j{}_i\,\vec{e}_j \qquad (4.12)$$

where $\omega^j{}_i$ is a set of differential coefficients expressing the mutual relations of unit vectors at different points (these forms have been denoted by e^{ji} in Prof. Hlawka's lectures).

In a Finsler space the $\omega^j{}_i$ are homogeneous of first degree in the dx^n, in an affine space the $\omega^j{}_i$ are differential forms linear in dx^k:

$$\omega^j{}_i = L^j{}_{ik}\,dx^k \qquad (4.13)$$

The functions $L^j{}_{ik}$ are the coefficients of the affine connection.

The differential of an arbitrary vector $\vec{v}$ can be expressed in the form

$$d\vec{v} = d(v^i\vec{e}_i) = dv^i\cdot\vec{e}_i + v^i d\vec{e}_i = (dv^j + L^j{}_{km}v^k dx^m)\vec{e}_j =$$

$$=: (Dv^j)\vec{e}_j =: v^j{}_{;k}\,dx^k\,\vec{e}_j \qquad (4.14)$$

In (4.14) the absolute differential Dv^j of the vector components and the covariant derivative

$$v^j{}_{;k} = v^j{}_{,k} + L^j{}_{mk}\,v^m \qquad (4.15)$$

have been defined.

The distance $d\vec{P}$ between two neighboring points on a manifold can be written heuristically as

$$d\vec{P} = dx^i\,\vec{e}_i \qquad (4.16)$$

since P and P+dP lie both in the tangent space of P. The condition

$$dd\vec{P} = 0 \qquad (4.17)$$

is not identity (since $d\vec{P}$ is no an exact differential of a quantity "$\vec{P}$") but expresses the fact that the manifold considered is torsion free. Writing (4.17) in the form

$$dd\vec{P} = \Omega^i e_i = d(dx^i \vec{e}_i) = dx^i d\vec{e}_i = dx^i \wedge \omega^j{}_i \vec{e}_j =$$

$$= dx^i \wedge L^j{}_{ik}\, dx^k \vec{e}_j = \tfrac{1}{2}(L^j{}_{ik} - L^j{}_{ki})\, dx^i \wedge dx^k \cdot \vec{e}_j =$$

$$=: T^j{}_{ik}\, dx^i \wedge dx^k\, \vec{e}_j \qquad (4.18)$$

introduces the torsion forms Ω^i and the torsion tensor $T^j{}_{ik}$ of the manifold. Heuristically speaking, $\Omega^i = 0$ means that one returns to the same point $\vec{P}$ after completing an infinitesimal "circle" on the manifold. (Manifolds with torsions can be visualized as crystals with screw type dislocations.) $d\vec{e}^i$ is no exact differential either, since the result of the parallel displacement will depend, in general, on the infinitesimal path taken from P to Q. (Fig. 5)

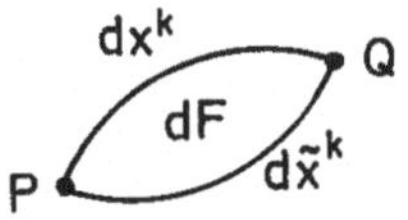

Figure 5 - Two different paths leading from P to Q.

The difference between the resulting $d\vec{e}_k$ will be proportional to the area dF between the two paths, i.e. of second order in dx^k. This is expressed by the fact that

$$dd\vec{e}_i \neq 0$$

We put

$$dd\vec{e}_i =: \Omega^j{}_i\, \vec{e}_j \qquad (4.19)$$

where $\Omega^j{}_i$ is the curvature form of the manifold.

A Riemannian manifold is a torsion free manifold on which a symmetric covariant tensor g of rank two is defined, i.e.

$$g(\vec{e}_i, \vec{e}_k) = g_{ik}(x) = g_{ki}(x) \qquad (4.20)$$

g_{ik} is the metric tensor. g_{ik} can be used to define a scalar product in the tangent space T

$$(\vec{e}_i,\vec{e}_k) = g_{ik}(x) \tag{4.21}$$

The distance between two points on the manifold is then given by

$$ds^2 = (d\vec{P},d\vec{P}) = dx^i\, dx^k \cdot g_{ik} \tag{4.22}$$

The connection coefficients $L^i{}_{jk} = \Gamma_{jk}^{\ i}$ are determined in a Riemannian space by the postulate that the space is torsion free (which implies that $\Gamma_{jk}^{\ i} = \Gamma_{kj}^{\ i}$; see (4.18)) and by the postulate that (4.21) is preserved during parallel transport (i.e. the angle between two vectors remains unchanged). Differentiating (4.21) gives

$$(d\vec{e}_i,\vec{e}_k) + (\vec{e}_i,d\vec{e}_k) = d\, g_{ik} \tag{4.23}$$

Inserting

$$d\vec{e}_i = \omega^j{}_i\, \vec{e}_j = \Gamma_{ik}^{\ j}\, dx^k\, \vec{e}_j$$

we obtain

$$\omega_{ik} + \omega_{ki} = d\, g_{ik} \tag{4.24}$$

with

$$\omega_{ik} = g_{i\ell}\, \omega^\ell{}_k \tag{4.25}$$

or

$$\Gamma_{ik}^{\ j}\, dx^k\, g_{jk} + \Gamma_{k\ell}^{\ j}\, dx^\ell\, g_{ji} = d\, g_{ik} \tag{4.26}$$

which is equivalent with the standard definition (1.4) of the Christoffel symbols.

Equation (4.21) shows that the natural basis vectors $\vec{e}_i = \partial/\partial x^i$ are not orthonormal since one cannot generally find coordinate lines which are orthonormal on the whole manifold.

Most calculations are simplified, however, when an orthonormal basis is used in tangent space, i.e. a basis $\vec{e}_i$ with

$$(\vec{e}_i,\vec{e}_j) = \eta_{ij} \tag{4.27}$$

The $\vec{e}_i$ can be written as a linear combination of arbitrary natural basis vectors $\vec{e}_{\bar{i}}$ as

$$\vec{e}_i = h_i^{\ \bar{i}}\, \vec{e}_{\bar{i}} \tag{4.28}$$

The coefficients $h_i^{\ \bar{i}}$ are called vierbein-components. The dual basis in T^* is then given by

$$\vec{e}^{\,i} = \omega^i = h^i_{\ \bar{i}}\, dx^{\bar{i}} \tag{4.29}$$

All the calculations performed before are independent of the base chosen in T.

In a general (it need neither be natural nor be orthonormal) basis $\vec{e}_i$ and $\vec{e}^{\,i} = \omega^i$, respectively, the relevant equations become

$$d\vec{P} = \omega^i\, \vec{e}_i \tag{4.16a}$$

and $dd\vec{P} =: (D\omega^i)\vec{e}_i = \Omega^i\, \vec{e}_i = 0$ (this defines the absolute differential $D\omega^i$ of ω^i). The torsion form Ω^i is now given by

$$\Omega^i = D\omega^i = d\omega^i + \omega^i_{\ k} \wedge \omega^k = 0 \tag{4.30}$$

Inserting $\omega^i_{\ k} = L^i_{\ k\ell}\omega^\ell$ this becomes (4.31)

$$d\omega^i + L^i_{\ k\ell}\omega^\ell \wedge \omega^k = 0 \tag{4.32}$$

Because of $d\omega^i \neq 0$ (except for a natural basis) we obtain $L^i_{\ k\ell} \neq L^i_{\ \ell k}$, i.e. the connection coefficients $L^i_{\ \ell k}$ are symmetric only when referred to a natural basis, in which case they become the Christoffel symbols. When we specialize to an orthonormal base (4.27) becomes

$$\omega_{ij} + \omega_{ji} = d\,\eta_{ij} = 0 \tag{4.33}$$

i.e. the forms ω_{ij} are anti-symmetric. In this case the $L^i_{\ jk}$ are called the Ricci-rotation coefficients (the most general relation between two orthonormal bases at different points is a rotation).

Finally we have to find an explicit expression for the curvature form $\Omega^i{}_j$ in a general base. We have

$$d(d\vec{e}_i) = \Omega^j{}_i\, \vec{e}_j = d(\vec{e}_j \omega^j{}_i) = d\vec{e}_j \wedge \omega^j{}_i + \vec{e}_j d\omega^j{}_i =$$

$$= \omega^k{}_j \wedge \omega^j{}_i\, \vec{e}_k + \vec{e}_j\, d\omega^j{}_i \tag{4.34}$$

or

$$\Omega^j{}_i = d\omega^j{}_i + \omega^j{}_\ell \wedge \omega^\ell{}_i \tag{4.35}$$

We define the curvature tensor by

$$\Omega^j{}_i = \frac{1}{2} R^j{}_{ik\ell}\, \omega^k \wedge \omega^\ell \tag{4.36}$$

By specializing to a natural base $\omega^k = dx^k$ one finds immediately that $R^j{}_{ik\ell}$ agrees with the curvature tensor defined in (1.5):

$$\Omega^j{}_i = d(\Gamma^j_{ik}\, dx^k) + \Gamma^j_{\ell m}\, \Gamma^\ell_{ik}\, dx^m \wedge dx^k =$$

$$= (\Gamma^j_{ik,m} + \Gamma^j_{\ell m}\, \Gamma^\ell_{ik}) dx^m \wedge dx^k = \tag{4.37}$$

$$= \frac{1}{2}(\Gamma^j_{ik,m} - \Gamma^j_{im,k} + \Gamma^j_{\ell m}\, \Gamma^\ell_{ik} - \Gamma^j_{\ell k}\, \Gamma^\ell_{im}) dx^m \wedge dx^k$$

Collecting our results we obtain the fundamental equations of Riemannian geometry in an arbitrary base:

$$d\vec{P} = \vec{e}_i \omega^i \qquad \omega_{ij} = g_{ik}\, \omega^k{}_j$$

$$(\vec{e}_i, \vec{e}_j) = g_{ij} \qquad \omega_{ij} + \omega_{ji} = d\, g_{ij}$$

$$ds^2 = (d\vec{P}, d\vec{P}) = g_{ij} \omega^i \otimes \omega^j \qquad d\omega^i + \omega^i{}_j \wedge \omega^j = 0$$

$$d\vec{e}_i = \omega^j{}_i \vec{e}^j \qquad \Omega^i{}_j = d\omega^i{}_j + \omega^i{}_k \wedge \omega^k{}_j$$

$$\Omega^i{}_j = \frac{1}{2} R^i{}_{jk\ell}\, \omega^k \wedge \omega^\ell \tag{4.38}$$

5. APPLICATION: SPHERICALLY SYMMETRIC SOLUTIONS OF THE EINSTEIN FIELD EQUATIONS

The preceding equations (4.38) can be applied to the explicit calculation of curvature tensors and turn out to be superior to the classical techniques. As an example let us apply these equations to a metric of the form

$$ds^2 = e^{2a}dt^2 - e^{2b}dr^2 - r^2\underbrace{(d\Theta^2 + \sin^2\Theta\, d\phi^2)}_{\equiv\, d\Omega^2} \tag{5.1}$$

where

$$a = a(r,t) \quad , \quad b = b(r,t) \tag{5.2}$$

It can be shown that every spherically symmetric line element can be written in this form, (5.1) will therefore be the metric to be used in the theory of spherically symmetric gravitational collapse.

We can introduce an orthonormal base in cotangent space by putting

$$\begin{aligned} \omega^0 &= e^a\, dt \\ \omega^1 &= e^b\, dr \\ \omega^2 &= r\, d\Theta \\ \omega^3 &= r\, \sin\Theta\, d\phi \end{aligned} \tag{5.3}$$

Denoting derivatives with respect to t and r by a dot and prime respectively we obtain

$$\begin{aligned} d\omega^0 &= e^a\, a'\, dr \wedge dt \\ d\omega^1 &= e^b\, \dot{b}\, dt \wedge dr \\ d\omega^2 &= dr \wedge d\Theta \\ d\omega^3 &= \sin\Theta dr \wedge d\phi + r\cos\Theta d\Theta \wedge d\phi \end{aligned} \tag{5.4}$$

Comparing this with

$$d\omega^i + \omega^i{}_j \wedge \omega^j = 0 \tag{5.5}$$

and taking into account

$$\omega_{ij} + \omega_{ji} = 0 \tag{5.6}$$

one obtains immediately

$$\omega^{o}_{1} \equiv e^{a-b}\, a'\, dt + e^{b-a}\, \dot{b}\, dr = \omega^{1}_{o}$$

$$\omega^{2}_{1} = e^{-b}\, d\Theta = -\, \omega^{1}_{2}$$

$$\omega^{3}_{1} = e^{-b}\, \sin\Theta\, d\phi = -\, \omega^{1}_{3}$$

$$\omega^{3}_{2} = \cos\Theta\, d\phi = -\, \omega^{2}_{3} \tag{5.7}$$

All other ω^{i}_{k} vanish. The exterior derivatives $d\omega^{i}_{k}$ become

$$d\omega^{o}_{1} = [-e^{-2b}(a'^{2}-a'b'+a'')+e^{-2a}(-\dot{a}\dot{b}+\dot{b}^{2}+\ddot{b})]\,\omega^{o}\wedge\omega^{1} \equiv$$

$$\equiv B\omega^{o}\wedge\omega^{1}$$

$$d\omega^{2}_{1} = -e^{-b}\,\dot{b}\; dt\wedge d\Theta - e^{-b}\, b'\; dr\wedge d\Theta$$

$$d\omega^{3}_{2} = -\sin\Theta\; d\Theta\wedge d\phi \tag{5.8}$$

We were able to omit $d\omega^{3}_{1}$ from this list, because of symmetry arguments. The 2 and 3 directions are the Θ and ϕ directions on the sphere. These directions are completely equivalent geometrically and the curvature tensor has to be symmetric when the 2 and 3 directions are interchanged. Note, however, that this argument is valid only in an orthonormal base. In the natural base $(d\Theta, d\phi)$ the basis vectors in these directions are of unequal length and the symmetry is destroyed. The fact that symmetries are preserved by the use of orthonormal basis vectors is an additional advantage of the method used here.

For the curvature form Ω^{i}_{j} we obtain:

$$\Omega^{o}_{1} = B\omega^{o}\wedge\omega^{1}$$

thus

$$R^{o}{}_{101} = B = - R^{o}{}_{110}$$

all other

$$R^{o}{}_{1ik} = 0$$

Furthermore

$$\Omega^{o}{}_{2} = \omega^{o}{}_{1} \wedge \omega^{1}{}_{2} = -(e^{-b}\, a'\, \omega^{o} + e^{-a}\, \dot{b}\, \omega^{1}) \wedge e^{-b}\, \omega^{2}/r \tag{5.11}$$

$$R^{o}{}_{202} = -e^{-2b}\, a'/r = R^{o}{}_{303}$$

$$R^{o}{}_{212} = -e^{-a-b}\, \dot{b}/r = R^{o}{}_{313}$$

$$\Omega^{1}{}_{2} = e^{-a-b}\, \dot{b}\, \omega^{o} \wedge \omega^{2}/r + e^{-2b}\, b'\, \omega^{1} \wedge \omega^{2}/r$$

$$R^{1}{}_{202} = e^{-a-b}\, \dot{b}/r = R^{1}{}_{303}$$

$$R^{1}{}_{212} = e^{-2b}\, b'/r = R^{1}{}_{313}$$

$$\Omega^{2}{}_{3} = \omega^{2} \wedge \omega^{3}/r^2 - e^{-b}\, \omega^{2} \wedge e^{-b}\, \omega^{3}/r^2$$

$$R^{2}{}_{323} = \frac{1}{r^2}(1-e^{-2b}) \tag{5.12}$$

This completes the list of nonvanishing $R^{i}{}_{k\ell m}$. The components of the Ricci tensor and of $G_{ik} = R_{ik} - \frac{1}{2}\eta_{ik} R$ can easily be obtained from these equations:

$$G_{oo} = -\frac{1}{r^2} - e^{-2b}\left(\frac{2b'}{r} - \frac{1}{r^2}\right) \tag{5.13}$$

$$G_{11} = \frac{1}{r^2} - e^{-2b}\left(\frac{2a'}{r} + \frac{1}{r^2}\right) \tag{5.14}$$

$$G_{22} = G_{33} = -e^{-2b}\left[a''+a'^2+(a'-b')/r - a'b'\right] + e^{-2a}\left(\ddot{b} + \dot{b}^2 - \dot{b}\dot{a}\right) \tag{5.15}$$

$$G_{o1} = -2\dot{b}\, e^{-a-b}/r \tag{5.16}$$

These components of G_{ik} refer to the orthonormal base ω^{i}, the standard components $G_{\bar{i}\bar{k}}$ referring to dx^{i} can be calculated from G_{ik} with the help of

$$G_{ik}\ \omega^{i} \otimes \omega^{k} = G_{\bar{i}\bar{k}}\ dx^{\bar{i}} \otimes dx^{\bar{k}} \tag{5.17}$$

GRAVITATIONAL COLLAPSE

6. THE TOLMAN-OPPPENHEIMER-VOLKOFF EOUATION

Before we can enter into a discussion of the dynamics of gravitational collapse we have to study the physics of static mass configuration of very high density.

In this case all time derivatives can be omitted in (5.13) - (5.16) and the energy-momentum tensor of matter becomes in an orthonormal frame

$$T_{ik} = \begin{pmatrix} \rho & & & \\ & p & & \\ & & p & \\ & & & p \end{pmatrix} \tag{6.1}$$

where ρ is the density and p the pressure within the mass distribution. Inserting (6.1), (5.13) and (5.14) into the Einstein equations we obtain

$$e^{-2b}\left(\frac{1}{r^2} - \frac{2b'}{r}\right) - \frac{1}{r^2} = \kappa\rho \tag{6.2}$$

$$e^{-2b}\left(\frac{1}{r^2} - \frac{2a'}{r}\right) - \frac{1}{r^2} = \kappa p \tag{6.3}$$

These two equations are obviously sufficient to determine a and b; their solution is

$$ds^2 = e^{2a}\ dt^2 - d\sigma^2 \tag{6.4}$$

$$d\sigma^2 = (1-2m(r)/r)^{-1}\ dr^2 + r^2\ d\Omega^2 \tag{6.5}$$

$$2m(r) = \kappa\int_0^r \rho\ r^2\ dr \tag{6.6}$$

$$a = -b + \int_\infty^r dr\ e^{2b}\ r(p+q).4\pi G \tag{6.7}$$

The geometry of the spacial part ($d\sigma^2$) of the line element is therefore determined by ρ only, p does not enter.

These equations determine the metric completely if p and ρ are known as functions of r.

We need, therefore, two additional equations. One is given by the equation of state

$$p = p(\rho) \tag{6.8}$$

The derivation of a reliable equation of state at high densities is a problem which is still unsolved, we shall come back to it later. The second equation which we need can be derived from (5.15) (we have not used the angular part of the Einstein equations yet) or simpler from

$$T_{l\ ;k}^{\ k} = 0 \tag{6.9}$$

Inserting (6.1) and (6.4) - (6.7) into (6.9) we obtain the Tolman-Oppenheimer-Volkoff (TOV) equation

$$-\frac{dp}{dr} = \frac{(p+\rho)\,[\,m(r)+4\pi Gpr^3]}{r[r-2m(r)]} \tag{6.10}$$

This is the relativistic generalization of the standard equation for hydrostatic equilibrium, which can be obtained from (6.10) by omitting the p-terms in the numerator and the 2m(r) term in the denominator. (6.10) has to be solved subject to the boundary condition that p(R)=0 at the boundary of the star (r=R).

Note that the pressure occurs twice at the r.h.s. of (6.10), this leads to the relativistic self-generation of pressure. At r=R the line element (6.4) - (6.7) can be joined smoothly to the exterior Schwarzschild solution (1.15), the mass M being given by

$$M = 4\pi\int_0^R dr\ r^2\rho \tag{6.11}$$

This equation for M looks like the standard definition of a mass in terms of density; it has, however, a number of remarkable properties:

a) M is positive if ρ is positive. If one compares this to the Newtonian equation (3.10) for the mass one sees that this fact is not trivial. According to (3.10) the mass of a highly collapsed object has no lower limit i.e. no ground-state exists since the gravitational self-energy can overcompensate all other (positive) energies.

b) In (6.11) the gravitational self-energy appears to be missing and no distinction seems to exist between the pre-assembly mass M_o and the actual mass of the body. The error in this argument is that $4\pi r^2 dr$ is not the volume element dV of curved space; dV follows from (6.5) to be

$$dV = 4\pi r^2\, dr\left(1-\frac{2m(r)}{r}\right)^{-1/2} \tag{6.12}$$

We can write (6.11) therefore in the form

$$M = \int dV\, \rho\left(1-\frac{2m(r)}{r}\right)^{1/2} \tag{6.13}$$

The fact that the integrand

$$\rho\left(1-\frac{2m(r)}{r}\right)^{1/2}$$

is smaller than ρ takes into account the effects of the gravitational self-energy, as one can show in detail. The pre-assembly mass M_o (which equals the baryon number A of the star if the unit of mass is suitably chosen) is given by

$$M_o = A = \int dV\, n(r) \tag{6.14}$$

where n(r) is the baryon density in the star. The connection between n, p and ρ is given by

$$p = -\frac{d(\text{Energy/Baryon})}{d(\text{Volume/Baryon})} = \frac{d(\rho/n)}{d(1/n)} = n\frac{d\rho}{dn} - \rho \tag{6.15}$$

or

$$p + \rho = n \frac{d\rho}{dn} \tag{6.16}$$

If the equation of state $p(\rho)$ is known, $n(\rho)$ can be determined from

$$\int \frac{d\rho}{\rho + p} = \int \frac{dn}{n} = \log n \tag{6.17}$$

7. INCOMPRESSIBLE MATTER

The only case in which the TOV-equation can be integrated analytically is for incompressible matter

$$T^o{}_o = \rho = \text{const.} \tag{7.1}$$

In this case $m(r)$ becomes

$$m = \frac{4\pi}{3} G \rho r^3 \tag{7.2}$$

Inserting this into (6.10) we obtain for the pressure p

$$p = \rho \frac{\sqrt{1-x^2} - \sqrt{1-X^2}}{3\sqrt{1-X^2} - \sqrt{1-x^2}} \tag{7.3}$$

where x is a dimensionless radial variable defined by

$$x = (8\pi \ G\rho/3)^{1/2} r \tag{7.4}$$

and X is the value of x at the surface of the star ($r=R$). For $X<<1$ (7.3) agrees with the ordinary nonrelativistic pressure distribution in a star. For $X=\sqrt{8/9}$ the pressure becomes infinite at $x=0$, i.e. all matter is crushed and collapse sets in. Stable mass configurations can therefore exist only for $X < \sqrt{8/9}$.

From (7.2) we have that the Schwarzschild radius 2M of the star is given by

$$2M = 8\pi \ G\rho/3 \cdot R^3 \tag{7.5}$$

X^2 can thus be written as

$$X^2 = 2M/R = \frac{\text{Schwarzschildradius}}{\text{radius}} \quad (7.6)$$

No star with a radius $R < 9/8(2M)$ can therefore be stable.

For incompressible matter the baryon density n agrees with the mass density, $n=\rho=\text{const}$. The pre-assembly mass of the star is thus (dV is given in (6.12))

$$A = n\int dV = \frac{4\pi\rho}{3} R^3 f(X) = Mf(X) \quad (7.7)$$

where

$$f(X) = \frac{3}{2}(\arcsin X - X\sqrt{1-X^2})/X^3 \quad (7.8)$$

$$f(X) = \begin{cases} 1+\frac{3}{10}X^2 & X<<1 \\ 3\pi/4 & X=1 \\ 1.374 & X=\sqrt{8/9} \end{cases} \quad (7.9)$$

The mass defect of the star becomes

$$\frac{\delta M}{A} = \frac{A-M}{A} = \frac{f(X)-1}{f(X)} \quad (7.10)$$

At the limit of stability ($X=\sqrt{8/9}$) we obtain $\delta M/A \approx 30\%$, i.e. 30% of the pre-assembly mass of the star is radiated during the evolution of the star leading to this configuration.

The constant X, which characterizes the degree of compaction of the star can be determined in a rather direct way by observing the red shift of light emanating from the stellar surface. The ratio of wave length λ at emission to the one observed at infinity, λ_∞, is given by

$$\frac{\lambda_\infty}{\lambda} \equiv 1+z = \frac{dt}{ds} = (1-\frac{2M}{R})^{-1/2} = (1-X^2)^{-1/2} \quad (7.11)$$

For $X=\sqrt{8/9}$ we obtain $z=2$, which is therefore the maximum redshift which can be explained by gravitational effects. Note that for $X=1$, $z=\infty$, i.e. light emanating from a source with $2M=R$ will be redshifted completely.

8. THE GEOMETRY OF THE SCHWARZSCHILD SOLUTION

In order to visualize the geometry of the Schwarzschild solution we have to restrict ourselves to a 2-dimensional surface which can then be embedded in 3-space.

We shall study here the geometry on a plane through the center of the star, i.e. $\Theta=\pi/2$, t=const., $d\Theta=dt=0$. The line element on this surface is given by (see (6.5))

$$d\sigma^2 = \left(1-\frac{2M}{r}\right)^{-1} dr^2 + r^2\, d\phi^2 \tag{8.1}$$

for r>R, i.e. in the exterior of the star.
It is easy to show that a parabola (z is an embedding coordinate)

$$z = \sqrt{8M}\,\sqrt{r-2M} \tag{8.2}$$

rotated around the axis r=0. (Fig. 6) gives a surface of the same intrinsic geometry as (8.1)

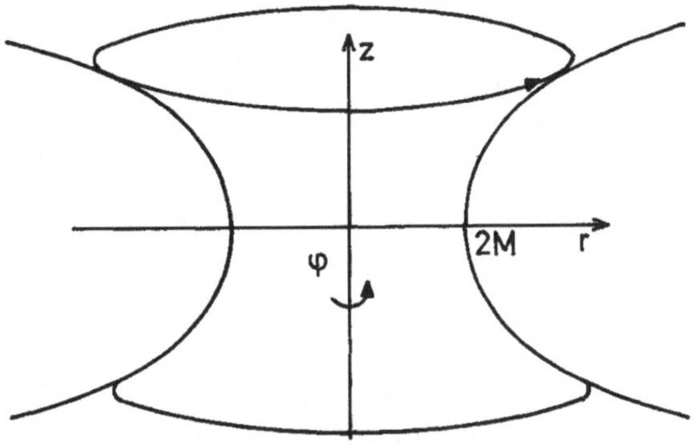

Figure 6 - Surface of revolution corresponding to Schwarzschild metric

In the interior of an incompressible mass distribution $d\sigma^2$ can be derived from (6.5) and (6.6) to be

$$d\sigma^2 = \frac{dr^2}{1-\frac{8\pi\, G\rho}{3}\, r^2} + r^2\, d\Omega^2 \tag{8.3}$$

This is the geometry on a space of constant curvature, i. e. of a 3-dimensional hypersphere with radius

$$a = \left(\frac{3}{8\pi\ G\rho}\right)^{1/2} \qquad (8.4)$$

(7.4) can therefore be written as x=r/a , X=R/a . The mass distribution on the hypersphere extends only to the maximum radius R as shown in Fig. 7

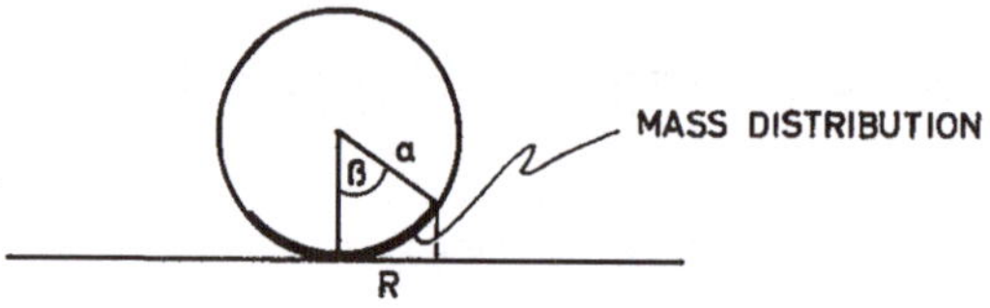

Figure 7 - The interior geometry

One can show that the hypersphere joins the exterior metric smoothly in r=R. (Fig. 8)

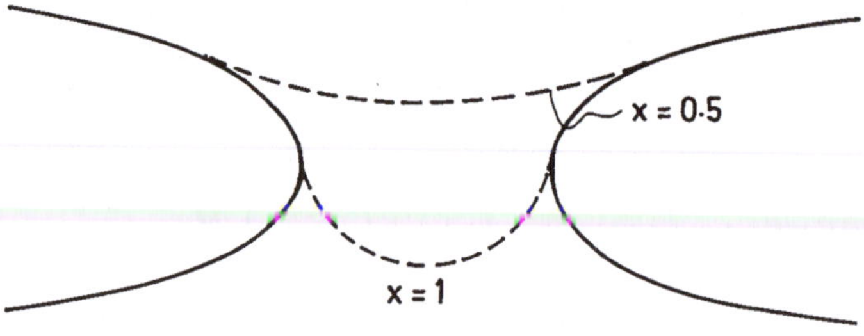

Figure 8 - The complete geometry of the Schwarzschild metric

The geometry on a plane through the center of a star with X=0.5 is therefore the same as on the surface shown above.

The figures show that the characterization of a star in terms of X is insufficient, since there are two mass distributions having the same value of X as shown in Fig. 9

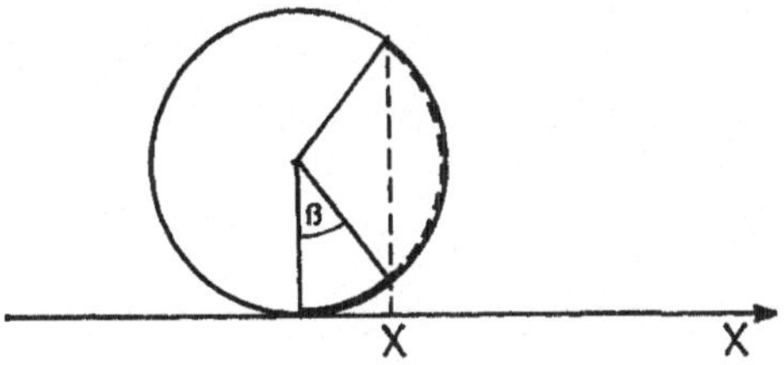

Figure 9 - Characterization of the mass distribution by β

A unique characterization can be achieved by putting

$$r = a \sin \beta \tag{8.5}$$

(7.5) and (7.7) are then rewritten in the form

$$M = \frac{a}{2} \sin^3\beta \qquad A = Mf(\beta)$$

$$f(\beta) = \frac{3}{2}(\beta - \sin\beta \cos\beta)/\sin^3\beta \tag{8.6}$$

In Fig. 10 we plot the mass defect $\frac{\delta M}{A} = \frac{A-M}{A}$ as a function of β

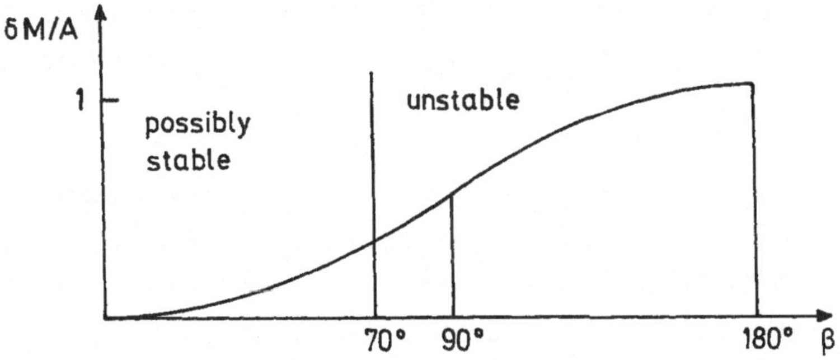

Figure 10 - Mass defect as a function of β

The figure shows that a star with $\beta=180^\circ$ has radiated all its mass during the collapse leading to this configuration.

In chapter 11 we shall show, however, that this conclusion is incorrect, because the quasistatic approximation used here is not applicable for $\beta > 70^{\circ}$.

9. THE EQUATION OF STATE AT HIGH PRESSURES

The only stability criterion used in the previous chapter was that the pressure has to be finite in the center of the star, which is a necessary, but not a sufficient condition for stability.

In this chapter we shall take into account furthermore, that the mass distribution has to be stable with respect to radial oscillations. All configurations with slightly lower or higher density (but equal baryon number A) have to have larger masses than the stable configuration.

In order to avoid the complication due to nuclear reactions we shall use as a starting point a mass consisting of (A/56) atoms of Fe^{56}, which is the endpoint of the thermonuclear evolution (it has the highest possible binding energy per nucleon). According to the remark made before the unit of mass has to be chosen therefore to be 1/56 of the mass of Fe^{56}.

Initially these atoms are widely dispersed; slowly they begin to contract (quasistatically) under the influence of their mutual gravitational attraction. We shall assume that the energy gained (gravitational binding energy) is radiated away in each instant, so that the contraction can continue quasistatically.

When the equilibrium density ρ_0 is reached the contraction stops. To compress the material further external forces would have to be applied, supplying energy to the system and thus increasing its mass. The shape of the mass density relation for the object considered is thus as shown

in Fig. 11

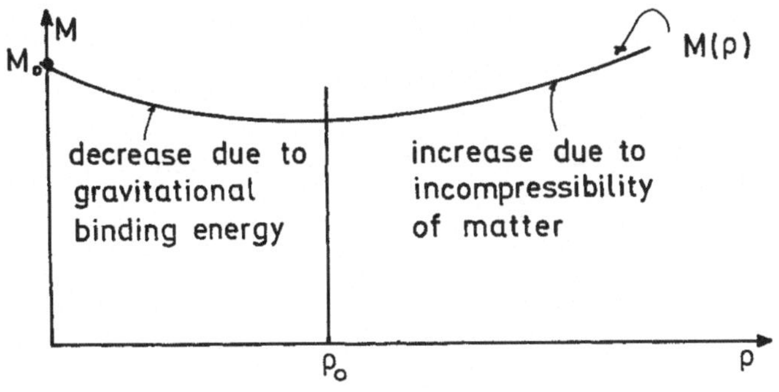

Figure 11 - Mass as a function of density

For A = 10^{50} (corresponding to the mass of earth) the equilibrium density is $\rho_o \sim 7 g/cm^3$ (density of iron), for $A=10^{57}$ (mass of the sun $\rho_o \approx 10^{15} g/cm^3$ (nuclear densities in a neutron star).

The $M(\rho)$ curve shown in Fig. 11 can be calculated from equation (8.6). Since compressional effects are important, it is clearly inadmissible to use the approximation $\rho \approx n$. To keep the mathematics as simple as possible we shall approximate energy and nucleon densities (ρ and n resp.) by constants within the star, but drop the assumption $\rho = n$.

The relevant equations are then

$$A = \int dV.n \approx n \int dV = 4\pi n\ R^3/3 \cdot f(\beta) \qquad (9.1)$$

$$M = 4\pi \int \rho r^2 dr = 4\pi\rho\ \ R^3/3 \qquad (9.2)$$

$$\beta = \text{arc sin } \left(\frac{3R^2}{8\pi G\rho}\right)^{1/2} \qquad (9.3)$$

where $f(\beta)$ is defined by (8.6). The function $M(\rho)$ can be calculated from (9.1) - (9.3), when $\rho(n)$ is known. The determination of the equation of state is therefore the central problem to be solved. There are two statements about $\rho(n)$ which can be made from general principles. Microstability demands that $p \geq 0$, i.e.

$$p = \rho\left(\frac{d\,\log\rho}{d\,\log n} - 1\right) > 0 \qquad (9.4)$$

and the condition that the velocity of sound v, has to be smaller than c(=1) implies

$$v^2 = dp/d\rho < 1 \qquad (9.5)$$

Assuming a simple power law

$$\rho = a \cdot n^{\gamma} \qquad (9.6)$$

we obtain the restrictions $1 \leq \gamma \leq 2$ from these two conditions.

At low densities ($\rho \lesssim 10^{15}$ g/cm^3=nuclear density) $\rho \sim n$ is a good approximation to the equation of state. At higher densities two possibilities arise:

a) Matter could be highly compressible at high densities, due to the fact that the Fermi sphere can be filled repeatedly with n, Λ, Σ... and other elementary particles. This hypothesis would imply $\gamma \gtrsim 1$.

b) If quarks exist the Fermi sphere can be filled only 3 times and matter behaves like a gas of neutrinos at high densities, when the Fermi-momentum is large compared to all rest masses. This implies $\gamma = 4/3$.

A detailed calculation of $\rho(n)$ has been performed by Harrison and Wheeler (see e.g. Harrison, Thorne, Wakano, Wheeler 1965). At very high densities these authors use an asymptotic law with $\gamma=4/3$. Fig. 12 shows schematically the $M(\rho)$ relation for three different values of A, calculated with the help of the Harrison-Wheeler equation of state. The figure shows that mass distributions with $A=5 \cdot 10^{55}$ nucleons and $7 \cdot 10^{56}$ reach equilibrium at densities $\sim 10^4$ g/cm^3 and $\sim 10^6$ g/cm^3, resp., while a mass with $A=2 \cdot 10^{57}$ collapses

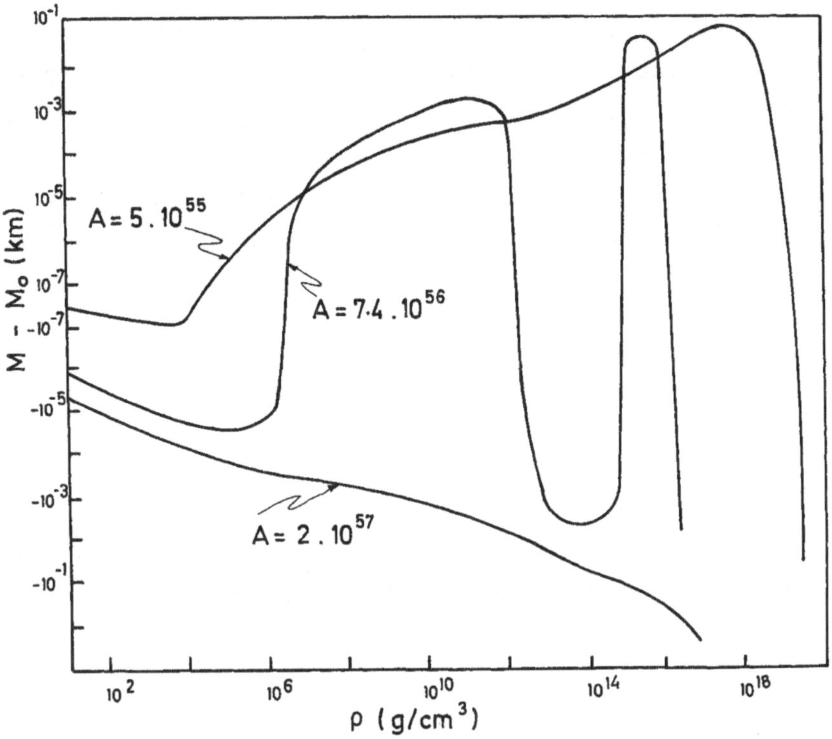

Figure 12 - Stability of various masses

indefinitely. (For the sun $A{\sim}1.2{\times}10^{57}$.) The collapse of smaller masses can be induced by compressing the mass to a sufficiently high density. In principle even the collapse of a bowling ball can be initiated by compressing it very hard.

The limit of stability $A{\sim}10^{57}$ can be explained roughly in the following way: If a collapsing mass configuration reaches nuclear densities before its radius R comes close to the Schwarzschild radius 2M, collapse will stop at this point (if it has not stopped at an even earlier state, as is the case for small values of A. An example for this is $A=5\cdot10^{55}$ where an equilibrium state is reached for $\rho=10^4 g/cm^3$). If the nuclear densities and the extremely strong repulsive forces which occur at these densities are reached only for $R<2M$ collapse cannot be stopped and will proceed indefinitely.

The critical value of A is thus the one for which nuclear densities are reached when the mass is compressed to a radius $R=2M$. Since the Schwarzschild radius of a body consisting of A nucleons (mass m) is given by

$$2M \sim A\, m\, G \tag{9.7}$$

and its radius R by

$$R \sim d \cdot A^{1/3} \tag{9.8}$$

where d is the average distance between particles. At nuclear densities $d \sim 1/m$ = Compton wave length of the nucleon $\sim 10^{-13}$cm. The critical value of A follows then from

$$R \simeq \frac{1}{m} A^{1/3} \;, \quad 2M \gtrsim A\, m\, G \tag{9.9}$$

or

$$A = \left(\frac{1}{m^2 G}\right)^{3/2} \sim 10^{58} \tag{9.10}$$

The most interesting question is therefore the ultimate fate of masses with $A>10^{58}$. The collapse of these masses seems inevitable, no equilibrium exists.

Before we can enter into a discussion of the dynamics of this collapse it will be necessary, however, to study the geometry of the Schwarzschild metric in more detail. This will be done in the next chapter.

10. KRUSKAL SPACE

In this chapter we shall continue the study of the geometry of the Schwarzschild metric. It will turn out that the simple discussion presented in chapter 8 leaves many questions to be answered: Is there no space at all for $r<2M$? Intuitively one feels that there should be a singularity at $r=0$, whereas there seems to be one at $r=2M$. If a test particle is dropped towards a Schwarzschild singularity where does it go once it reaches $r=2M$? Can an object collapse beyond its own Schwarzschild radius? In this case the interior and exterior metrics could not be joined together in the simple manner shown in Fig. 8.

To answer these questions we have to understand the geometry of the Schwarzschild metric in detail. It will turn out that the modern manifold concept is useful even in the simple situation discussed here.

The key step towards the understanding of the full geometry of the Schwarzschild solution is a coordinate transformation $(t,r,\Theta,\phi) \rightarrow (u,v,\Theta,\phi)$ introduced by Kruskal (1960) which is defined in an implicit way by:

$$ds^2 = f^2(u,v)(dv^2-du^2)-r^2(u,v)\; d\Omega^2 \tag{10.1}$$

$$u^2-v^2 = ((r/2M)-1)\; e^{r/2M} \tag{10.2}$$

$$2uv/(u^2+v^2) = \mathrm{tgh}(t/2M) \tag{10.3}$$

$$f^2(u,v) = \frac{32M^3}{r(u,v)}\; e^{-r/2M} \tag{10.4}$$

The geometrical significance of the transformation becomes most transparent in a (u,v) diagram, from which one can read off the space-time behaviour of radial trajectories (see Fig. 13). The radial variables Θ, ϕ are omitted in this diagram.

Remark: It is important to distinguish clearly between the different types of diagrams used here. In Fig. 8 a surface is shown which has the same geometry as a (*space-like*) cross section through the center ($r=0$) of the mass configuration. Fig. 13 is a *space-time* diagram (analogous to those used in special relativity) permitting a study of radial geodesics light cones etc. In Fig. 16 both types of diagrams are used together.

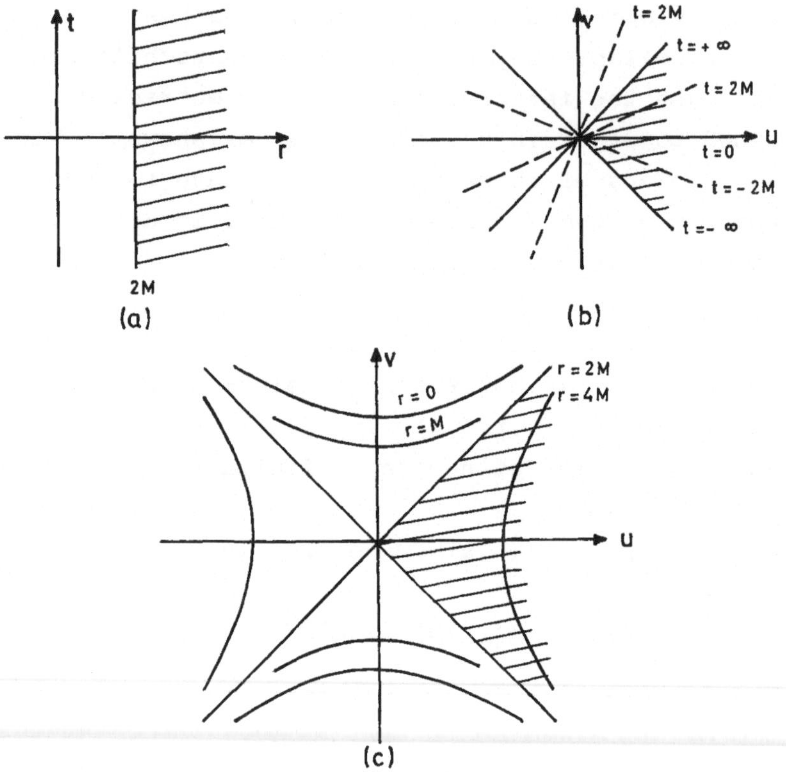

Figure 13 - Kruskal space – space time diagrams.

The physical region r>2M of the Schwarzschild space is mapped onto one quadrant of the Kruskal space by the transformation (10.1) - (10.4). These equations show, however, that the line element is regular even in the other three quadrants up to the line r=0, which is a hyperbola in the (u,v) plane. This means that (t,r,Θ,ϕ) is a coordinate patch (chart) which can be used only on one part of the whole manifold.

One can show that Kruskal space is the maximum analytic extension of the Schwarzschild manifold, i.e. the Kruskal coordinates (u,v,Θ,ϕ) are an admissible chart on the whole manifold which can be constructed by an analytical extension of Schwarzschild space. This situation is illustrated in Fig. 14

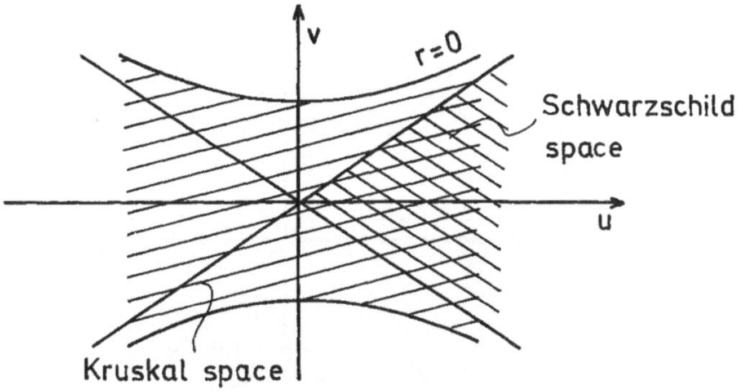

Figure 14 - Kruskal space

In (10.1) v is obviously a time like coordinate, proportional to the proper time element of an observer located at a point u=const.

This observer starts at a finite radius $r>2M$ for $v=0$, begins to fall towards smaller values of r later on and reaches the singularity $r=0$ at a finite proper time.

A special advantage of the Kruskal metric is that radial light rays are simply given by

$$du = \pm dv \tag{10.5}$$

The light cone has therefore the same form as in special relativity, which simplifies the discussion of causality problems. The light cone $u=\pm v$ (Fig. 15) separates Kruskal space into four regions, the catastrophic ones (I and III, $r<2M$) and the non-catastrophic ones (II and IV, $r>2M$). Both branches of all light cones emanating from the points within I reach the singularity $r=0$; every observer within the catastrophic region I is therefore bound to fall into the singularity. It is possible to send light rays into I, but no signal can come back from there; similarly it is possible to send signals from III to the non-catastrophic regions, but no signal can be sent from II or IV to III.

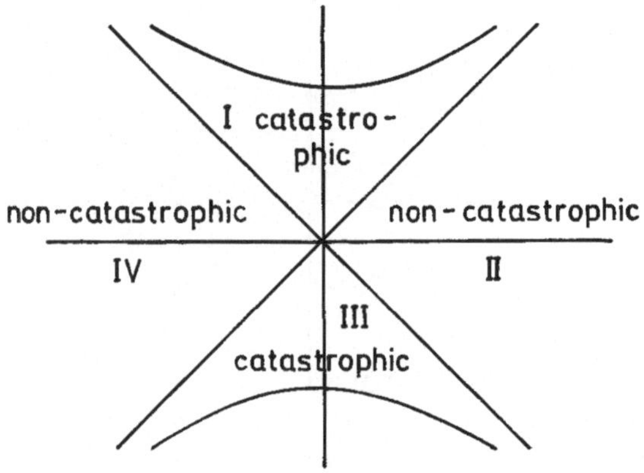

Figure 15 - Kruscal space

Fig. 13c shows, furthermore, that the lines r=const.<2M are not time-like. Only a small part of Kruskal space (the regions II + IV) can thus be explored with the help of static (r=const) observers.

In chapter 7 we have studied the geometry on a (space-like) cross section through the center of the star and have constructed an embedding diagram for this surface. Fig. 13b shows, however, that the cross-sections t=const., $\Theta=\pi/2$ do not give a complete picture of the manifold, since the catastrophic regions I and III cannot be obtained in this way. We have to choose, therefore, a different family of cross sections through the center of the mass distribution; the only restrictions being that the surfaces should actually be space-like (slope $<45^{\circ}$ in the u-v diagram) and that taken together they should give a complete picture of the manifold. The simplest choice fulfilling these conditions are the surfaces v=const, $\Theta=\pi/2$.

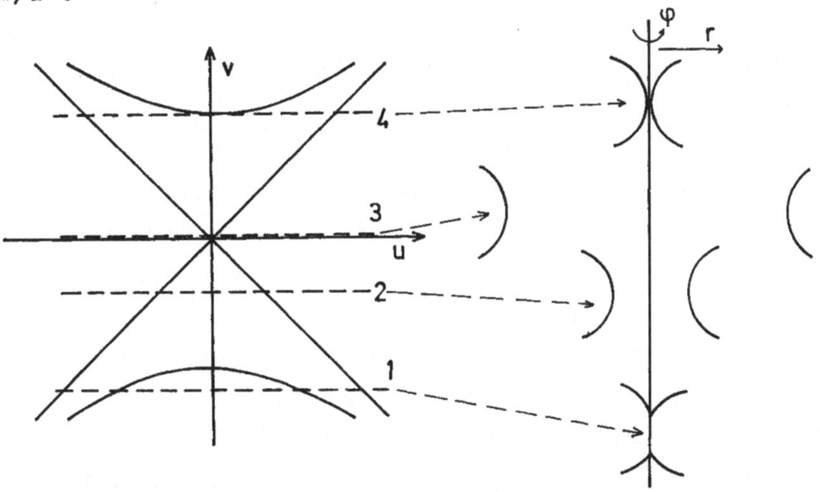

Space-time diagram

Space-like slices along the geometry.

Figure 16 - Kruskal space

Fig. 16 shows four typical cross sections through Kruskal space, obtained in this way. For v=0 (cross section 3) we obtain the paraboloid shown in Fig. 6 (v=0 implies t=0= const.). For positive values of v the Schwarzschild throat begins to close slowly and in v=1 the two surfaces begin to touch. It seems slightly surprising at first that the Schwarzschild metric shows a dynamical behaviour; one has to bear in mind, however, that the static observers (r= const.) can not see the whole manifold and dynamic ones (e.g. u=const.) have to be used to explore all of space time.

11. THE DYNAMICS OF GRAVITATIONAL COLLAPSE

We are now in a position to discuss the dynamics of spherically symmetric gravitational collapse. The simplest situation - which we shall consider here - results when an object with a nucleon number $A \gg 10^{58}$ collapses. For $A \sim 10^{68}$ (this is the typical size of a galaxy) the Schwarzschild radius is $2M = 10^{16}$ cm and the density of the massive sphere is about 10^{19}nucleons/cm^3 when the radius R equals the Schwarzschild radius. In this case pressure effects can be neglected for $R \gtrsim 2M$ (at very small values of R when matter reaches nuclear densities they have to be taken into account, see below) and the collapse of the object can be approximated by a free fall. In this approximation we can also neglect (thermal) electromagnetic radiation, which would lead to a possible mass loss of the system. Thus the total mass C(as seen from the outside) will be constant. The exterior metric is therefore given by the Kruskal metric as before.

As the initial configuration of the collapsing object we choose a static, homogeneous mass distribution of radius R, density ρ at $t=v=0$. For $t=0$ the cross section through the center of the star has one of the forms shown in Fig. 8, i.e. space is given by the Schwarzschild metric outside and a (momentarily static) space of constant curvature on the inside. These two spaces are joined together smoothly at the surface of the star $r=R$.

The subsequent dynamical evolution of the exterior metric can be read off Fig. 17. The stellar surface (in Fig. 17 the atom which is at $r=R$ for $t=0$) begins to collapse freely along the geodesic shown in Fig. 17 .

The figure shows that the stellar surface crosses the Schwarzschild radius $r=2M$ at a finite proper time and reaches $r=0$ also within a finite time, as measured by an

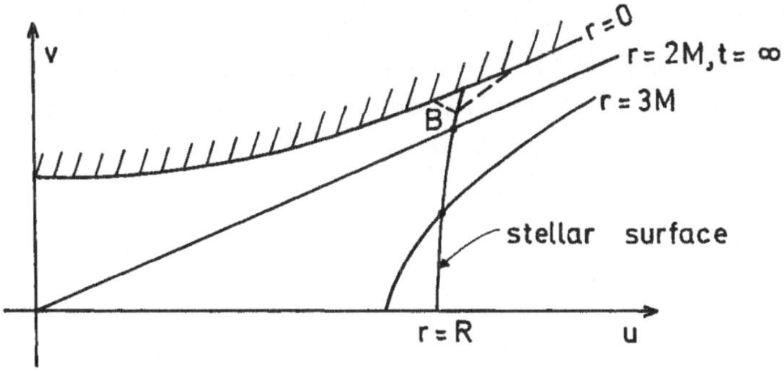

Figure 17 - Gravitational collapse - exterior metric

observer sitting on the surface. For an exterior observer, located at r>2M the collapse takes infinite time and the Schwarzschild radius is crossed by the collapsing object only a t=∞ . The object will lose its luminosity, however, quite rapidly (exponentially), so that the metaphor of a "black hole" is not totally unjustified.(This loss of luminosity is due to the rapidly increasing red shift of the radiation emitted by the system, as (7.11) shows .)

Fig. 17 shows, furthermore, that pressure effects which become relevant after the object has crossed its Schwarzschild radius will not be able in principle to stop the collapse to r=0. The reason for this is that the both parts of the light cone emanating from point B (Fig. 17) reach r=0 and pressure effects can change the trajectory of the stellar surface - once it has reached B - only to any other world line within this light cone. We conclude, therefore, that once an object has crossed its own Schwarzschild radius it cannot be stopped from collapsing completely (to r=0) by any type of force law.

We turn now to the interior solution which will turn out to be well known. Let us repeat that for t=0 the interior solution is given by a space of constant curvature (density) which is momentarily static; pressure has been neglected. This geometry and energy momentum tensor agrees exactly with a closed Friedmann universe at the moment of time symmetry (maximum expansion). Let us briefly summarize the essential facts about this universe. The line element is given by

$$ds^2 = a^2(\eta)[d\eta^2 - d\alpha^2 - \sin^2\alpha(d\theta^2 + \sin^2\theta d\phi^2)] \qquad (11.1)$$

where η is a time-like coordinate and $\alpha (0 \leq \alpha < \pi)$ is the radial coordinate. The expansion law is given by (a is the radius of the universe)

$$a(\eta) = \frac{a_o}{2}(1+\cos\eta) \qquad (11.2)$$

$$a_o = \left(\frac{3}{8\pi G\rho_o}\right)^{1/2} \qquad (11.3)$$

where ρ_o is the density and a_o the radius at the moment of maximum expansion of the universe (Fig. 18)

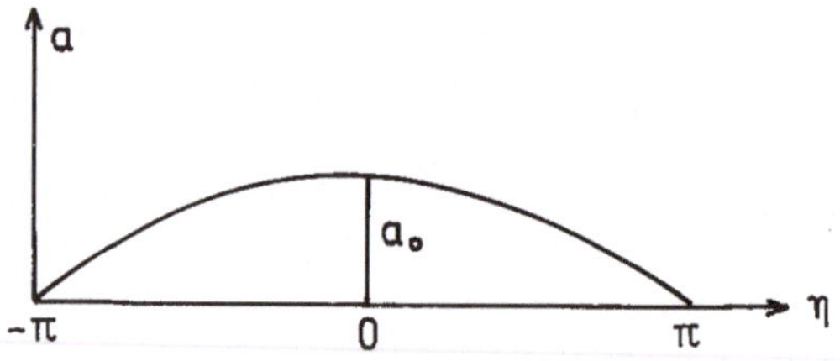

Figure 18 - Radius of the Friedmann universe as a function of η

This universe originates in a big bang at $\eta = -\pi$, reaches its maximum expansion at $\eta = 0$ and collapses back into a singularity for $\eta = \pi$. The time which it takes the universe to collapse is given by

$$T = \int_0^\pi a(\eta)\, d\eta = \frac{a_o}{2} \qquad (11.4)$$

(T is approximately equal to the time it takes for a light ray to circle the universe once at maximum expansion).

Since the interior geometry of the Schwarzschild

metric agrees with the one of the closed Friedmann universe for $t=0$ ($\eta=0$) (The first time derivatives of both geometries agree too, i.e. they are both static), the subsequent evolution of both geometries has to be the same too. The only additional fact which has to be taken into account is that the interior Schwarzschild geometry at $t=0$ does not agree with the full Friedmann universe, but only with the sector with $\alpha<\beta$. (β is the angle shown in Fig. 7. Only the part of the hypersphere with $\alpha < \beta$ is used as interior metric in Fig. 8.) Fig. 19 shows a space-time diagram (η-α-diagram) for the Friedmann universe.

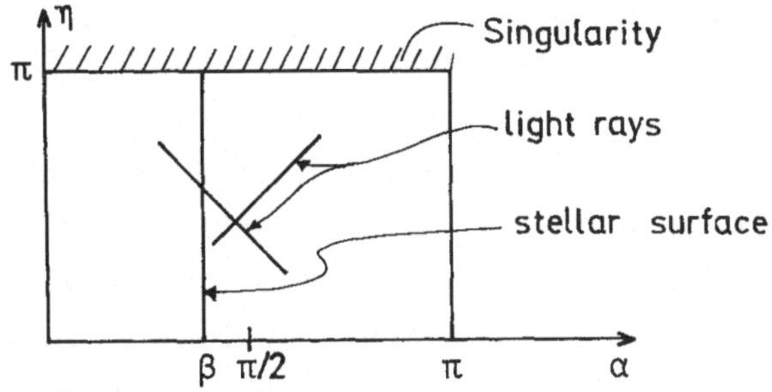

Figure 19 - Gravitational collapse - interior geometry

The advantage of the coordinates (η,α) is that light rays appear as straight lines with slope ± 1 in Fig. 19. An atom on the stellar surface is represented by the point $\alpha=\beta$ for $t=\eta=0$. At later times it moves along the geodesic α=const. shown in Fig. 19.(In the Friedmann universe each atom of the "substratum" remains at constant values of α,Θ,ϕ during the evolution of the universe, i.e. α,Θ,ϕ are comoving coordinates.) Thus the region $\alpha<\beta$, $0\leq\eta\leq\pi$ represents the history of the interior of the collapsing mass distribution, while the exterior geometry is contained in Fig. 17.

Oppenheimer and Snyder (1939) have shown that the two metrics can actually be joined smoothly at the "stellar surface" shown in Fig. 17 and 19 respectively. The complete geometry is hence the one shown in Fig. 20

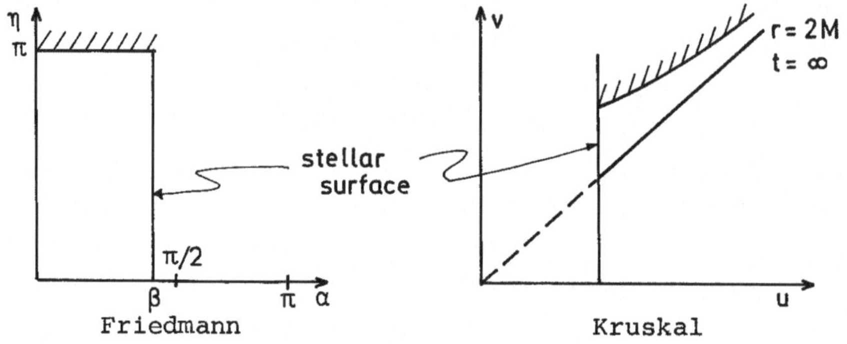

Figure 20 - Gravitational collapse - complete geometry.

We are now in a position to answer the questions raised before about spherically symmetrical gravitational collapse. For an outside observer the star will never vanish completely, since the Schwarzschild radius $r=2M$ will be reached only at $t=\infty$. The light emitted by the star will, however,be redshifted more and more so that the outside appearance of the star will resemble more and more a "black hole" in space.

For the inside observer the collapse will be completed in finite time. No communication is possible with the outside once the Schwarzschild radius is crossed.

In order to visualize the geometry of the Schwarzschild metric we have to construct from the space-time diagrams (Fig. 20) purely spacial diagrams corresponding to cross sections through the center of the collapsing object and to find surfaces of revolution which have the same intrinsic geometry as these cross sections through the object.

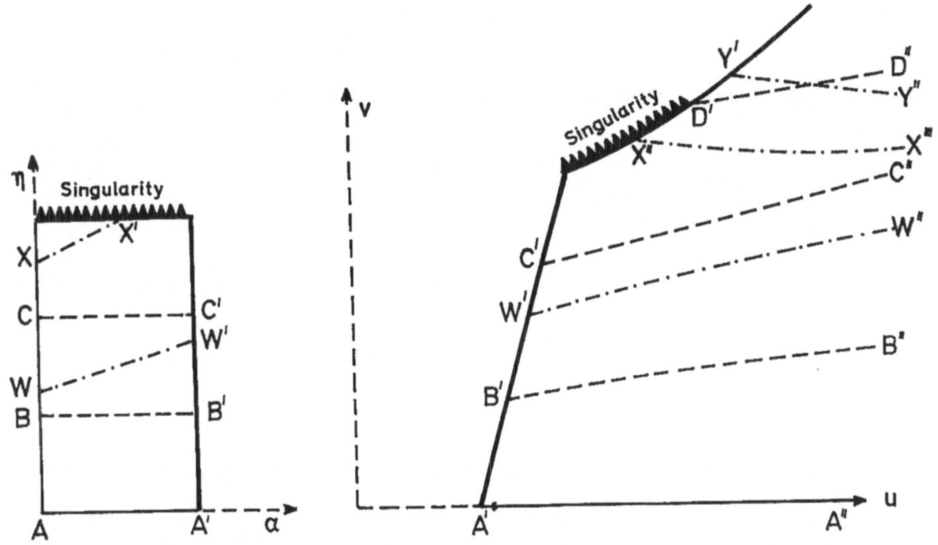

Space - time diagrams with various (space-like) cross sections.

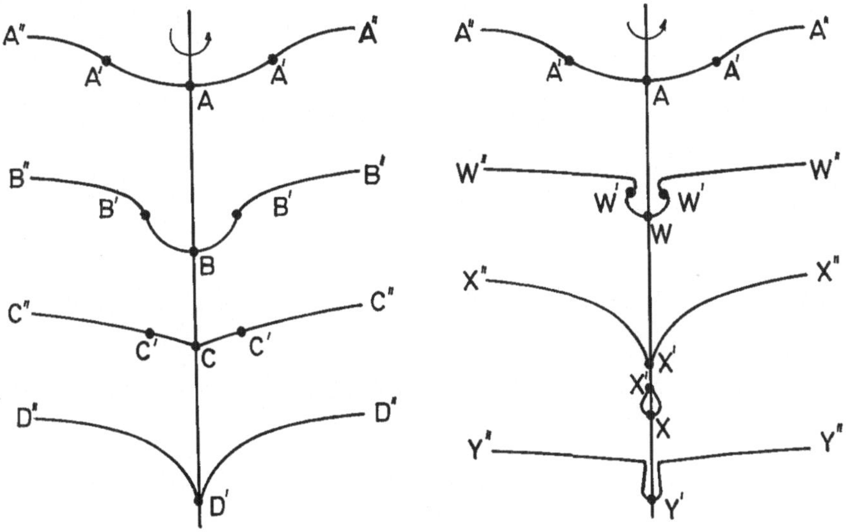

Two different families of embedding diagrams for space-like slices

Fig. 21

REFERENCES

Dicke, R. H. (1964) in "Relativity, Groups and Topology", C.+B.de Witt eds., Gordon and Breach, New York.

Harrison, K., Thorne, K., Wakano, M., Wheeler, J. (1965) "Gravitational Collapse", University of Chicago Press.

Kruskal, M.D. (1960) Phys. Rev. 119, 1743

Oppenheimer J. R., Snyder, H. (1939), Phys. Rev. 56, 455

Pound, R. V., Rebka, G. A. (1960), Phys. Rev. Lett. 4, 337

Shapiro, I. I. (1969) Phys. Rev. Letters 20, 1265 (1968) and Bericht auf der ESRO-Tagung, Interlaken 1969.

Thorne, K. (1970), Astrophys. Journal (to be published)

Weber, J. (1970), Phys. Rev. Letters 24, 945.

Acta Physica Austriaca, Suppl. VII, 355–391 (1970)

THE CONSTRUCTION OF PHYSICAL STATES IN QUANTUM FIELD THEORY*

BY

L. STREIT
Bell Telephone Laboratories, Inc.
Murray Hill, New Jersey

1. INTRODUCTION

There were the years when quantum field theorists and the dynamical problem sat there and looked at each other in despair. Nonlinear theories would either give colorful nonsense answers (infrared, ultraviolet catastrophes) or else they would be butchered beyond resemblance to anything that the purists would call field theory. In those days it used to be a party game to guess at what the miracle cure might be, the "totally new idea" that was needed to lead out of this mess. Now, after a thousand days of rapid progress and of most fascinating results, it has become transparent that such "gifts from heaven" were not required. The ingredients of the success story were to a large extent "more of the same": space and ultraviolet cutoffs, theories "in a box", path integrals, perturbation expansions etc., etc.

The crucial difference is that now things are under

---

* Lecture given at IX. Internationale Universitätswochen für Kernphysik, Schladming, February 23 - March 7, 1970.

control. Models are still being butchered to make them tractable, but this now is reversible butchering. One is fascinated by the skill that it took to extract the grain of truth that was hidden in the usual more or less formal procedures and by the courage of attempting to blend the miscellaneous bits of information thus obtained into a workable program.

The results were of two kinds: First of all existence theorems and various general properties were established for specific models.

Best known, and most advanced, is the work of A. Jaffe and J. Glimm on quartic selfinteraction of scalar "mesons" in one-dimensional space (GJ 1-3, CJ 1, He 1, Ja 1) for short*

$$(\phi^4)_2$$

and to a lesser degree on the Yukawa interaction

$$(\phi\bar{\psi}\psi)_2 \quad \text{(Gl 1, GJ 4, 5, Ja 2, He 1, Pa 1)}$$

and on $(\phi^4)_3$ (Gl 3, He 1) .

This sequence reflects a hierarchy of increasing difficulties on which we plan to shed some light in the course of these lectures. Other models of recent interest are (Os 1,He 1)

$$(\phi^3)_{s+1} \qquad s \geq 1$$

and higher even powers of a scalar field in one dimension (Gl 2, Ro 1). Apart from statements about specific models there has emerged, as a second type of result, a much better understanding of Hamiltonian quantum field theory in general and especially of the approximations involved, their domains of validity and their limitation; thus foundations for future work are laid and a giant step is made toward reliable numerical calculations. On these talks we

* The subscript denotes the member S+1 of space time dimensions.

shall focus on some such general methods rather than on a specific model, for their intrinsic interest as well as for contrast: work on the $(\phi^4)_2$ model was reviewed by J. R. Klauder (Kl 1) in last year's Schladming school.

2. STATING THE PROBLEM

Let us begin by clarifying what we shall mean by physical states in quantum field theory. While the asymptotic multiparticle states (Jo 1, p.119) would be prime candidates to carry this label, a particle interpretation has yet to be established in the nonlinear theories under consideration. As a first step we shall concentrate on a more modest goal, namely to find states ψ in Fock space H such that

$$H_{ren}\,\psi \equiv (H_o+V(g)+R)\,\psi\ \varepsilon\ H$$

Here R stands for renormalization terms that may be required to make H_{ren} well defined and a large part of our effort will be devoted to finding a consistent renormalization; H_o is the appropriate kinetic energy operator, e.g. for scalar mesons

$$H_o = \frac{1}{2}\int d^sx\ :\pi^2(x)+(\nabla\phi(x))^2+m^2\phi^2(x)\ :$$

and V the interaction term with g symbolizing the spatial cutoff that is required by Haag's theorem (SW 1, St 2), e.g.:

$$V(g) = \int d^sx\ g(x)\ :\ \phi^4(x)\ :$$

with supp g a bounded set in space. We shall not have much to say about the limiting case g→const. This removal of the spacial cutoff is a much later step, a future step for most of the models in question, while on the other hand our endeavour is clearly the very first step in the discussion of Hamiltonian dynamics: H must remain rather meaningless as an operator as long as

$$H\psi = \infty$$

for any nonzero ψ that one can think of.

For S=1 the above H_o,V would be exactly those that Glimm and Jaffe have studied, and one might object to our program that at least in the $(\phi^4)_2$ model those physical states as we understand them are nothing very mysterious: any Fock vector ψ_n describing n ("bare") particles with a decent wave function has finite energy. More generally, if also more technically it is fairly easy to establish for the domain D(H) of $H=H_{ren}$ that

$$D(H) \supset \bigcap_{n=1}^{\infty} D(H_o^n)$$

i.e. H may see any vector ψ for which $H_o^n\psi$ is finite for all n, among them most prominently the ("bare") vacuum $|0\rangle$ for which

$$H_o|0\rangle = 0 .$$

Such simplicity, however, singles out $(\phi^n)_2$; it is not shared by the other local non-linear interactions under consideration, where to find such "physical" or "dressed" states is a nontrivial part of the dynamical problem.

While at this point the reader may want to discover by a simple and straightforward computation why in the case of $(\phi^4)_{S+1>2}$

$$(H_o + V(g))\psi = \infty$$

(in the sense that the length $||(H_o+V)\psi||$ is infinite) for all the "obvious" states (bare vacuum, n-particle states, etc.), we shall now turn to a very few general remarks on selfadjoint operators and their perturbations to provide a certain amount of perspective and background for our subsequent discussion of the problem as it is posed by quantum field theory.

3. WAVE OPERATORS, THE FRIEDRICHS METHOD

As for the importance of finding states in the domain of the Hamiltonian operator H one just has to recall that the existence of a dense domain D(H)

$$\overline{D(H)} = \mathcal{H}$$

is a necessary first step to establish selfadjointness which means

$$H\psi = H^{+}\psi \qquad \text{for all} \quad \psi \in D(H) = D(H^{+})$$

While we have this definition before us it might be worth pointing out that the devil, as usual, hides in the small detail. Assume we have what we want, namely a dense D(H) such that

$$H\psi = H^{+}\psi \qquad \text{for all} \qquad \psi \in D$$

- we shall call such operators symmetric - then we "only" have to worry about

$$D(H) = D(H^{+})$$

Recall the definition of the adjoint H^{+} of an operator H. We want

$$(\phi, H\psi) = (H^{+}\phi, \psi) \equiv (\phi', \psi)$$

for all $\psi \in D(H)$.
Hence $D(H^{+}) = \{\phi : \exists \phi' \text{ such that } (\phi, H\psi) = (\phi', \psi) \, \forall \psi \in D(H)\}$
and for a symmetric operator we thus have only

$$D(H) \subseteq D(H^{+})$$

and not, in general, equality. From the formula for $D(H^{+})$ it is clear that if we manage to increase D(H), $D(H^{+})$ can at most become smaller, in favourable situations - e.g. when H is semibounded (Yo 1, p317) - one can blow up D(H) and shrink $D(H^{+})$ until they become equal, obtaining what is called a selfadjoint extension of H. In general, however, no selfadjoint extension need exist and even if it exists it need not be unique (Yo 1, p.349). Since self-adjointness is so hard to establish for the Hamiltonian it is probably worth while to recall what its relevance is from the physicist point of view. Depending on ones taste and preference one might observe that (Ma 1)

a) H is supposed to be the "energy" operator and it is one of the fundamental postulates of quantum mechanics that the operators corresponding to such "observables" be self adjoint.

b) the temporal development of the physical system is supposed to obey the Schrödinger equation

$$i \, \partial_t \, \psi(t) = H \, \psi(t)$$

In this context selfadjointness supplies us with an existence and uniqueness theorem for solutions of this equation:

$$\psi(t) = e^{-iHt} \, \psi(0)$$

c) One may want to "diagonalize" H, find the energy spectrum (or in a relativistic theory the mass spectrum), find eigenvectors of H like bound states, scattering states etc. Through the "spectral theorem" (Yo 1, p.313) this becomes automatic once selfadjointness is established:

any selfadjoint operator H has a spectral decomposition

$$H = \int_{-\infty}^{+\infty} E \, P(dE)$$

with P projecting onto states with energy in the interval dE.

Now in nonrelativistic quantum mechanics of a finite number of particles typically

$$H = H_o + gV$$

where

$$H_o = \frac{1}{2} \sum_{i=1}^{n} \frac{p_i^2}{2m_i} \quad , \; V = V(x_1, \ldots, x_n)$$

and selfadjointness of H can be established in a number of ways for the usual choices of V(x), like e.g. through Kato's theorem (Ka 1) for potentials that are "relatively bounded" with respect to H_o

$$||V\psi|| \leq \varepsilon||H_o\psi|| + \text{const.}||\psi||$$

$$0\leq\varepsilon<1 \quad , \quad \text{all} \quad \psi\varepsilon D(H_o)$$

or through the construction of Møller operators (wave operators) $\Omega_\pm$ that intertwine between H_o and H :

$$H\Omega_\pm = \Omega_\pm H_o$$

If Ω is unitary then

$$H = \Omega H_o \Omega^{-1}$$

is selfadjoint with

$$D(H) = \Omega D(H_o)$$

i.e. Ω maps bare states into physical states. Formally,if

$$H_o|E>_o = E|E>_o$$

it follows for $|E> \equiv \Omega|E>_o$ that

$$H|E> = E|E> \quad .$$

More rigorously Faddeev (Fd 1) showed in this context that $\Omega_\pm$ can be written as an integral operator in the space where H_o is diagonal (kinetic energy space, angular variables suppressed, V acting as an integral operator). I.e.

$$(H_of)(E) = E\ f(E)$$

$$(\Omega_\pm f)(E) = \int\omega_\pm(E,E')f(E')dE'$$

where $\omega_\pm(.,E')$ is an eigenfunction of H with eigenvalue E'

$$H\omega_\pm(.,E') = E'\omega_\pm(.,E')$$

from which the formal version follows by setting

$$f(E') = \delta(E-E')$$

Note however, that the usual definition (Sw 1, p.325)

$$\Omega_{\pm} = \lim_{t\to\pm\infty} e^{itH} e^{-itH_o}$$

would not serve our purpose of exploring the properties of H : we do not know enough about H to judge whether e^{itH} and the limit make sense. Instead we shall follow Friedrichs (Fr 1) and construct Ω as follows. (For another construction that has been applied to nonsymmetric perturbations cf. (Ka 1, 2)).

Rewriting

$$(H_o+gV)\Omega = \Omega H_o$$

in the form

$$[H_o,\Omega] = - gV\Omega$$

we see that Ω is a solution to an equation of the general type

$$[H_o,X] = A$$

for the solution of which we shall use the notation

$$X = \Gamma(A)$$

If A acts as an integral operator in momentum space

$$(Af)(k) = \int a(k,k')f(k')dk' ,$$

$$(H_o f)(k) = E(k)f(k)$$

then so does $\Gamma(A)$ with the kernel

$$(\gamma a)(k,k') = \frac{a(k,k')}{E(k)-E(k')}$$

since

$$\begin{aligned} [H_o,\Gamma(A)] &= H_o\Gamma(A) - \Gamma(A) H_o \\ &\sim E(k) \frac{a(k,k')}{E(k)-E(k')} - \frac{a(k,k')}{E(k)-E(k')} E(k') \\ &= a(k,k') \\ &\sim A \qquad \text{q.e.d.} \end{aligned}$$

Clearly such definition of $\Gamma(A)$ is not unambiguous: if

$a(k,k') \neq 0$ for $E(k) = E(k')$ a suitable convention, like

$$(\gamma a)(k,k') = a(k,k')(E(k)-E(k')-i\varepsilon)^{-1}$$

has to be chosen (Fr 1,p.22). Furthermore not only $\Gamma(A)$ but also

$$R + \Gamma(A)$$

obey the commutator equation as long as

$$[H_o,R] = 0$$

Thus the solution of the equation

$$[H_o,\Omega] = -gV\Omega$$

for the Møller operator reads

$$\Omega = R - g\Gamma(V\Omega)$$

Since we want

$$\Omega \Big|_{g=o} = 1$$

we shall set $R = 1$ to obtain the "Friedrichs equation"

$$\Omega = 1 - g\Gamma(V\Omega)$$

with the iteration solution

$$\Omega = \sum_{n=o}^{\infty} g^n \Omega^{(n)}$$

where

$$\Omega^{(o)} = 1 \quad , \quad \Omega^{(n)} = (-1)^n \Gamma(V\Omega^{(n-1)}) \quad .$$

Convergence can be proven for suitable ("gentle") classes of perturbations (Fr 1, p.21), cf. also (Fd 1, Ka 2), one verifies isometry

$$\Omega^+ \Omega = 1$$

and finally, if the usual Møller operators exist

$$\Omega = \lim_{t\to-\infty} e^{iHt} e^{-iH_o t} = \Omega_-$$

as a consequence of the $-i\varepsilon$ convention in the definition of Γ.

While clearly the perturbed operator H must have the same spectrum as the unperturbed H_o for unitary equivalence one usually makes the further restriction that H_o should not have discrete spectrum. Since we shall ultimately be concerned with quantum field theory where the bare vacuum is an eigenstate of H_o

$$H_o|0\rangle = 0$$

a closer look at this question is worth while. An exact statement of the situation is given by the following Lemma (Ka 1, p.528):

Let $$\underset{t\to-\infty}{\text{s-lim}}\ e^{iHt}\, e^{-iH_ot} = \Omega_-$$

exist and let

$$H_o\psi = E\psi .$$

Then

$$H\,\psi = E\,\psi .$$

I.e. the perturbation may change neither eigenvector nor eigenvalue (while clearly

$$H|0\rangle \neq 0$$

for the usual quantum field theories with local interactions). The argument is very simple: the existence of the strong limit implies that

$$\|(e^{iH(t+a)}\, e^{-iH_o(t+a)} - e^{iHt}\, e^{-iH_ot})\psi\| \underset{t\to-\infty}{\to} 0$$

for arbitrary a . Since ψ is an eigenvector of H_o we can rewrite the norm as

$$\|e^{-iEt}\, e^{iHt}(e^{iHa}\, e^{-iH_oa} - 1)\,\psi\|$$

$$= \|(e^{iHa}\, e^{-iH_oa} - 1)\,\psi\|$$

which does not depend on t and hence must be identically zero, i.e.

$$e^{i(H-E)a}\,\psi = \psi ,$$

$$(H-E)\psi = 0 \quad \text{q.e.d.}$$

Note, however, that this does not exclude the existence of an intertwining operator Ω such that $H\Omega = \Omega H_o$ even if discrete eigenvectors are being perturbed: they need not be strong limits of

$$e^{iHt}\, e^{-iH_o t}$$

4. EXAMPLE OF A PERTURBED GROUND STATE: THE LEE MODEL

It has been pointed out by Friedrichs (Fr 1, p.31) - and we follow him in this paragraph - that the Lee model (Sw 1, p.352) with static "nucleons" v,n of mass M,m and a Θ-meson of mass μ furnishes a well-known example of the situation where the unperturbed Hamiltonian

$$\hat{H}_o \equiv M\int d^s p\ v^*(p)\ v(p) + m\int d^s p\ n^*(p)\ n(p) + \int d^s k\ \omega(k)\ \Theta^*(k)\ \Theta(k) \qquad \omega(k) \equiv \sqrt{\mu^2+k^2}$$

has discrete and continuous spectrum. In the n-Θ sector the ground state is that of a (bare) v-particle and the continuum is produced by the n-Θ states.

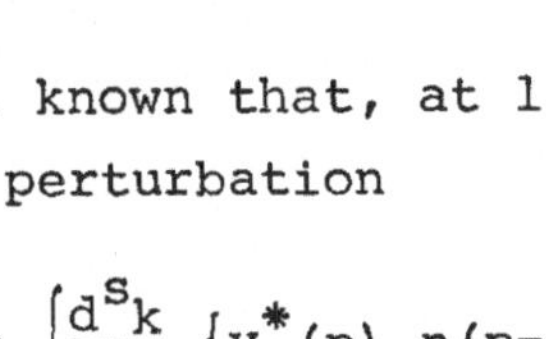

Energy Spectrum of the n-Θ Sector

It is well known that, at least with an ultraviolet cutoff, the perturbation

$$\hat{V} \equiv g\int d^s p \int \frac{d^s k}{\sqrt{2\omega}} \{v^*(p)\ n(p-k)\ \Theta(k) + h.c.\}$$

preserves this situation qualitatively, however, to keep the width of the energy gap unchanged one introduces the so-called mass renormalization

$$\hat{R} = -\delta M \int d^s p \; v^*(p) \; v(p)$$

An arbitrary state in the n-Θ sector may be written as

$$\psi = \int d^s p \; \chi(p) \{\phi_\nu \; v^*(p) |0> + \int d^s k \; \phi(k) \; n^*(p-k) \theta^*(k) |0>\}$$

with the inner product

$$(\psi,\psi') = (\chi,\chi') \{\phi_\nu^* \cdot \phi_\nu' + (\phi,\phi')\}$$

i.e.

$$\psi = \chi \otimes (\phi_\nu \oplus \phi) \; .$$

Since $\hat{H}_o$, $\hat{V}$, $\hat{R}$ act trivially on the first factor we shall focus on the second, using the notation

$$\hat{H}_o \Big|_{n-\Theta} = 1 \otimes H_o \qquad \text{etc.}$$

Thus

$$H_o \; \Phi = \begin{pmatrix} m \; \phi_\nu \\ E(k) \phi(k) \end{pmatrix}$$

where

$$E(k) = m + \sqrt{\mu^2 + k^2}$$

A general integral operator will act on Φ as

$$A\Phi = \begin{pmatrix} a_{11} & a_{12} \\ a_{21} & a_{22} \end{pmatrix} \begin{pmatrix} \phi_\nu \\ \phi \end{pmatrix} = \begin{pmatrix} a_{11} \; \phi_\nu + \int a_{12} \; (k') \phi(k') dk' \\ a_{21} \; (k) \phi_\nu + \int a_{22} \; (k,k') \phi(k') dk' \end{pmatrix} \; .$$

With this notation and with $\omega' = \omega(k')$, $E' = E(k')$ we have

$$H_o = \begin{pmatrix} M & 0 \\ 0 & E(k) \delta(\vec{k} - \vec{k}') \end{pmatrix} \qquad V+R = \begin{pmatrix} -\delta M & g/\sqrt{2\omega'} \\ g/\sqrt{2\omega} & 0 \end{pmatrix}$$

and

$$[H_o, A] = \begin{pmatrix} 0 & (M-E') a_{12} \\ (E-M) a_{21} & (E-E') a_{22} \end{pmatrix} \equiv B \qquad \text{such that for}$$

$$B = \begin{pmatrix} 0 & b_{12} \\ b_{21} & b_{22} \end{pmatrix} \quad \ldots. \quad \Gamma(B) = A = \begin{pmatrix} 0 & b_{12}/M-E' \\ b_{21}/E-M & b_{22}/E-E' \end{pmatrix}$$

(Note that we can solve the commutator only for operators B with $b_{11} = 0$).

With this definition of Γ we obtain, just as in the preceding paragraph the Friedrichs equation

$$W = 1 - \Gamma\{(V+R)W\}$$

for the intertwining operator W:

$$(H_o+V+R)W = WH_o$$

Making the ansatz

$$W = \begin{pmatrix} w_{11} & w_{11} \\ w_{21} & w_{22} \end{pmatrix}$$

a first consistency requirement is $\{(V+R)W\}_{11} = 0$, hence $w_{11} = 1$, so that the Friedrichs equation reads:

$$\begin{pmatrix} 1 & w_{12}(k') \\ w_{21}(k) & w_{22}(k,k') \end{pmatrix} = \begin{pmatrix} 1 & \dfrac{\delta M w_{12} - g\int \frac{d^s k}{\sqrt{2\omega}} w_{22}}{M - E'} \\ -\dfrac{g}{\sqrt{2\omega}(E-M)} & \delta(\vec{k}-\vec{k}') - \dfrac{g w_{22}}{\sqrt{2\omega}(E-E')} \end{pmatrix}$$

We can immediately read off

$$w_{21} = -\frac{g}{\sqrt{2\omega}(E-M)}$$

and from the above condition on $\{(V+R)W\}_{11}$ we obtain

$$0 = \{(V+R)W\}_{11} = v_{11} w_{11} + v_{12} w_{21} = -\delta M - g^2 \int \frac{d^s k}{2\omega(E-M)}$$

Insertion of w_{22} into the equation for w_{12} yields

$$w_{12} = \frac{g}{\sqrt{2\omega'}} \left\{E' - M - \delta M - g^2 \int \frac{d^s k}{2\omega} \frac{1}{E-E'}\right\}^{-1}$$

in which we want to eliminate

$$\delta M = -g^2 \int \frac{d^s k}{2\omega(E-M)}$$

such that

$$w_{12} = \frac{g}{\sqrt{2\omega'}} \left\{ (E'-M)\left(1-g^2\int\frac{d^sk}{2\omega}\,\frac{1}{(E-M)(E-E')}\right\}^{-1}$$

w_{22} can then be read off as an expression in terms of w_{12}. After this derivation of W two remarks are in order

1) To get the Friedrichs construction to work we had to set $w_{11}=1$ which in turn produced the mass renormalization δM as given above. This makes good sense: recall that the same δM is usually obtained by imposing that M be an eigenvalue of H_o as well as of H_{ren} (Sw 1, p.358)

2) To make an important point let us take closer look at the integrals in question. Disregarding the present popularity of s=1 we shall assume for the moment that the world is not one- but two-dimensional, since in this case (as in the old-fashioned three-dimensional world) simple power counting reveals that

$$\delta M = -\infty$$

However, noting how δM cancels against another divergent integral in w_{12} we see that W is perfectly well defined and so is the renormalized Hamiltonian H_{ren} through

$$H_{ren}W = WH_o$$

We recognize that W is a fairly stable object, finite even in the case of theories that are singular enough to require infinite renormalizations, such that, through equations like the above, the wave operator can serve to give rigorous meaning to H_{ren}.

Similarly we shall see how in a fullfledged field theory renormalization will emerge very naturally from the construction of approximate "dressing transformations".

5. OF CREATION AND ANNIHILATION

In quantum field theory we typically encounter free Hamiltonians

$$H_o = \int d^s k \; \omega(k) \; a^+(k) \; a(k)$$

and perturbations of the form

$$V = \sum_{\ell,m} \int d^s k_1 \ldots d^s k_\ell \; d^s p_1 \ldots d^s p_m \; v_{\ell m}(k_1,\ldots,k_\ell,p_1,\ldots,p_m) \cdot$$

$$\cdot \; a^+(k_1)\ldots a^+(k_\ell) \; a(p_1)\ldots a(p_m)$$

where a^+, a are creation and annihilation operators obeying the canonical commutation relations (CCR):

$$[a(k), \; a^+(k')] = \delta(k-k') \quad .$$

(It is only for simplicity that we disregard the possibility of more than one kind of excitations and of their obeying Fermi statistics .)

Now the creation and annihilation operators of quantum field theory are not at all uniquely characterized by the CCR: there are uncountably many inequivalent irreducible representations of this algebra. While we shall start out from the Fock representation which is characterized by the existence of a cyclic vector $|0\rangle$ for which

$$a(k)|0\rangle \equiv 0$$

we shall also be confronted with the situation suggested by the van Hove model (Ho 1): for sufficiently singular perturbations

$$H_{ren} = H_o + V + R$$

will have no domain at all in Fock space. It turns out though that there are other representations of the a^+,a by the use of which ultraviolet divergences can be removed and H_{ren} be made a well defined operator*.

Already the van Hove model indicates that such strange representations of the CCR have to be tailored to fit the Hamiltonian under consideration and to me it is one of the most fascinating aspects of constructive field theory that one now begins to understand how this can be done for non-trivial models of quantum field theory. Returning now to the expression for V we note that

$$V = \sum V_{\ell m}$$

is a sum of normal ordered terms (all creation operators to the left!) It turns out to be useful to depict $V_{\ell m}$ by a vertex with ℓ lines to the left and m to the right, e.g.

V_{23} 

These graphs are helpful when it comes to normal ordering a product of two such operators

$$U_{k\ell}\, V_{mn} = :U_{k\ell}\, V_{mn}: + U_{k\ell} -\!\ominus\!- V_{mn}$$

where the contraction symbol $-\!\ominus\!-$ stands for the usual normal ordering procedure:

a) write the a^+, a in U.V in normal order and eliminate at least one $a(p_i)$ from U and one $a^+(k_j)$ from V, replacing them by $\delta^s(p_i - k_j)$

b) sum over all possible ways of doing this (i.e. of "making contractions", the term with no contractions is the one between colons); graphically, with interior lines standing for δ-functions:

---

* Similarly for the infrared problem. While we shall not discuss it in these lectures there is good reason for optimism that the methods discussed here are also useful there.

$U_{23}\,V_{21}$ = 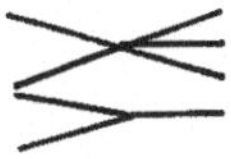 +  + 

$$= \; :UV: \; + \; \underbrace{UV}_{1} \; + \; \underbrace{UV}_{2}$$

A graph in the normal ordering of

$$V : W_1 \ldots W_n :$$

will be called connected if there is at least one line from V to each of the W_i. For the sum over all connected graphs we shall use the Friedrichs notation

$$V \angle : W_1 \ldots W_n :$$

With the convention

$$V \angle : \; 1 \; : = V$$

we can compute (Fr 1, p.61)

$$\begin{aligned} V : e^W : &\equiv V \sum_n \frac{1}{n!} : W^n : \\ &= \sum_n \frac{1}{n!} \sum_{m=o}^{n} \binom{n}{m} : (V \angle :W^m:)\, W^{n-m} : \\ &= \sum_{m,n} \frac{1}{m!} \frac{1}{(n-m)!} : (V \angle :W^m:)\, W^{n-m} : \\ &= :(V \angle :e^W:)(:e^W:) : \end{aligned}$$

While this expression may not look pretty at first sight, it will turn out to be very useful. Note that the first factor really contains only a finite number of terms since V has only a finite number of "lines" to contract with.

Furthermore, since within a normal ordered expression factors may freely be permuted we may write

$$V : e^W : = :(:e^W:)(V \angle :e^W:) :$$

Hence this is a "pull through" formula that will be handy to construct relations similar to our

$$(H_o + V)\,\Omega = \Omega H_o \;.$$

6. APPROXIMATE DRESSING TRANSFORMATIONS

To characterize the singular nature of the perturbations

$$V = \sum_{\ell m} V_{\ell m}$$

we might note that $V_{\ell m}$ is of the order of magnitude $N^{\frac{1}{2}(\ell+m)}$ (N the number operator).

To make this plausible recall that in the case of the one dimensional oscillator

$$a|n\rangle = n^{\frac{1}{2}}|n-1\rangle$$

so that $a(N+1)^{-\frac{1}{2}}$ is bounded.

Similarly one can show (Gl 4, prop.2.1): "If $v_{\ell m}(\underline{k},\underline{p})\in L^2$ and $a+b+\ell+m \leq 0$ it follows that

$$||(N+1)^{a/2} V_{\ell m}(N+1)^{b/2}|| \leq \text{const}||v_{\ell m}||_2 \quad "$$

By the same rule of thumb

$$H_o \sim N$$

hence in field theories like (ϕ^4) where

$$V \sim N^2$$

the perturbation is in this sense much larger than H_o in the realm of large particle numbers, the relative boundedness of V with respect to H_o that is so useful in the context of ordinary quantum mechanism ("Kato's theorem", cf. sec. 3) is far from being applicable here. Let us go one step further in this qualitative analysis and find out which of the $V_{\ell m}$, given V, causes the worst trouble. The motivation here is that hopefully if we can deal with just the most singular part of the perturbation, the rest will cause no further trouble in the sense that physical states of the approximate Hamiltonian are also in the domain of the full operator. If we take as a trial state one of n bare particles with wave function

$$f(k_1,\ldots,k_n)$$

i.e. $|f\rangle = F_{no}|0\rangle$ we can study the norm of $V|f\rangle$ by our graph technique

$$||V|f\rangle||^2 = (F_{no})^* V^*V F_{no}\Big|_{oo}$$

we have to take the sum of all graphs of $F^*V^*V F$ with no external legs. Let us look at two extreme cases:

a) $V = V_{om}$ has only annihilation operators

(m=3,n=4)

all lines emerging from V go to the right hence they all are contracted with F.

b) $V = V_{\ell o}$ has only creation operators here we have (if e.g. $\ell=3$, $n=4$)

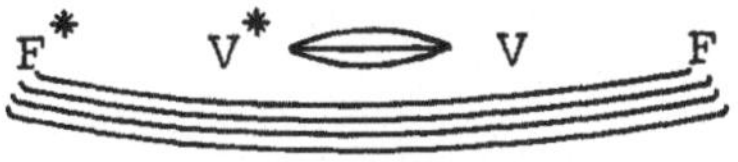

+ ...

plus graphs where some lines go from V^* to F and from F^* to V. In terms of kernel functions the first graph would read

$$\int f_n(k,q)\; v^*_{om}(k)\; v_{om}(p)\; f_n(p,q)\; d^{sm}k\; d^{sm}p\; d^{s(n-m)}q$$

Consequently the kernel v_{om} may be pretty singular if we restrict the states f to be smooth, like e.g.

$$v \in \phi' \;,\; f \in \phi \,.$$

On the other hand the second graph amounts to

$$\int v^*v \;\cdot\; \int f^*f$$

So as soon as $v_{\ell o} \notin L^2(R^{s\ell})$ this will blow up: no bare n-particle state can be seen by V and hence not by H_o+V. Except for scalar selfinteracting fields in one dimensional space all the local polynomial interactions that are usually considered give rise to such singular kernels! To find a domain for

$$H_o + V_{no}$$

in these cases we shall make use of the Friedrichs formalism. We need not be deterred here by the fact that H_o+V_{no} is not symmetric since no use is made of e^{iHt} as in the time dependent method, intertwining operators can in principle be obtained this way (Ka 2). A problem of different nature will, however, prevent us from completing the construction: the series

$$\Omega = \sum g^n \Omega^{(n)}$$

will be divergent in general . We shall see later how this can be repaired. Let us begin by seeing how Γ acts on $V_{\ell m}$. $\Gamma(V)$ obeying

$$[H_o, \Gamma(V)] = V$$

is obtained very simply through replacing the kernels $v_{\ell m}$ by

$$\gamma v_{\ell m}(k_1,\ldots,k_\ell,p_1,\ldots,p_m) = \frac{v_{\ell m}}{E(k)-E(p)}$$

with

$$E(k) = \sum_{i=1}^{\ell} \omega(k_i) \quad , \quad E(p) = \sum_{i=1}^{m} \omega(p_i)$$

Recall that insertion of $\Omega = \sum g^n \Omega^{(n)}$ into $(H_o+gV)\Omega = \Omega H_o$ yields in k-th order of g

$$\Omega^{(k)} = - \Gamma(V\Omega^{(k-1)}) .$$

For $V = V_{no}$ this is solved by

$$\Omega^{(k)} = \frac{(-1)^k}{k!} \{\Gamma(V_{no})\}^k$$

so that at least formally

$$\Omega = e^{-g\Gamma(V_{no})}$$

intertwines between H_o+V_{no} and H_o .

7. RENORMALIZATION

Before we take up the question of convergence let us see whether we may multiply Ω from the left by the full Hamiltonian (g=1):

$$H = H_o + V_{no} + \sum_{\ell=o}^{n-1} V_{\ell,n-\ell}$$

$$H\Omega = \Omega H_o + \sum_{\ell=o}^{n-1} V_{\ell,n-\ell}\,\Omega$$

To control the second term we shall normal order it with the help of Friedrichs' formula:

$$V_{\ell m}\Omega = V_{\ell m}\, e^{-\Gamma(V_{no})} = V_{\ell m} : e^{-\Gamma(V_{no})} :$$

$$= :(:e^{-\Gamma(V_{no})}:)(V_{\ell m} \prec :e^{-\Gamma(V_{no})}:):$$

$$= e^{-\Gamma(V_{no})}(V_{\ell m} \prec e^{-\Gamma(V_{no})}) = \Omega(V_{\ell m} \prec \Omega)$$

(note that we are free to set or to omit the normal ordering colons around expressions that are by definition normal ordered). Thus

$$H\Omega = \Omega(H_o + \sum_{\ell=o}^{n-1} V_{\ell,n-\ell} \prec \Omega) \quad .$$

To see graphically what the connected product amounts to let us take a very simple example (Ea 1, also He 1, p.117)

$$V = V_{20} + V_{11} + V_{02} \quad .$$

Here V_{02} contributes

$$V_{02} + \frac{(-1)}{1!}\underbrace{V_{02}\Gamma(V_{20})}_{1,2} + \frac{(-1)^2}{2!} V_{02} \prec (\Gamma(V_{20}))^2$$

V_{02} : V_{02} 

$\underbrace{V_{02}\Gamma(V_{20})}_{1}$: V_{02}  $\Gamma(V_{20})$

$\underbrace{V_{02}\Gamma(V_{20})}_{2}$:

$V_{02} \prec (\Gamma(V_{20}))^2$:

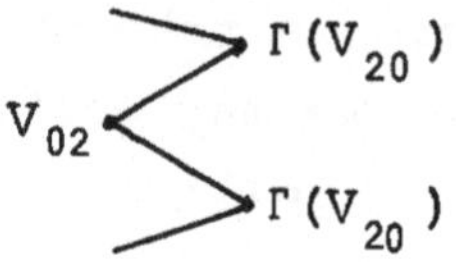

V_{11} can connect in only one way to $\Gamma(V_{20})$ and contributes

$$V_{11} - \underbrace{V_{11}\Gamma(V_{20})}_{1}$$

V_{11} :

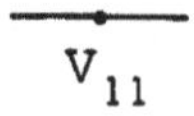

$\underbrace{V_{11}\Gamma(V_{20})}$:

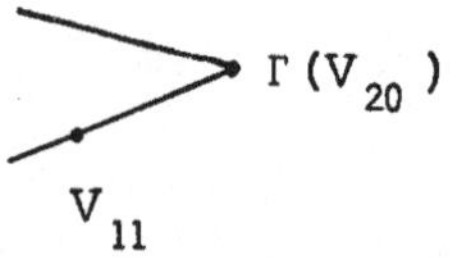

Among all these there is one ominous bubble graph and indeed it turns out to be the most dangerous one. Take e.g. $(\phi^2)_4$, where

$$V = \int d^3x\ g(x)\ :\phi^2(x):$$

Then

$$R_{oo} \equiv V_{02}\Gamma(V_{20}) \sim \int d^3k_1 d^3k_2 \frac{|\tilde{g}(k_1+k_2)|^2}{\omega_1(\omega_1+\omega_2)\omega_2}$$

which according to the rule of thumb (He 1, p.83) "replace $|g(p)|^2$ by $\delta(p)$" behaves like $\int d\omega/\omega$, hence diverges.

This, I claim, is a divergence of the fascinating kind. If we are so bold as to subtract it on both sides of our equation, we are left with

$$H_{ren}\,\Omega \equiv \Omega(H_o + \text{finite terms})$$

where

$$H_{ren} = H + R_{oo}$$

is obtained from H by an infinite renormalization (of the ground state energy) and makes sense since the r.h.s. does.

Recall for analogy the Lee model where the existence

of an intertwining operator necessitated the introduction of an additive renormalization to match the spectra of H_o and H_{ren}. Here, where an approximate dressing transformation links H_{ren} to a finite expression we will not expect the addition of R_{oo} to equalize the spectra. However, R_{oo} cancels divergences such that H_{ren} is well defined while H is not. To see this clearly let us take a closer look at

$$H = \frac{1}{2}\int d^3x \;:\pi^2(x) + (\nabla\phi)^2(x) + m^2\phi^2(x)\;: + $$
$$+ \frac{1}{2}\int d^3x \; g(x) \;:\phi^2(x)\;:$$

We have

$$i[H,\phi(f)] = \pi(f)$$

$$i[H,\pi(f)] = -\phi(\omega^2 f) - \phi(gf) \equiv -\phi(\hat{\omega}^2 f)$$

where

$$(\omega^2 f)(x) = (-\Delta + m^2)f(x) \qquad (gf)(x) = g(x)f(x)$$

and

$$\hat{\omega}^2 \geq 0 \qquad \text{if} \qquad g(x) \geq -m \quad \forall \; x$$

The same canonical equations would result if we replaced H by

$$H' \equiv \hat{a}^+ \omega \hat{a} = \sum_{m,n} \hat{\omega}_{m,n} \; \hat{a}_m + \hat{a}_n \quad ,$$

where the creation and annihilation operators $\hat{a}^+_m$ and $\hat{a}_n$ are the usual functions of ϕ, π, and $\hat{\omega}$.

Since the ϕ, π are irreducible the canonical equations determine H up to a c-number:

$$H' = H + C$$

Now the following facts are important (Fa 1):

C is infinite

Hence, since H' makes sense, H does not.

R_{oo} - C is finite

Hence the renormalized Hamiltonian H_{ren} differs from the well defined H' by a finite constant only, i.e. we have indeed a well-defined renormalized Hamiltonian H_{ren}. By the way, note that

$$(\psi_m, H\psi_n)$$

is finite for sufficiently smooth m- resp.n-particle wave functions ψ. Hence $H' = H+C$ is not allowed to see them as a bilinear form, let alone as an operator when C or equivalently R_{oo} iis infinite. This illustrates that the domain problem does not go away through the renormalization. For the $(\phi^4)_3$ interaction where one deals with

$$\sum_{\ell=0}^{3} V_{\ell,4-\ell} \;-\!\!\!< \; e^{-\Gamma(V_{40})}$$

the corresponding divergent bubble graph would be

$$\underbrace{V_{04}\,\Gamma(V_{40})}_{4} \quad : \quad V_{04}$$  $$\Gamma(V_{40})$$

but there are others, e.g.

$$\underbrace{V_{04}\,\Gamma(V_{40})}_{3} \quad : \quad V_{04}$$ 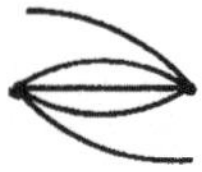 $$\Gamma(V_{40})$$

which turn out to be divergent (Gl 3, He 1, p.98). Note the exterior lines here: they suggest that divergences might here be cancelled by setting

$$H_{ren} = H + E \cdot 1 + \delta m^2 \int g(x) \; :\phi^2(x): \; d^2x$$

where E and δm^2 are infinite ground state and mass renormalizations to cancel the divergent graphs. It was demonstrated by Glimm (Gl 3) and Hepp (He 1) that such (local!) renormalization is indeed possible but to appreciate this Herculean effort we must first return to our quest for physical states, i.e. essentially for a well-defined dressing transformation .

8. A CLOSER LOOK AT THE APPROXIMATE DRESSING TRANSFORMATIONS

To see whether the formal

$$\Omega = e^{-g\Gamma(V_{no})}$$

exists let us investigate

$$\| \Omega |0> \|^2 = \Omega^+ \Omega \Big|_{00}$$

$$= \sum_{\nu=0}^{\infty} \left(\frac{g^\nu}{\nu!}\right)^2 (\Gamma(V_{no})^+)^\nu (\Gamma(V_{no}))^\nu \Big|_{00}$$

While $(g^\nu/\nu!)^2$ looks like a pretty good convergence factor, optimism is shattered immediately when you note that there are $(n\nu)!$ ways to form the $n\nu$ contractions of normal ordering, so that one really deals with

$$\frac{(n\nu)!}{\nu!\ \nu!} g^{2\nu}$$

and there is little hope that Ω might exist except in the case n=2, as in $(\phi^2)_4$ where closer investigation (see below and (Ea 1, He 1)) shows convergence for sufficiently small g . A solution for this dilemma lies in the observation that it is really only the ultraviolet limit which requires a dressing transformation. More precisely, let us decompose the kernel $v_{no}(k_1,\ldots,k_n)$ into a sum of functions involving and larger momenta

$$v_{no} = \sum_{j=0}^{\infty} v_{no}^j \quad , \quad v_{no}^j = \begin{cases} v_{no} & \text{if } \max_{1\le i\le n} |k_i| \varepsilon I_j \\ 0 & \text{otherwise} \end{cases} ,$$

with $\{I_j\}$ a suitably chosen decomposition of the positive reals into halfopen intervals

$$I_j = [a_j, a_{j+1}) ,$$

such that

$$e^{-\Gamma(V_{no})} = e^{-\sum_j \Gamma(V_{no}^j)} = \prod_j e^{-\Gamma(V_{no}^j)}$$

In keeping with the fact that dressing transformations are important only at high momenta (large j) one replaces the infinite series $e^{-\Gamma(V^j)}$ by the finite truncation

$$e_j^{-\Gamma(V^j_{no})} \equiv \sum_{k=o}^{j} \frac{(-\Gamma(V^j_{no}))^k}{k!}$$

There are two conditions on the specific truncation (choice of $\{I_j\}$) : it must be drastic enough to ensure the existence of the resulting truncated Ω and mild enough to preserve

$$H_{ren}\Omega = \Omega(H_o + \text{"finite"})$$

It is a matter of detailed and quite elaborate estimates that the two conditions are indeed compatible for models of interest (Fa 1, Gl 3, He 1).

After this discussion of series convergence let us now focus on the lowest order term

$$\underbrace{\Gamma(V_{no})^+ \; \Gamma(V_{no})}_{n}$$

There we encounter the third type of bubble graph, let us recall in which context they occured:

A : $\underbrace{V_{on} \; V_{no}}_{n} < \infty$: $D(H) \cap D(H_o)$ is dense

$V_{no} \varepsilon L^2$

B : $\underbrace{V_{on} \; \Gamma(V_{no})}_{n} < \infty$: H_{ren} is a bilinear form on a dense subset of $D(H_o)$

$V_{on} \gamma V_{no} \varepsilon L^1$

C : $\underbrace{\Gamma(V_{no})^+ \; \Gamma(V_{no})}_{n} < \infty$: $D(H_{ren})$ can be constructed (in Fock space).

$\gamma V_{no} \; \varepsilon L^2$

(Note that $A \Rightarrow B \Rightarrow C$ because of the damping action of Γ.)

It has been conjectured by Glimm (Gl 5) that the two sides

of this list might be equivalent (at least for the usual local interactions, nonlocal quadratic perturbations have been found by Eachus (Ea 1) which violate this pattern). For all the usual models that have been investigated it works. The most famous among them is of type A: For $(\phi^4)_2$, we have

$$V_{40} \sim \prod_{i=1}^{4} \left(\int \frac{dk_i}{\sqrt{\omega_i}} a^+(k_i)\right) \tilde{g}\left(\sum_1^4 k_i\right)$$

hence

$$\underbrace{V_{04} \quad V_{40}}_{4} \sim \prod_{i=1}^{4} \left(\int \frac{dk_i}{\omega_i}\right) |\tilde{g}(\sum k_i)|^2 < \infty$$

and indeed Glimm and Jaffe have shown that the conjectured domain property is true (G J 1). In fact it holds for a larger class of polynomial selfinteractions in one dimension that have been studied by Glimm (Gl 2) and by Rosen (Ro 1) and are of type A :

$$V = \int dx\, g(x) \sum_{\nu}^{2n} c_\nu :\phi^\nu(x): \qquad c_{2n} > 0$$

An example of type B is provided by $(\phi^2)_3$. Since H is a bilinear form on smooth n-particle states and

$$R_{oo} = \underbrace{V \quad \Gamma(V_{02})}_{2}$$

is finite, $H_{ren} = H + R_{oo}$ is one, too. For $(\phi^2)_4$ which is not in B but of type C this argument fails, as we have seen earlier (sec.7).One verifies,however,that H_{ren} does indeed have a domain in Fock space, either by constructing an approximate dressing transformation or by taking a closer look at the exact solution. Recall that H was diagonalized by a canonical transformation

$$a_n \to \hat{a}_n$$

To show that there is a unitary operator U with

$$U\, a_n\, U^+ = \hat{a}_n$$

we use the formalism of generating functionals (Ar 1)

$$E(f,g) \equiv (\psi, e^{i\phi(f)+i\pi(g)}\psi)$$

If ψ is chosen to be $|0\rangle$, the ground state of H_o, we readily compute E to be

$$E_o(f,g) = \langle 0|e^{i\phi(f)+i\pi(g)}|0\rangle = e^{-1/4(f,\omega^{-1}f)-1/4(g,\omega g)}$$

If we choose the ground state of H' instead we must replace ω by $\hat{\omega}$ and obtain

$$\hat{E}(f,g) = e^{-1/4(f,\hat{\omega}^{-1}f)\ -1/4(g,\hat{\omega}g)}$$

The crucial question whether this state lies in Fock space can be decided through the following Lemma (Sh 1, KMP 1, Ea 1): $E_o(f,g)$ and $\hat{E}(f,g)$ correspond to the same CCR representation if

$$\omega^{-1/2}\hat{\omega}\,\omega^{-1/2} - 1$$

is a Hilbert Schmidt operator. For sufficiently small g this can indeed be verified by using a binomial expansion for

$$\hat{\omega} = (\omega^2 + g)^{1/2}$$

cf. (Ea 1). Another vastly more difficult model of type C is the Yukawa interaction $(\phi\bar{\psi}\psi)_2$ in one dimensional space. For it there obviously is no solution in closed form and as expected for a non-B theory

$$\langle\phi,\ H_{ren}\,\psi\ \rangle$$

is infinite for all the usual bare states ϕ, ψ, so that here the construction of physical states becomes a crucial first step in the investigation. As in the case of boson selfinteractions that we have discussed above the method of approximate dressing transformations implies infinite renormalizations for the ground state energy and the boson mass

$$R_{oo}(g) = \infty \qquad \delta m^2(g) = \infty$$

and produces a dense set of vectors on which the renormalized Hamiltonian is defined (Gl 1, He 1). Recent results (G J 4, 5, Ja 2) based on this procedure are

Selfadjointness of $H_{ren}(g)$
Locality of the fields
Existence of a vacuum vector $H_{ren}(g)\Phi_o(g)=0$
Equations of motion for ϕ and ψ .

9. INFINITE WAVE FUNCTION RENORMALIZATION

Up to this point we have only considered models that obey the condition C

$$\gamma v_{no} \in L^2(R^{sn})$$

While this is badly violated in all realistic situations one will first try to understand interactions where γv_{no} is "almost in L^2 ", i.e. where the integral over $|\gamma v_{no}|^2$ diverges no more than logarithmically. As examples in point physical states for $(\phi^4)_3$ were constructed by Glimm (Gl 3) and by Hepp. In keeping with Glimm's conjectures they do not correspond to vectors in Fock space. The resulting non-Fock representations of the canonical commutation relations have been studied systematically by Fabrey (Fa 1). In the following sketch of their techniques we recall that the violation of condition C is related to a term by term explosion of the power series for Ω. Hence to simplify our argument we shall disregard in this paragraph the truncation required to ensure series convergence. The blow up occurs in expressions like

$$\begin{aligned} ||\Omega|f>||^2 &= (F_{ro})^{+}\Omega^{+}\Omega\, F_{ro}\big|_{oo} \\ &= \underbrace{F^{+}\,F}_{r} + \underbrace{F^{+}\,F}_{r} \cdot \underbrace{\Gamma(V_{no})^{+}\quad\Gamma}_{n}(V_{no}) + \ldots \\ &= \underbrace{F^{+}\,F}_{r}\,(1+B+cB^2+\ldots) \end{aligned}$$

$$+ \text{ other graphs} . \qquad B \equiv \underbrace{\Gamma^+(V)\,\Gamma(V)}_{n}$$

The trick will be to recognize that those other graphs can also be grouped into classes which differ only by the number of bubbles B, so that the term in brackets is attached to each such class as an over all factor by which matrix elements may be renormalized.
To see this let

$$\frac{1}{k!}\,\frac{1}{\ell!}\,G_{o,k\ell}$$

be the graph of $||\Omega|f>||^2$ which contains none of the dangerous bubbles B and is of the orders k and ℓ in $\Gamma(V_{no})^+$ and $\Gamma(V_{no})$ *. Denoting the same graph but with m bubbles by

$$\frac{1}{(k+m)!}\,\frac{1}{(k+\ell)!}\,G_{m,k\ell}$$

the objective is to sum over all members of the above mentioned class, i.e. to take

$$\sum_{m=o}^{\infty}\frac{1}{(k+m)!}\,\frac{1}{(\ell+m)!}\,G_{m,k\ell} \equiv G_{k\ell}$$

To express $G_{m,k\ell}$ in terms of $G_{o,k\ell}$ and B a look at its graphical representation may be helpful:

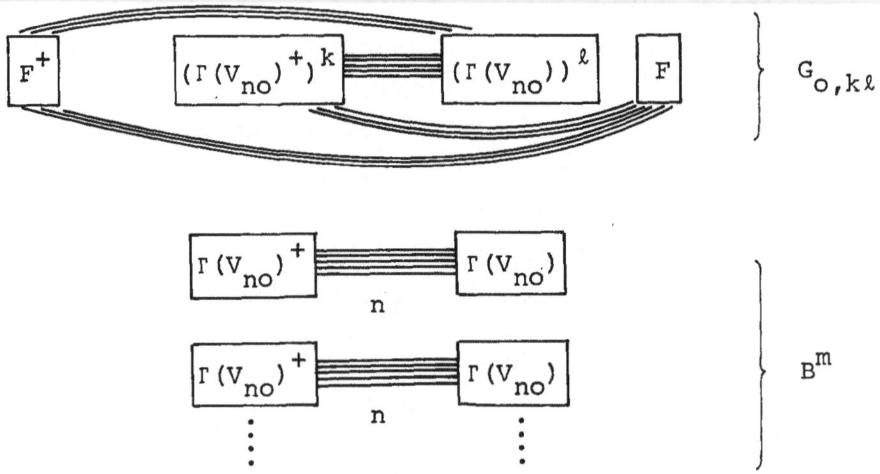

* $k \neq \ell$ will contribute if $\Omega^+\Omega$ is sandwiched between states of different particle number .

How many ways are there to form $G_{m,k\ell}$, given $G_{o,k\ell}$?

There are $\binom{m+k}{k}$ ways to select m of the m+k $\Gamma(V)^+$

There are $\binom{m+\ell}{\ell}$ ways to select of the +k $\Gamma(V)$

There are m! ways to contract them to bubbles B^m .

Hence

$$G_{k\ell} = \sum_m \frac{1}{(k+m)!} \frac{1}{(\ell+m)!} \binom{k+m}{k} \binom{\ell+m}{\ell} m! \, B^m \; k! \, \ell! \, G_{o,k\ell} =$$

$$= \sum_m \frac{B^m}{m!} G_{o,k\ell}$$

$$= e^B \, G_{o,k\ell}$$

This shows e^B to be a universal factor on all graphs of all matrix elements, and we recognize that a new inner product can be defined by

$$\langle f|f'\rangle_{ren} = \lim_{c\to\infty} e^{-B_c} \langle f|\Omega_c^+ \, \Omega_c|f'\rangle$$

with c some high energy cut-off to make B_c finite as long as $c<\infty$. A more detailed look at this procedure is required to prove positive definiteness

$$\langle g|g\rangle_{ren} > 0 \qquad \text{if} \qquad g \neq 0$$

for a sufficiently large set of states g and to verify that

$$\lim_{c\to\infty} \langle f|\Omega_c^+ \, H_c \, \Omega_c|g\rangle \equiv \langle f|H_\infty|g\rangle_{ren}$$

defines a symmetric operator H_∞ . This has been done for $(\phi^4)_3$ by Glimm (Gl 3) and by Hepp (He 1), cf. also (Fa 1), and marks the point to which work on these models has progressed.

In keeping with our effort to illustrate things with the help of simple models we shall focus once more on the quadratic perturbation

$$V \sim \int d^s x \; g(x) \; :\phi^2(x):$$

for which

$$\gamma V_{20} \sim \frac{\tilde{g}(k_1+k_2)}{\sqrt{\omega_1\omega_2}\,(\omega_1+\omega_2)}$$

is not in L^2 for s=4 so that we will expect $(\phi^2)_5$ to require an infinite wave function renormalization. For a suitable class of g(x) we can cast $\Gamma(V_{20})$ into the form

$$\Gamma(V_{20}) = \sum_n \lambda_n (a_n^+)^2$$

so that

$$\Omega = \bigotimes_n e^{-\lambda_n (a_n^+)^2}$$

and

$$B = \underbrace{\Gamma(V_{20})^+ \; \Gamma(V_{20})}_{2} = 2 \sum \lambda_n^2 = \infty$$

Now

$$\| e^{-\lambda (a^+)^2} |0\rangle \|^2 = (1 - 4|\lambda|^2)^{1/2}$$

showing that for convergence of the exponential series the coupling must be sufficiently small to ensure $|\lambda_n| < \frac{1}{2}$ for all n. In that case the renormalized

$$e^{-B/2} \Omega |0\rangle = \bigotimes_n e^{-\lambda_n^2} e^{-\lambda_n (a_n^+)^2} |0\rangle$$

$$= \bigotimes_n |\lambda_n\rangle$$

is a well defined product vector with the following properties

i) $\prod_n |\langle 0|\lambda_n\rangle| = \infty$

By the Klauder, McKenna and Woods theorem (KMW 1, St 1) this means that

$$\bigotimes_n |\lambda_n\rangle$$

generates a CCR representation inequivalent to Fock's, as suggested by Glimm's conjecture and in accordance with the nature of the exact solution

$$\hat{E}(f,g) = e^{-\frac{1}{4}(f,\hat{\omega}^{-1} f) \; -\frac{1}{4}(g,\hat{\omega} g)}$$

since (Ea 1) $\omega^{-1/2}\hat{\omega}\,\omega^{-1/2} - 1$ is not Hilbert Schmidt for s=4 (c.f. the lemma for equivalence in sec. 8).

ii) The above immediately suggests the question whether the approximately dressed vacuum appears as a vector in the space of the exact solution $\hat{E}$. This is not trivial: we are only guaranteed to obtain some representation of the CCR in which H_{ren} makes sense, while what we call the exact solution is a very special case, namely the one where H_{ren} is semibounded (Sr 1). Luckily the expectation value of

$$e^{i\phi(f) + i\pi(g)}$$

in the state

$$\bigotimes_n |\lambda_n\rangle$$

is, up to a finite normalization, of the form (Ea 1, KMP 1):

$$\hat{\hat{E}}(f,g) \equiv e^{-1/4(f,\hat{\hat{\omega}}^{-1}f)-1/4(g,\hat{\hat{\omega}}g)}$$

and $\hat{\hat{\omega}}$ is such that

$$\hat{\omega}^{-1/2}\;\hat{\hat{\omega}}\;\hat{\omega}^{-1/2} - 1$$

is a Hilbert Schmidt operator, thus by our lemma the approximate ground state gives rise to the same representation as the exact solution E.

iii) By the same techniques one verifies further that for f,g restricted to bounded subsets of momentum space $\hat{\hat{E}}(f,g)$ is equivalent to the Fock representation. In this sense $\hat{\hat{E}}$ is non-Fock only in the ultraviolet limit.

10. SUMMARY

Starting from the observation that for most local relativistic quantum field theory models

$$H\,\psi = \infty$$

for all the obvious states ψ, we have used the Friedrichs method to construct mappings

$$\Omega : D(H_o) \longrightarrow D(H_{ren})$$

How far $D(H_{ren})$ of a given model is removed from the domain $D(H_o)$ of the free Hamiltonian can be found from Glimm's perturbation theoretic classification which we reproduce here for $(\phi^n)_{s+1}$, (with L denoting the class discussed in 9.).

n \ s	1	2	3	4
1	A	A	A	A
2	A	B	C	L
3	A	C		
4	A	L		

We have seen how renormalizations of ground state energy, masses and wave-functions are imposed by consistency requirements and how for theories as singular as $(\phi\bar{\psi}\psi)_2$ or $(\phi^4)_3$, the renormalized Hamiltonian can thus be established as a densely defined, symmetric operator and a first step toward the construction of the corresponding dynamics is done.

Discussions with W. J. Eachus and J. R. Klauder are gratefully acknowledged.

REFERENCES

(Ar 1) H. Araki: Hamiltonian Formalism and the CCR in Quantum Field Theory, Journ. Math. Phys. 1, 492 (1960).

(CJ 1) J. Camon and A. Jaffe: Lorentz Covariance of the $\lambda(\phi^4)_2$ Quantum Field Theory. To appear.

(Ea 1) W. J. Eachus: A Solvable Model in Quantum Field Theory. Thesis, Syracuse, 1970.

(Fa 1) J. D. Fabrey: Exponential Representations of the Canonical Commutation Relations, Thesis, MIT, 1969.

(Fd 1) L. D. Faddeev: On the Friedrichs Model in the Theory of Perturbations of a Continuous Spectrum. Trudy Mat. Inst. Steklova 73, 292 (1964).

(Fr 1) K. O. Friedrichs: Perturbation of Spectra in Hilbert Space. Providence, Amer. Math. Soc. 1965.

(Gl 1) J. Glimm: Yukawa Coupling of Quantum Fields in Two Dimensions I, II. Comm. Math. Phys. 5, 343 (1967), 6, 120 (1967).

(Gl 2) J. Glimm: Boson Fields with Nonlinear Selfinteraction in Two Dimensions. Comm. Math. Phys. 8, 12 (1968).

(Gl 3) J. Glimm: Boson Fields with the $:\phi^4:$ Interaction in Three Dimensions. Comm. Math. Phys. 10, 1 (1968).

(Gl 4) J. Glimm: The Foundations of Quantum Field Theory; preprint.

(Gl 5) J. Glimm: Models for Quantum Field Theory (Varenna Lectures). Academic Press, New York, 1969.

(GJ 1) J. Glimm and A. Jaffe: A $\lambda\phi^4$ Quantum Field Theory without Cutoffs I. Phys. Rev. 176, 1945 (1968).

(GJ 2) J. Glimm and A. Jaffe: $\lambda(\phi^4)_2$ Quantum Field Theory Without Cutoffs II: The Field Operators and the Approximate Vacuum. Ann. of Math., to appear.

(GJ 3) J. Glimm and A. Jaffe: The $\lambda(\phi^4)_2$ Quantum Field

Theory without Cutoffs III: The Physical Vacuum. To appear.

(GJ 4) J. Glimm and A. Jaffe: A Model for Yukawa Quantum Field Theory. Phys. Rev. Letters 23, 1362 (1969).

(GJ 5) J. Glimm and A. Jaffe: Self-Adjointness of the Yukawa Hamiltonian; preprint 1969.

(He 1) K. Hepp: Theorie de la renormalization, Springer-Verlag, Berlin 1969.

(Ho 1) L. van Hove: Les difficultés de divergences pour un modèle particular de champ quantifié. Physica 18, 145 (1952).

(Ja 1) A. Jaffe: Constructive Quantum Field Theory; Lecture notes,Zürich, 1968.

(Ja 2) A. Jaffe: Princeton seminar talk, unpublished, 1969.

(Jo 1) R. Jost: The General Theory of Quantized Fields. Providence, Amer. Math. Soc. 1965.

(Ka 1) T. Kato: Perturbation Theory for Linear Operators; Springer-Verlag, Berlin, 1966.

(Ka 2) T. Kato: Wave Operators and Similarity for some Non-selfadjoint Operators; Math. Ann. 162, 258 (1966).

(Kl 1) J. R. Klauder: Hamiltonian Approach to Quantum Field Theory; Acta Phys. Austriaca, Suppl. VI, 167 (1969).

(KMW 1) J. R. Klauder, J. McKenna, and E. J. Woods: Direct Product Representations of the CCR: Journ. Math. Phys. 7, 822 (1966).

(KMP 1) P. Kristensen, L. Mejlbo and E. Poulsen: Tempered Distributions in Infinitely Many Dimensions III; Comm. Math. Phys. 6, 29 (1967).

(Ma 1) G. W. Mackey; Mathematical Foundations of Quantum Mechanics; Benjamin, New York, 1963.

(Os 1) K. Osterwalder: Thesis, Zürich (1969).

(Pa 1) S. Parrott: Uniqueness of the Hamiltonian in Quantum Field Theories; Comm. Math. Phys. 13, 68 (1969).

(Ro 1) L. Rosen: A $\lambda\phi^{2n}$ Field Theory without Cutoffs; Comm. Math. Phys. 16, 157 (1970).

(Sr 1) B. Schroer: Infrateilchen in der Quantenfeldtheorie; Fortschritte d. Phys. 11, 1 (1963).

(Sw 1) S. Schweber: Relativistic Quantum Field Theory; Row and Peterson, New York, 1961.

(Sh 1) D. Shale: Linear Symmetries of Free Boson Fields; Trans. Am. Math. Soc. 103, 149 (1962).

(SW 1) R. F. Streater and A. S. Wightman: PCT, Spin and Statistics, and All That; Benjamin, New York 1964

(St 1) L. Streit: Test Function Spaces for Direct Product Representations of the CCR; Comm. Math. Phys. 4, 22 (1967).

(St 2) L. Streit: A Generalization of Haag's Theorem; Nuovo Cim. 62A, 673 (1969).

(Yo 1) K. Yosida: Functional Analysis; Springer-Verlag, Berlin, 1966.

Acta Physica Austriaca, Suppl. VII, 392—442 (1970)

THE O(3,1) ANALYSIS OF CURRENTS AND CURRENT COMMUTATORS[*]

BY

W. RÜHL
CERN - Geneva

SUMMARY

A current density operator $j_\mu(x)$ between one-particle states

$$\langle p_2, q_2 | j_\mu(o) | p_1, q_1 \rangle$$

is the simplest example of a covariant (in this case: vector) operator connecting two (in this case: reducible unitary) representations of the group SL(2,C). In these lectures we review first the representation theory of the group SL(2,C). This enables us to derive a decomposition theorem for covariant operators with heuristic arguments. By this theorem we can decompose any covariant operator into a direct integral of irreducible covariant operators, which are defined to connect two irreducible representations of SL(2,C). This formalism is then applied to several problems of physical interest.

a. We analyse current matrix elements. The interpretation of the singularities of the Fourier transforms leads

---

[*] Lecture given at IX. Internationale Universitätswochen für Kernphysik, Schladming, February 23 - March 7, 197o.

us to the concept of dynamical groups. We obtain asymptotic expansions of the form factors.

b. We analyse diagonal matrix elements of current commutators

$$\int d^4x\, e^{ikx} \langle p,q | [j^a_\mu(x), j^b_\nu(o)] | p,q \rangle$$

The singularities of the Fourier transforms are interpreted as Lorentz poles and cuts. We obtain asymptotic expansions in $\nu = kp/M$. The photon mass is fixed.

c. We analyse the same matrix elements in a different fashion such that asymptotic expansions in powers of ν for fixed

$$k^2/2kp$$

result. An interpretation of the singularities of the Fourier transforms leads to a novel picture of inelastic electron-proton scattering.

1. THE REPRESENTATION THEORY OF THE MATRIX GROUP SL (2,C)

1.1 Homogeneous functions

Let $F(z_1,z_2)$ be not necessarily analytic functions of two complex variables z_1,z_2. We call them homogeneous if for any complex $\alpha \neq 0$ the relation

$$F(\alpha z_1, \alpha z_2) = \alpha^{\mu} \bar{\alpha}^{\mu'} F(z_1,z_2) \tag{1.1}$$

is satisfied. μ and μ' are complex numbers. In order that (1.1) reduces to an identity for $\alpha = \exp 2\pi i$, we must have

$$\mu' - \mu = m \quad \text{integral} \tag{1.2}$$

We define the symbol χ, the type of the homogeneity,

by the pair of numbers

$$\chi = \{n_1, n_2\}$$
$$n_1 = \mu+1$$
$$n_2 = \mu'+1 \tag{1.3}$$

We shall later use also the notation

$$\chi = (M,\lambda)$$

where

$$n_1 = -M+\lambda$$
$$n_2 = +M+\lambda \tag{1.4}$$

From (1.2), (1.3) and (1.4) we have m=2M.

We define a linear vector space D_χ as follows.

a. D_χ is a linear vector space of functions $F(z_1,z_2)$ of the type χ;

b. Every function $F(z_1,z_2) \varepsilon D_\chi$ is infinitely differentiable with exception of the point $z_1 = z_2 = 0$;

c. D_χ is a topological vector space whose topology is defined by sequences. A sequence $\{F_n(z_1,z_2)\}$ is said to converge to zero, if it converges together with all its derivatives to zero in every compact domain not containing the point $z_1 = z_2 = 0$.

One can show that the space D_χ is closed.

In D_χ we define an operator T_a^χ for every $a\varepsilon SL(2,C)$ by

$$(z_1', z_2') = (z_1, z_2)\begin{pmatrix} a_{11} & a_{12} \\ a_{21} & a_{22} \end{pmatrix} \tag{1.5}$$

(this is a matrix multiplication) and

$$T_a^\chi F(z_1,z_2) = F(z_1',z_2') \tag{1.6}$$

These operators T_a^χ are continuous on D_χ. In addition they satisfy the group law

$$T^{\chi}_{a_1} T^{\chi}_{a_2} = T^{\chi}_{a_1 a_2} \tag{1.7}$$

Nevertheless we do not want to speak yet of representations.

1.2 Other realizations of the spaces D_{χ}.

Because of the homogeneity of the functions $F(z_1,z_2)$ we may consider one of the variables z_1,z_2 as superfluous and eliminate it. It is, for example, sufficient to give the values of F for fixed $z_2=1$. Let us define therefore

$$f(z) = F(z,1) \tag{1.8}$$

Due to the homogeneity (1.1) we then have

$$F(z_1,z_2) = z_2^{n_1-1} \bar{z}_2^{n_2-1} f(\frac{z_1}{z_2}) \tag{1.9}$$

It follows that $f(z)$ allows an asymptotic expansion at $z=\infty$

$$f(z) \cong z^{n_1-1} \bar{z}^{n_2-1} \sum_{j,k\geq o}^{\infty} c_{jk} z^{-j} \bar{z}^{-k} \tag{1.1o}$$

This expansion obviously depends on χ. Since the correspondence between $f(z)$ and $F(z_1,z_2)$ is one-to-one due to (1.8) and (1.9), we can realize the space D_{χ} by functions $f(z)$. Note that the expansion (1.1o) guarantees the infinite differentiability of F at $z_2=0$, the asymptotic expansion (1.1o) is therefore indispensable. The topology is simply carried over, though we do not want to explicitly formulate what this amounts to.

In this new realization the operators T^{χ}_{a} take on the form

$$T^{\chi}_{a} f(z) = \alpha^{\chi}(z,a) f(z_a) \tag{1.11}$$

where we use the notations

$$z_a = \frac{a_{11}z+a_{21}}{a_{12}z+a_{22}}$$

$$\alpha^{\chi}(z,a) = |a_{12}z+a_{22}|^{2\lambda-2} \exp[-2i\, M \arg(a_{12}z+a_{22})] =$$

$$= (a_{12}z+a_{22})^{n_1-1} (\bar{a}_{12}\bar{z}+\bar{a}_{22})^{n_2-1} \qquad (1.13)$$

For some matrices of triangular shape we introduce the notations

$$\zeta = \begin{pmatrix} 1 & 0 \\ z & 1 \end{pmatrix}, \quad \zeta_a = \begin{pmatrix} 1 & 0 \\ z_a & 1 \end{pmatrix}, \quad k = \begin{pmatrix} \delta^{-1} & \gamma \\ 0 & \delta \end{pmatrix} \qquad (1.14)$$

where z, z_a, γ, δ are complex numbers. With the notations (1.12) to (1.14) we have the transformation formula (for right cosets of the subgroup K of matrices of the shape k)

$$\zeta a = k\zeta_a \qquad (1.15)$$

and

$$\delta(k) = \delta(z,a) = a_{12}z+a_{22}$$

$$\alpha^{\chi}(z,a) = \delta(z,a)^{n_1-1}\, \overline{\delta(z,a)}^{n_2-1} \qquad (1.16)$$

which can be verified easily by direct computation.

Among the infinity of other possibilities to realize the spaces D_{χ} there is a third one of some practical importance. It is obviously also sufficient to know the functions $F(z_1,z_2)$ on a unit sphere

$$z_1 = (|z_1|^2 + |z_2|^2)^{1/2} u_{21}$$

$$z_2 = (|z_1|^2 + |z_2|^2)^{1/2} u_{22}$$

$$|u_{21}|^2 + |u_{22}|^2 = 1 \qquad (1.17)$$

Setting $u_{11}=\bar{u}_{22}$ and $u_{12}=-\bar{u}_{21}$, we obtain a matrix $u\varepsilon SU(2)$

$$u = \begin{pmatrix} u_{11} & u_{12} \\ u_{21} & u_{22} \end{pmatrix}$$

which is therefore uniquely fixed by z_1 and z_2. We define

$$\phi(u) = F(u_{21}, u_{22}) \tag{1.18}$$

such that the homogeneity implies

$$F(z_1, z_2) = (|z_1|^2+|z_2|^2)^{\lambda-1} \phi(u(z_1, z_2)) \tag{1.19}$$

The functions $\phi(u)$ satisfy a constraint following from (1.1) and (1.18) which reduces the number of independent real parameters on which they depend to two, namely

$$u(\psi) = e^{\frac{i}{2}\psi\sigma_3}$$

$$\phi(u(\psi)u) = F(e^{-\frac{i}{2}\psi} u_{21}, e^{-\frac{i}{2}\psi} u_{22})$$

$$= e^{iM\psi} \phi(u) \tag{1.2o}$$

We say that $\phi(u)$ is covariant on right cosets of the group U(1) of matrices $u(\psi)$ in SU(2).

Since the functions $\phi(u)$ are defined on a compact manifold which can be regarded as a compact submanifold of the manifold of variables z_1 and z_2, the spaces D_χ have now been realized by infinitely differentiable functions $\phi(u)$ which are submitted to the single constraint (1.2o). This realization is therefore of particular importance if we have to deal with linear continuous functionals on D_χ. These are distributions in a standard sense. The operators T^χ_a can be written

$$T^\chi_a \phi(u) = \alpha^\chi(u,a)\phi(u_a) \tag{1.21}$$

where u_a is obtained from

$$u\,a = k\,u_a$$

$$k = \begin{pmatrix} \delta^{-1}(u,a) & \gamma \\ o & \delta(u,a) \end{pmatrix} \qquad (1.22)$$

i.e., similar as in (1.15) u is a representative of a right coset of the subgroup K of triangular matrices k on which SL(2,C) acts by right translation. In addition we have

$$\alpha^{\chi}(u,a) = \delta(u,a)^{n_1-1}\,\overline{\delta(u,a)}^{n_2-1} \qquad (1.23)$$

The matrix k is determined up to a factor $u(\psi)$. This ambiguity cancels, however, in (1.21) because of the constraint (1.2o).

1.3 The completion of D_χ and the principal series

We complete the spaces D_χ, $\chi=(M,\lambda)$ with the scalar product norm

$$||\phi||^2 = \int |\phi(u)|^2 d\mu(u) \qquad (1.24)$$

where $d\mu(u)$ is the normalized Haar measure on SU(2). In this manner we obtain a Hilbert space $L^2_M(U)$ [U=SU(2)]. In the norm (1.24) the operators T^χ_a are continuous on D_χ. With the factorization

$$a = u_1 d\, u_2,\; u_{1,2} \in SU(2)$$

$$d = e^{\frac{1}{2}\eta\sigma_3}, \eta \geqq o \qquad (1.25)$$

the norm of T^χ_a can easily be computed to be

$$||T^\chi_a|| = e^{\eta|\mathrm{Re}\,\lambda|} \qquad (1.26)$$

The operators T_a^χ can therefore be extended unambiguously and norm conservingly onto the Hilbert space $L_M^2(U)$. We see from (1.26) that the operators thus obtained have unit norm on the imaginary λ axis. We can show indeed that the scalar product implied by (1.24) is invariant in this case, so that the operators T_a^χ are isometric. Since the relation (1.7) can also be extended on $L_M^2(U)$, the operator T_a^χ has an inverse $T^\chi_{a^{-1}}$. It is unitary if λ is imaginary. We have found the

Theorem

For every χ the equations (1.21) to (1.23) define a representation in the Hilbert space $L_M^2(U)$. These representations are unitary at least if λ is purely imaginary.

The unitary representations obtained in this fashion are denoted the "principal series". This does not mean that all other complex λ lead to non-unitary representations. It means only that a scalar product of the form (1.24) is invariant only for the principal series. There are other representations (λ is real in that case) for which a different invariant scalar product can be constructed, which leads to another series of unitary representations, the so-called "supplementary series". For the purposes of harmonic analysis of L^2 functions on SL(2,C) the principal series is, however, sufficient and we concentrate our interest on it. For non-unitary representations an operator norm like (1.26) depends on the realization. If we use the point transformation between the two realizations of D_χ in terms of functions $\phi(u)$ or functions $f(z)$ to rewrite the norm (1.24) we find

$$||f||^2_\lambda = \int |f(z)|^2 (1+|z|^2)^{-2\,\mathrm{Re}\,\lambda}\, dx\, dy \qquad (1.27)$$

$$z = x + i\, y$$

which is finite for every function f(z) of D_χ because of the expansion (1.1o). By completion we obtain a Hilbert space $L^2_\lambda(Z)$. The point transformation between the two realizations yields an isomorphism between the Hilbert spaces $L^2_\lambda(Z)$ and $L^2_M(U)$. For the half plane Re $\lambda \leqq 0$ we can also perform the completion of D_χ with the norm

$$||f||^2 = \int |f(z)|^2\, dx\, dy \qquad (1.28)$$

From this discussion we should keep in mind that for a non-unitary representation no a priori principle exists which selects a norm. For our purposes the norm (1.24), (1.27) turns out to be particularly useful, and we shall therefore use it throughout.

As a consequence of our definition of a norm (1.24), the space $L^2_M(U)$ is independent of λ. All representations χ with a fixed M can therefore be realized on the same carrier space $L^2_M(U)$. This allows us to consider the operators T^χ_a as operator functions of λ. For any pair $\phi_{1,2}(u) \in L^2_M(U)$ the matrix element

$$\int \overline{\phi_1(u)}\, T^\chi_a\, \phi_2(u) d\mu(u)$$

is in fact entire analytic in λ. Infinite differentiability in $\underline{a}$ can, however, only be guaranteed if $\phi_{1,2} \in D_\chi$.

1.4 Representation functions for the canonical basis

The theorem of Peter and Weyl for SU(2) (this is a special case of a Plancherel theorem) supplies us with an orthonormal basis in $L^2_M(U)$. In fact, this theorem tells us that the matrix elements

$$(2j+1)^{1/2} D^j_{q_1 q_2}(u)$$

of representations of SU(2) to the spin j [say with the conventions of Edmonds[1]] form an orthonormal basis in the Hilbert space $L^2(U)$ of measurable functions on SU(2) with finite scalar product norm (1.24).

The sub-basis

$$\phi^j_q(u) = (2j+1)^{1/2} D^j_{M,q}(u) \tag{1.29}$$

for $-j \leq q \leq +j$, $j = |M|, |M|+1, \ldots$ satisfies the covariance constraint (1.2o) and lies therefore in $L^2_M(U)$. We call it the canonical basis. Its elements are polynomials in the matrix elements of u and, therefore, lie also in the dense subspace D_χ. If we realize the space D_χ by functions $f(z)$ the canonical basis consists of functions $f^j_q(z)$ which can easily be computed. With the norm (1.27) we find

$$f^j_q(z) = \\ = \left[\frac{2j+1}{\pi} \frac{(j+M)!(j-M)!}{(j+q)!(j-q)!}\right]^{1/2} (1+z\bar{z})^{\lambda-j-1} G^j_q(z) \tag{1.3o}$$

where $G^j_q(z)$ is a polynomial in z and $\bar{z}$

$$G^j_q(z) \begin{cases} = z^{q-M} \binom{j+q}{q-M} {}_2F_1(-j+q,-j-M,q-M+1;-z\bar{z}) \\ \qquad \text{for } q \geq M \\ = (-\bar{z})^{M-q} \binom{j-q}{M-q} {}_2F_1(-j-q,-j+M,M-q+1;-z\bar{z}) \\ \qquad \text{for } M \geq q \end{cases} \tag{1.31}$$

The matrix elements of an operator A in the canonical basis are written

$$<\chi;j_2q_2|A|\chi;j_1q_1> = \int \overline{\phi_{q_2}^{j_2}(u)} A\, \phi_{q_1}^{j_1}(u)d\mu(u) \qquad (1.32)$$

The matrix elements of T_a^χ

$$D^\chi_{j_2q_2\;j_1q_1}(a) = <\chi;j_2q_2|T_a^\chi|\chi;j_1q_1> \qquad (1.33)$$

are called representation functions. In the case that we use a realization of D_χ in terms of functions $f(z)$, we may present (1.32), (1.33) in the alternative form

$$<\chi;j_2q_2|A|\chi;j_1q_1> = (-1)^{q_2-M} \int f_{-q_2}^{j_2}(z)^{-\chi}\, A\, f_{q_1}^{j_1}(z)^{\chi}\, dx\, dy \qquad (1.34)$$

where $f_{-q_2}^{j_2}(z)^{-\chi}$ belongs to the canonical basis in $D_{-\chi}$,

$$-\chi = (-M,-\lambda) \text{ if } \chi = (M,\lambda) \qquad (1.35)$$

We can verify (1.34) easily by means of (1.27), (1.3o), (1.31). In this context we want to mention that the invariance of D_χ under T_a^χ implies that the representation function (1.33) tends to zero faster than any power of $j_2(j_1)$ if $j_2(j_1)$ goes to infinity and q_1, q_2, j_1 (j_2) are fixed. The group law $T_{a_2}^\chi T_{a_1}^\chi = T_{a_2a_1}^\chi$ expressed by the representation functions (1.33) involves a summation which consequently converges rapidly. In (1.34) we made use of the fact that the spaces D_χ, $D_{-\chi}$ form a dual pair in the (not quite usual) sense that any $f_2 \varepsilon D_{-\chi}$ generates an invariant continuous linear functional on D_χ and vice versa. For $f_1 \varepsilon D_\chi$ this linear functional has the form

$$(f_2,f_1) = \int f_2(z)f_1(z)dx\, dy \qquad (1.36)$$

The invariance of (1.36) is easily verified.

With the factorization (1.25) the representation functions (1.33) can be reduced to functions $d^{\chi}_{j_2 j_1 q}(\eta)$ defined by

$$D^{\chi}_{j_2 q_2\, j_1 q_1}(a) = \sum_q D^{j_2}_{q_2 q}(u_1) d^{\chi}_{j_2 j_1 q}(\eta) D^{j_1}_{q q_1}(u_2) \qquad (1.37)$$

namely

$$D^{\chi}_{j_2 q_2\, j_1 q_1}(d) = \delta_{q_2 q_1}\, d^{\chi}_{j_2 j_1 q_1}(\eta)\,,\quad d = e^{\frac{1}{2}\eta\sigma_3}$$

$$D^{\chi}_{j_2 q_2\, j_1 q_1}(u) = \delta_{j_2 j_1}\, D^{j_1}_{q_2 q_1}(u) \qquad (1.38)$$

If we restrict the representation χ of SL(2,C) to SU(2) it reduces into irreducible representations of SU(2) with spins $j=|M|,\ |M|+1,\ \ldots$ The multiplicity of these representations is one. The canonical basis can therefore be characterized (up to a normalization) by the reduction chain

$$SL(2,C) \supset SU(2) \supset U(1)$$

Inserting the reduced rotation functions $d^{j}_{q_1 q_2}(z)$ into the matrix element (1.33) of the representation function yields the integral representation

$$d^{\chi}_{j_1 j_2 q}(\eta) = (2j_1+1)^{\frac{1}{2}}(2j_2+1)^{\frac{1}{2}} \times$$

$$\times \frac{1}{2}\int_{-1}^{+1} dt\, d^{j_1}_{Mq}(t)\, d^{j_2}_{Mq}(t_d)\,[\mathrm{ch}\,\eta - t\,\mathrm{sh}\,\eta]^{\lambda-1} \qquad (1.39)$$

where

$$t_d = \frac{t-\mathrm{th}\,\eta}{1-t\,\mathrm{th}\,\eta} \qquad (1.40)$$

Inserting the known Jacobi polynomials $d^j_{Mq}(z)$ into (1.39) allows us to perform the integration. We obtain

$$d^{\chi}_{j_1 j_2 q}(\eta) =$$

$$= \{(2j_1+1)(2j_2+1)(j_1+M)!(j_1-M)!(j_2+M)!(j_2-M)! \times$$
$$\times (j_1+q)!(j_1-q)!(j_2+q)!(j_2-q)!\}^{1/2} \times$$
$$\times (2\mathrm{sh}\,\eta)^{-j_1-j_2-1} \sum_{\nu\mu} C_{\nu\mu} \frac{2\mathrm{sh}(\lambda-j_2+\nu)\eta}{\lambda-j_2+\nu} e^{(j_1-\mu)\eta} \tag{1.41}$$

where the coefficients $C_{\nu\mu}$ are certain rational numbers. They are defined by the sum

$$C_{\nu\mu}(j_1 j_2 q M) = (-1)^{\mu} \sum_{n_1 n_2} [n_1!(j_1+q-n_1)!(n_1-q+M)! \times$$
$$\times (j_1-n_1-M)!n_2!(j_2+q-n_2)!(n_2-q+M)!(j_2-n_2-M)!]^{-1} \times$$
$$\times \binom{j_1+j_2-n_1-n_2+q-M}{-n_1+\frac{1}{2}(\nu+\mu+q-M)} \binom{n_1+n_2-q+M}{n_2-\frac{1}{2}(\nu-\mu+q-M)} \tag{1.42}$$

The labels ν and μ run over a domain of integers restricted by

$$\nu+\mu = \max(\nu+\mu) - 2s = \min(\nu+\mu) + 2t$$
$$\nu-\mu = \max(\nu-\mu) - 2s' = \min(\nu-\mu) + 2t'$$

s, s', t, t' are integers. The extrema of the domain are easily computed to be

$$\max(\mu-\varepsilon\nu) = (1-\varepsilon)j_2+2j_1-|q+\varepsilon M|$$
$$\min(\mu-\varepsilon\nu) = -(1+\varepsilon)j_2+|q+\varepsilon M| \tag{1.43}$$

The computation of the coefficients $C_{\nu\mu}$ simplifies considerably, if we take three symmetry relations into account which are easily read off (1.42). These relations are

$$
\begin{aligned}
C_{\nu\mu}(j_1 j_2 qM) &= C_{\nu\mu}(j_1 j_2 Mq) \\
&= (-1)^{2j_1} C_{\nu,2j_1-\mu}(j_1 j_2 q,-M) \\
&= (-1)^{j_1+j_2} C_{j_1+j_2-\nu,j_1+j_2-\mu}(j_2 j_1 qM)
\end{aligned}
\tag{1.44}
$$

In addition these coefficients satisfy certain sum rules which follow from the fact that the limit

$$\lim_{\eta\to 0} \eta^{-|j_1-j_2|} d^{\chi}_{j_1 j_2 q}(\eta)$$

must be finite. A few of these sum rules and a big set of coefficients are tabulated in the Appendix.

Besides the representation functions $d^{\chi}_{j_1 j_2 q}(\eta)$ which we call more specifically "representation functions of the first kind" from now on, representation functions of the second kind are of a certain importance for asymptotic expansions. They lack a direct group theoretical interpretation. Instead we define them directly by

$$e^{\chi}_{j_1 j_2 q}(\eta) =$$

$$= \{\ldots\}^{1/2} (2\mathrm{sh}\ \eta)^{-j_1-j_2-1} \sum_{\nu\mu} C_{\nu\mu}(\lambda-j_2+\nu)^{-1} \exp(\lambda+j_1-j_2+\nu-\mu)\eta \tag{1.45}$$

where the curly bracket is the same as in (1.41). These functions fall off exponentially in one half plane for $\eta\to\infty$ or $\lambda\to\infty$. This is a property also known from

Legendre functions of the second kind and cylinder functions of the third kind (Hankel functions). The exact asymptotic behaviour of the functions (1.45) can be obtained easily by means of (1.43).

If we make use of the combination

$$C_{\nu\mu}(j_2 j_1 q,-M) = (-1)^{j_1-j_2} C_{j_1+j_2-\nu, j_1-j_2+\mu}(j_1 j_2 qM) \quad (1.46)$$

of the symmetry relations (1.44), we obtain the following relation connecting functions of the first and the second kind

$$d^{\chi}_{j_1 j_2 q}(\eta) = e^{\chi}_{j_1 j_2 q}(\eta) + (-1)^{j_1-j_2} e^{-\chi}_{j_2 j_1 q}(\eta) \quad (1.47)$$

It is more difficult to prove the relation

$$d^{\chi}_{j_1 j_2 q}(\eta) = e^{\chi}_{j_1 j_2 q}(\eta) + \beta^{j_1}(-\lambda)\beta^{j_2}(\lambda) e^{-\chi}_{j_1 j_2 q}(\eta) \quad (1.48)$$

where $\beta^{j}(\lambda)$ is defined by

$$\beta^{j}(\lambda) = \frac{\Gamma(j+\lambda+1)}{\Gamma(j-\lambda+1)} \quad (1.49)$$

1.5 Reducible and equivalent representations

Symmetry relations for the functions $d^{\chi}_{j_1 j_2 q}(\eta)$ can be directly deduced from (1.47), (1.48), the most important one of which is

$$d^{\chi}_{j_1 j_2 q}(\eta) = \beta^{j_1}(-\lambda)\beta^{j_2}(\lambda) d^{-\chi}_{j_1 j_2 q}(\eta) \quad (1.5o)$$

Let us denote

$j_o' = \lambda, j_o = M, j_o' - j_o > o$ integral

so that both n_1, n_2 (1.4)

$$n_{1,2} = \mp M+\lambda$$

are positive integers. Since the functions $d^{\chi}_{j_1 j_2 q}(\eta)$ are entire analytic in λ, the poles of the Γ functions in (1.5o) must be cancelled by first order zeros of the d functions. This yields

$$d^{\chi}_{j_1 j_2 q}(\eta) = o \quad \text{for} \quad j_o \leqq j_2 < j_o', \; j_1 \geqq j_o'$$

$$d^{-\chi}_{j_1 j_2 q}(\eta) = o \quad \text{for} \quad j_o \leqq j_1 < j_o', \; j_2 \geqq j_o'$$

This means that the subspace E_χ of D_χ spanned by the basis elements with

$$j_o \leqq j < j_o'$$

is an invariant subspace. Its dimension is finite

$$\dim E_\chi = \sum_{j=j_o}^{j_o'-1} (2j+1) = j_o'^2 - j_o^2 = n_1 n_2 \tag{1.51}$$

As we see from (1.3o), (1.31) this subspace consists of polynomials in z and $\bar{z}$, precisely it consists of all polynomials admitting the asymptotic expansion (1.1o). Moreover we have that the basis elements of $D_{-\chi}$ with

$$j \geqq j_o'$$

span an invariant subspace $F_{-\chi}$ of $D_{-\chi}$ with infinite dimension. This result can be summarized in the following theorem (the rigorous proof of which is tedious):

Theorem

The representations χ are completely irreducible except in the case that n_1, n_2 are both positive or both negative integers.

We define an operator A_χ on the canonical basis of D_χ onto the canonical basis in $D_{-\chi}$ by

$$A_\chi \phi_q^j(u)^\chi = C(\chi)\,\beta^j(\chi)\,\phi_q^j(u)^{-\chi} \tag{1.52}$$

where $C(\chi)$ is a normalization constant. For $j\to\infty$ $\beta^j(\chi)=\beta^j(\lambda)$ is bounded by a power of j

$$|\beta^j(\lambda)| \approx j^{2\mathrm{Re}\,\lambda} \tag{1.53}$$

Since the expansion coefficients of an element of D_χ in the canonical basis decrease rapidly A_χ can be continued on D_χ and establishes a continuous map from D_χ into $D_{-\chi}$ whenever both χ and $-\chi$ are completely irreducible. It possesses the continuous inverse $A_{-\chi}$. Because of the symmetry relation (1.5o) for the representation functions from which we started, it intertwines the representations on D_χ and $D_{-\chi}$

$$T_a^{-\chi} A_\chi = A_\chi T_a^\chi \tag{1.54}$$

For the principal series and the norm (1.24) we can make A_χ an isometric operator by an appropriate adjustment of $C(\chi)$. It can therefore be continued on the Hilbert space $L_M^2(U)$, say. Using the standard notion of equivalence of unitary representations, we end up with the

Theorem

Two representations χ and $-\chi$ of the principal series are equivalent.

If χ is not completely irreducible, say, n_1, n_2 are both positive integers, the operator A_χ defined by (1.52) annihilates the subspace E_χ and yields a continuous map of the quotient space D_χ/E_χ on $F_{-\chi}$. It intertwines the representations induced by χ on these spaces. If n_1, n_2 are both negative integers, A_χ is defined on the whole space D_χ only if we let $C(\chi)$ assume a first order zero in an appropriate fashion. Then A_χ annihilates F_χ and establishes a continuous map of D_χ/F_χ on $E_{-\chi}$. It intertwines the representations on these spaces. Taking other limits of $C(\chi)$ it is possible to obtain continuous intertwining operators A'_χ which are only defined on the invariant subspaces E_χ respectively F_χ. They map the invariant spaces on the quotient spaces and represent the inverses of the former operators A_χ.

This is the complete scheme of equivalences (strictly: intertwining operators) which is connected with the symmetry relation (1.5o) for the representation functions. It is not yet complete in general. In the case that n_1, n_2 are positive integers there exists in addition a bicontinuous intertwining operator B_χ from D_χ/E_χ onto $D_{\chi'}$ (and similarly onto $D_{-\chi'}$) where

$$\chi' = \{-n_1, n_2\}$$

is completely irreducible. The relation between the representation functions following from the existence of this operator B_χ is

$$d^{\chi}_{j_1 j_2 q}(\eta) = \delta^{j_2}(\delta^{j_1})^{-1}\, d^{\chi'}_{j_1 j_2 q}(\eta) \tag{1.55}$$

with

$$j_1, j_2 \geqq \tfrac{1}{2}(n_1+n_2) = \lambda > |M|$$

$$\delta^j = \left[\frac{(j-M)!\,(j+\lambda)!}{(j+M)!\,(j-\lambda)!}\right]^{1/2} \tag{1.56}$$

Contrary to (1.5o), the symmetry relation (1.55) is valid only for a discrete set of half-integral λ's.

We finally mention that all the intertwining operators can be given in the form of pseudo-differential operators on D_χ realized by functions $f(z)$. In this form they have been studied first by Gelfand and his co-workers(2).

2. COVARIANT OPERATORS AND THEIR REDUCTION

2.1 A survey of the problems

Finite dimensional irreducible representations of SL(2,C) are often characterized by two half integers j_1, j_2 such that j_1 denotes the spin of the proper spinor part and j_2 the spin of the conjugate spinor part. These representations can be identified with the finite dimensional representations on the spaces of polynomials E_χ obtained in section 1.5 by setting

$$\chi = \{n_1, n_2\} = [j_1, j_2]$$

with

$$n_1 = 2j_1 + 1, \; n_2 = 2j_2 + 1$$

Let us consider the two two-dimensional representations $\chi = [½, o]$ and $\hat{\chi} = [o, ½]$. Vectors in the respective spaces E are often given as spinors ξ_α, $\eta_{\dot{\alpha}}$, $\alpha = \pm ½$ [note that we can in general define a spinor basis which differs from the canonical basis (1.29), (1.3o)]. With the help of the Pauli matrices we can map the tensor product $E_\chi \times E_{\hat{\chi}}$ into the space $E_{\chi'}$, $\chi' = [½, ½]$, which carries the four-vector representation, by

$$x_\mu = \sum_{\alpha\beta} (\sigma_\mu)^{\dot\alpha\beta} \eta_{\dot\alpha}\xi_\beta$$

This map is covariant in the following sense: A simultaneous transformation T_a^χ, $T_a^{\hat\chi}$ in E_χ, $E_{\hat\chi}$ induces a transformation $T_a^{\chi'}$ in $E_{\chi'}$.

We want to generalize this concept in this chapter by admitting tensor products $D_{\chi_1} \times D_{\chi_2}$ which are mapped by an "irreducible covariant operator" into a given space $E_{\chi'}$. Of course, we could also admit image spaces $D_{\chi'}$ such that χ' is completely irreducible. This would simplify the problem in general. The problem to construct such irreducible covariant operators is obviously connected with the problem to reduce a tensor product representation. In the case that all three χ_1, χ_2, χ' belong to the principal series both problems are in fact equivalent. A complete solution has been found for them by Naimark. If all three representations are completely irreducible but do not all belong to the principal series, we restrict them to the spaces D. In this case the covariant operators can be obtained from Naimark's solution by analytic continuation. If χ' is continued into a point where $D_{\chi'}$ possesses an invariant subspace $E_{\chi'}$, these irreducible covariant operators become singular. But with a method similar to the one used to obtain the intertwining operators, we can pick out that part which has the image space $E_{\chi'}$. It turns out to be a differential operator(3).

Our next task is to reduce a covariant operator, like the matrix element of the current density between single-particle states

$$\langle p_2, q_2 | j_\mu(o) | p_1, q_1 \rangle$$

into a direct integral of irreducible covariant operators. First we must define rigorously what we mean by a covariant operator in general. We shall find a definition which is general enough to cover all cases of interest for us. In the case that all three representations χ_1, χ_2, χ' belong to the principal series the problem of decomposing a covariant operator into irreducibles is easily reduced to Naimark's theorem on the tensor product decomposition. In turn Naimark's theorem can be deduced from the Plancherel theorem. We have used a method of analytic continuation to establish the decomposition for also four-vector operators and for antisymmetric and symmetric tensor covariance[(3),(4),(5)]. The singularities met in the analytic continuation were evaluated in the canonical basis. For this reason we did not establish a formula for the decomposition of a general covariant operator but only for the cases mentioned. The major ingredients of the general formula can, however, be guessed easily, as we shall see later. An essential complication in proving the decomposition formula is due to the fact that this formula involves contributions from a "discrete series" of irreducible covariant operators in addition to the principal series whenever

$$\max(j_1, j_2) > \frac{1}{2}$$

A simple trick allows us to deal with this discrete series when only the principal series is known. In the case $j_1=j_2=0$ our covariant operators can always be represented by one distribution on the group SL(2,C) and its reduction is equivalent with the Plancherel decomposition of this distribution. The discrete series is known to be absent in this special case, indeed.

2.2 Irreducible covariant operators

We define an irreducible covariant operator $T(z|z')$ to be a map from $D_{\chi_1}(Z)$ into the space $E_{\chi'}(Z')$ of polynomials in the variables z', $\bar{z}'$, such that the coefficients of these polynomials lie in $D_{\chi_2}(Z)$. Explicitly

$$f(z)\varepsilon D_{\chi_1} \to T(z|z')f(z) = \sum_{i=o}^{2j_1} \sum_{j=o}^{2j_2} c_{ij}\, z'^{i}\, \bar{z}'^{j}$$

$$c_{ij} = c_{ij}(z)\varepsilon D_{\chi_2} \qquad (2.1)$$

We require in addition that the coefficients depend continuously on the element of $D_{\chi_1}(Z)$ and that

$$T^{\chi_2}_{a^{-1}} T(z|z') T^{\chi_1}_{a} = \qquad (2.2)$$

$$= (a_{12}z'+a_{22})^{2j_1}(\bar{a}_{12}\bar{z}'+\bar{a}_{22})^{2j_2} T(z|z'_a)$$

Due to the "duality" of the pair of spaces $D_{\chi_2}, D_{-\chi_2}$ this definition is essentially equivalent to the definition of an irreducible covariant operator as a map of $D_{\chi_1} \times D_{-\chi_2}$ into $E_{\chi'}$ with separate continuity in either argument and a covariance property similar to (2.2). We call χ' the covariance of the operator. Such operators exist only if χ_1 and χ_2 are related as

$$\pm\chi_2 = \chi_1 + \alpha \qquad (2.3)$$

$$\alpha = (\Delta M, \Delta\lambda)$$

where ΔM and $\Delta\lambda$ are half integers. We are interested only in the upper sign of (2.3) since the covariant operator for the lower sign can be obtained from the operator with the upper sign by left multiplication with the intertwining operator A_{χ_2} which maps D_{χ_2} into $D_{-\chi_2}$. The covariant operators satisfying all constraints of our definition turn out to be differential operators[6]

$$T_{\nu\mu}(z|z') =$$

$$= (z-z')^{A_1}(\bar{z}-\bar{z}')^{B_1}\left(\frac{\partial}{\partial z}\right)^{\nu}\left(\frac{\partial}{\partial \bar{z}}\right)^{\mu}(z-z')^{A_2}(\bar{z}-\bar{z}')^{B_2} =$$

$$= \sum_{k=o}^{\nu}\sum_{\ell=o}^{\mu} k!\ell!\binom{\nu}{k}\binom{\mu}{\ell}\binom{A_2}{k}\binom{B_2}{\ell} \times \tag{2.4}$$

$$\times\ (z-z')^{2j_1-k}(\bar{z}-\bar{z}')^{2j_2-\ell}\left(\frac{\partial}{\partial z}\right)^{\nu-k}\left(\frac{\partial}{\partial \bar{z}}\right)^{\mu-\ell}$$

For one covariance $\chi'=[j_1,j_2]$ we obtain $(2j_1+1)(2j_2+1)$ different irreducible covariant operators labelled by ν and μ which range over the integers in the intervals

$$o \leqq \nu \leqq 2j_1$$

$$o \leqq \mu \leqq 2j_2 \tag{2.5}$$

The powers $A_{1,2}$ and $B_{1,2}$ are defined as

$$A_1 = -M_1+\lambda_1+\nu$$

$$B_1 = +M_2+\lambda_2+\mu$$

$$A_2 = 2j_1-A_1$$

$$B_2 = 2j_2-B_1 \tag{2.6}$$

and for α (2.3) we find

$$\alpha = (\nu-\mu+j_2-j_1, j_1+j_2-\nu-\mu) \tag{2.7}$$

Instead of using the parameters ν and μ we shall label the different covariant operators for one covariance from now on by their α. It is sometimes useful to plot the α's in the ΔM $\Delta\lambda$ plane in order to visualize which types of irreducible covariant operators exist. The Figure[*] presents an example. We shall call such a plot an α-diagram. It is invariant under inversion at the origin.

The operators $T_\alpha(z|z')$ (2.4) can be expanded in the canonical basis of E_χ, (1.3o)

$$T_\alpha(z|z') =$$

$$= \sum_{|j_1-j_2|\leq j\leq j_1+j_2} \sum_{-j\leq Q\leq j} T^j_Q(z|\alpha) f^j_Q(z')^{\chi'} \tag{2.8}$$

Matrix elements of these operators defined as by (1.32), (1.34) can be reduced by the Wigner-Eckart theorem of SU(2) as

$$\langle\chi;J_2Q_2|T^J_Q(\alpha)|\chi-\alpha;J_1Q_1\rangle = \tag{2.9}$$

$$= (-1)^{2J}(2J_1+1)^{-\frac{1}{2}}(J_2Q_2;JQ|J_1Q_1)\langle\chi;J_2||T^J(\alpha)||\chi-\alpha;J_1\rangle$$

The number of independent matrix elements which have actually to be computed can further be reduced by three kinds of symmetry relations. The most important symmetry is "Weyl's symmetry" which relates the matrix elements for two equivalent representations χ and $-\chi$

$$\langle-\chi;J_2||T^J(-\alpha)||-\chi+\alpha;J_1\rangle =$$

$$= (-1)^{2j_2}\beta^{J_2}(\lambda)\beta^{J_1}(-\lambda+\Delta_\alpha\lambda)\langle\chi;J_2||T^J(\alpha)||\chi-\alpha;J_1\rangle \tag{2.1o}$$

The covariance is the same on both sides. Another symmetry is of importance if we extend SL(2,C) by parity, we call it the symmetry under "parity conjugation". If we set

---

[*] See appendix

$$\hat{\chi} = (-M,\lambda) \quad \text{for} \quad \chi = (M,\lambda)$$

$$\hat{\alpha} = (-\Delta M,\Delta\lambda) \quad \text{for} \quad \alpha = (\Delta M,\Delta\lambda)$$

we have

$$<\hat{\chi};J_2 || T^J(\hat{\alpha}) || \hat{\chi}-\hat{\alpha};J_1> =$$

$$= (-1)^{J_2+J-J_1+M'+\Delta_\alpha M} <\chi;J_2 || T^J(\alpha) || \chi-\alpha;J_1> \qquad (2.11)$$

The covariant operators belong to covariances χ' and $\hat{\chi}'$, respectively. Finally, we have the symmetry under "transposition"

$$<\chi;J_2 || T^J(\alpha) || \chi-\alpha;J_1> =$$

$$= (-1)^{J-M'} <-\chi+\alpha;J_1 || T^J(\alpha) || -\chi;J_2> \qquad (2.12)$$

which, in some cases, can be connected with time reversal. In general we can express the matrix elements of the irreducible covariant operators (2.4) by interpolated 9j symbols. For the simple cases of physical interest (four-vector and tensors of rank two) they are all tabulated in(4),(5).

2.3 Covariant operators and their decomposition

We consider the space C^∞ of infinitely differentiable functions f(a) on SL(2,C) which have compact support. A covariant operator can be defined as a linear map T(z) from the tensor product space $C_1^\infty \times C_2^\infty$ into the space $E_{\chi'}(Z)$ so that under all right translations in C^∞

$$T(z)(T_a^r f_1 \times T_a^r f_2) =$$

$$= (a_{12}z+a_{22})^{2j_1}(\bar{a}_{12}\bar{z}+\bar{a}_{22})^{2j_2}T(z_a)(f_1\times f_2) \tag{2.13}$$

In addition we require continuity of T(z) in both arguments f_1 and f_2. Such covariant operator can formally be presented as

$$T(z)(f_1\times f_2) =$$
$$= \int \Gamma(z;a_2,a_1)f_2(a_2)f_1(a_1)d\mu(a_2)d\mu(a_1) \tag{2.14}$$

where Γ is a polynomial in z, $\bar{z}$ with distributions as coefficients. $d\mu(a)$ is the Haar measure on SL(2,C) which we normalize as follows. We project SL(2,C) on the flat space R^6 by

$$\begin{pmatrix} a_{11} & a_{12} \\ a_{21} & a_{22} \end{pmatrix} = \begin{pmatrix} \delta^{-1} & \gamma \\ o & \delta \end{pmatrix}\begin{pmatrix} 1 & o \\ z & 1 \end{pmatrix} \tag{2.15}$$

$$a_{22} \neq o$$

$$\delta = \delta_1+i\delta_2,\ \gamma = \gamma_1+i\gamma_2,\ z = x+iy$$

and set

$$d\mu(a) = (2\pi)^{-4}\, d\delta_1 d\delta_2 d\gamma_1 d\gamma_2 dx\, dy \tag{2.16}$$

If we expand the function $\Gamma(z)$ in the canonical basis of $E_{\chi'}(Z)$ we get

$$\Gamma(z;a_2,a_1) = \sum_{JQ} \Gamma^J_Q(a_2,a_1)f^J_Q(z)^{\chi'} \tag{2.17}$$

where the coefficients $\Gamma^J_Q(a_2,a_1)$ are distributions satisfying the covariance constraint

$$\Gamma^J_Q(a_2a^{-1},a_1a^{-1}) = \sum_{J'Q'} D^{\chi'}_{JQJ'Q'}(a)\Gamma^{J'}_{Q'}(a_2,a_1) \tag{2.18}$$

which replaces (2.13).

For the remainder of this chapter we restrict our discussion to the class of "smooth" covariant operators, which are generated by infinitely differentiable functions $\Gamma^J_Q(a_2,a_1)$. We assume further that the functions $\Gamma^J_Q(a,e)$ (e is the group unit) fall off faster than any power of $|a|$

$$|a|^2 = \mathrm{Tr}(aa^+) \tag{2.19}$$

for $|a|\to\infty$. In physical applications we are in general not dealing with smooth covariant operators in this sense. Consequently, we are forced to extend our formalism.

We can make a further reduction of the class of covariant operators. Applying a Fourier transformation on SU(2) we get

$$\Gamma^J_Q(a_2,a_1)\begin{matrix} S_2r_2 \\ q_2\end{matrix},\begin{matrix} S_1r_1 \\ q_1\end{matrix} =$$

$$= \int D^{S_2}_{q_2r_2}(u_2^{-1})D^{S_1}_{r_1q_1}(u_1)\Gamma^J_Q(u_2a_2,u_1a_1)d\mu(u_2)d\mu(u_1) \tag{2.20}$$

where $d\mu(u)$ is the normalized Haar measure on SU(2). For a fixed set of superscripts these projections behave as

$$\Gamma^J_Q(u_2a_2,u_1a_1)_{q_2q_1} =$$

$$= \sum_{q_1'q_2'} D^{S_2}_{q_2q_2'}(u_2)D^{S_1}_{q_1'q_1}(u_1^{-1})\Gamma^J_Q(a_2,a_1)_{q_2'q_1'} \tag{2.21}$$

We can invert the projection (2.2o) by

$$\Gamma^J_Q(a_2,a_1) =$$

$$= \sum_{q_1q_2S_1S_2}(2S_1+1)(2S_2+1)\Gamma^J_Q(a_2,a_1)\begin{matrix} S_2q_2 \\ q_2\end{matrix},\begin{matrix} S_1q_1 \\ q_1\end{matrix} \tag{2.22}$$

In the applications we shall see that the projection

(2.2o) is most appropriate for dealing with matrix elements of operators between simple-particle states with timelike momentum.

However, this is only a practical argument. In principle it suffices completely to decompose operators with the property (2.21) due to the inversion formula (2.22). In the sequel we omit the superscripts in (2.2o) as we have done already in (2.21).

We observe that the matrix elements

$$<\chi;S_2q_2|T^{\chi}_{a_2}T^J_Q(\alpha)T^{\chi-\alpha}_{a_1^{-1}}|\chi-\alpha;S_1q_1> \qquad (2.23)$$

have exactly the same covariance properties (2.18), (2.21) as the functions $\Gamma^J_Q(a_2,a_1)_{q_2q_1}$ provided the irreducible covariant operators $T^J_Q(\alpha)$ belong to the covariance χ'. We suggest therefore the decomposition formula

$$\Gamma^J_Q(a_2,a_1)_{q_2q_1} =$$

$$= +i\int_{-i\infty}^{+i\infty} d\lambda \sum_M \sum_\alpha F_\alpha(M,\lambda)<\chi;S_2q_2|T^{\chi}_{a_2}T^J_Q(\alpha) \times$$

$$\times\ T^{\chi-\alpha}_{a_1^{-1}}|\chi-\alpha;S_1q_1> + \text{discrete series} \qquad (2.24)$$

We call the functions $F_\alpha(M,\lambda)$ the Fourier transforms. The main part of the expansion theorem consists in expressing these Fourier transforms by the functions Γ^J_Q. If we investigate this Fourier transformation [consequently, we might call (2.24) the inverse Fourier transformation] we find that the Fourier transforms consist in fact of two factors. One is the proper Fourier transform, which, due to the assumed smoothness of the covariant operator is an entire analytic function of λ which goes to zero faster than any power of $|\lambda|$ on any line parallel to the imaginary λ axis. The other factor is a function $C_\alpha(M,\lambda)$ which is rational in λ and plays the same role as the weight function

$$(2J+1)\pi \text{ ctg } \pi(J-\varepsilon), \ \varepsilon = o, \ \frac{1}{2}$$

in the Plancherel theorem for the group SU(1,1). In particular, its poles are connected with the discrete series.

In order to incorporate the discrete series in the integral (2.24) we use the Weyl symmetry of the Fourier transforms

$$F_{-\alpha}(-\chi) = (-1)^{2j_2} \beta^{S_2}(-\lambda)\beta^{S_1}(\lambda-\Delta_\alpha\lambda)F_\alpha(\lambda) \tag{2.25}$$

the Weyl symmetry of the matrix elements of $T_Q^J(\alpha)$ (2.1o), and the relation between functions of the first and second kind (1.48).

We find

$$\Gamma_Q^J(d(\eta),e)_{q_2q_1} = 2i \int_{C_\pm} d\lambda \sum_M \sum_\alpha F_\alpha(M,\lambda) \times$$

$$\times \sum_{J_2} e^\chi_{S_2,J_2q_2}(\eta) \langle\chi;J_2q_2|T_Q^J(\alpha)|\chi-\alpha;S_1q_1\rangle \tag{2.26}$$

where the contour C_+ (C_-) is defined as follows. It consists of two infinite intervals on the imaginary axis that are connected by a half circle. The radius and position of this half circle is chosen such that it goes around the poles of the functions $C_\alpha(M,\lambda)$ and the functions of the second kind, which are all placed at half integral or integral points on the real axis, in the positive (negative) sense. It is directed upwards. Because of the rapid fall-off of $F_\alpha(M,\lambda)$ along the lines $\text{Re } \lambda = \lambda_o$ and its holomorphy to the right of C_+ (to the left of C_-) we may displace the contour to the right (to the left). We assume that after such displacement the contour lies on Re $\lambda=\lambda_o$ with its infinite parts. We can then estimate the right-hand side of (2.26) by

$$c_1(\lambda_o)|\text{ch }\eta|^{-|\lambda_o|+c_2}, \eta\rightarrow-(\text{sign }\lambda_o)\infty$$

where c_2 is independent of λ_o. It decreases therefore faster than any power of ch η as required by the left-hand side.

The form (2.26) of the expansion formula does not only present the easiest way to deal with the discrete series, but is also the starting point for asymptotic expansions of Γ_Q^J in powers of ch η when the covariant operator is not smooth.

3. THE O(3,1) ANALYSIS OF CURRENT OPERATORS

3.1 Introductory remarks

In this part of our lectures we are concerned with applications of our formalism to the O(3,1) analysis of current operators. We do not intend to go through the details of the calculations but rather concentrate on the most interesting points. First, we have to prepare the matrix elements such that they coincide with matrix elements of certain covariant operators in the fashion as they appear in our formalism. This is a procedure which is almost always the same and we present it therefore only once. Second, we want to know which Fourier transforms actually are non-zero and how their analytic form is restricted by the additional inputs as: parity invariance, time reversal invariance, current conservation, and crossing symmetry. Third, we give a physical model interpretation of the singularities of the Fourier transforms. This alone can lead us to predictions. It differs in all three cases discussed. Finally, we present some of the most interesting implications.

3.2 Current matrix elements

We sandwich the current density operator $j_\mu(x)$ between one-particle states

$$\Gamma_\mu(p_2,p_1)_{q_2q_1} = N_1^{-1}N_2^{-1}\langle p_2,q_2|j_\mu(o)|p_1,q_1\rangle \tag{3.1}$$

where $p_{1,2}$ are the four-momenta, $q_{1,2}$ are spin projections on a fixed third co-ordinate axis (Wigner spin, no helicities), and $N_{1,2}$ are normalization factors which normalize the vertex function in a standard fashion. First we replace the vector basis by the canonical basis in E_χ, for $\chi'=[\frac{1}{2}, \frac{1}{2}]$. This is a trivial issue. Next, we want to replace the four-momenta in the arguments by group elements of SL(2,C). We define boosts a(p) by

$$\underset{\sim}{p}^R = a(p)\ \underset{\sim}{p}\ a(p)^\dagger \tag{3.2}$$

with

$$\underset{\sim}{p} = p_o e + p_1\sigma_1 + p_2\sigma_2 + p_3\sigma_3 \tag{3.3}$$

In order to make the solution of (3.2) unique we can in addition impose the constraints

$$a(p) = a(p)^\dagger,\ \mathrm{Tr}\ a(p) > o \tag{3.4}$$

This permits us to define

$$\Gamma'^J_Q(a(p_2),a(p_1))_{q_2q_1} = \Gamma^J_Q(p_2,p_1)_{q_2q_1} \tag{3.5}$$

(we shall later drop the prime). Next we want to extend the definition of Γ'^J_Q from the submanifold of Hermitian positive matrices to the whole manifold SL(2,C). We know that any matrix of SL(2,C) admits a polar decomposition

$$a = u\, a(p) \tag{3.6}$$

where $u \in SU(2)$. In view of (3.2) we may say that a acts on the particle first by rotating it at rest with the rotation u^{-1} and then boosting it up by $a(p)^{-1}$ to the momentum p. This leads us to the definition

$$\Gamma'^J_Q(u_2 a(p_2), u_1 a(p_1))_{q_2 q_1} =$$

$$= \sum_{q_1' q_2'} D^{S_2}_{q_2 q_2'}(u_2) D^{S_1}_{q_1' q_1}(u_1^{-1}) \Gamma'^J_Q(a(p_2), a(p_1))_{q_2' q_1'} \tag{3.7}$$

which is already equivalent to the covariance relation (2.21). The correct SL(2,C) covariance under right multiplication of a_2, a_1 by a^{-1} (2.18) can be verified easily.

These functions generate a covariant operator in the general sense but not a smooth covariant operator. In fact we have

$$q^2 = (p_2 - p_1)^2 = M_1^2 + M_2^2 - 2M_1 M_2 \operatorname{ch} \eta \tag{3.8}$$

with the same variable η as in (2.26). Infinite differentiability can therefore be assumed if the masses M_1 and M_2 are equal. But in general we have both to cut off the vertex function at $q^2 \to -\infty$ and to regularize it.

If the regularization and the cut-off are removed, the Fourier transforms show up singularities which may impinge on our contour $C_\pm$ of (2.26) and force us to deform it. The hypothesis that in all our applications a deformation of the contour (that may depend on M and α) is sufficient to take account of the singularities is crucial for our approach. In addition, we do not want to get contributions from infinity if we perform a parallel displacement of the contours $C_{\pm}$ to the right or left. From the Riemann-Lebesgue lemma it can be seen that local integrability of the form factors is sufficient for this to hold whenever the singularities do not accumulate at $\pm i\infty$ in any strip of finite width parallel to the imaginary λ axis.

The single-particle states span a Hilbert space which carries an irreducible unitary representation of the Poincaré group. If we restrict the Poincaré group to the homogeneous Lorentz group, this representation reduces into a direct integral of irreducible representations of the homogeneous Lorentz group which all belong to the principal series. Each vector of the single-particle space is mapped onto a finite set of L^2 functions defined on the imaginary axis, the functions of the set labelled by $|M| \leq S$. The current density operator is submitted to a simultaneous decomposition into a direct integral of irreducible four-vector operators. These irreducible four-vector operators do not all act between pairs of representations of the principal series, but some connect also non-unitary representations with the representations of the principal series. This puzzle is solved by the fact that the domain of the current density operator is not the whole single-particle Hilbert space but only a dense subspace of it. A vector of this domain does not yield an arbitrary L^2 function under the reduction but only functions which possess analytic continuations into a strip of sufficient width to contain the non-unitary representations required by the irreducible four-vector operators.

The homogeneous Lorentz group plays a second role as the smallest possible dynamical symmetry group which is compatible with Lorentz covariance. If our Fourier transforms were rational functions, the vertex functions resulting from (2.26) would reproduce one of Barut's ansatzes for a dynamical group SL(2,C). However, it is easy to see that rational functions cannot imply the correct analytic structure of the vertex functions. This is one of the known short-comings of Barut's model indeed.

Nevertheless, we want to accept the idea of a dynamical group, but in the weaker version, where we do not look for a global representation of the Fourier transforms by rational functions, but are interested instead only in poles of the Fourier transforms close to the imaginary axis. This means that we abandon a global representation of the vertex functions, too, but instead obtain asymptotic expansions at $q^2 \to -\infty$. The dynamical group model is linked to the concept of towers of particles and resonances. We could make use of the idea of such towers by assuming that the poles in the Fourier transformed vertex function of the proton are correlated with poles in the Fourier transformed vertex function for the transition of the proton into resonances, provided these resonances and the proton can be fitted into one tower. We have done this [7]. But the predictions are not specific enough to be compared with the presently still rather poor experimental data on the asymptotic behaviour of the resonance production form factors.

We are therefore forced to apply our formalism in this case on a purely phenomenological level [8].

In the case of the electromagnetic current between nucleon states we have six non-vanishing Fourier transforms

$$F_{(o,1)}{}^{(\pm\frac{1}{2},\lambda)}, F_{(o,-1)}{}^{(\pm\frac{1}{2},\lambda)}, F_{(1,o)}{}^{(\frac{1}{2},\lambda)}, F_{(-1,o)}{}^{(-\frac{1}{2},\lambda)} \tag{3.9}$$

They are connected by Weyl's symmetry (2.25). From parity invariance follows

$$F_{(o,\pm 1)}(\tfrac{1}{2},\lambda) = F_{(o,\pm 1)}(-\tfrac{1}{2},\lambda)$$

$$F_{(1,o)}(\tfrac{1}{2},\lambda) = -F_{(-1,o)}(-\tfrac{1}{2},\lambda) \qquad (3.1o)$$

Current conservation implies for the "smoothed" vertex function

$$F_{(o,1)}(\tfrac{1}{2},\lambda)-F_{(o,1)}(\tfrac{1}{2},\lambda+1)+F_{(o,1)}(\tfrac{1}{2},-\lambda) -F_{(o,1)}(\tfrac{1}{2},-\lambda+1) = o \qquad (3.11)$$

which we assume to hold true also after the cut-off and the regularization is removed. Time reversal invariance and the Hermiticity of the current imply finally

$$F_{(o,1)}(\tfrac{1}{2},\lambda)-F_{(o,1)}(\tfrac{1}{2},-\lambda+1) = o \qquad (3.12)$$

We assume that a pole in one of the independent Fourier transforms lies closest to the imaginary axis and compute how the form factors behave. For the pole (to the left of C_-)

$$F_{(1,o)}(\tfrac{1}{2},\lambda) \sim \frac{\beta_1}{\lambda-\lambda_1} + \frac{\bar{\beta}_1}{\lambda-\bar{\lambda}_1} \qquad (3.13)$$

+ mirror poles induced by Weyl's symmetry

we find the asymptotic behaviour

$$G_E \simeq C_E^1\left(-\frac{q^2}{M^2}\right)^{\lambda_1 - \frac{3}{2}}$$

$$G_M \simeq C_M^1\left(-\frac{q^2}{M^2}\right)^{\lambda_1 - \frac{3}{2}}$$

$$C_E^1 : C_M^1 = \left(\lambda_1 - \frac{1}{2}\right)^{-1} \qquad (3.14)$$

provided λ_1 is real. If λ_1 is complex,

$$\lambda_1 = \lambda_1' + i\lambda_1'' \tag{3.15}$$

we would get in addition to (3.14) oscillating factors of the kind

$$\cos[\lambda_1'' \lg(-\frac{q^2}{M^2}) + c]$$

A pole in $F_{(o,1)}(\frac{1}{2},\lambda)$ of the kind

$$F_{(o,1)}(\tfrac{1}{2},\lambda) \sim \frac{\beta_2}{\lambda-\lambda_2-1} + \frac{\bar{\beta}_2}{\lambda-\bar{\lambda}_2-1} \tag{3.16}$$

+ mirror poles induced by Weyl's symmetry

implies for real λ_2 [otherwise the situation is as in (3.14)]

$$G_E \simeq C_E^2(-\frac{q^2}{M^2})^{\lambda_2 - \frac{1}{2}}$$

$$G_M \simeq C_M^2(-\frac{q^2}{M^2})^{\lambda_2 - \frac{3}{2}}$$

$$C_E^2 : C_M^2 = -\frac{1}{2}(\lambda_2 + \frac{1}{2}) \tag{3.17}$$

We want to leave the discussion of these results to our readers.

3.3 Commutators of current density operators; photon mass fixed

We consider the matrix elements

$$M_{\mu\nu}^{ab}(k,p)_{q_2 q_1} =$$

$$= N^{-2} \int d^4x\, e^{ikx} \langle p,q_2|[j_\mu^a(x),j_\nu^b(o)]|p,q_1\rangle \tag{3.18}$$

which are diagonal in the momentum p of the timelike particle. k is a spacelike four-momentum

$$k^2 = -\mu^2, \mu > o \tag{3.19}$$

This amplitude describes the absorptive part of the forward scattering of a spacelike "charged" photon. The labels "a" and "b" denote a "charge" in a quite general fashion, for example an intrinsic parity of the current and an SU(3) component. The total cross-section for electron-proton scattering in the approximation of one-photon exchange is proportional to a matrix element of the type (3.18) if we set a=b equal the charge component of the SU(3) octet, the parity is that of a proper four-vector, and the momentum p belongs to the proton. The spins $q_1=q_2$ are summed over in that case.

The only new ingredient in the formal treatment of (3.18) is due to the spacelike momentum k. This leads to some technical complications. We define first ($s=\frac{1}{2}$ for example, in the case of the proton)

$$\Gamma^{ab}_{\mu\nu}(k,p)^S_q = \sum_{q_1 q_2 q_2'} D^S_{q_2 q_2'}(-i\sigma_2)(sq_2, sq_1)Sq) \times$$

$$\times M^{ab}_{\mu\nu}(k,p)_{q_2' q_1} \tag{3.2o}$$

Next, we map k and p on their respective boosts a(k), a(p) in SL(2,C). We define the boost a(k) by

$$\underset{\sim}{k}^R = a(k)\ \underset{\sim}{k}\ a(k)^\dagger$$

$$k^R = (o,o,o,\mu) \tag{3.21}$$

With $v \in SU(1,1)$ and $u \in SU(2)$ we obtain in this fashion amplitudes with the properties

$$\Gamma^{ab}_{\mu\nu}(va_2, ua_1)^S_q = \sum_{q'} D^S_{q'q}(u^{-1})\Gamma^{ab}_{\mu\nu}(a_2,a_1)^S_{q'}$$

$$\Gamma^{ab}_{\mu\nu}(a_2a^{-1},a_1a^{-1})^S_q = \sum_{\mu'\nu'}(\Lambda_a)^{\mu'}_{\mu}(\Lambda_a)^{\nu'}_{\nu}\Gamma^{ab}_{\mu'\nu'}(a_2,a_1)^S_q \tag{3.22}$$

Here Λ_a denotes the matrix $\Lambda \in SO(3,1)$ corresponding to $a \in SL(2,C)$. The covariance on the right cosets of SU(1,1) is according to the trivial representation of SU(1,1).

Finally, we can transform from the tensor basis to the canonical basis. We obtain contributions in all four irreducible representations $\chi'=[j_1,j_2]$ with

$\chi' = \begin{cases} [1,1]: & \text{symmetric traceless tensor of rank two;} \\ [1,o]: & \text{antisymmetric self-dual tensor;} \\ [o,1]: & \text{antisymmetric antiself-dual tensor;} \\ [o,o]: & \text{trivial symmetric tensor proportional to } g_{\mu\nu}. \end{cases}$

In the case of the proton matrix elements and two currents of the same SU(3) multiplet the total spin S is restricted to

$S = 0$ for the symmetric and

$S = 1$ for the antisymmetric tensors

If the spins q_2,q_1 are summed over in (3.18) we get $S=0$ automatically and only contributions from the symmetric tensors.

Matrix elements of the irreducible covariant operators in the canonical basis only are not appropriate to take account of the SU(1,1) covariance in the argument a_2. The correct recipe is to take the canonical basis on one side such that the SU(2) covariance in the argument a_1 can be taken care of, and to use a new basis, the "pseudo-basis" on the other side. This pseudobasis is obtained if we restrict the homogeneous functions

$F(z_1,z_2)$ (section 1.1) to a two-shell hyperboloid, each shell of which can be identified with the cosets $SU(1,1)/U(1)$. Applying the Plancherel theorem of $SU(1,1)$ to these homogeneous spaces gives the desired basis, which can therefore be characterized by the chain of restrictions

$$SL(2,C) \supset SU(1,1) \supset U(1)$$

The decomposition of the regularized matrix elements (3.22) looks then for each of the four χ'

$$\Gamma^J_Q(a_2,a_1)^S_q = i \int_{-i\infty}^{+i\infty} d\lambda \sum_{\alpha,\zeta} F_\alpha(\lambda,\zeta) \times$$

$$\times \lim_{J'\to o} \langle o\lambda;J'\zeta o|T^{(o,\lambda)}_{a_2} T^J_Q(\alpha) T^{\chi-\alpha}_{a_1^{-1}}|\chi-\alpha;Sq\rangle$$

$$+ \text{ discrete series} \tag{3.23}$$

The invariant M is necessarily zero and has therefore been dropped as an argument of the Fourier transforms. $\zeta=\pm 1$ labels the two shells of the hyperboloid. The limit $J'\to 0$ has to be performed as an analytic continuation from any point of the principal series of $SU(1,1)$ ($\mathrm{Re}\ J'=-\frac{1}{2}$) since the trivial representation of $SU(1,1)$ is not contained in the pseudobasis. It has Plancherel measure zero. The Fourier transforms are Weyl symmetric as

$$F_\alpha(\lambda,\zeta) = \frac{\Gamma(\lambda)\Gamma(S-\lambda+\Delta_\alpha\lambda+1)}{\Gamma(-\lambda)\Gamma(S+\lambda-\Delta_\alpha\lambda+1)} F_{-\alpha}(-\lambda,-\zeta) \tag{3.24}$$

It allows us to restrict the summation over ζ in (3.23) to one value. Since for a fixed ζ the representation functions in the mixed basis have already an exponential decrease in a half-plane depending only on the sign of η, we have instead of (3.23)

$$\Gamma^J_Q(d(\eta),e)^S_Q =$$

$$= 2i \int_{C_\pm} d\lambda \sum_\alpha F_\alpha(\lambda,+1) \lim_{J'\to o} \langle o\lambda;J',+1,o| T^{(o,\lambda)}_{d(\eta)} \times$$

$$\times T^J_Q(\alpha)|\chi-\alpha;SQ\rangle \qquad (3.25)$$

such that the discrete series is incorporated. Equation (3.25) can be used for asymptotic expansions when the cut-off is removed.

Our amplitude describes the elastic forward scattering of a photon. We should therefore expect that such asymptotic expansion is of the type invented by Toller[(9)]. If we set $a_2=e$ in (3.23) we can interpret the matrix elements in (3.23) as follows. The momentum of the photon and the tensor indices appear on the left-hand side, the momentum and spin of the time-like particle appear on the right-hand side. They are coupled by a "propagator" which propagates an "object" with zero four-momentum and with a spin transforming as the representation $\chi-\alpha$ of SL(2,C). It would be mathematically equivalent to set $a_1=e$ and to let the "object" propagated have the spin $\chi=(o,\lambda)$. However, it makes a difference whether the quantum numbers of the Lorentz poles are identified with $\chi-\alpha$ or with χ. In the first case we speak of a natural exchange of Lorentz poles, only this case gives reasonable results.

Before we discuss results we want to add that the two values of ζ are also connected by crossing symmetry. If we denote by ($\pm$) the symmetric respectively antisymmetric part of the amplitude $M^{ab}_{\mu\nu}$ (3.18) under the exchange $a\leftrightarrow b$, the Fourier transforms are related as

$$F^{(\pm)}_\alpha(\lambda,\zeta) = \mp(-1)^{j_2-j_1} F^{(\pm)}_\alpha(\lambda,-\zeta), \chi'=[j_1,j_2] \qquad (3.26)$$

We study the case of the proton and currents of equal parities in more detail. We use the invariant amplitudes of Dietz and Kupsch[10]

$$M_{\mu\nu}(k,p)_{q_2q_1} =$$

$$= \tilde{u}_{q_2}(p)[p_\mu p_\nu \rho_1 + k_\mu k_\nu \rho_2 + \frac{1}{2}(k_\mu p_\nu + k_\nu p_\mu)\rho_3 +$$

$$+ g_{\mu\nu}\rho_4 + \varepsilon_{\mu\nu\sigma\tau}\, p^\sigma \gamma_5 \gamma^\tau \rho_5 + \varepsilon_{\mu\nu\sigma\tau}\, k^\sigma \gamma_5 \gamma^\tau \rho_6 +$$

$$+ (\varepsilon_{\nu\lambda\sigma\tau}\, k_\mu - \varepsilon_{\mu\lambda\sigma\tau}\, k_\nu) p^\sigma k^\lambda \gamma_5 \gamma^\tau \rho_7] u_{q_1}(p) \qquad (3.27)$$

with

$$\varepsilon_{0123} = +1, \quad \gamma_5 = i\gamma_0\gamma_1\gamma_2\gamma_3$$

If the currents are conserved we have the constraints

$$2M\nu\rho_1 - \mu^2\rho_3 = 0$$

$$2\mu^2\rho_2 - M\nu\rho_3 - 2\rho_4 = 0$$

$$\rho_5 + \mu^2\rho_7 = 0$$

$$\nu = \frac{kp}{M} \qquad (3.28)$$

The invariant amplitudes ρ_i, i=1,2,3,4, possess the independent Fourier transforms

$$F_{(0,0)}(\lambda,+),\ F_{(0,2)}(\lambda,+),\ F_{(0,-2)}(\lambda,+) \text{ for } \chi'=[1,1]$$

$$F(\lambda,+) \text{ for } \chi'=[0,0] \qquad (3.29)$$

The spin S=0 state of the proton has parity +1 and $\chi-\alpha=(0,\lambda-\Delta_\alpha\lambda)$ has M=0. The singularities of the Fourier transforms (3.29) belong therefore to class I in the

Freedman-Wang-Toller classification [M=O and parity $\pi=(-1)^{S}$]. A simple pole at

$$\lambda_I = \lambda - \Delta_\alpha \lambda$$

implies for $\nu\to\infty$, μ fixed

$$M^2\rho_1 \simeq c_1 (\frac{2\nu}{\mu})^{\lambda_I -3}$$

$$\mu^2\rho_2 \simeq c_2 (\frac{2\nu}{\mu})^{\lambda_I -1}$$

$$M\mu\rho_3 \simeq c_3 (\frac{2\nu}{\mu})^{\lambda_I -2}$$

$$4\rho_4 \simeq c_4 (\frac{2\nu}{\mu})^{\lambda_I -1} \qquad (3.3o)$$

Since all terms in the constraints (3.28) occur with the same power, current conservation to highest order in ν can be achieved by a mere adjustment of the coupling constants (residues). It turns out, however, that this adjustment of the coupling constants is also sufficient to cancel the lower powers in the constraints (3.28).

The standard relation between the position of a Lorentz pole and the intercept of the parent Regge trajectory is

$$\lambda_I = \alpha(o)+1 \qquad (3.31)$$

The Walecka functions W_1, W_2 behave then asymptotically like

$$2\pi W_1 = \frac{\nu^2 M^2}{\mu^2}\rho_1 - \mu^2\rho_2 \simeq (\tfrac{1}{4}c_1 - c_2)(\frac{2\nu}{\mu})^{\alpha(o)}$$

$$2\pi W_2 = M^2\rho_1 \simeq c_1 (\frac{2\nu}{\mu})^{\alpha(o)-2} \qquad (3.32)$$

so that Pomeranchuk exchange implies that

$$\lim_{\nu\to\infty} \nu W_2 = \text{const.} \qquad (3.33)$$

The antisymmetric tensors behave exactly equal. They obtain contributions from class II poles $[M=0, \pi=(-1)^{S+1}]$ and class III poles $[M\neq 0$, in particular $M=\pm 1]$. The poles of class II satisfy current conservation automatically. For the details of all the results we refer to the original article(11).

3.4 Commutators of current density operators; photon mass tends to infinity

We study the same matrix element (3.18) as before. We define a new four-momentum vector Q by

$$Q = k+\alpha p$$

$$Q^2 = \ell^2 > 0 \quad \text{fixed}$$

$$Q_0 > 0 \qquad (3.34)$$

The manifold spanned by the variables

$$p,k,\mu^2 \ : \ p^2 = M^2, k^2 = -\mu^2, -\infty < k_0 < \infty$$

$$0 < \mu^2 < \infty$$

can be mapped one-to-one on the manifold of variables

$$p,Q,x^{-1} \ : \ p^2 = M^2, Q^2 = \ell^2, x = \frac{\mu^2}{2M\nu}$$

$$-\infty < x^{-1} < \infty$$

This enables us to define the new function

$$M'_{\mu\nu}(Q,p,x)_{q_2 q_1} = M_{\mu\nu}(k,p)_{q_2 q_1} \ F(\mu^2,x)^{-1} \qquad (3.35)$$

where $F(\mu^2)$ is an unknown function of μ^2 for the moment (see below). For fixed x this new function can be considered formally as a forward scattering amplitude $M'_{\mu\nu}$ for a timelike photon of momentum Q and be submitted to an O(3,1) analysis. Current conservation uses the vector k and not Q and entails therefore typical complications: correlations between different Lorentz poles. In order that an O(3,1) analysis leads to predictions we want to make the hypothesis of Lorentz poles for $M'_{\mu\nu}$. This hypothesis can be justified by the following model.

Solving (3.34) for α yields

$$\alpha M = -\nu+(\ell^2+\mu^2+\nu^2)^{1/2} \tag{3.36}$$

and for $\nu\to\infty$, x fixed

$$\alpha \approx x + O(\tfrac{1}{\nu}) \tag{3.37}$$

We can therefore interpret (3.34) such that the photon picks up a fraction x of the proton and forms with it a resonance of mass ℓ. This resonance is scattered off the remainder of the proton $(1-x)p$ in forward direction by an exchange of a Lorentz pole. The function $F(\mu^2)$ in (3.35) is to take the coupling of the photon to the resonance into account. This picture interpolates therefore between pure resonance production at $x\to 1$ and classical Lorentz pole exchange at $x\to o$. In the following we assume $x\neq o$, $x\neq 1$. Again we consider the proton and a pair of currents of equal parities.

The symmetric tensors have contributions from the Fourier transforms

$$F_{(o,o)}(\lambda), F_{(o,2)}(\lambda), F_{(o,-2)}(\lambda) \text{ for } \chi' = [1,1]$$

and

$$F(\lambda) \quad \text{for} \quad \chi' = [o,o]$$

and only class I Lorentz poles. Due to current conservation, a Lorentz pole at

$$\lambda-\Delta_\alpha \lambda = \lambda_I$$

entails another pole at λ_I-2. We make therefore the ansatz

$$F_{(o,o)}(\lambda) = \sum_{\nu=o}^{N} [\frac{A_\nu}{\lambda-\lambda_I+2\nu} + \frac{A'_\nu}{\lambda-\lambda_I+1+2\nu}] + \text{remainder}$$

$$F_{(o,2)}(\lambda) = \sum_{\nu=o}^{N} [\frac{B_\nu}{\lambda-\lambda_I-2+2\nu} + \frac{B'_\nu}{\lambda-\lambda_I-1+2\nu}] + \text{remainder}$$

$$F_{(o,-2)}(\lambda) = \sum_{\nu=o}^{N} [\frac{C_\nu}{\lambda-\lambda_I+2+2\nu} + \frac{C'_\nu}{\lambda-\lambda_I+3+2\nu}] + \text{remainder}$$

$$F(\lambda) = \sum_{\nu=o}^{N} [\frac{D_\nu}{\lambda-\lambda_I+2\nu} + \frac{D'_\nu}{\lambda-\lambda_I+1+2\nu}] + \text{remainder} \quad (3.38)$$

The constraints of current conservation imply

$$B_o = o$$

$$[1-(\frac{Mx}{\ell})^2]A_o - \frac{1}{\lambda_I-1}(\frac{Mx}{\ell})^2 B_1+\lambda_I(\lambda_I-1)(\lambda_I-2)C_o = o$$

$$D_o + \frac{1}{2}(\lambda_I-1)^2 A_o = o$$

$$D'_o + \frac{1}{2}(\lambda_I-2)^2 A'_o = \frac{2Mx}{\ell}\frac{\lambda_I-1}{\lambda_I}[(\lambda_I-1)A_o+B_1] \quad (3.39)$$

The asymptotic formulae following from (3.39) are

$$M^2\rho_1 \approx c_1 \; F(\mu^2)\,(\frac{2\nu}{\ell})^{\lambda_I - 2}$$

$$\ell^2\rho_2 \approx c_2 \; F(\mu^2)\,(\frac{2\nu}{\ell})^{\lambda_I - 3} \qquad (3.4o)$$

For the Walecka functions this implies

$$2\pi W_1 \approx c_1 \; \frac{1}{4}(\frac{Mx}{\ell})^{-1} \; F(\mu^2)\,(\frac{2\nu}{\ell})^{\lambda_I - 1}$$

$$2\pi W_2 \approx c_1 \; F(\mu^2)\,(\frac{2\nu}{\ell})^{\lambda_I - 2} \qquad (3.41)$$

The Pomeranchuk exchange implies now that both W_1 and νW_2 scale if and only if

$$F(\mu^2) \approx c\mu^{-2} \quad \text{for} \quad \mu^2 \to \infty$$

In any case $F(\mu^2)$ cancels in the ratio

$$\lim_{\substack{\nu\to\infty \\ x \text{ fixed}}} \frac{\nu W_2}{W_1} = 2Mx \qquad (3.42)$$

The last result is equivalent to

$$\lim_{\substack{\nu\to\infty \\ x \text{ fixed}}} \frac{\sigma_\ell}{\sigma_t} = o \qquad (3.43)$$

where σ_ℓ and σ_t denote the cross-sections of photo-production induced by a longitudinal, respectively transverse spacelike photon.

Again the antisymmetric tensors exhibit similar properties. Current conservation implies that class II and III poles are correlated. For the details we refer again to the original article[12].

APPENDIX

Tables of the coefficients $C_{\nu\mu}(j_1,j_2,q,M)$ for $M=0,\frac{1}{2},1,\frac{3}{2}$

Let n be defined by

$$n = 2\max(j_1,j_2)$$

Then, the following sum rules (1) to (n) are valid

(1) $\sum_{\nu,\mu} C_{\nu\mu} = 0$

(2) $\sum_{\nu,\mu} \mu C_{\nu\mu} = 0$

(3a) $\sum_{\nu,\mu} \nu C_{\nu\mu} = 0$

(3b) $\sum_{\nu,\mu} (\nu^2+3\mu^2) C_{\nu\mu} = 0$

(4a) $\sum_{\nu,\mu} \nu\mu C_{\nu\mu} = 0$

(4b) $\sum_{\nu,\mu} \mu(\nu^2+\mu^2) C_{\nu\mu} = 0$

(5a) $\sum_{\nu,\mu} (3\nu^2+5\mu^2) C_{\nu\mu} = 0$

(5b) $\sum_{\nu,\mu} \nu(\nu^2+5\mu^2) C_{\nu\mu} = 0$

(5c) $\sum_{\nu,\mu} (\nu^4+10\nu^2\mu^2+5\mu^4) C_{\nu\mu} = 0$

etc., $n \geq 6$. The summation extends over all ν and μ.

(A) $M = 0$

$C_{00}(0,0,0,0)$:

ν	0
μ	0
C	1

$C_{\nu\mu}(1,0,0,0)$:

ν	0	0	1
μ	0	2	1
C	1	1	-2

$C_{\nu\mu}(0,1,0,0)$:

ν	0	1	1
μ	0	-1	1
C	2	-1	-1

$C_{\nu\mu}(1,1,0,0)$:

ν	0	0	1	1	1	2	2
μ	0	2	-1	1	3	0	2
C	2	2	-1	-6	-1	2	2

$C_{\nu\mu}(1,1,1,0)$:

ν	0	1	1	2
μ	1	0	2	1
C	-1	1	1	-1

(B) $M = \frac{1}{2}$

$C_{\nu\mu}(\frac{1}{2},\frac{1}{2},\frac{1}{2},\frac{1}{2})$:

ν	0	1
μ	0	1
C	1	-1

$C_{\nu\mu}(\frac{1}{2},\frac{1}{2},-\frac{1}{2},\frac{1}{2})$:

ν	0	1
μ	1	0
C	-1	1

$C_{\nu\mu}(\frac{1}{2},\frac{3}{2},\frac{1}{2},\frac{1}{2})$:

ν	0	1	1	2	2
μ	0	-1	1	0	2
C	$\frac{3}{2}$	-1	-2	1	$\frac{1}{2}$

$C_{\nu\mu}(\frac{1}{2},\frac{3}{2},-\frac{1}{2},\frac{1}{2})$:

ν	0	1	1	2	2
μ	1	0	2	-1	1
C	$-\frac{3}{2}$	2	1	$-\frac{1}{2}$	-1

$C_{\nu\mu}(\frac{3}{2},\frac{1}{2},\frac{1}{2},\frac{1}{2})$:

ν	0	0	1	1	2
μ	0	2	1	3	2
C	$\frac{1}{2}$	1	-2	-1	$\frac{3}{2}$

$C_{\nu\mu}(\frac{3}{2},\frac{1}{2},-\frac{1}{2},\frac{1}{2})$:

ν	0	0	1	1	2
μ	1	3	0	2	1
C	-1	$-\frac{1}{2}$	1	2	$-\frac{3}{2}$

$C_{\nu\mu}(\frac{3}{2},\frac{3}{2},\frac{1}{2},\frac{1}{2})$:

ν	0	0	1	1	1	2	2	2	3	3
μ	0	2	-1	1	3	0	2	4	1	3
4C	3	6	-2	-17	-8	8	17	2	-6	-3

$C_{\nu\mu}(\frac{3}{2},\frac{3}{2},-\frac{1}{2},\frac{1}{2})$:

ν	0	0	1	1	1	2	2	2	3	3
μ	1	3	0	2	4	-1	1	3	0	2
4C	-6	-3	8	17	2	-2	-17	-8	3	6

(C) M = 1

$C_{\nu\mu}(1,1,0,1)$:

ν	0	1	1	2
μ	1	0	2	1
C	-1	1	1	-1

$C_{\nu\mu}(1,1,1,1)$:

ν	0	1	2
μ	0	1	2
4C	1	-2	1

$C_{\nu\mu}(1,1,-1,1)$:

ν	0	1	2
μ	2	1	0
4C	1	-2	1

(D) $M = \frac{3}{2}$

$C_{\nu\mu}(\frac{3}{2},\frac{3}{2},\frac{3}{2},\frac{3}{2})$:

ν	0	1	2	3
μ	0	1	2	3
36C	1	-3	3	-1

$C_{\nu\mu}(\frac{3}{2},\frac{3}{2},-\frac{3}{2},\frac{3}{2})$:

ν	0	1	2	3
μ	3	2	1	0
36C	-1	3	-3	1

$C_{\nu\mu}(\frac{3}{2},\frac{3}{2},\frac{3}{2},\frac{1}{2})$:

ν	0	1	1	2	2	3
μ	1	0	2	1	3	2
4C	-1	1	2	-2	-1	1

$C_{\nu\mu}(\frac{3}{2},\frac{3}{2},-\frac{3}{2},\frac{1}{2})$:

ν	0	1	1	2	2	3
μ	2	1	3	0	2	1
4C	1	-2	-1	1	2	-1

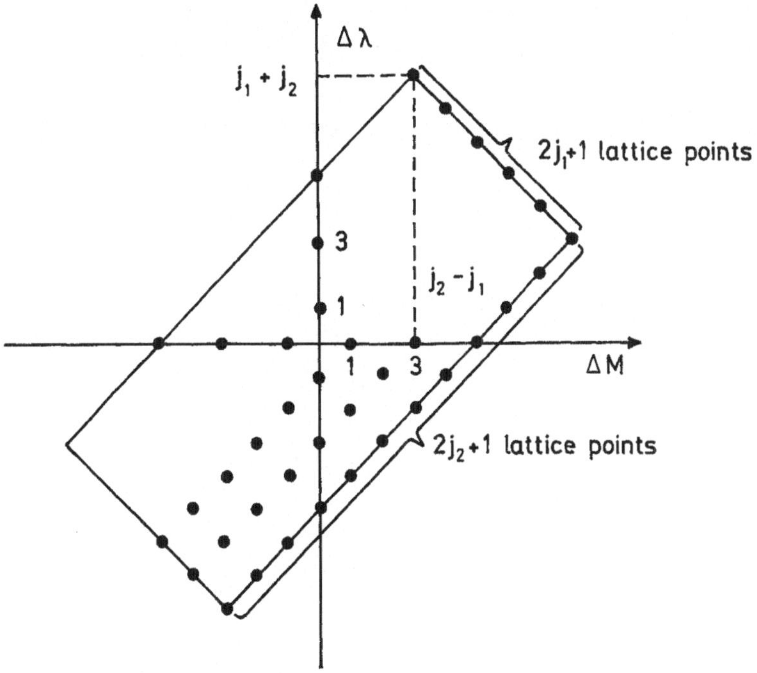

REFERENCES

1) A. R. Edmonds, Angular momentum in quantum mechanics, Princeton University Press, Princeton (1960)

2) I. M. Gelfand, M. I. Graev and N. Ya Vilenkin, Generalized functions, Academic Press, New York, Vol. 5 (1966)

3) W. Rühl, Nuovo Cimento 63A, 1131 (1969)

4) W. Rühl, Nuovo Cimento 63A, 1163 (1969)

5) W. Rühl, Tensor operators of rank two for the homogeneous Lorentz group, CERN preprint TH. 1125 (197o)

6) J. Wess, Lectures in Theoretical Physics. Vol. loB, edited by A. O. Barut and W. E. Brittin, p. 325, New York (1968)

7) J. Kupsch and W. Rühl, Nuovo Cimento 64A, 991 (1969)

8) W. Rühl, Nuclear Phys. B11, 5o5 (1969)

9) M. Toller, Nuovo Cimento 53A, 671 (1968)

lo) K. Dietz and J. Kupsch, Nuclear Phys. B2, 581 (1967)

11) W. Rühl, CERN preprint TH. 1132 (197o)

12) W. Rühl, CERN preprint TH. 1146 (197o).

Acta Physica Austriaca, Suppl. VII, 443–547 (1970)

DUALITY*†

BY

M. KUGLER

Department of Nuclear Physics
Weizmann Institute of Science, Rehovot, Israel

1. INTRODUCTION

Before starting this series of lectures on duality, I decided to look up the word duality in the dictionary. Here is how Fowler's Modern English Usage (1) treats duality. The words dual, dualistic, are explained by: "both words are of the learned kind and better avoided when such ordinary words as two, two-fold, double, connected, divided, half and half, ambiguous will do". In contrast to what appears in the dictionary, duality has turned out to be a fruitful and promising way to view strong interaction. It is only a few years since Dolen, Horn and Schmid (2) have proposed the idea. It is already hard to write a paper about strong interaction physics without using the subject duality.

---

*Lecture given at IX. Internationale Universitätswochen für Kernphysik, Schladming, February 23 - March 7, 197o.

†This research has been supported in part by the Air Force Office of Scientific Research through the European Office of Aerospace Research, OAR, United States Air Force under Contract F-61052-68-C0070.

Several good lecture notes and review articles on the subject of duality exist in the literature. Among these we mention the lectures of Jacob[3] and Horn[4] at this Winter School a year ago, the lectures of Lipkin, Jackson and Jacob at the Lund Conference[5] and the lectures of Harari at the Brookhaven Summer School[6]. This just goes to prove that duality is a very popular subject in the last few years. I will not assume that my audience knows all that was written on the subject. Otherwise, this lecture would be superfluous. I will, therefore, start at the beginning and try to present the developments of duality in the last few years. Some overlap between my lecture notes and other reviews is, thus, unavoidable. In this presentation I will try to stress the basic physical ideas and, whenever possible, the connections with experiment. I will also try to show that duality is indeed a very useful tool in investigating strong interaction.

Let me begin with an example which illustrates two physical ways of developing intuition about Regge poles. Consider first electron-positron scattering. To lowest order in α this process is given by the diagrams of fig. 1. Diagram 1a can be considered as an s-channel bound state, namely, the photon. Diagram 1b is a t-channel particle exchange . In quantum electrodynamics, one is supposed to add the diagrams. People who believe, that Regge poles resemble ordinary elementary particle exchanges, would therefore add the Regge pole to the direct channel resonances. This gives rise to the so called "interference model", whose only justification comes from intuition. When the Van Hove[7] model is considered similar intuitive pictures may arise. Positronium on the other hand is a real e^+e^- bound state. It certainly includes part of diagram 1b. The intuition is therefore based on lowest order perturbation theory only.

A different type of intuition concerning Regge poles can be developed by considering the multi-peripheral model of Amati, Fubini and Stanghellini[8]. In this model a Regge pole is given by an infinite sum of diagrams (see fig. 2). When all the ladder diagrams in fig. 2 are added, they give rise to Regge behaviour in the high energy limit ($s^{\alpha(t)}\beta(t)$). The s-channel bound state in this model is an integral part of the t-channel Regge pole and should therefore not be added to the t-channel Regge pole.

We see here two types of intuition, both based on field theoretical models, but giving completely different results. In spirit, duality is close to the second type of intuition and even more extreme than that.

2. FINITE ENERGY SUM RULES. A MATHEMATICAL INTRODUCTION

After discussing intuitive approach in the introduction, we will now develop one of the main mathematical tools to help us discuss duality.

Consider a scattering amplitude $A(\nu, t)$. We assume that A has the following expansion for large ν:

$$A(\nu, t) = \sum_i \beta_i(t)\frac{1 \pm e^{-i\pi\alpha_i(t)}}{\sin \pi\alpha_i(t)}\nu^{\alpha_i(t)} \tag{2.1}$$

The $\pm$ sign depends on the crossing properties of the amplitude. Consider now the following integral.

$$\oint_C \nu^n \mathrm{Im}\, A(\nu, t)d\nu = 0 \tag{2.2}$$

The integration is carried out over the contour C in fig. 3. Since all possible singularities of A are outside the contour of integration, the right hand side of equation (2.2) will vanish. If we now assume that equation (2.1) holds on the large semi-circle of radius N, we can evaluate the integral on the semi-circle and obtain the standard formula of finite energy sum rules (FESR) (9)

$$\int_0^N \mathrm{Im}\ \nu^n\ \mathrm{Im}\ A(\nu,\ t) = \sum_i \beta_i(t) \frac{N^{\alpha_i(t)+n+1}}{\alpha_i(t)+n+1} \tag{2.3}$$

Equation (2.2) holds. Provided we have the following crossing property

$$A(\nu,\ t) = (-)^{n+1} A(-\nu,\ t) \tag{2.4}$$

For even functions even under $\nu \to -\nu$ we get a trivial identity o = o. The equation of finite energy sum rules has been derived as a consequence of analyticity and the assumption of asymptotic Regge behaviour.

Mathematically, the finite energy sum rule equation is not very profound. If we knew a solution of a scattering amplitude, we could just use FESR as a consistency check to investigate Regge behaviour. Physics comes into the problem when we start discussing the following questions:

1) What are the allowed values of N?
2) What Regge trajectories contribute to the right hand side of the equation?
3) What should be included in the left hand side of the equation?

Physically, all this can be summarized by the following question. Given low energy data to be included on the left hand side of the equation, how well can this help us to determine α_i and β_i and through them the high energy behaviour of the amplitude? To answer the physical question arising from finite energy sum rules, we will study a few case histories which I find very illuminating.

3. DOLEN-HORN-SCHMID AND πN CHARGE EXCHANGE

The first extensive application of finite energy sum rules has been carried out by Dolen, Horn and Schmid[(2)]. The figs. 4, 5, 6 will illustrate how one can use the low energy phase shifts to calculate the high energy behaviour as determined by the functions β_i and α_i. πN charge exchange is of course an ideal example because high energy Regge fits show that the process is dominated by the exchange of the ρ trajectory. There is of course a cloud on this horizon and that is the πN charge exchange polarization measurement. If only ρ exchange were to dominate the high energy behaviour, the polarization is predicted to vanish. Yet, experimentally, it is nonvanishing. On the other hand one may claim that the polarization is very sensitive to other exchanges, and ρ exchange is still the predominant feature of our reaction.

What have Dolen-Horn-Schmid achieved? They have used low energy phase shifts to saturate the left hand side of the finite energy sum rule. They have put the cutoff N to be at medium energy. (This is defined by the point to which phase shift analysis existed). Using the phase shifts they could predict the high energy behaviour of πN charge exchange, reasonably well. This by itself is no minor achievement, especially in a field like high energy physics, where very little predictive power is available.

The major achievement of Dolen-Horn-Schmid comes from their asking the question: "What should be included on the left hand side of the finite energy sum rule?" The answer they gave is everything. All the scattering amplitudes at low energies including the resonances should be used in the calculation. This is just what they did. This is certainly in contrast with the interference model based on the analogy with quantum electrodynamics, which we have discussed in the introduction. According to Dolen-Horn-Schmid, Regge behaviour is some average of the amplitude including resonances. The interference model would say that Regge behaviour is the background and resonances sit on top of this background. The authors even went one step further. They said that, in their case at least, everything on the left hand side is resonances and no background appears at all (fig. 4). On the basis of this observation they formulated the idea of duality which says as follows: Regge poles and resonances are two alternative manners of describing the scattering amplitude. Each of them may be used and the criterion for use is the physical simplicity. In the region where a single resonance dominates, the Regge pole description would be very complicated. At high energy the Regge pole description is simple, yet a huge number of resonances would be needed to describe the amplitude. Finite energy sum rules are then the mathematical tool of connecting the two approaches. This idea, though striking at first, has become the basic idea on which duality is based. All our lectures will deal with consequences, generalizations and the physical analysis of this idea.

A very nice example of how this idea is used is given by Dolen-Horn-Schmid. This concerns the zeros in $\beta(t)$. From high energy Regge analysis we know that the $A'^{(-)}$ amplitude has a zero at t=-o.2 and the $B^{(-)}$ amplitude has a zero at t=-o.6. These are zeros of $\beta(t)$. When one considers the contribution of various resonances to A' and B, one notes that each resonance separately seems to have a zero at -o.2 for the amplitude A' and at -o.6 for the amplitude B (fig. 7). Two alternative explanations of the zero are therefore available: (i) The Regge explanation says that $\beta(t)$ vanishes. (ii) The resonance explanation says that each resonance contributes a zero at the proper point and this is the reason for the vanishing of the amplitude. These two explanations were viewed as contradictory in the pre-duality days. According to duality the explanations are not contradictory. They are complementary descriptions of the same fact. It must be emphasized that duality, so far, does not predict that each resonance will have a zero at the right point. Only the sum of resonance contribution to the FESR will be zero. It so happens that the sum of resonance contribution vanishes not because of cancellation between various resonances. In this case each term vanishes separately. This observation will become very important when we will consider further developments of duality.

Once we learn that resonances are sufficient to describe the left hand side of the finite energy sum rule equation, we can use the resonance parameters, as given by phase shift analysis, and calculate from them the amplitude at unphysical positive values of t. The amplitude at positive values of t can then be used to calculate the finite energy sum rule for t>o. Thus, we have a hope to calculate the finite energy sum rule at a neighborhood of the ρ pole. (Roughly, t=o.5 BeV^2). Dolen-Horn-Schmid have done this (fig. 5) and have succeeded in calculating the parameters of the ρ particle from knowledge of the π-nucleon resonances. This calculation is obviously of the bootstrap type. It suggests a completely new bootstrap scheme based on finite energy sum rules.

In the next section we will give a simple and interesting application of this bootstrap idea.

4. $\pi\pi \to \pi\omega$

In this section we will describe a particularly simple and attractive application of the FESR bootstrap idea. The work is due to Ademollo, Rubinstein, Veneziano and Virasoro[10]. It will serve us to illustrate the basic ideas and techniques and also as an introduction to the Veneziano formula which we will discuss later. Historically, it is indeed this work which has led to the Veneziano formula.

Only a small number of trajectories and particles can contribute to $\pi\pi\to\pi\omega$. This makes the reaction particularly attractive. The contributing particles in all channels must have I=1, G=+ and natural parity. The only particles having such quantum numbers are those on the ρ trajectory. In a finite energy sum rule calculation the left hand side of the equation will be dominated by the ρ trajectory and the right hand side will include the same trajectory. The calculation can be viewed, therefore, as a self-consistent bootstrap calculation of the ρ trajectory. The process $\pi\pi\to\pi\omega$ has only one amplitude

$$\varepsilon_\alpha T_\alpha(\nu, t) = \varepsilon_\alpha P_{1\mu} P_{2\nu} P_{3\lambda} \varepsilon_{\lambda\mu\nu\alpha} A(\nu, t) \tag{4.1}$$

where ε_α is the ω polarization and P_i the π momenta. The amplitude A has the following asymptotic behaviour

$$A(\nu, t) \to \beta(t) \frac{e^{-i\pi\alpha(t)} - 1}{\sin \pi\alpha(t)} \left(\frac{\nu}{\nu_o}\right)^{\alpha(t)-1} \tag{4.2}$$

where α is that of the ρ trajectory. In order to calculate $\beta(t)$ it was parametrized as

$$\beta(t) = \frac{\bar{\beta}}{\Gamma(\alpha(t))} \tag{4.3}$$

This insures that the proper zeros will appear at nonsense points. Substituting only the ρ resonance on the left hand side of the finite energy sum rule the authors obtain the result in fig. 8. Saturating the same sum rule with the ρ and its recurrence the $g(3^-)$ particle in the resonance side the authors obtain even better results (fig. 9). Calculating the third moment sum rule with the inclusion of the ρ and the g meson the authors obtain fig. lo. This figure shows a remarkable success in obtaining the Regge parameters from the resonance parameters and this reaction. The ingredients of this success are:

1) Finite energy sum rules with resonance saturation and a very low cutoff N.
2) The choice of a particularly simple reaction where only a small number of mesons contribute.
3) The use of linear trajectory and the proper parametrization of residues. These ingredients are important for future developments and we will return to discuss them in the following sections.

The authors of this calculation note that when they use higher values of N, their agreement is no longer good. Troubles begin already after the inclusion of the 5^- particle on the ρ trajectory (fig. 11). One is thus forced to conclude that a single linear trajectory cannot support itself in the sense of finite energy sum rules. The reason for this fact is obvious: The resonances lie on a trajectory where $\ell \propto s$. On the other hand one expects on the basis of a semi-classical argument that the dominant partial waves are at $\ell=kR$. Higher partial waves with $\ell>>kR$ are outside the range of interaction and should be very small. Therefore, additional trajectories are needed in order to sustain the t-channel ρ trajectory in a finite energy sum rule calculation.

There are some clues which we should get from the success of finite energy sum rule with low values of N. Let us recall that in the derivation of finite energy sum rules we had to assume that the Regge behaviour is an adequate description of the scattering amplitude at a semi-circle of radius N. If N can be small then the Regge behaviour is a good description even at low energies. Moreover, if the finite energy sum rule equation holds with two limits of integration N_1 and N_2 we can consider the difference of two FESR's and obtain:

$$\int_{N_1}^{N_2} d\nu \; \nu^n \; \mathrm{Im}\, A(\nu, t) = \sum \beta_i(t) \frac{1 \pm e^{-i\pi\alpha_i(t)}}{\sin \pi\alpha_i(t)} \frac{N_2^{\alpha_i+n+1} - N_1^{\alpha_i+n+1}}{\alpha_i+n+1} \tag{4.4}$$

Whenever N_1 and N_2 can be taken close to each other, the Regge formula should be a good approximation to the scattering amplitude not only in an average (or global) sense, but in the local sense. That is, point by point. So far we have investigated duality in the global sense by means of finite energy sum rules, let us next turn to the investigation of duality in the local sense.

5. SCHMID CIRCLES

We have seen that the Regge representation in terms of a few poles may be a reasonable approximation for the scattering amplitudes. As such, it can be used for an approximate phase shift analysis. This analysis may differ slightly from the experimental one, but it should reproduce the main features of the scattering amplitude. Schmid[11] was the first to show that upon analysing the t channel Regge poles into s channel partial waves, structures may appear in the Argand diagram of partial waves. These resemble closely resonances. This very surprising result led to a flood of papers investigating whether such circles should be identified as resonances, and re-examining the criteria for the "identification" of resonances. There is no final conclusion on this subject. Opinions may still be divided. I will return to this point in the next section. For the moment I would like to ask the reader to keep this question in mind, while we investigate some of the mathematical details which give rise to loops in the Argand diagram, and some of the physical consequences which will arise from this investigation.

The first question to be answered is why a Regge representation which varies smoothly with energy gives rise to partial waves which have a rapid energy variation. This should not come as a total surprise to us. We all know that the Roper resonance at 148o MeV cannot be seen in a total cross section or angular distribution investigation, yet, when we consider the P_{11} partial wave in πN scattering the presence of the Roper resonance is obvious. (That is how a resonance was first discovered by Roper.) Our picture of a Regge pole must contain many partial waves, each varying with energy, yet the total sum is smooth, somewhat like fig. 12.

Our next point concerns the reason why a Regge behaved amplitude gives rise to circles or loops in the Argand diagram. This is simply understandable if we consider how a partial wave amplitude a_ℓ is determined, from the total amplitude

$$\text{Im } a_\ell = \frac{1}{2} \int_0^{-4k^2} P_\ell(1+ \frac{t}{2k^2})\, \beta(t)\, \nu^{\alpha(t)} \frac{dt}{2k^2} \qquad (5.1)$$

The function $\beta(t)$ will have zeros at fixed t. (At points where $\alpha(t)$ goes through an integer). $P_\ell(1+t/2k^2)$ will have zeros at fixed $t/2k^2$. We have to integrate the product of these functions. The integral will be maximal when the two functions oscillate in phase. When we consider fig. 13, we conclude that at $k^2=k_2^2$ P_ℓ and β oscillate in phase and the integral will be maximal. The behaviour of Re a_ℓ is given near $k^2=k_2^2$ by dispersion relations. We can summarize this behaviour by fig. 14. In an Argand diagram this will look like a resonance in fig. 15.

The resonance structure is given by the zeros of $\beta(t)$ which appear as dips in the angular distribution. We thus correlate the presence of dips in the angular distribution with the presence of resonances [12]. Amplitudes having low energy resonances such as $\bar{p}p$, πp, K^-p should have dips in the angular distribution, whereas amplitudes which do not have resonances such as pp and K^+p, should have no dips in the angular distribution at high energy. This is in striking agreement with experiment as can be seen from figs. 16, 17, 18. The next question which can be answered now is the functional dependence of the resonance energy on ℓ. On the basis of our simple discussion before, we conclude that k^2 should be such that

$$P_\ell(1+ \frac{t_o}{2k^2}) = o \qquad (5.2)$$

where t_o is defined by

$$\beta(t_o) = o \qquad (5.3)$$

These equations can be easily solved when we use the approximation used in any eikonal calculation.

$$P_\ell(1+\frac{t}{2k^2}) \sim J_o((2\ell+1)\sqrt{\frac{t_o}{2k^2}}) \qquad (5.4)$$

This approximation is valid for $t/k^2 << 1$ and $\ell >> 1$. The solution of this equation is:

$$(\ell + \frac{1}{2}) = \text{const} \sqrt{k^2} \qquad (5.5)$$

In the standard Chew-Frautschi plot this would give rise to a trajectory which starts off linearly and then curves. There are two possible ways of interpreting this result.

1) We have a single curving Regge trajectory[13].
2) We have an "effective trajectory" which is composed of many parallel linearly rising trajectories. The important resonances on each trajectory sit on the curve ℓ=ck. The second view has two supporting points.

(a) The width of the resonance on the curve ℓ=k is very large and it increases like ℓ for large ℓ.
(b) The Veneziano model favors the second point of view. We will return to discuss this point when we consider the Veneziano model.

We will now turn to investigate the usefulness of phase shift analysis of Regge poles. Fig. 19 shows a rather remarkable agreement of the Regge projection with experimental resonances. This is taken from an analysis by Collins, Johnson and Squires[14]. We should not be over-impressed by this fit. If one changes slightly the high energy parametrization of the data the resonance structure resulting from the new high energy parametrization may be different. Thus some care should be exercised before we use high energy analysis data to really predict resonances[15].

Next, let us turn to the example of $\pi\pi\to\pi\omega$. The partial wave analysis for this process gives[16] very striking results. These may be summarized as follows:

a) The high energy Regge parametrization gives very good prediction for the resonances in the low energy region. (The ρ meson of course cannot be predicted since it lies below threshold.)

b) In this analysis new resonances appear. They do not lie on the leading Regge trajectory but could be placed on daughter trajectories (fig. 2o).

c) The inclusion of such resonances will improve the agreement of finite energy sum rules with experiments at higher cutoffs. We recall that at the 5^- resonance the ρ trajectory could no longer sustain itself and additional trajectories were needed.

The list of successes we have discussed raises a very important question to high energy theorists and experimentalists. What is a resonance and what should be identified with a resonance?

6. WHAT IS A RESONANCE?

In treating any experimental result, the theoretician's definition of a resonance, as a pole of the scattering amplitude on an unphysical sheet in the complex energy plane, is of very little practical use. We do not have any absolute criterion to identify a resonance. Lacking absolute criteria for this purpose we will demand a few necessary, but not sufficient, conditions for circles in the Argand diagram to be defined as resonances[17].

1. Resonances have well defined quantum numbers in the <u>s-channel</u> (spin, isospin etc.)
2. <u>Resonances factorize</u> in the s-channel.
3. Resonances can be observed in production experiments, (i. e., $\pi N \to \pi N^*$) in addition to formation experiments, (i. e., $\pi N \to N^* \to \pi N$).

After introducing the important necessary conditions for the identification of a resonance, we now introduce a fourth condition which is not based on any theoretical grounds but is strongly based on present experience. We will assume that

4. <u>there are no exotic resonances or exotic Regge poles</u>. By an exotic resonance we mean a resonance which is not described by $\bar{q}q$ or qqq in the quark model. Resonances which have I=2, I=5/2, B=2 etc., will be called exotics of the first kind. Exotics of the second kind appear due to unusual spin parity properties, such as odd CP natural parity mesons, 0^{--} mesons, etc.

Regge trajectories have well defined quantum numbers in the t-channel whereas resonances have well defined quantum numbers in the s-channel. Duality which connects t-channel exchanges with s-channel resonances will thus give rise to interesting constraints. The following sections will be devoted to the investigation of the constraints.

For the experimentalists the question of the definition of a resonance is rather crucial. The important question in analyzing experiments is often how to separate resonances from something else. Duality teaches us a lesson in that the possibility of describing data in terms of t-channel exchanges may not be in contradiction with the resonance description. This point was stressed by Chew and Pignotti[18] when they discussed the separation of the A_1 resonance from the Deck effect. If duality holds the Deck effect may just be a t-channel description of the A_1. Before we attempt a separation of the A_1 we must gain a better understanding of duality. To avoid raising false hopes I want to emphasize at this point that we do not yet possess such understanding.

To my mind the crucial property in identifying a resonance is factorization. This property can be experimentally verified. It really boils down to the old rule of thumb "resonances can be produced in several reactions and the branching ratio for decay should not depend on the production mechanism".

In discussing the Veneziano model we will encounter several new developments in the factorization of resonances.

7. LOCAL DUALITY

In the previous sections we have discussed two aspects of duality.

1. Resonances generate t-channel Regge trajectories via finite energy sum rules.
2. t-channel Regge trajectories generate s-channel resonances via partial wave projection.

The basic philosophy of duality is that resonances and Regge poles are two equivalent descriptions of the scattering amplitude. One of the descriptions may, however, be much more complicated than the other. Consider π-nucleon scattering at the energy of the first resonance (1238 MeV); the resonance description is extremely simple but a Regge pole description would be very complicated. Many Regge poles would have to be added in order to reproduce the experimental rapid energy variation of the cross section. These Regge poles should have very low intercept, as such they cannot be observed at high energy.

Clearly, such a complicated description offers no advantages. At high energies, the Regge pole description is simple whereas the resonance description is very complicated. Very many resonances are needed to produce a sharp forward peak in the angular distribution. The intermediate energy region may be described simply in both ways.

There are, in many processes, energy regions where both the resonance and the Regge pole descriptions, are equally simple. In these regions it is useful to propose local duality. Only in these regions will we demand that a few Regge poles <u>or</u> a few resonances describe correctly the amplitude point by point. There may be a small error in the description, and we should always be careful to check whether we are dealing with phenomena which are larger than the errors we have committed. Obviously local duality should be assumed for the imaginary part of the amplitude only. The real part of a resonance contribution vanishes at the point of a resonance and has a very long tail outside the resonance. A resonance could therefore influence the real part of a scattering amplitude far away from the point of the resonance. We will thus make unreasonable approximations if we approximate the real part of the scattering amplitudes by resonances and include only a small number of these.

It is not very useful to go into the mathematical details of assuming local duality in all energy regions. If we were to go into this, we would have had to answer such questions as: "Is the Regge description of an amplitude in terms of an infinite number of Regge poles a convergent expansion or just an asymptotic expansion?" Obviously the expansion cannot converge everywhere. If we were to assume that a Regge expansion in terms of $\nu^{\alpha(t)}$

is convergent, we could never get a pole in ν variable. The expansion must therefore diverge at the points where the amplitude has poles in s. Several mathematical analyses of this type exist in the literature. To my opinion they are not useful physically and I will not go into their details.

We have said that the local duality idea is useful only in an intermediate energy region. There are two important exceptions to our rule. First, a reaction which has exotic quantum numbers in the s-channel is very simply described in terms of resonances; all resonances vanish. Such reactions are pp, K^+p etc. Second, there are reactions in which the leading Regge poles do not contribute because of exotic quantum numbers. Such a reaction is $K^-p \to K^+\Xi^-$. The Regge description for this case is extremely simple. No t-channel Regge pole can be exchanged. Only u-channel exchanges are possible and therefore the amplitude must be small. We must therefore see how the amplitudes which have simple physical description from one point of view behave. These reactions will give us a clue to what can happen in nature.

The K^+p reaction shows a flat total cross section and no dips in the angular distribution. The last point has already been discussed. As to the flat total cross section must we abandon the idea of duality? The answer is negative. We will have to investigate the flat part of the total cross section separately and see that the flat part due to the Pomeranchuk exchange is different from all other trajectories. This will be done in the next section.

The reaction $\pi^-p \to K^+\Sigma^-$ is dominated in the s-channel by N^* resonances(19), but can be generated by no known (non exotic) Regge exchange. In order for this reaction to vanish resonances have to cancel each other and a strong correlation between resonances in the s-channel must exist.

The examples we have discussed show that the ideas of local duality lead to an amusing picture of strong scattering amplitudes. Many resonances must exist but these are strongly correlated so that they construct Regge poles. On the other hand relations between various Regge poles must exist so that these can describe the s-channel resonances. Moreover, the Regge poles are analytic continuations of t- and u-channel resonances. Extremely strong constraints of the bootstrap type must hold for strong interactions in order that duality be a reasonable picture. The next sections will be devoted to the investigation of all the constraints resulting from duality and their comparison with experiment.

8. THE TWO COMPONENT THEORY OF THE SCATTERING AMPLITUDE

We have already noted that some channels lack resonances. Such channels K^+p and pp exhibit a different high energy behaviour than other channels. Whereas πp and K^-p and $\bar{p}p$ cross sections decrease towards their asymptotic limit, the total cross sections for K^+p, K^+n and pp are remarkably flatter (fig. 21). If local duality is to hold the flat part of the cross section in K^+p cannot be built from resonances locally since such resonances do not exist. We are thus forced, along with Harari[20] and Freund[21], to assume that the strong scattering amplitude can be divided into two components.

1. The Pomeranchuk part contributes to the flat total cross section and is built from non resonating background.
2. The other Regge trajectories contribute to the decreasing

part of the total cross sections. These are built from resonances only. Before going into the detailed predictions of this conjecture we must add a historical footnote. We see that the two component idea is a natural consequence of local duality. The Harari-Freund conjecture was formulated before the idea of local duality took hold. Indeed it is this conjecture which contributed towards the establishment of the local duality idea.

In the two component assumption of the scattering amplitude we can write two sum rules

$$\int_0^N \nu^n \operatorname{Im} A_B(\nu, t)d\nu = \beta_P(t)\frac{N^{\alpha_P(t)+n+1}}{\alpha_P(t)+n+1} \qquad (8.1)$$

$$\int_0^N \nu^n \operatorname{Im} A_{Res}(\nu, t)d\nu = \sum_{i\neq Pom} \beta_i\frac{N^{\alpha_i+n+1}}{\alpha_i+n+1} \qquad (8.2)$$

where A_B is the non resonating background and A_{Res} includes the resonances. We consider an amplitude A which has no definite crossing symmetry and we assume that the sum rule can be written from zero to N only. That is for instance only over the physical region of K^+p scattering not including the K^-p region. In K^+p scattering the left hand side of equation (8.2) vanishes for all N. We let us lead to the constraint

$$\sum_i \beta_i \frac{N^{\alpha_i}}{\alpha_i+1} = 0$$

These constraints are strong exchange degeneracy predictions. They demand both equality of β and equality of the trajectory α_i.

Another prediction of the two component theory is that all total cross sections for elastic processes which have resonances should decrease with energy to their asymptotic value. This follows from the fact that resonances have positive coupling to elastic processes. This prediction seems to hold in all measurable reactions. The model also predicts that all inelastic reactions which contain no Pomeranchuk exchange such as $pp \to p\Delta$, $K^+p \to K^+\Delta^+$, $K^+p \to K^{*+}p$ should be purely real. It is not easy to verify these predictions at the moment.

To test the two component theory, Gilman, Harari and Zarmi[(22)] have undertaken to calculate the P' trajectory parameters from finite energy sum rules using resonances only. This cannot be done in the usual approach because we do not know how to disentangle the P' trajectory from the Pomeranchuk trajectory. The results obtained for the P' trajectory are reasonable.

A word of warning must now be added. We assumed that the P' trajectory is generated from resonances. We must check whether the success depends on the parametrization of the resonances. The answer that Gilman, Harari and Zarmi give is that the parametrization is not crucial as long as the resonances do not have a long tail towards high energies, and a proper threshold behaviour is included. Taking a resonance Breit-Wigner parametrization with a long tail, may result in a resonance which influences the scattering amplitude far away from the resonance location. This is certainly against the spirit of local duality, and against simple common sense, since a Breit-Wigner parametrization should not be trusted far away from the resonance.

As a further check for the two component idea, Harari and Zarmi[(23)] have plotted the partial wave amplitude in πN scattering in a slightly unusual way. They have used partial wave amplitudes with well defined isospin in the t-channel. Taking the result of phase shift analysis, these amplitudes show very beautiful circles in the Argand diagram for I_t=1, and circles distorted by the presence of a positive imaginary background for I_t=0. They also show that resonances alone reproduce ρ, the I_t=1 part of the scattering amplitude, and reproduce badly the I_t=0 part of the scattering amplitude. A positive additive background is needed to reproduce the I_t=0 amplitude. All this is shown in figs. 22 to 25 which speak for themselves.

This analysis teaches us two things:

a) The two component theory looks reasonable in this reaction.

b) All non-Pomeranchuk amplitudes can be completely described by resonances. This last statement is closely related to the statements of local and global duality. This statement is stronger than local duality because it says that resonances are a good description of the amplitudes which have no Pomeranchuk exchange even at energies where Regge poles are a bad description. This resonance dominance will be shown to underlie some of the successes of duality.

This two component theory of scattering amplitudes has taught us a lot about the ordinary Regge trajectories and their relation to resonances. It has, however, taught us very little about the Pomeranchuk trajectory. All we deduced are negative statements about the Pomeranchuk, namely, it is not built from resonances, it is built from something else but we do not know what this something else is. The problem of the Pomeranchuk trajectory and its nature thus remains open. It is to my opinion one of the most interesting problems in high energy hadron dynamics. The physical idea describing the Pomeranchuk as some diffraction process due to the shadow of many inelastic reactions is very appealing. Also the droplet model seems to point in the same direction. Yet, no final answer to the nature of the Pomeranchuk trajectory is available. The question concerning the nature of the Pomeranchuk trajectory becomes even more crucial in view of the latest high energy scattering data from Serpukhov. In this experiment the K^-p, π^-p, π^-n cross sections seem to flatten the region of 45-7o BeV/c[24]. For a discussion of this subject see R. J. N. Phillips[25].

9. EXCHANGE DEGENERACY AND CROSSING

We have seen in previous sections that exchange degeneracy appears naturally as a consequence of the local duality picture and the two component theory of the amplitude. In this section we will discuss other cases, generalize the results to higher symmetries SU(2) and SU(3) and see the difficulties arising in such a picture. We will also derive some of the nicest successes of the duality approach. It should be emphasized that our derivation does not completely rely on duality. Another important ingredient has been added. That is the absence of exotic trajectories. An exotic trajectory is most easily defined in terms of the quark model. It is a trajectory or resonance which cannot be constructed from three quarks or quark-antiquark. This definition is valid whether quarks exist or not, since it relies on quantum numbers only. The physical meaning of the exotic states is however rather obscure if real quarks do not exist. One may take the attitude that quarks represent a mathematical structure underlying the theory and such mathematical quarks have no connection with the existence of real spin 1/2, 1/3 e charge particles, but in the absence of a satisfactory theory of mathematical quarks even this is a bit obscure.

How well is the absence of exotic states established, and how strongly do our results depend on the complete absence of exotic states? These two questions will play an important role in what follows. We must, therefore, clarify them here. There are several established exotic facts in the literature. The first are the bumps and the K^+p total cross section at $\sim$1900 MeV[26]. These may be due to K^+p resonances but even if these resonances exist, they are coupled very weakly to K^+p. The other evidence for presence of exotic exchanges comes from γD experiments at SLAC[27].

In the absence of exotic resonances, and neglecting deuteron corrections, we have the following predictions

$$R_1 = \frac{\sigma(\gamma D\to K\Sigma N)}{\sigma(\gamma P\to K\Sigma)} = 3 \qquad (9.1)$$

$$R_2 = \frac{\sigma(\gamma D\to \pi^+\Delta N)}{\sigma(\gamma P\to \pi^+\Delta)} = 4 \qquad (9.2)$$

Experimentally $R_1=2.37\pm.11$, $R_2=3.0\pm.5$ with slight t dependence. Another ratio which depends only on the absence of deuteron corrections is:

$$R_3 = \frac{\sigma(\gamma D\to K^+\Lambda N)}{\sigma(\gamma P\to K^+\Lambda)} = 1 \qquad (9.3)$$

This ratio is well satisfied experimentally. It should be emphasized that the γD experiments are sensitive to the interference of the exotic exchange to non exotic exchange. These tests may thus be more sensitive than reaction checking directly for exotic exchange such as: $pn\to N^{*-}N^{*++}$[28], $\bar{p}p\to\bar{\Sigma}^+\Sigma^-$[29], $\bar{p}p\to K^+K^-$[30] which are observed to vanish very rapidly at high energy. In the next paragraph we will discuss the results obtained from assuming duality and the absence of exotic resonances.

Consider first $\pi\pi$ scattering[31]. $\pi^+ \pi^+$ is exotic, and $\pi^+ \pi^-$ is not exotic. The amplitudes are dominated by I=O(P') and I=1(ρ) exchanges. If local duality is to hold, the $\pi^+ \pi^+$ cross section must be flat, and the $\pi^+ \pi^-$ cross section must decrease. This can be obtained only if

$$\alpha_{P'} = \alpha_\rho \qquad (9.3)$$

and

$$\beta_{P'\pi^+\pi^+} = \beta_{\rho\pi^+\pi^+} \qquad (9.4)$$

Formally, this is seen if we consider the s-channel isospin amplitudes S_o, S_1, S_2 (S_i denotes isospin i in the s-channel, similarly for T_i). These are connected by a crossing matrix to the channel amplitudes.

$$S_i = X_{ij}T_j \qquad (9.5)$$

The non exotic assumption means

$$S_2 = 0 \quad \text{and} \quad T_2 = 0 \qquad (9.6)$$

We have thus one equation to relate the two quantities: T_o and T_1. The only solution is:

$$T_o = c\, T_1 \qquad (9.7)$$

i. e. exchange degeneracy between t channel trajectories.

In the following paragraphs I will list a few of the results obtained by crossing alone.

1) PP→PP (P denotes a pseudoscalar meson). We demand that resonances or Regge poles behaving like the 27, lo and $\overline{10}$ representation of SU(3) do not exist in any channel. This imposes three relations between 8_{SS}, 8_{AA} and 1 in SU(3) notation. We thus have a unique solution demanding

exchange degenerate octets with odd ℓ odd signature and nonet, even ℓ even signature. This holds for both the t-channel and the s-channel. It would seem that we cannot satisfy three equations with three quantities. This is however, misleading since the 8×8 crossing matrix in SU(3) is singular.

2) $\bar{\Delta}\Delta \rightarrow \bar{\Delta}\Delta$ with isospin symmetry. In this case we have four s-channel isospin amplitudes: S_0, S_1, S_2, S_3 and four t-channel amplitudes. If we demand that no I=2 or I=3 mesons exist, we find that there is only one solution. All amplitudes vanish. This follows from the fact that we try to solve two homogeneous equations with two unknowns and the crossing matrix is non-singular. This surprising result is one of the main difficulties arising from our model. We will return to discuss these difficulties.

3) Similar cases where no solutions exist, arise in other reactions involving baryon-antibaryon. Consider the reaction $\bar{B}\Delta \rightarrow \bar{B}\Delta$ where B denotes the octet of baryon and Δ the decuplet. We demand that no resonances appear in the 35, 27 and lo in the s-channel and no Regge trajectory in the 27. This gives one equation for one parameter (the octet in the s-channel). Again only a null solution exists.

4) Similarly, a unique null solution exists in $\bar{B}B \rightarrow \bar{B}\Delta$.

5) When considering the $\bar{B}B \rightarrow \bar{B}B$ in SU(3), we do not encounter any difficulty when treating the crossing matrix. In this case we assume that the crossed channel $BB \rightarrow BB$ has no resonances. Trajectories must appear in exchange degenerate pairs with even and odd charge conjugation so as to add in $\bar{B}B$ and cancel in BB. The solution is non-unique due to the fact that the crossing matrix is

singular. We obtain two linearly independent solutions. The additional solution appears because in $\bar{B}B$, we allow amplitudes which behave like 8_{FD} and 8_{DF}. These are forbidden in PP reactions because of charge conjugation. In the next section we will see that by imposing factorization we are led into trouble, because both solutions will vanish.

6) The most important reaction from many points of view is $MB \to MB$, where M notes the octet of mesons and B the baryon octet. This reaction along with $MB \to M\Delta$ are the best measured reactions and their investigation is therefore extremely important. In $MB \to MB$ we encounter a new situation. This time, the s-channel decuplet is allowed and the solution is highly non-unique. We get a three parameter solution from the crossing matrix. Any linear combination of solutions is also a solution. Physics will have to tell us which of the many solutions is preferred by nature.

7) $MB \to M\Delta$ has a unique nonvanishing solution.

In this section we have discussed the algebraic constraints imposed on the Regge trajectory by crossing and local duality. Note that we have discussed mainly s-t crossing. In meson-baryon, also s-u crossing may present constraints, this point has not been conclusively investigated. The striking features of the solutions we have obtained are exchange degeneracies between trajectories of different charge conjugation and signature and different SU(3) quantum numbers. In particular, the vector and tensor trajectories are exchange degenerate and the octet and singlet of tensor and vector trajectories are also exchange degenerate. This appearance of trajectories in nonets is a remarkable feature reminiscent, in a way, of the quark model. Here we obtained the quark model structure without making complete quark model assumptions. Only our definition of exotic channels is related to quarks. All our treatment in this section was based on symmetry considerations (SU(2) or SU(3)). The next section will use the duality assumptions to investigate how the SU(3) symmetry is broken.

10. DUALITY AND SU(3) BREAKING

In this section we will follow the discussion of Lipkin[32] in order to obtain results which were obtained earlier, by Chiu and Finkelstein[33]. We will first discuss meson-meson scattering without SU(3) assumptions. We will, however, assume factorization. This assumption may raise some difficulties of its own. In the meson-meson case factorization is not essential. It is just a short-cut. The derivation can be given without assuming factorization. In baryon-antibaryon factorization is the essential ingredient of the difficulties which arise.

In this section we will have to treat meson trajectories having I=0 or 1 and C=+ or C=-1. Let us denote these by the names of the mesons having the required quantum numbers. f, ρ, A_2 and ω. In $\pi\pi$ scattering only the ρ and f contribute. These contributions are

$$\pi^+\pi^- : \; f_{\pi\pi} + \rho_{\pi\pi} \tag{10.1}$$

$$\pi^+\pi^+ : \; f_{\pi\pi} - \rho_{\pi\pi} \tag{10.2}$$

From the fact that $\pi^+\pi^+$ has no resonances, we find

$$f_{\pi\pi} = +\rho_{\pi\pi} \tag{10.3}$$

Next, consider $K\pi$ scattering. The relative contributions are

$$K^+ \pi^+ : \quad f_{K\pi} - \rho_{K\pi} \tag{1o.4}$$

$$K^+ \pi^o : \quad f_{K\pi} + \rho_{K\pi} \tag{1o.5}$$

demanding now that $K^+ \pi^+$ has no resonances, we find that

$$f_{K\pi} = \rho_{K\pi} \tag{1o.6}$$

When we consider KK scattering, we get the following four equations:

$$K^+ K^- : \quad f_{KK} + A_{2_{KK}} + \omega_{KK} + \rho_{KK} \tag{1o.7}$$

$$K^+ K^+ : \quad f_{KK} + A_{2_{KK}} - \omega_{KK} - \rho_{KK} \tag{1o.8}$$

$$K^+ K^o : \quad f_{KK} - A_{2_{KK}} - \omega_{KK} + \rho_{KK} \tag{1o.9}$$

$$K^+ \bar{K}^o : \quad f_{KK} - A_{2_{KK}} + \omega_{KK} - \rho_{KK} \tag{1o.1o}$$

When we demand that neither $K^+ K^+$ nor $K^+ K^o$ have resonances we find that

$$f_{KK} = \omega_{KK} \tag{1o.11}$$

and

$$A_{2_{KK}} = \rho_{KK} \tag{1o.12}$$

Using factorization we find that

$$f_{KK}\, f_{\pi\pi} = f^2_{K\pi} \tag{1o.13}$$

and similar equations for other couplings. These can be

combined with eqs. (10.3), (10.6) to show that

$$f_{KK} = \rho_{KK} \qquad (10.14)$$

If this is substituted into (10.10) using (10.11) and (10.12) we obtain $\sigma_{tot}(K^+ \bar{K}^o)=0$. This is a very undesired solution. The resonances in this reaction all contribute with positive sign since we are treating an elastic process and no cancellation can occur. The vanishing of this cross section means that all resonances decouple from $K^+ \bar{K}^o$. The only way to avoid this unpleasant feature is to assume that an extra pair of trajectories exist which do not couple to pions but couple to kaons. These will make $K^+ \bar{K}^o$ non-vanishing. A pair of these trajectories is needed so that $K^+ \bar{K}^o$ will be non-vanishing but $K^+ K^o$ will still vanish. The two trajectories can be identified with the f' and ϕ which are known experimentally to decay into $K\bar{K}$ and not into pions.

If we translate this into SU(3) we have proven that the mixing angle of the singlet and octet components in $\phi\omega$ and $f^o f^{o'}$ obeys $tg^2\Theta=1/2$. This is the canonical SU(6) or quark model mixing angle. Obtaining this result from a duality argument is a very pleasant surprise. One should not be confused at this point to assume that there is no solution to the duality equations which obeys SU(3). This is incorrect. If the f^o and f' and ϕ and ω were exchange degenerate, any mixing angle would have been possible, since we could not distinguish between mixing angles. What the duality calculations show is, that once we break the SU(3) symmetry via the breaking of intercept, only one mixing angle is possible.

The discussion of the baryon-antibaryon reaction follows closely the line of our discussion for meson-meson. There are however, two major differences.

1) The BB reactions are exotic. Therefore all trajectories coupling to baryons must appear in exchange degenerate pairs of opposite charge conjugations. These can cancel in the BB and add coherently in $\bar{B}B$.

2) Baryons are not eigenstates of charge conjugation or G parity. Thus all mesons can couple to $\Sigma\Sigma$ when in the analogous $\pi\pi$ reaction only those with positive G parity could contribute. We can find a solution to the BB case which is identical to that of the meson solution, because the $\bar{B}B$ reactions cannot be distinguished from MM reactions. If we assume however factorization this solution will vanish and only null solutions exist[34, 35].

When we discuss baryon-antibaryon systems where one of the baryons is in the decuplet we run into trouble immediately. This we have already mentioned. Theory predicts complete vanishing of all couplings when decuplets are concerned. It should be emphasized that when decuplets are involved the troubles depend neither on factorization, nor on local duality. This is most easily seen[36] if we consider the reaction $\Delta^{++}\bar{\Delta}^{o}\to\bar{\Delta}^{+}\Delta^{+}$. This charge exchange reaction has only I=1 in the t-channel. In the s-channel it has only I=2 or 3 states. The reaction is exotic in the s-channel, the u-channel has B=2 and thus it also is exotic. Therefore, if we write a finite energy sum rule no s-channel resonances or u-channel resonances exist and the I=1 coupling in the t-channel vanishes trivially.

The troubles with baryon-antibaryon and decuplets are a major difficulty in the theory of duality. Several attempts at answering the question have been made. These can be classified into three groups. The first keeps the duality idea but changes the definition of exotic states[34]. This group suggests that there exist $\bar{B}B$ resonances with quantum numbers outside the octet. These may be in the 27 or any system with quantum numbers of two quarks and two antiquarks. These resonances do not decay into mesons. They are not seen because nobody has been able to go to high enough energies. The second group suggests that something other than resonances should be used in saturating the sum rules. These may be quark-antiquark annihilation processes[32], Regge cuts[37] or various other unspecified structures. A third approach[38] says that we have to abandon the ideas of duality when treating $\bar{B}B$ reactions. We will return to this point.

In concluding this section, I would like to emphasize the importance of duality in providing a framework for the investigation of SU(3) breaking. The dynamical assumption of duality provides for nonets of mesons and for a strong mixing between the singlets and octets. This mixing is strong, in spite of the relative weakness of SU(3) breaking interactions, because the mixed states should be degenerate in the limit of exact SU(3). The prediction of the octet-singlet mixing angle is considered one of the nice successes of the duality approach. In the next section we will discuss the graphical way of presenting the duality results we have obtained so far.

11. DUALITY DIAGRAMS

The clue to describing the duality properties of strong interaction comes from the fact that we want to forbid a set of resonances, namely the exotic resonances. Since exotic resonances are defined in terms of the quark model, our graphic description will have to be based on the quark model. In diagrammatic[39] notations all particles (incoming, outgoing or intermediate states) are described by either three quarks or by quark-antiquark. Thus no particle can have exotic quantum numbers, and this is the main ingredient of our calculations. The rules for drawing diagrams are stated as follows:

(1) All baryons are made out of three quarks.

(II) All mesons are made out of quark-antiquark.

(III) A quark line cannot double back on itself and end up as an antiquark in the same particles.

(IV) When we cut a diagram at any $B=\pm 1$ channel, we cut only through three quark lines.

(V) When we cut a diagram through any B=O channel we cut only through a $q\bar{q}$ pair (figs. 26, 27).

These rules for constructing diagrams insure that all mesonic channels will have the quantum numbers of 1 and 8 in SU(3) and all baryonic numbers will have the quantum numbers of 1, 8 or 1o in SU(3). Any duality diagram is therefore a solution to the crossing problem involved in the local duality game. The reverse of this statement is false. The duality diagrams do not constitute the most general solution to the crossing problem. Moreover, the impossibility of drawing a diagram in certain cases does not prove that no solution exists. We will return to this point. It should also be emphasized that the Pomeranchuk trajectory is completely excluded from the duality diagram picture which deals only with trajectories which are dual to resonances.

The usefulness of duality diagrams is best proven by considering a few examples. Diagram (a) of fig. 27 describes the meson-meson scattering process. In particular, the ϕ meson is a $\lambda\bar{\lambda}$ combination and cannot couple to pions which are pure non-strange quarks. The reaction $\bar{K}^o K^+ \to \bar{K}^o K^+$ is dominated by $\lambda\bar{\lambda}$ in the t-channel. This is in agreement with our proof of exchange degeneracy which shows that the A_2, ρ, ω, f do not contribute to this process (see fig. 26).

The diagrams for meson-baryon scattering with meson exchange and meson-baryon scattering with baryon exchange are presented in figs. 27b and c.

In the diagrammatic approach it is also easy to see that baryon-antibaryon cannot be drawn. The only baryon-antibaryon diagram is given in fig. 27d. It is this figure which tempted many people to assume that the baryon-antibaryon process is dominated by mesons having the quantum numbers of two quarks, two antiquarks. Note, that the solution which we found for $\bar{B}B$ is illegal from the duality diagrams point of view. This is a simple proof that duality diagrams are not the most general solution.

The beauty of duality diagrams lies in the ease of using them for making predictions. We should remember that the predictions obtained from the diagrams hold only for the imaginary part of the amplitude, since this is the part which obeys local duality. We cannot draw diagrams for the following processes

$$\begin{aligned} &K^- \; B \to \pi^- \; B' \\ &\pi^+ \; B \to \bar{K}^o \; B' \\ &K^+ \; B \to K^o \; B' \\ &\bar{K} \; B \to M^o \; B' \end{aligned} \qquad (11.1)$$

where B and B' are any nonexotic baryon or resonance, M^o is a neutral nonstrange meson which does not contain $\lambda\bar{\lambda}$, (such as π, ρ, ω, A_1, f^o, A_2, etc.). The amplitudes for these reactions such as $K^-p \to \pi^-\Sigma^+$ or $K^-p \to \pi^o\Lambda$ are predicted to be purely real. This can be checked directly from dispersion relations. If we assume that duality applies equally to both amplitudes appearing in meson-baryon scattering, then both amplitudes are real and the polarization in these reactions is predicted to vanish. Duality diagrams give rise to more predictions than we have listed. The reader can work them out for himself, or look them up in the original papers.

Next we must face a crucial question. What do duality diagrams mean? The two authors who have independently and simultaneously proposed these diagrams take different attitudes. The attitude adopted by Harari is that we do not know how to calculate diagrams, especially since they should include SU(3) breaking effects and are therefore more general than SU(3). Taking this attitude he is forced to make his predictions on the basis of the only calculable diagrams. Those which cannot be drawn are predicted to vanish. Rosner, on the other hand, assumes both SU(3) and factorization of Regge pole residua and obtains the diagrams as a solution for the meson-meson and meson-baryon scattering problem. His assumption suffers from the drawback of assuming SU(3) even though the diagrams are more general, and of assuming factorization, which is certainly a dubious assumption.

The next question which we have to answer is how general are the solutions which are predicted by the diagrams? Even when factorization is assumed, the meson-baryon problem has two linearly independent solutions. Rosner accommodates these solutions by giving different weights to diagrams where the inert quarks in the baryon are symmetric or antisymmetric. This is not to the liking of many quark model enthusiasts, because they want to assume that the inert quarks play no role whatsoever in the scattering process. In the standard quark model all trajectories couple to baryons via F-coupling. Rosner's trick allows some D coupling which is needed for experimental purposes.

The main advantage of duality diagrams, in my opinion, is their simplicity. They are very simple to use and very nice to look at, and to visualize results of calculations. Mathematically they may however be misleading. In section 15 we will argue that the assumption of factorization is much too strong and has very little physical foundation. It also seems to disagree with experiments in a few cases. When the assumption of factorization is relaxed, some of the processes which are predicted to be completely real by duality diagrams may acquire an imaginary part. In particular, the process $K^- p \to \pi^o \Lambda$ is predicted by duality diagrams to be completely real and to have vanishing polarization. When the polarization of the Λ is measured it is not small[40] in some ranges of t. This result is inconsistent with the diagrams. If we believe the polarization, the diagrams have to be abandoned and factorization should be given up. There are of course ways to argue about the significance of the result; the experiment is still at a low energy and at higher energies where local duality holds everything will be all right. We will return to the question of polarization and agreement with experimental data in the next section.

When discussing duality diagrams, we have used them mainly as a symmetry tool or a quark counting tool. These diagrams are, however, very intuitive so that one is tempted to assume that some deeper dynamical reason underlies them. Such a dynamical reason has not yet been found. The diagrams have however found use in several dual calculations which may, in the future, when their difficulties are overcome, be a basis for a completely dual theory of strong interaction.

12. EXOTICS OF THE SECOND KIND

So far we have discussed exotic resonances only when their internal quantum numbers were exotic. In the quark model there are also forbidden spin-parity assignments which are called exotics of the second kind. Such are 0^{--} mesons and mesons of natural parity and odd CP. This follows from the fact that mesons are built out of a fermion-antifermion system. (Generalizations of the quark model such as the Gell-Mann–Zweig[41] proposal have been able to introduce such states.) In this section we will discuss a paper by Schwimmer[42] which I find very intriguing.

The basic assumption of this work is that no natural parity odd CP trajectories or resonances exist. Schwimmer considers the $\pi^0\eta$ system, in this case odd CP reduces to odd ℓ. Therefore the assumption of no exotic states leads to the conclusion that all resonances in this reaction must have even angular momentum. The angular distribution should thus be symmetric around 90^0. In the high energy region this means that the trajectory which can be exchanged in the t-channel must have the same α and coupling as the trajectory which can be exchanged in the u-channel. We have therefore deduced exchange degeneracy between the P' trajectory and the A_2 trajectory.

When other reactions are considered in the same way one obtains also the ϕ ω and f f' mixing angle. One remark should be made. If SU(3) symmetry is imposed, the odd CP resonances must be in a 10 or $\overline{10}$ representation of SU(3). Thus, the condition for avoiding odd CP trajectories is consistent with the demand that no exotic internal quantum numbers appear. The remarkable thing about this work is that by demanding conditions which follow from spin and parity of the quark-antiquark system, one obtains physical results. This approach may be telling us that the dynamics underlying duality may have something to do with quarks. What such a dynamical theory is, we do not know at the moment. It is a challenging and interesting problem to try and find out.

13. DUALITY AND LOW ENERGY PARAMETERS

In this section we will use the absence of exotic resonances, combine it with an unorthodox but perfectly correct decomposition of the scattering amplitude to obtain the following results[43]:

(a) The contribution of the background to the scattering amplitude is strongly constrained.

(b) In some reactions, satisfying conditions to be specified later, <u>the low energy parameters are determined by resonances only</u>.

We will make use of the standard assumptions:

(1) Regge trajectories are generated by resonances.

(2) Non-resonant background generates the Pomeranchuk trajectory only.

(3) In channels where no Pomeranchuk can be exchanged the <u>imaginary part</u> of the amplitude is completely described by resonances.

(4) Exotic resonances are absent in the sense that $\alpha_{exotic} < 0$ for t near the forward direction.

By assumption (3) we will use resonances to describe the <u>imaginary part</u> when no Pomeranchuk exchange is possible. The <u>real part</u> of the amplitudes will be calculated from dispersion relations. Whenever no subtractions are required in the dispersion relations the real part will be completely determined by the resonances.

To explain our treatment we consider $\pi\pi$ scattering. In $\pi\pi$ scattering only the I=2 channel is exotic. By considering exotic states in two channels simultaneously the usual exchange degeneracy is deduced. We will consider three exotic amplitudes simultaneously one in each channel: S_2, T_2 und U_2 (in our notation T_2 means that the amplitude has I=2 in the t-channel). By assumption (4) exotic states couple weakly in the sense that $\alpha_2(t)<o$ for the leading t-channel Regge pole (or cut). Then T_2 will obey a fixed t, unsubtracted dispersion relation:

$$T_2(\nu, t) = \frac{2}{\pi} \int \frac{\nu' \, d\nu'}{\nu'^2 - \nu^2} \, \text{Im} \, T_2(\nu', t) \tag{13.1}$$

Similarly S_2 will obey an unsubtracted, fixed s dispersion relation; and U_2 an unsubtracted, fixed u dispersion relation. By assumption (3), we can saturate these dispersion relations with resonances only. The background, Pomeron, contribution does not appear in any of these integrals.

S_2, T_2 and U_2 are a complete, though not orthogonal, set of isospin amplitudes and we can therefore write:

$$T_o = \frac{3}{2}(U_2 + S_2) - \frac{1}{2} T_2 \tag{13.2}$$

$$T_1 = S_2 - U_2 \tag{13.3}$$

It is usually remarked that the dispersion relation for T_o needs a subtraction. This statement is misleading. Eqs. (13.1) and (13.2) show that T_o can indeed be calculated from unsubtracted dispersion relations, provided we use three simultaneous dispersion relations one in each channel. We have thus succeeded in determining all the amplitudes in $\pi\pi$ scattering in terms of resonance contributions to the imaginary part. The use of simultaneous dispersion relations enabled us to determine the real part of all amplitudes. The Veneziano formula which describes reasonably well the imaginary part of resonances will thus give a good result for the real part, especially for the Adler consistency condition (see discussion in section 21).

At first glance our results seem surprising. Obviously we cannot hope to calculate T_o from resonances at all energies. At large s, for instance, T_o is dominated by the background due to Pomeranchuk exchange. The resolution of this paradox is obvious.

Our ability to calculate all amplitudes depends on the range of validity of assumptions (3) and (4) (i. e. $\alpha_2 < o$ and resonance dominance). These assumptions cannot be rigorous everywhere. At large s the amplitude S_2 (namely $\pi^+\pi^+$ scattering) cannot be built of resonances only because of the t-channel Pomeranchuk background. At large s, $\alpha_2(s)$ may also become positive.

We thus assume that there is a range in t where T_2 is dominated by resonances and $\alpha_2(t) < o$. This range is drawn schematically as region 1 in fig. 28. By crossing our assumptions hold for S_2 and U_2 in regions 2 and 3 respectively. Only in the intersection of regions 1, 2 and 3, (the shaded region in fig. 28) can we calculate all amplitudes.

The size and shape of the shaded region are of great interest and we will return to this point. If our purpose is the evaluation of the $\pi\pi$ amplitude at the Adler point $s=t=u=m_\pi^2$, which is very close to the physical region. Our use of the simultaneous dispersion relations is most probably justified, since the shaded region extends probably that far.

A discussion similar to $\pi\pi$ can be carried out in $K\pi$ scattering. In this case the exotic channels are $S_{3/2}$ and $U_{3/2}$. T_1 also obeys a convergent dispersion relation. Again we find that a complete set of amplitudes can be calculated from knowledge of resonances only. Our approach cannot be used in $\pi\eta$ or πX^0 scattering, where no exotic channels exist.

14. EXCHANGE DEGENERACY AND EXPERIMENT

Exchange degeneracy while being very old is one of the main predictions of the duality approach. It is therefore very useful to investigate where and how closely exchange degeneracy agrees with experiments. This investigation can be carried out in two regions. For positive t we can plot the resonances on a Chew-Frautschi plot and see the extent to which two trajectories seem to overlap. This approach looks rather promising and straight line exchange degenerate trajectories are rather common. The ρ A_2 ω f trajectories look reasonably degenerate. The same holds for the f' ϕ trajectory. Of course in this case we have only two points and we can always draw a straight line between any two points. If we believe that the slopes of Regge trajectories should be equal then the ϕ f' slope agrees with that of the ρ trajectory. Baryon exchange degeneracies do not look as good, except for the Y^* trajectory.

In the literature, one distinguishes two types of exchange degeneracy, usually referred to as weak and strong exchange degeneracy. Weak exchange degeneracy refers to the equality of $\alpha(t)$ whereas strong exchange degeneracy implies also a relation between the couplings of the exchange degenerate trajectories.

The simplest way to proceed in testing exchange degeneracy would be to analyze data in terms of Regge poles and extract the trajectories and couplings from the data. This procedure is, however, not very trivial. The analysis of high energy data is often not sufficient in order to give accurate values of α and β. Much depends at the moment on the particular way of parametrizing the data. On the particular mechanism chosen at nonsense points and on the availability of small error data. Luckily, there are other and sometimes better ways of checking exchange degeneracy. The total cross section in exotic channels constitutes by itself a good test of exchange degeneracy. Here it should be remarked that whereas the K^+p and K^+n cross-sections are extremely flat this is not the case in pp scattering (fig. 21). In the latter reaction the total cross-section shows a decrease with energy. The decrease is significantly smaller than the decrease in πN, K^-p or $\bar{p}p$, yet it is definitely present. This effect must be attributed to the breaking of strong exchange degeneracy and to the breakdown of one of our assumptions. It is tempting to associate decrease of $\sigma(pp)$ to the u-channel bound states. The $\pi\rho$ etc. are, because of proton-proton kinematics, very close to the s-channel threshold. Nevertheless, this breaking seems to be a problem.

A further test of exchange degeneracy follows from a simple observation. Namely, two exchange degenerate trajectories of opposite signature cannot interfere in their contribution to the differential cross-section. This is best illustrated graphically.

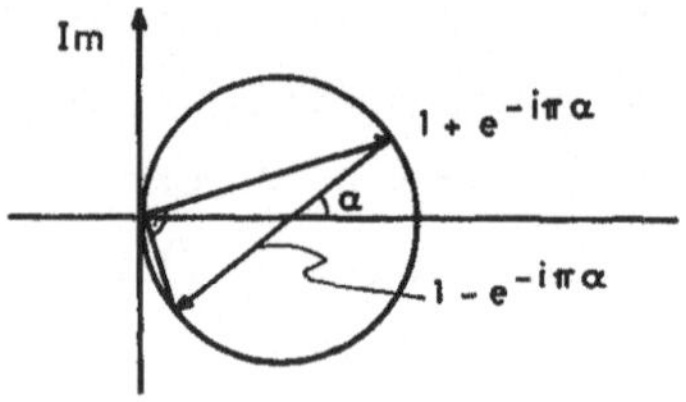

It is easy to see from this picture that the two contributions $1+e^{-i\pi\alpha}$ and $1-e^{-i\pi\alpha}$ are perpendicular and cannot interfere. This may therefore be a check for weak exchange degeneracy. Consider now two processes such as $K^-p\to\pi^-\Sigma^+$ and $\pi^+p\to K^+\Sigma^+$, these are related to each other by crossing. The two K^* trajectories contribute to both processes. The difference between the processes comes from the relative sign of K^* versus K^{**}. Since the K^* and K^{**} contributions are 9o$^{\circ}$ out of phase this change of sign cannot influence the cross section.

Weak exchange degeneracy predicts that these processes should therefore have equal cross section. Fig. 29 shows the experimental data at 5.5 GeV/c. Disregarding the possibility of incorrect normalization in one of the experiments, the two curves do not agree well with each other. Several other comparisons of the same kind are shown in figs. 3o to 35(44). These pictures show unpleasant disagreement with the data.

In KN charge exchange the discrepancy seems to decrease with energy and at 5.5 GeV there is no discrepancy. Also, at small values of t the discrepancy seems to be small. Only at $t=-.25$ does the discrepancy begin to be considerable.

How do we interpret these failures? Should we abandon the aesthetically beautiful scheme of duality in favor of something else, and if so, what part of the scheme should we abandon? We could also argue that duality is an approximate scheme and these experiments measure the breaking of duality. The questions relating to the breaking of duality are very relevant and we shall try to tackle them in this and the next sections.

Several points are worth noticing. The equality of differential cross-sections holds only if one pair of exchange degenerate trajectories is present. If more than one pair is present, interference between the leading and non-leading pair could break the equality of differential cross-sections. We thus conclude that only at high energy do we expect the differential cross sections to be equal.

The agreement with experiment is the most important criterion for any theory in strong interactions. Therefore, I strongly urge experimentalists to provide us with an extensive check of the duality assumptions and of exchange degeneracy. The more checks the better. Theoreticians are badly in need of an idea and any experimental data will be of great help. The data could help us investigate which is the way out of the difficulty. In the next sections we will try to describe a few of these ways. Unfortunately, we do not yet have a definite answer to the problem.

15. CUTS AND FACTORIZATION

Duality, as discussed so far, holds as a relation between resonances and Regge poles. Most of our results will, however, not be changed if poles are really "effective poles"(2) or some combination of Regge cuts and poles.

It is important to find out the role which cuts play in the duality game. Are they generated by resonances or do they correspond to background?
Resonances predict well the high energy behaviour. If cuts play an important role in the high energy behaviour we must conclude that they are generated by resonances. This does not exclude the idea that cuts are partially responsible for duality breaking(45).

The theory of complex angular momenta teaches us that Regge cuts should exist. We know something about their intercepts, but we cannot predict the strength of their coupling. Nor can we predict therefore which cuts are important and which are not.
In analyzing high-energy data we cannot distinguish between a cut and a pole. The difference between the two contributions is just logarithmic and that is hard to notice. Phenomenological cuts, based mainly on the absorption idea, have had a number of successes in the last few years. Even in these models the correct amount of cut contribution is not very clear.
We can distinguish between "weak absorbers" whose cuts are just a small correction to improve Regge pole fits, and "strong absorbers" whose cuts play a predominant role(46). The two approaches also differ in their parametrization of the Regge pole. The two fits to the data are roughly of the same quality, so that in general we do not know how important the cuts are.

One of the striking examples where Regge cuts are needed is the photo-production of charged π mesons. The amplitude for $\gamma p \to \pi^+ n$(47) involves spin flip both at the nucleon vertex and at the $\gamma\pi$ vertex. An ordinary Regge pole should give a vanishing contribution near t=o. Naive pole theory predicts a dip in the forward direction. Experimentally this dip is not present and we observe a sharp spike with a typical width of t=o.o2 BeV^2 which is of the order of m_π^2. We can produce such a dip from Regge poles only if we assume that the pion is an M=1 trajectory and a conspiring trajectory is present. This conspiring trajectory should not give rise to a pole, since nobody has observed a 0^+ particle with G=- and $m=m_\pi$. This particle can be eliminated by adding a zero in the residue of the pole. If we take this model seriously, we run into additional trouble. To show this, we must use the crucial property in which poles differ from cuts, namely Regge poles factorize whereas cuts need not do so. Le Bellac[48] has shown that factorization would impose a forward zero in the reaction $\pi^+ p \to \rho^o \Delta^{++}$ where the ρ^o has zero helicity. The data show the opposite behaviour, namely a very strong forward peak. One could try to avoid the factorization difficulties by introducing a few other trajectories such as the A_1, but this just pushes the trouble into other reactions[47]. On the other hand, $\gamma p \to \pi^+ n$ is very easily and very well fitted with π exchange plus absorption. This predicts the right peaking at t=o and the right angular distribution near t=o. What does all this have to do with duality? It turns out that a finite energy sum rule calculation for $\gamma p \to \pi^+ n$ gives excellent results and enables us to predict the angular distribution, and the results which can be obtained from polarized photon experiments as well[47]. The necessary input into this finite energy sum rule calculation are the low lying resonances. This success of finite energy sum rules proves that duality holds between resonances and the effective high energy behaviour, which includes both poles and cuts. It also proves that the factorization assumption must probably be abandoned.

There are two more reasons which can show that factorization probably does not hold for Regge poles. The first is the experimentalist's evidence that the A_2 is doubled, it corresponds to two resonances. There is no reason to believe that such doubling should not effect the trajectory. We must conclude that the trajectory does not behave like a single Regge pole and thus need not factorize. A more convincing argument against factorization comes from the crossover point in nucleon-nucleon scattering[47]. This point at t=-.15 is the point where the $\bar{p}p$ and pp cross-sections are equal. This cross-over point must be attributed to the vanishing of the ω trajectory coupling to nucleons. Factorization will then imply that the ω contribution should vanish for all other processes. In reactions like $K^+p \to K^* p$ and $\pi p \to \rho p$ where the ω contribution can be extracted, there is no dip at the cross-over point and thus factorization is in trouble.

The presence of Regge cuts will alter little the discussions which we have carried out before. This point is clarified by investigating the nature of cuts which should appear. The cuts can be classified into two groups. First, the exchange of a Pomeranchuk trajectory and an ordinary trajectory. Second, the exchange of two ordinary trajectories. The first group cannot have exotic quantum numbers; it has $\alpha(o) \sim 1/2$. These are presumably the cuts which are given by the absorption model. The second group of cuts may have exotic quantum numbers. Although we cannot estimate quantitatively the importance of the exotic contributions, we can estimate the intercept of these cuts, and these should have $\alpha(o) < o$. Thus, if only objects with high intercepts are discussed all results based on the absence of non-exotic trajectories still hold, but factorization is ruined. This argument reduces the trouble $\bar{B}B$ reaction which has a non factorizing solution. The only reactions in contradiction with the duality assumption remain those in which Δ's are involved. We have therefore a physical reason to examine those reactions separately. This will be the subject of the next section dealing with the breaking of duality.

16. THE BREAKING OF DUALITY

A physical theory is usually an approximation based on neglecting some of the non-relevant features. When the first successes of a theory become clear the next step should be to investigate where the things neglected come back to haunt us and where the predictions of the theory do not hold. Hopefully, one can correlate the two cases and understand how the physical limitations follow from the features the theory has neglected. This is being attempted now in the case of duality.

One proposal for the mechanism of duality breaking is by the Cal-Tech group[38]. They suggest that the crucial assumption in the duality calculation is the existence of an intermediate energy region. The intermediate energy region is defined as the region where both the Regge description and the resonant description of the scattering amplitude are relatively simple. Next they assume that the intermediate energy region starts and ends at some well-defined energy. Therefore they argue that processes involving high external masses will not have an intermediate energy region. The region where both descriptions are simple should be below the threshold of the reaction or at least too close to it to make a Regge description reasonable. These authors classify the exchange degeneracy results by the processes from which this exchange degeneracy can be obtained. They argue that the higher the external mass is the worse will the exchange degeneracy be. The best exchange degeneracy according to this argument is between vector and tensor trajectories since this can be obtained by Pseudoscalar-Pseudoscalar scattering. Progressively worse are the exchange degeneracies obtained from baryonic channels and from higher meson channels. The baryon-antibaryon or $\bar{\Delta}\Delta$

problem according to this approach is not a problem at all. The external masses are so high that no intermediate region exists and we cannot use the local duality assumption in the absence of an intermediate region. Supporting evidence for this point of view can be seen in the fact that K^+p cross-sections are much flatter than pp cross-sections and in the experimental pattern of exchange degeneracy breaking. They do not stress the fact that the $\Delta\Delta$ problem follows without local duality, see section lo.

The Cal-Tech group investigates mainly the breaking of exchange degeneracy in the positive region of s. They consider channels for which the exchange is exotic and deduce the constraint imposed by this on the resonances. One could also investigate the opposite procedure, namely take channels where resonances are exotic and consider the relations imposed by this fact on the Regge trajectory. This way we investigate the breaking of exchange degeneracy in the negative t region.

What should a priori be expected for a pattern of breaking? Any approximate theory contains some errors due to its assumptions. In our case these errors may come from the presence of exotic resonances which couple weakly, the presence of exotic exchanges or cuts, the presence of non-resonant background needed for the finite energy sum rule and several other reasons. Using our approximate theory of duality, we ask ourselves where should the prediction be most accurate. The answer to that is that the large effects predicted by our theory should be relatively accurate whereas when we discuss small quantities the relative error may be much larger. Since we deal with Regge exchanges, the large quantities are due to trajectories with high intercepts and the small quantities are due to trajectories with low intercepts. Our general rule is therefore that <u>high intercept</u>

trajectories will obey exchange degeneracy better than low intercept trajectories. It is remarkable that the results of this demand agree with the Cal-Tech approach. This happens because the low-lying trajectories can be isolated only if we consider high external masses. The high intercept vector and tensor trajectories couple to low external masses such as two pseudoscalars.

The $\bar{B}B$ and $\bar{\Delta}\Delta$ reactions are one of the important problems in duality. We have already mentioned several approaches to this problem. The Cal-Tech solution prefers to abandon duality while other groups suggest new saturation mechanisms. Here we would like to offer a different solution to the problem.

We have already discussed factorization[(49)], and our conclusion was that this feature of Regge poles should probably be abandoned. If we abandon factorization the $\bar{B}B$ reactions are no longer a problem. We have seen that non factorizing solutions exist. The $\bar{\Delta}\Delta$ problem is still with us. The solution I would like to propose to the $\bar{\Delta}\Delta$ problem is that we trust duality, and the absence of exotic states. When duality predicts a zero solution we believe the result and conclude that the coupling of leading trajectories to the troublesome channels vanishes or is very small. Lower trajectories for which the duality predictions do not hold so well may still contribute. If we are correct the following consequences hold.

(1) The couplings of the dominant Regge trajectories to the forbidden channels are small.

(2) The leading resonances do not decay into troublesome channels.

(3) No large exotic resonances[(34)] are needed to save the situation. This is nice because no such resonances have yet been observed.

Our support for this point of view is the following. There are two experimentally observed reactions which belong to the troublesome class. This is $\bar{B}B\to\bar{B}\Delta$ and $\bar{B}B\to\bar{\Delta}\Delta$. If we solve the SU(3) crossing problem we find that only null solutions are possible. Concerning the experimental data $\bar{p}p\to\bar{p}\Delta$ is observed but not at very high energy. If we conclude that $p\bar{p}\to\bar{p}\Delta$ cannot have contributions from the leading vector and tensor trajectories the same conclusion holds for $pp\to p\Delta$. Since only I=1 trajectories couple to this reaction they cannot cancel in one channel and add coherently in the second. Therefore the real part of the amplitude $pp\to p\Delta$ as well as the imaginary part should have a small contribution from vector and tensor exchange. This reaction has been carefully investigated and it is consistent with having no contribution from vector and tensor trajectories. Everything seems to be dominated by π exchange. This is observed both in the energy dependence of the cross-section (see fig. 36) and in the angular distribution. Both can be fitted well with π exchange plus absorption up to 22 GeV/c. Our prediction holds therefore in one case. In the reaction $\pi p\to\pi\Delta$ or $Kp\to K\Delta$ which are not troublesome according to duality, the contributions of the ρ and A_2 trajectories have been observed. This is a comfortable point but certainly not a conclusive proof.

If we abandon factorization the meson baryon solution may also change. We no longer have the prediction from duality diagram, we also have a more general solution than the one imposed by factorization. Instead of a two parameter solution, we get a three parameter solution.

The other test of exchange degeneracy which looks somewhat disasterous can be explained by the fact that they do not check exchange degeneracy alone. The presence of low lying exchange degenerate pair of trajectories could change the result. We are therefore still hopeful that duality is a reasonable description of nature. It is not a perfectly rigorous description and we should always keep in mind that breaking of duality exists. We made some progress in understanding something about the nature of duality breaking but we cannot claim to have a theory of duality breaking.

After treating duality in a phenomenological way we will now pass to the concrete example of a function obeying duality and investigate it. We turn to the Veneziano formula.

17. THE VENEZIANO FORMULA

The dream of an S-matrix theory is to find, obtain, or guess a function which satisfies the following requirements:

1) analyticity in s, t, u
2) crossing
3) unitarity
4) Regge behaviour
5) duality in the global and local sense

Such a function could be a candidate for the solution of the strong interaction problem. Though the Veneziano formula[50] has not achieved this dream, but it has come rather close to it. In the next few sections we will discuss briefly this formula. We will try to avoid the use of cumbersome mathematics and to stress the simple, physical ideas.

The starting point for obtaining the Veneziano function are a few ingredients of the research on the reaction $\pi\pi\rightarrow\pi\omega$ which we have already discussed. These are (a) the parametrization of the $\beta(t)$ in terms of gamma functions; (b) the success of finite energy sum rules in this approximation; (c) the presence of daughter trajectories parallel to the leading trajectory; (d) the use of narrow resonances as an input to the finite energy sum rules; (e) the use of linear trajectories. (This point is of course connected with the narrowness of the resonances.) All these features have been incorporated into the Veneziano function for $\pi\pi\to\pi\omega$ by a remarkable guess. Veneziano proposes that the scattering amplitude for the process $\pi\pi\to\pi\omega$ be given by:

$$A(s, t, u) = \bar{\beta}(B(s, t) + B(s, u) + B(u, t)) \qquad (17.1)$$

where

$$B(s, t) = \frac{\Gamma(1-\alpha(s))\Gamma(1-\alpha(t))}{\Gamma(2-\alpha(s)-\alpha(t))} \text{ and } \alpha(s) = as+b \qquad (17.2)$$

The following features of this function are immediately obvious:

1) The function has a pole whenever $\alpha(s)=n$. The same holds true for $\alpha(t)$ and $\alpha(u)$.

2) The function obeys crossing symmetry. This is obvious from inspection.

3) Whenever both $\alpha(s)$ and $\alpha(t)$ approach an integer, there is no double pole at the two variables because the denominator contributes a zero. This is a necessity since otherwise the residue of a pole in s would be singular in t, a highly unwanted situation. The pole and zero structure of the amplitude B(s, t) is given in fig. 37. It is important to notice that whenever two pole lines intercept a zero line will appear to cancel the unwanted double pole.

We can now use the Stirling approximation

$$\Gamma(z) \underset{z\to\infty}{\to} \sqrt{2\pi}\, e^{-z}\, z^{z-1/2} \qquad (17.3)$$

to investigate the high energy behaviour. This expansion holds on any ray in the complex z plane, except the negative z axis where the gamma function has poles. We will also have to use

$$\Gamma(z)\Gamma(1-z) = \frac{\pi}{\sin \pi z} \tag{17.4}$$

This equation will help us to convert the gamma function of a negative integer to a gamma function of positive integers. Given these equations anybody can investigate the asymptotic behaviour of the Veneziano model. Leaving the details as an exercise for the reader, I would like to quote just the results.

1) If s goes to infinity on the real axis there is no simple asymptotic expansion of the Veneziano formula. This is obvious because an infinite number of poles in s exist.

2) Outside the real axis we find the following asymptotic behaviour. As $s \to +\infty$, t fixed

$$B(s, t) \to \frac{1}{\Gamma(\alpha(t))}(as)^{\alpha(t)-1} e^{-i\pi\alpha(t)/\sin \pi\alpha(t)} \tag{17.5}$$

$$B(u, t) \to \frac{1}{\Gamma(\alpha(t))}(as)^{\alpha(t)-1/\sin \pi\alpha(t)} \tag{17.6}$$

$$B(s, u) \to O(e^{-cs)} \tag{17.7}$$

A few remarks should be made. a) $B(s, t)$ which has poles for positive s is the function which contributes the imaginary part of the high energy asymptotic behaviour. Both $B(s, t)$ and $B(u, t)$ contribute the real part of the asymptotic behaviour. b) $B(s, u)$ vanishes exponentially provided <u>the trajectories in the s- and u-channel have the same slope</u>. This remark is very important because the consistency of the Veneziano formula demands that all trajectories appearing in this problem should have equal slopes. This is in remarkable agreement with experiment. (For the $\pi\pi\to\pi\omega$ example this is of course trivial.) It is also important to remark that both B(s, t) and B(u, t) have asymptotic expansion in powers of s which are associated to Regge behaviour.

We have already seen that the Veneziano formula has narrow resonance poles and Regge asymptotic behaviour. These are two ingredients of duality. Let us now investigate duality of the formula more closely. Consider the function B(s, t). This can be written as either of two expansions.

$$B(s, t) = \sum_n \frac{1}{\alpha(s)-n} Q_n(\alpha(t)) \tag{17.8}$$

or

$$B(s, t) = \sum_n \frac{1}{\alpha(t)-n} Q_n(\alpha(s)) \tag{17.9}$$

where

$$Q_n(x) = \frac{\Gamma(1-x)}{\Gamma(n)\Gamma(2-x-n)} \tag{17.1o}$$

$Q_n(\alpha(t))$ is a polynomial of degree N-1 in $\alpha(t)$. These expansions are easily interpreted in terms of duality. The function can be expressed as either a sum of s-channel poles whose residues are polynomials in t or as a sum of poles in the t-channel whose residua are polynomials in s. Taking two such sums would involve severe double counting. The two expansions converge in different regions. The expansion in s-channel poles converges whenever $\alpha(t)<1$. The t-channel pole expansion converges whenever $\alpha(s)<1$. This divergence of the expansion is necessary in order to have poles in the t-channel in spite of the fact that Q_n are analytic in t. The two alternative expansions give a nice illustration to the idea of duality.

Our amplitude has Regge pole behaviour and s-channel resonances. The expansions we have discussed prove that the Veneziano amplitudes will rigorously satisfy finite energy sum rules. This insures the existence of global duality. Local duality in the resonance region is insured by the fact that the s-channel discontinuity of the amplitude is given correctly by the residua of the s-channel resonance poles.

The residue of each s-channel pole is a polynomial in $\alpha(t)$ and therefore a polynomial in $\cos \theta_s$. This polynomial is not a Legendre polynomial. Therefore, the s-channel pole describes a combination of resonances with many values of j. This produces the parallel daughters which we have already discussed. The Veneziano formula can be easily generalized to include different particles with isospin complication, the inclusion of particles with high spin is a much more difficult problem.

Let us turn to $\pi\pi$ scattering. In this reaction the amplitude S_2 is given by

$$S_2 = \frac{\Gamma(1-\alpha(t))\Gamma(1-\alpha(u))}{\Gamma(1-\alpha(t)-\alpha(u))} \qquad (17.11)$$

This automatically insures that S_2 which is exotic in the s-channel will have no resonances in this channel. From S_2 we can calculate all other isospin amplitudes. This has already been discussed in section 14. Using this we find that $\alpha(t)$ describes an exchange degenerate pair of trajectories, the ρ and f^0. A series of narrow resonance has to appear. This is summarized in fig. 38. The Veneziano formula makes therefore a list of predictions for $\pi\pi$ scattering.

These predictions include:

1) the appearance of an infinite number of resonances. Among them we list: The σ or ε meson with $\ell=0$, I=0 and $m=m_\rho$, the ρ'-meson with the quantum numbers of the ρ and $m=m_f$ and many other mesons. 2) At $\alpha(s)=n$ there will be n+1 mesons. This remarkable infinite spectrum is very intriguing. One would like to find out whether such mesons exist. Experimentally, there are good indications for the existence of the ε and some evidence against the existence of a ρ'. Higher mesons have not been seriously investigated.

Some aspects of local duality show up in an interesting way in the Veneziano formula. Consider for instance $\pi^+\pi^-$ scattering. This amplitude has I=2 in the u-channel and can therefore be described by the amplitude $U_2(s, t)$. Since the amplitude has I=2 in the u-channel, the scattering process should have no backward peak. At the point $s=m_\rho^2$ for instance there are two resonances of opposite angular momentum, the ρ and the σ. These add in the forward direction but, because of the behaviour of P_1 vs. P_0, cancel each other in the backward direction. Similarly, at the mass of the f^0, there should be a ρ' and an ε' such that the resonances cancel each other in the backward direction. Of course, such a cancellation can be exact only at one point, and the backward direction does not vanish identically. Only when we go to high enough energy, the number of resonances begins to be so large that they can cancel each other well in the backward direction.

18. THE STRUCTURE OF RESONANCES IN THE VENEZIANO MODEL

In the last section we have seen how local duality gives a complicated resonance structure in the Veneziano model. We will continue to investigate this point further. Our investigation will center around two points: 1) How do the many resonances reproduce the "quasi phase shift analysis" of t-channel Regge poles. 2) The investigation of the factorization property of resonance.

In our discussion of local duality, section 5, we have seen that the t-channel Regge poles may serve as an approximate amplitude for the purpose of an approximate phase shift analysis. We have also seen that when we analyse the amplitude at high energy, each partial wave has a maximum at $\ell \propto k$. The Veneziano model on the other hand has linearly rising trajectories. The only way to reconcile the two results is to assume that the resonances that are lying on linear trajectories have a maximum at $\ell \propto k$. This is indeed the case[51]. In fig. 39 we see that many narrow resonances cooperate in producing the big bump at $\ell \propto k$. The partial wave projection of a Regge pole just gives the envelope of all the narrow resonances. We are faced with a "giant resonance" situation like those occurring in nuclear physics. Fig. 4o shows the location of this giant resonance and how many resonances will participate in constructing each giant resonance. To convince you that such a situation is not too far fetched, I will show an analogous situation in nuclear physics. The figs. 41 and 42 show how a bad resolution experiment gives rise to a giant resonance and when the resolution is increased narrow resonances making up the giant structure begin to appear[52]. Note that we have a clue for identifying the giant resonances from this figures: two reactions measured the maximum of the resonance appears shifted. Such a thing can also happen in the partial wave projection of Regge pole. Good resolution experiments will provide us with the more intimate structure of the resonance.

In our discussion of resonances, we have reached a conclusion that a major condition for the identification of a resonance is the factorization property. We will now turn to investigate how the factorization property is achieved in the Veneziano model. We will use a simplified discussion based on four point functions only[53]. The most general discussion should be based on the n point function, but this will no longer have the physical simplicity we want to achieve here.

By direct projection of partial waves we can calculate Im a_ℓ from the Veneziano formula. From

$$A(s, t) = \sum \frac{Q_n(s, t)}{s-s_n} \tag{18.1}$$

we obtain (for $\ell<\sqrt{s_n}$) the residue of a partial wave amplitude. This is given by

$$g(s_n) = c\frac{s_n^{\alpha(t=o)-1}}{\log s_n} + (t\to u) \tag{18.2}$$

Using these results we can impose our factorization constraint. This will be done by considering three reactions (1) a→a, (2) a→b and (3) b→b. The reaction will in general be dominated by different Regge poles in the t- and u-channels. These will be denoted by 1, 2, 3 respectively. The factorization condition takes the form

$$g_\ell^{aa}(s_n)g_\ell^{bb}(s_n) = [g_\ell^{ab}(s_n)]^2 \tag{18.3}$$

When we substitute eq. (18.2) into (18.3) we obtain

$$\alpha_1+\alpha_3 = 2\alpha_2 \tag{18.4}$$

where α_i denotes the intercept of the dominant Regge pole in reaction i. This will constitute our main constraints in imposing factorization.

As an example we will consider π and σ mesons. (The σ has $J^{PC}=0^{++}$.) Our three reactions will be:

$$\text{(i)} \quad \pi^0\pi^0 \to \pi^0\pi^0$$

$$\text{(ii)} \quad \pi^0\pi^0 \to \sigma\,\sigma$$

$$\text{(iii)} \quad \sigma\,\sigma \to \sigma\,\sigma \qquad (18.5)$$

Reaction (ii) is dominated by the exchange of unnatural parity trajectories. The leading candidates for exchange in this reactions are the π and A_1 trajectories. Both have $\alpha_{A_1}(o)\sim\alpha_\pi(o)\sim o$. If we now apply eq. (18.4) we obtain

$$\alpha_3(o) \approx -1/2 \qquad (18.6)$$

What do we conclude from this result? One possible conclusion could be that the P' trajectory does not couple to $\sigma\sigma$. This can be done in our case but when we generalize it to other reactions we are led into trouble. The other conclusion we shall have to make is that a pole in a partial wave amplitude of the Veneziano formula does not correspond to a single resonance, because it does not factorize. Our factorization condition has therefore to be abandoned.

The usual discussion of factorization due to Fubini and Veneziano(54) starts from an n point Veneziano formula, where all particles have no isospin and are identical and have positive parity. In this somewhat unrealistic model, the factorization difficulties arise because of the many amplitudes appearing in the coupling of high spin mesons. The conclusion from this model is that an enormous degeneracy exists. The number of states with $m^2=n$ increases like e^{cm}. Some of the states also are ghosts in that they have negative coupling. When the generalization of the n point function to physical particles is attempted, it seems that the degeneracy will even become worse[55]. The full solution of this problem for physical particles is not yet known.

How do we explain the difficulties we are encountering in factorization? One way of saying this is that there are no difficulties, just an unusual situation. Another way out of the dilemma would be to assume that we are trying to use a narrow resonance approximation in a region where this approximation does not work. Nature revenges itself on our crudeness of approximation and forces us to live with an unpleasant spectrum of resonances. The crucial question will be, what will happen to the Veneziano formula when we improve it to make it consistent with unitarity. The answer to this question is unclear because we do not know how to make the Veneziano function consistent with unitarity. This will be the subject of our next section.

19. ATTEMPTS AT UNITARIZING THE VENEZIANO MODEL

The Veneziano formula is in obvious contradiction with experiment and with theory because it contains zero width resonances, therefore poles in the physical region of a scattering process. Many attempts in overcoming this contradiction have been made but none of them is very successful. We will list and discuss a few of these attempts.

The obvious way in giving resonances a finite width is to give the Regge trajectories $\alpha(s)$ an imaginary part[56]. We can make this consistently so that $\alpha(s)=n$ will be satisfied only for values of s on the unphysical sheet. This approach runs immediately into trouble. We have noticed that in the linear trajectory approach the residue of a pole in s is a polynomial in $\alpha(t)$. This is also a polynomial in t and in $\cos\Theta_s$ provided the trajectory is linear. When we give the trajectory an imaginary part we also introduce a nonlinearity, and therefore at a given point in s all partial waves will have poles. The poles at very high ℓ have become known as ancestors. The existence of ancestors is very unpleasant since we do not believe for instance resonances with ℓ=2o at 1 BeV. One may take the attitude that the ancestors are small and may be neglected. This is often the case but we have lost the simplicity of the formula.

In order to get rid of ancestors for phenomenological use one may include a width only in one trajectory, say s-channel trajectory, and leave the t-channel trajectories linear. This will give resonances of finite width in the s-channel and no ancestors. While for a particular use such an approach may not be bad, it destroys crossing symmetry which is one of the achievements of the Veneziano formula. Suzuki(57) has succeeded in adding successfully a width in both channels by modifying an integral representation for the Veneziano formula. This approach is complicated mathematically and I will not discuss it here. Even in his approach, he has not yet succeeded in proving Regge behaviour along the physical region, because of the technical difficulties his approach encountered.

Another approach would consider the Veneziano function as a distribution and use standard smearing techniques to widen the resonance(58). The difficulties in such an approach are two-fold. a) The simplest smearing procedure gives resonances at the same energy identical width. This is somewhat unpleasant since the Veneziano formula predicts that the σ meson should have a stronger coupling to $\pi\pi$ than the ρ meson. If we introduce equal width for both mesons, we ruin the connection between the strength of coupling and the total width, which is an important ingredient of unitarity. b) The high energy usually is no longer dominated by Regge poles. In this case Regge cuts will make a major contribution. c) The smearing procedure is very arbitrary and we may lose many of the physical predictions of the Veneziano formula.

A different approach to the Veneziano formula is to consider it as a Born approximation and try to use various techniques to unitarize this approximation. Such techniques can be the N/D calculations[59], the use of the K-matrix[60] to unitarize a Born term, the insertion of width by hand etc. While this is not an attractive way to treat the Veneziano formula, it may be a rather useful one in performing phenomenology.

Mathematically, the most beautiful and ambitious program in unitarizing the Veneziano formula is that started by Kikkawa, Sakita and Virasoro[61]. These authors want to consider the Veneziano formula as a Born term in some field theoretical sense. The obvious procedure for unitarizing such a Born term is the inclusion of higher diagrams. This is not a simple program. First, the authors had to overcome the difficulty of writing a dual Feynman diagram, namely a Feynman diagram whose s-channel poles also correspond to t-channel poles. Once this had been achieved, the next purpose was to calculate the correction to pole diagrams. This of course is a very ambitious program. The authors had to tackle the ancient and unsolved problem of calculating all the diagrams in a field theory and, in their case, an extremely complicated field theory. Moreover, their field theory was not a weak interaction field theory. The expansion parameter in this case is of the order of strong interaction coupling constants. The authors hoped that the inclusion of a Born diagram which already contained many of the desired features of the scattering amplitude such as resonances, Regge behaviour and so on, would make the expansion rapidly converging in spite of the apparent large value of the expansion parameter. When the high degeneracy of resonances was included in the so-called Feynman diagram method, a very unpleasant feature appeared: The integrands are exponentially divergent and this infinity is much worse than that encountered even in bad field theories[62]. The "renormalization" of this program is still unclear[63]. Much of the work in this direction is mathematically beautiful and is carried out with immense technical skill. The connections of this approach with experiment are not yet evident, since the attempt had to use the n point Veneziano function which can be written in general only for physically unrealistic cases. I do not know whether the difficulties in this program will be overcome and how. Possibly the person to lecture here in a few years on the subject duality, if the subject survives at all, will give you an answer.

2o. SOME APPLICATIONS OF THE VENEZIANO MODEL

So far, we have discussed both the theoretically pleasing and the unpleasant aspects of the Veneziano formula but we have kept away from experiment. One of the reasons we have done so is that the Veneziano formula as such needs modifying before we can confront it with data. Lack of space prohibits me from going into details about this subject. Much has already been written about it. One will find useful material in the reviews mentioned already and some enthusiastic reviews by Lovelace[64] on the subject.

Before discussing in great detail the phenomenological applications we should clarify the point concerning the uniqueness of the Veneziano formula. It turns out that we can always add to a Veneziano formula terms of the type

$$B_{nm\ell}(s, t) = \frac{\Gamma(1-\alpha(s)+n)\,\Gamma(1-\alpha(t)+m)}{\Gamma(1-\alpha(s)-\alpha(t)+\ell)} \qquad (2o.1)$$

Such terms are known as satellites. It has been shown by Khuri(65) and by Tiktopoulos(66) that the most general functions obeying duality and having zero width resonances can be written as an infinite sum of such terms.

$$A(s, t) = \sum c_{nm\ell}\, B_{nm\ell}(s, t) \qquad (2o.2)$$

We have therefore an enormous freedom in writing down the Veneziano formulae. The four point function in itself cannot give an answer to what the coefficients are. One would hope that the lowest term, i. e., the term without satellites, could be sufficient to discuss everything. This is, however, not the case. When problems with spin are encountered, one has to invoke satellites. There are indications from the n point functions that sometimes, for the consistency of the problem, satellites are also needed(67). This occurs when the external particles are not on the leading Regge trajectory. Obviously, we do not have yet a selfconsistent scheme in order to define the coefficients $c_{nm\ell}$ and we must use them as parameters to fit the data. Mostly, we will try to manage with as few terms as possible to make life simpler. Some care should be exercised when using the Veneziano formula and this concerns the Pomeranchuk contribution. We already know from the two-component theory of scattering amplitudes that the Pomeranchuk contribution is connected to s-channel background and not to resonances. The Veneziano function, on the other hand, is just the resonant part of the amplitude, which is connected with ordinary Regge trajectories. In applying the Veneziano formula, we should be careful not to apply it to processes where the Pomeranchuk can be dominant. A typical example of this case is one of the nice successes of the Veneziano formula for meson-meson scattering. When we use the Veneziano formula for $\pi\pi$ scattering which we have already written down, we can, following Lovelace[68], investigate whether this formula obeys the Adler self-consistency condition. At the point $s=t=u=m_\pi^2$ the amplitude should vanish. In $\pi\pi$ scattering, this vanishing happens if and only if

$$\alpha_\rho (m_\pi^2) = 1/2 \tag{2o.3}$$

This value α_ρ determines the intercept of the ρ trajectory to be roughly 1/2 in remarkable agreement with experiment. When we continue this procedure to πK scattering we obtain

$$\alpha_{K^*} (m_K^2) = 1/2 \tag{2o.4}$$

This condition, too, is in agreement with experiment. When we use the same procedure to investigate the $\pi\eta$ scattering or πX^o scattering we get into contradiction with experiment. The Adler condition demands

$$\alpha_{A_2} (m_\eta^2) = 1/2 \tag{2o.5}$$

$$\alpha_{A_2} (m_{X^o}^2) = 3/2 \tag{2o.6}$$

These results do not agree with experiment. In view of our discussion about the importance of exotic channels to determine the Adler consistency condition from resonances, we understand this disagreement and claim that it is due to the contribution of the Pomeranchuk trajectory(43). The connection between the Adler consistency condition, current algebra results and the Veneziano formula has been investigated also by Ademollo, Weinberg and Veneziano(69). They have deduced a spacing rule for trajectories which resembles the Lovelace results. Trajectories of opposite normality which can decay into each other by pion emission have intercepts which differ by one-half. The difficulty in this prediction is the use of single-term Veneziano formulae which may be suspect when higher spins are involved.

The connection between the Veneziano formula and current algebra is still being investigated. In a mysterious way, people hope that finding the proper Veneziano formula will help them understand the representation of current algebra. I do not know whether this is true. On the other hand, if one is interested in current algebra representation, one might as well try Veneziano functions because they are nice.

The spectrum implied by the Veneziano formula, namely, the existence of an infinite number of mesons of each spin parity gives rise to a few interesting speculations about current coupling to vector mesons. The usual vector dominance model can be modified in the presence of an infinite number of vector mesons by introducing direct couplings between the photon and each vector meson. We do not have any physical insight to tell us what these couplings would be. A V.D.M. enthusiast would assume that all couplings except the first vanish. Other people would try to introduce a coupling to many vector mesons in order to improve the agreement of vector dominance with experiment and in order to try and fit form factors. Some attempts in writing Veneziano's formulae for the scattering of photons on particles have been made. See for instance a beautiful example in the lecture of Landshoff[70]. All examples of this kind suffer from enormous arbitrariness.

Whenn a theoretician discusses duality and the Veneziano formulae with an experimentalist, the most important remark he will hear sounds something like the following: "That is all very nice, but tell me, how do I describe my data?" If the extreme local duality picture is correct, each of many reaction mechanism is an alternative description of the data. Only some may be simpler than the rest. How, then, is an experimentalist able to disentangle resonance production from background, phase space, or resonance production from another channel. The answer to this is not simple. To be frank about it, I do not know how to do it in general, and the Veneziano formula may give us a clue in some cases.

Application of the Veneziano formula to πN scattering is not very successful. Firstly since the baryon trajectories α are functions of $\alpha(s)$ parity doubling occurs because of the Mac Dowell symmetry. To overcome this and to fit low energy resonances many satellite terms are needed[71]. These predict correctly the forward πN charge exchange scattering, but fail badly in describing the backward scattering[72].

The next example I will discuss concerns the reaction $\bar{p}n \to 3\pi$. The remarkable feature in this case is the very complicated structure of the Dalitz plot. The initial state of the $\bar{p}n$ is a $^1S_0$ state i. e. it has the quantum numbers of the π meson. Lovelace[68] was the first to use a $\pi\pi \to \pi\pi$ Veneziano formula to fit the complicated structure of the Dalitz plot. In this fit $\alpha(s)$ was given an imaginary part.

Lovelace's fit looks remarkable. When satellites are introduced the fit can even be improved. The appearance of pole and zero lines is the essential ingredient which makes the fit of a complicated Dalitz plot possible.

21. CONCLUSIONS

We have followed the story of duality from infancy through childhood up to the Veneziano formula. My discussion is incomplete, but sufficient to introduce some of the major ideas and problems of the subject. It is my hope that this introduction is sufficient to arouse your curiosity and interest in the subject. I may be accused of following closely the Rehovot or Israeli party-line. To these accusations I plead guilty. As I could not cover the whole topic extensively, my choice of subjects was limited, somewhat, by personal interest.

I have not discussed some of the most recent developments of duality. The n point functions. The Feynman like diagrams. The applications of 5 point functions to phenomenology. The attempts to construct n point functions for physical particles. The operator formalism for discussing n point functions. Some of these are still in a rather fluid stage. The reasons for omitting these subjects are technical, i. e. space and time limitations.

Whatever the results of these new investigations duality is here to stay. It has revolutionized our way of thinking about strong interactions. It has however not answered all questions we are asking of it. The questions which still remain open have an appeal to every taste. For the mathematically minded theoreticians, there are many questions in generalizing the Veneziano formula.

For the phenomenologist the analysis of data in terms of duality. For the experimentalist the many checks still needed to confirm or disprove duality predictions. And last for all of us as high energy physicists the challenge of understanding duality in all it's implications with the hope that more interesting physics is still hiding under the surface. This is my hope and I am sure we will know much more in the future.

ACKNOWLEDGMENT : The author is grateful to his colleagues at the Weizmann Institute and Tel Aviv University for many helpful discussions. These discussions have set the tone and approach used in the present series of lectures.

REFERENCES

1. H. W. Fowler, Modern English Usage, Clarendon Press Oxford.
2. R. Dolen, D. Horn and C. Schmid, Phys. Rev. 166, 1768 (1968).
3. M. Jacob, VIII. Internationale Universitätswochen für Kernphysik, P. Urban, Editor.
4. D. Horn, ibid.
5. H. J. Lipkin, J. D. Jackson, M. Jacob, Proceedings of the 1969 Lund Conference on Elementary Particles.
6. H. Harari, Proceedings of the Brookhaven Summer School, 1969.
7. L. Van Hove, Phys. Letters 24B, 183 (1967).
8. D. Amati, S. Fubini, A. Stanghellini, Nuovo Cimento 26, 896 (1962).
9. K. Igi and S. Matsuda, Phys. Rev. Letters 18, 625 (1967)

A. Logunov, L. D. Soloviev and A. N. Tavkhelidze, Phys. Letters 24, 181 (1967), and Ref. 2.

1o. M. Ademollo, H. R. Rubinstein, G. Veneziano and M. A. Virasoro, Phys. Letters 27B, 92 (1969) and Phys. Rev. 176, 19o4 (1968).

11. C. Schmid, Phys. Rev. Letters 2o, 689 (1968).

12. M. Kugler, Phys. Rev. 18o, 1538 (1969).

13. M. Kugler, Phys. Rev. Letters 21, 57o (1968).

14. P. D. B. Collins, R. C. Johnson and E. J. Squires, Phys. Letters 27B, 23 (1968).

15. H. Harari, Rapporteur Talk, Proceedings of the Vienna Conference.

16. H. R. Rubinstein, A. Schwimmer, G. Veneziano, M. A. Virasoro, Phys. Rev. Letters 21, 491 (1968).

17. We follow the discussion of Ref. 12 and Ref. 15.

18. G. F. Chew and A. Pignotti, Phys. Rev. Letters 2o, 1o78 (1967).

19. V. A. Alessandrini and E. J. Squires, Phys. Letters 27B, 3oo (1968).

2o. H. Harari, Phys. Rev. Letters 2o, 1395 (1968).

21. P. G. O. Freund, Ibid. 2o, 235 (1968).

22. F. J. Gilman, H. Harari and Y. Zarmi, Phys. Rev. Letters 21, 323 (1968).

23. H. Harari and Y. Zarmi, Phys. Rev. 187, 223o (1969).

24. J. V. Allaby et al., Phys. Letters 3oB, 5oo (1969).

25. R. J. N. Phillips, These proceedings.

26. Particle Data Group K^+N Compilation.

27. H. Harari, Proceedings of the Liverpool Conference on Electrons and Photons.

28. A. Shapira et al., Weizmann Institute preprint.

29. H. W. Atherton et al., fast antiproton group, CERN preprint.

3o. H. Nicholson et al., Phys. Rev. Letters 23, 6o3 (1969).

31. We follow the discussion of reference 12.
32. H. J. Lipkin, Nucl. Phys. B9, 349 (1969).
33. C. N. Chiu and J. Finkelstein, Phys. Letters 27B, 576 (1968).
34. J. L. Rosner, Phys. Rev. Letters 21, 95o (1968).
35. M. Kugler, Phys. Letters, to be published.
36. H. J. Lipkin, Weizmann Institute preprint.
37. S. Pinsky, Phys. Rev. Letters 22, 677 (1969).
C. Michael, Nucl. Phys. B13, 644 (1969).
38. J. Mandula, J. Weyers and G. Zweig, Phys. Rev. Letters 23, 266 (1969).
39. H. Harari, Phys. Rev. Letters 22, 562 (1969).
J. L. Rosner, Ibid. 22, 689 (1969).
4o. D. J. Crennel et al., Phys. Rev. Letters 23, 1347 (1969).
41. M. Gell-Mann and G. Zweig, unpublished. Discussed in Ref. 15.
42. A. Schwimmer, Phys. Rev. 184, 15o8 (1969).
43. M.Kugler, Phys. Letters 31B, 372 (197o).
44. F. J. Gilman, Phys. Letters 29B, 673 (1969);
K. W. Lai and J. Louie, Paper submitted to the Irvine Conference on Regge Poles.
45. See ref. 37.
46. G. Hite, These Proceedings.
47. J. D. Jackson, Ref. 5 and Rev. Mod. Phys. 42, 12 (197o). Further references are quoted in this paper.
48. M. Le Bellac, Phys. Letters 25B, 524 (1967).
49. M. Kugler, Ref. 35.
5o. G. Veneziano, Nuovo Cimento 57A, 19o (1968).
51. A. Schwimmer, Weizmann Institute Thesis (unpublished).
52. P. P. Singh et al., Nucl. Phys. 65, 577 (1965).
I thank S. S. Hanna for a discussion of his data.
53. M. Kugler and M. Milgrom, Nucl. Phys. B13, 294 (1969).
54. S. Fubini and G. Veneziano, Nuovo Cimento 64A, 811 (1969).

55. M. Bishari and V. Rittenberg, Weizmann Institute Preprint.
56. R. Z. Roskies, Phys. Rev. Letters 21, 1851 (1968).
57. M. Suzuki, Phys. Rev. Letters 23, 2o5 (1969).
58. A. Martin, Phys. Letters 29B, 431 (1969).
K. Huang, Phys. Rev. Letters 23, 9o3 (1969).
N. F. Bali et al., Phys. Rev. Letters 23, 9oo (1969).
59. D. Atkinson et al., Phys. Letters 29B, 423 (1969).
6o. C. Lovelace, Proceedings of the ANL Conference on $\pi\pi$ and πK Interactions.
61. K. Kikkawa, B. Sakita and M. A. Virasoro, Phys. Rev. 184, 17o1 (1969).
62. G. Frye and L. Susskind, Phys. Letters 31B, 537 (197o); A. Neveu and J. Sherk, Phys. Rev., to be published.
63. See, however, D. J. Gross et al., Phys. Letters 31B, 592 (197o).
64. C. Lovelace, Review talk at the Irvine Conference on Regge Poles (197o).
65. N. N. Khuri, Phys. Rev. 185, 1876 (1969).
66. G. Tiktopoulos, Phys. Letters 31B, 138 (197o).
67. H. R. Rubinstein, E. J. Squires and M. Chaichian, Phys. Letters 3oB, 189 (1969).
68. C. Lovelace, Phys. Letters 28B, 265 (1968).
69. M. Ademollo, G. Veneziano and S. Weinberg, Phys. Rev. Letters 22, 83 (1969).
7o. P. Landshoff, These proceedings.
71. R. Carlitz and M. Kisslinger, Phys. Rev. Letters 24, 186 (197o).
72. See: S. Fenster and K. C. Wali, Phys. Rev., in press. Further references are quoted there.
73. G. Altarelli and H. R. Rubinstein, Phys. Rev. 178, 2165 (1969).

FIGURE CAPTIONS

Fig. 1 e^+e^- scattering to lowest order in α.

Fig. 2 A Regge pole in the multiperipheral model.

Fig. 3 The contour C used in deriving FESR.

Fig. 4 Plot of $\mathrm{Im}A'^{(-)}$ at $t=0$. Comparison between different models. Taken from ref. 2.

Fig. 5 $\mathrm{Im}\ \nu B^{(-)}$ and $\mathrm{Im}\ A'^{(-)}$ at $t=m_\rho^2$. Taken from ref. 2.

Fig. 6 Comparison of the Regge residue functions from (I) high-energy fits and (II) the FESR, under the one-pole assumption ($\alpha_\rho=0.57+0.96t$). The error from neglecting the background integral in the j plane is included. From ref. 2.

Fig. 7 First zeros of the prominent resonances on the Mandelstam plot for the πN problem. From ref. 2.

Fig. 8 Saturation of the $\pi\pi\to\pi\omega$ sum rules with the ρ resonance alone. Dashed line represents the resonance side and full line the Regge side. Ordinates in arbitrary units. From ref. lo.

Fig. 9 Saturation of the same sum rule as in fig. 8 with the ρ and the $\rho_3(3^-)$ in the resonance side. On the upper left side the most important region is shown on a larger scale. Here the Regge side is represented by the dashed line. From ref. lo.

Fig.lo Third moment sum rule for $\pi\pi\to\pi\omega$. Saturation with ρ and $\rho_3(3^-)$. Here the Regge side is represented by the full line. From ref. lo.

Fig.11 Same as in figs. 8 and 9 with the $\rho_5(5^-)$ included in the resonance side. Here the Regge side is represented by a dashed line. From ref. lo.

Fig.12 A Regge pole built from resonances (schematic).

Fig.13 $\beta(t)\nu^{\alpha(t)}$ and $P_\ell(1+t/2k_1^2)$. The overlap of these functions is maximum for k_2^2.

$$① \ldots P_\ell(1 + \frac{t}{2k_1^2}) \ , \quad ② \ldots P_\ell(1 + \frac{t}{2k_2^2}) \ ,$$

$$③ \ldots P_\ell(1 + \frac{t}{2k_3^2}) \ .$$

Fig.14 Im a_ℓ and Re a_ℓ vs k^2. At $k^2=k_2^2$ Im a_ℓ has a maximum.

Fig.15 Argand diagram for $a_\ell(k^2)$.

Fig.16 K^-p elastic scattering. From D.R.O. Morrison, Rapporteur talk at the Lund conference.

Fig.17 K^+p elastic scattering. From the same source as fig. 16.

Fig.18 Angular distributions for elastic reactions. Exotic channels do not show dips in angular distribution while non-exotic channels have dips. From same source as fig. 16.

Fig.19 Argand diagrams for partial waves in which the projected t- and u-channel trajectories in πN scattering correspond most closely to the "experimental" phase shifts. From ref. 14.

Fig.2o Argand diagrams for $\pi\pi\to\pi\omega$. From ref. 16.

Fig.21 Comparison of the negative particle total cross-sections with the Regge-pole model predictions of Barger et al. Taken from ref 24.

Fig.22 Argand diagrams for amplitudes with well defined I_t. From ref. 23.

Fig.23 Same as fig. 22.

Fig.24 Reconstruction of the πN amplitude from resonances (dashed line). From ref. 23.

Fig.25 Same as fig. 24.

Fig.26 Duality diagram for $\bar{K}^oK^+\rightarrow\bar{K}^oK^+$.

Fig.27 Diagrams for

a) meson-meson scattering

b) forward meson-baryon scattering

c) backward meson-baryon scattering, and

d) an "illegal" diagram for baryon-antibaryon scattering.

The s- and t-channel intermediate states are marked by dashed lines. From ref. 39.

Fig.28 The regions where resonance dominance and $\alpha_2<o$ hold. In the shaded region all amplitudes can be calculated from resonances. From ref. 43.

Fig.29 $\sigma(K^-p\rightarrow\Sigma^+\pi^-)$and $\sigma(\pi^+p\rightarrow\Sigma^+K^+)$. From ref. 44.

Fig.3o $\sigma(K^-p\rightarrow\pi^-\Sigma^+)$ and $\sigma(\pi^+p\rightarrow K^+\Sigma^+)$. From ref. 44.

Fig.31 $\sigma(K^-n\rightarrow\bar{K}^o\Delta)$ and $\sigma(K^+p\rightarrow K^o\Delta^{++})$. From ref. 44.

Fig.32 Taken from ref. 44.

Fig.33 The differential cross section for $K^-p\rightarrow\pi^-\Sigma^+$ at 4.1 GeV/c and two times the differential cross section for $\pi^-p\rightarrow K^o\Sigma^o$ at 4.0 GeV/c. From ref. 44.

Fig.34 From ref. 44.

Fig.35 $\sigma(K^+p\rightarrow K^o\Delta^{++})$ and $\sigma(K^-n\rightarrow\bar{K}^o\Delta^-)$. From ref. 44.

Fig.36 $\sigma(pp\rightarrow n\Delta^{++})$ and $\sigma(\bar{p}n\rightarrow\bar{\Delta}^{--}p)$. The straight line is p_{lab}^{-2} as predicted by π exchange.

Fig.37 Lines of zeros and poles in a single term Veneziano formula. $B(\alpha(s), \alpha(t))$.

Fig.38 Chew-Frautschi plot of resonances appearing in the $\pi\pi$ Veneziano formula.

Fig.39 Im a_{21} in $\pi\pi\rightarrow\pi\omega$. From ref. 51.

Fig.4o Trajectories in the Veneziano model for $\pi\pi\rightarrow\pi\omega$. The horizontal lines denote the widths of giant resonances. From ref. 51.

Fig.41 Giant resonance in nuclear physics for various experimental resolutions.

Fig.42 Same as fig. 41.

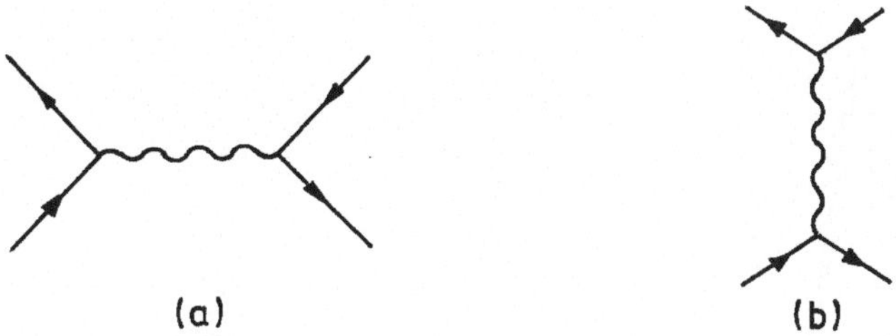

Fig. 1

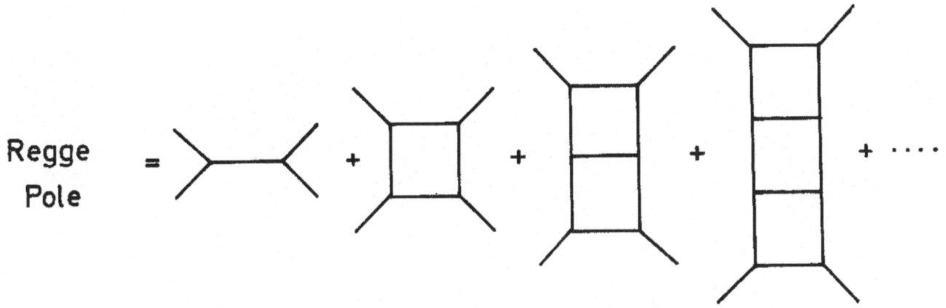

Fig. 2

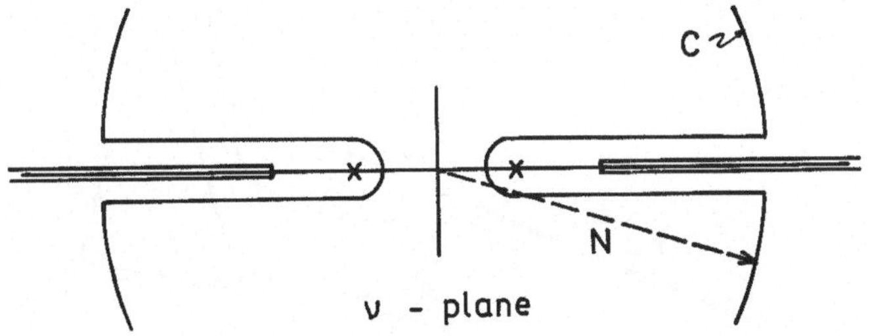

Fig. 3

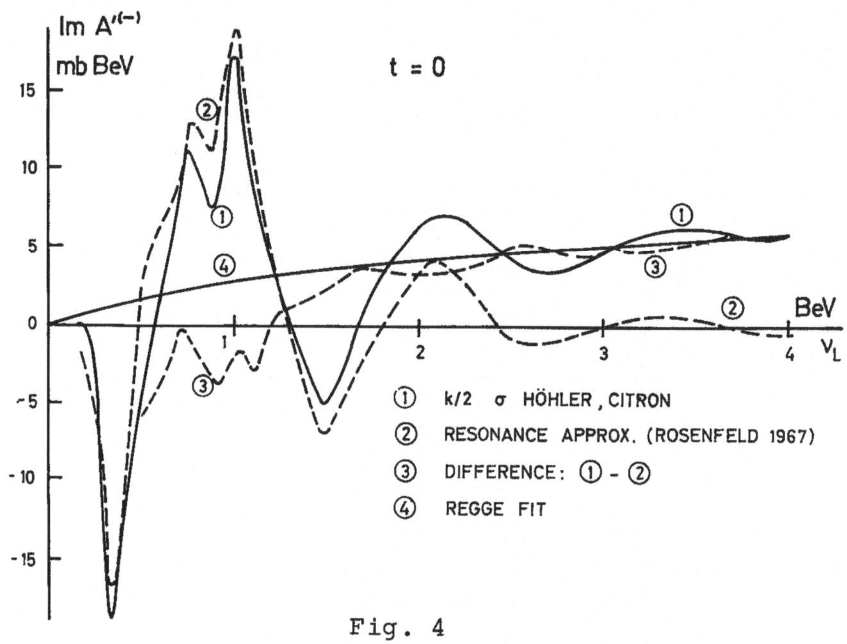
Im A'(−)
mb BeV
t = 0
BeV
ν_L
① k/2 σ HÖHLER, CITRON
② RESONANCE APPROX. (ROSENFELD 1967)
③ DIFFERENCE: ① - ②
④ REGGE FIT

Fig. 4

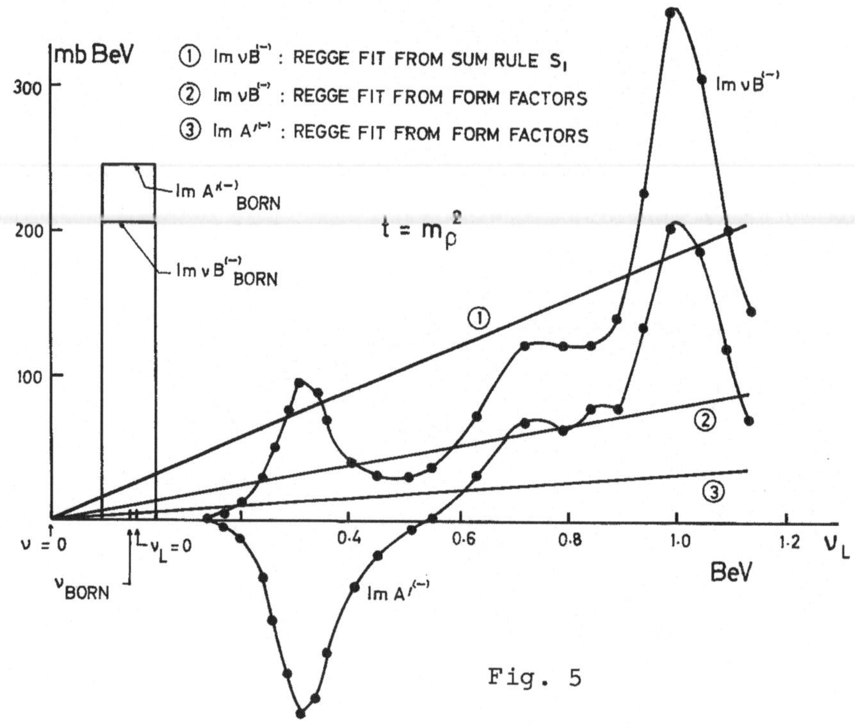
mb BeV
① Im νB(−) : REGGE FIT FROM SUM RULE S1
② Im νB(−) : REGGE FIT FROM FORM FACTORS
③ Im A'(−) : REGGE FIT FROM FORM FACTORS
Im A'(−) BORN
Im νB(−) BORN
t = m_ρ^2
Im νB(−)
Im A'(−)
ν = 0
ν_L = 0
ν_{BORN}
BeV
ν_L

Fig. 5

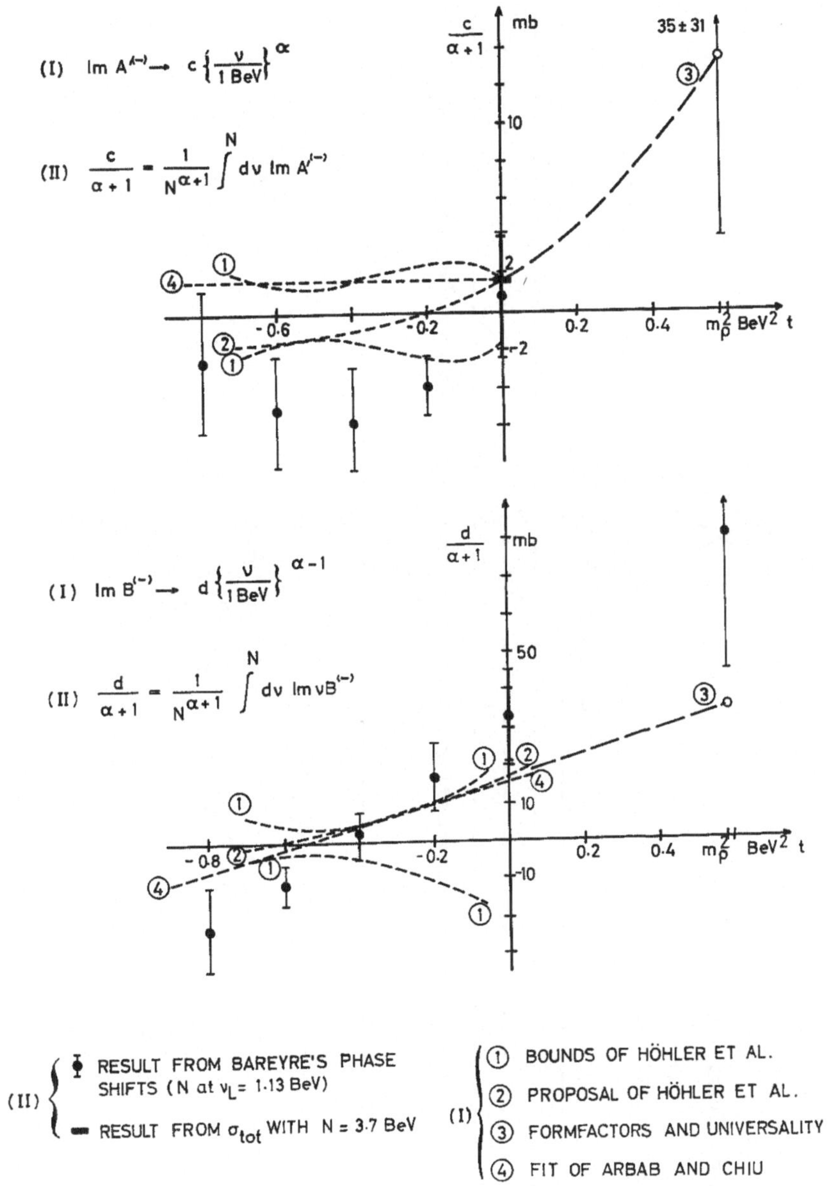

(I) $\text{Im } A^{(-)} \longrightarrow c \left\{ \dfrac{\nu}{1 \text{ BeV}} \right\}^{\alpha}$

(II) $\dfrac{c}{\alpha + 1} = \dfrac{1}{N^{\alpha + 1}} \displaystyle\int^{N} d\nu \ \text{Im } A'^{(-)}$

(I) $\text{Im } B^{(-)} \longrightarrow d \left\{ \dfrac{\nu}{1 \text{ BeV}} \right\}^{\alpha - 1}$

(II) $\dfrac{d}{\alpha + 1} = \dfrac{1}{N^{\alpha + 1}} \displaystyle\int^{N} d\nu \ \text{Im } \nu B^{(-)}$

(II) $\begin{cases} \text{RESULT FROM BAREYRE'S PHASE} \\ \text{SHIFTS (N at } \nu_L = 1.13 \text{ BeV)} \\ \text{— RESULT FROM } \sigma_{tot} \text{ WITH } N = 3.7 \text{ BeV} \end{cases}$

(I) $\begin{cases} ① \text{ BOUNDS OF HÖHLER ET AL.} \\ ② \text{ PROPOSAL OF HÖHLER ET AL.} \\ ③ \text{ FORMFACTORS AND UNIVERSALITY} \\ ④ \text{ FIT OF ARBAB AND CHIU} \end{cases}$

Fig. 6

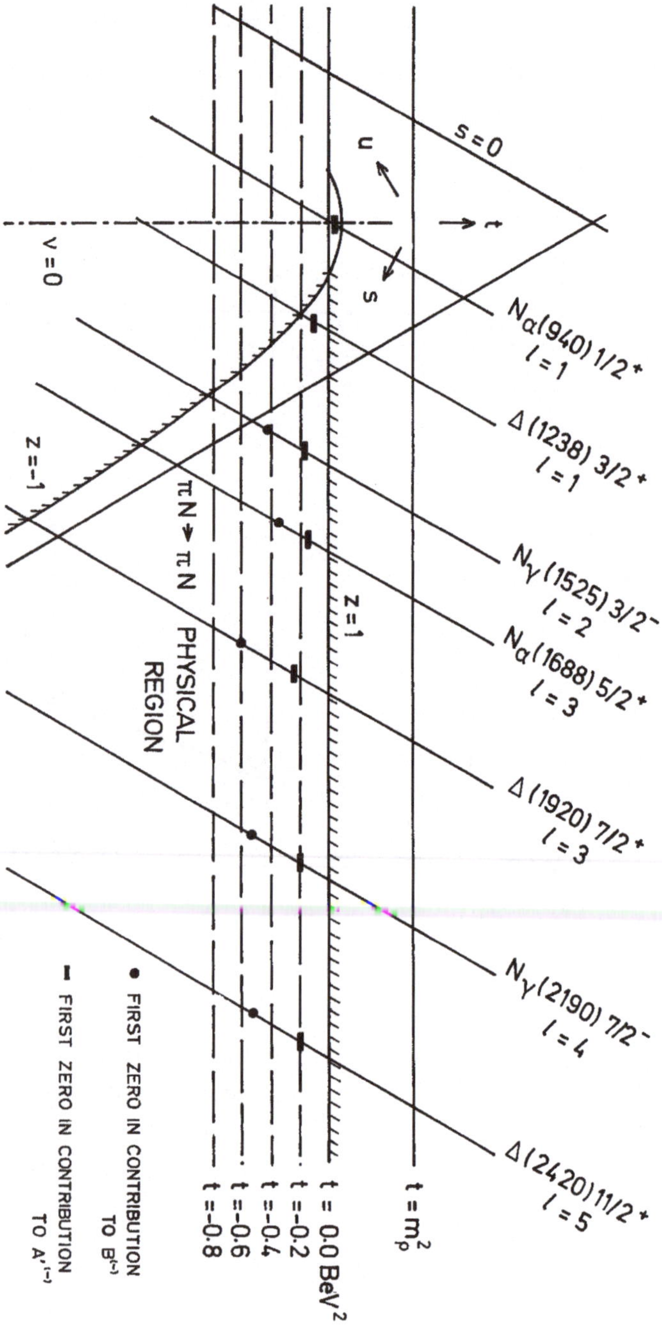
s = 0
u
t
s
ν = 0
z = -1
z = 1
πN → πN
PHYSICAL REGION
Nα(940) 1/2+ l = 1
Δ(1238) 3/2+ l = 1
Nγ(1525) 3/2− l = 2
Nα(1688) 5/2+ l = 3
Δ(1920) 7/2+ l = 3
Nγ(2190) 7/2− l = 4
Δ(2420) 11/2+ l = 5
t = m_ρ^2
t = 0.0 BeV^2
t = -0.2
t = -0.4
t = -0.6
t = -0.8
● FIRST ZERO IN CONTRIBUTION TO B^(-)
▬ FIRST ZERO IN CONTRIBUTION TO A'^(-)

Fig. 7

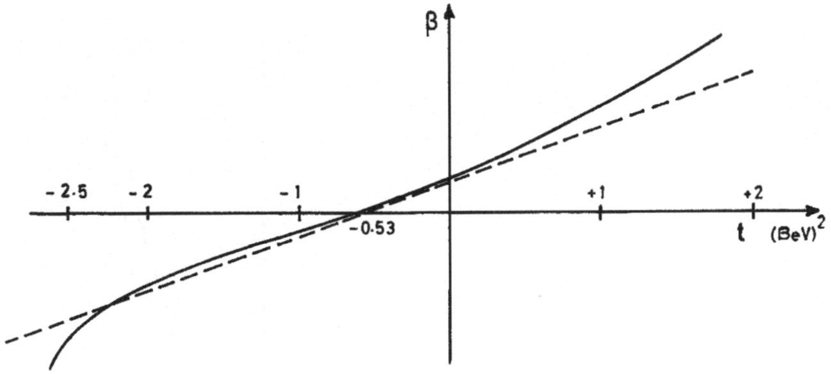

Fig. 8

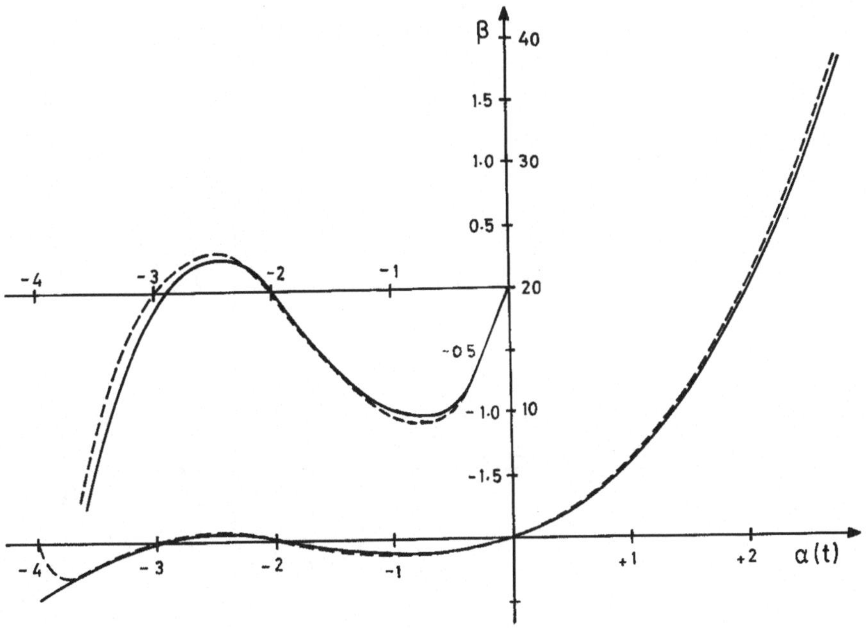

Fig. 9

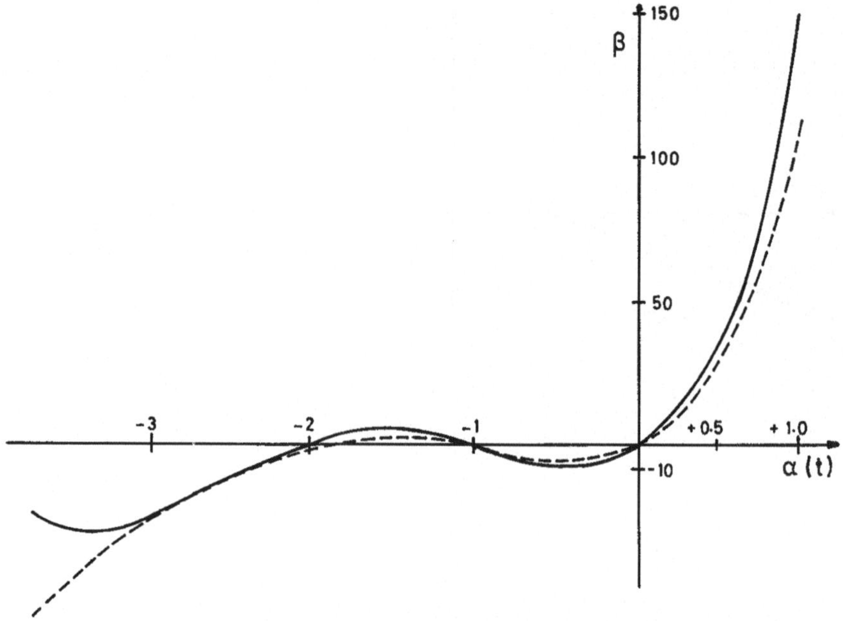

Fig. 1o

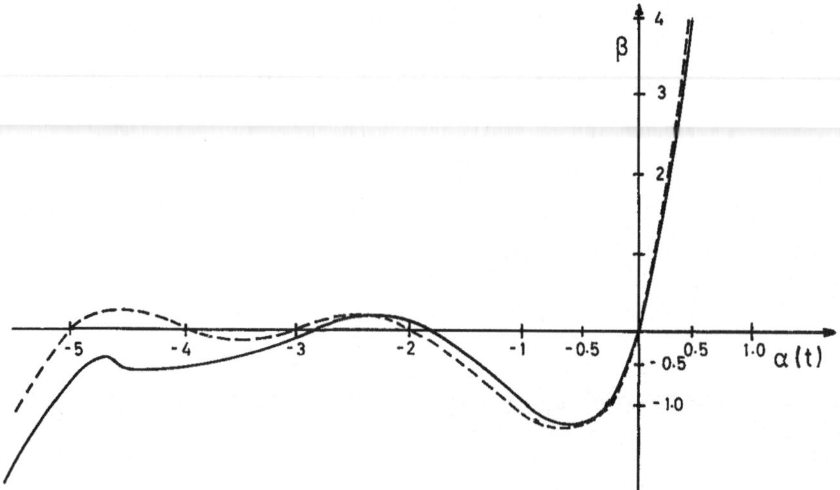

Fig. 11

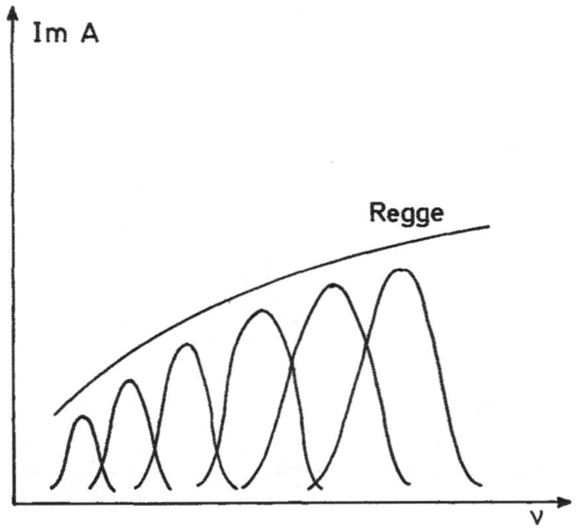

Fig. 12

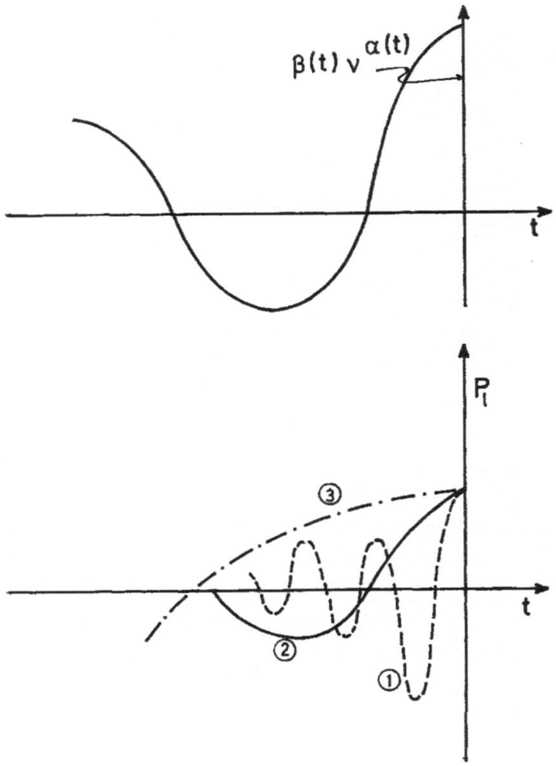

Fig. 13

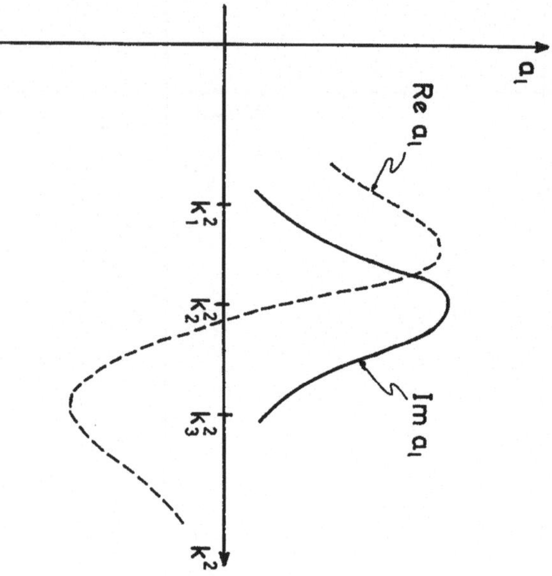

Fig. 14

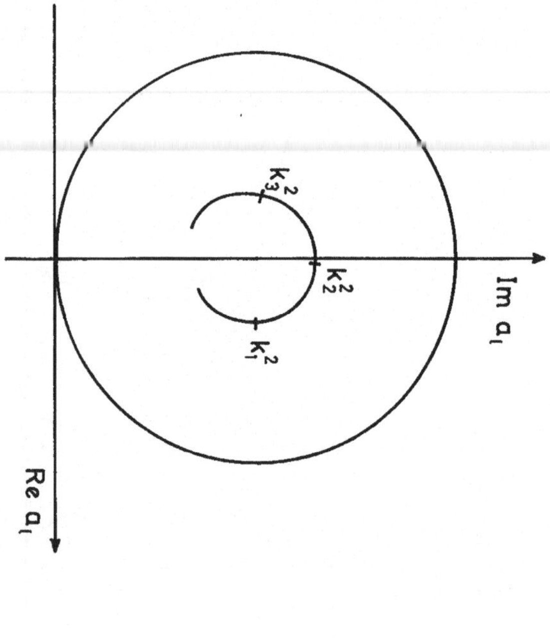

Fig. 15

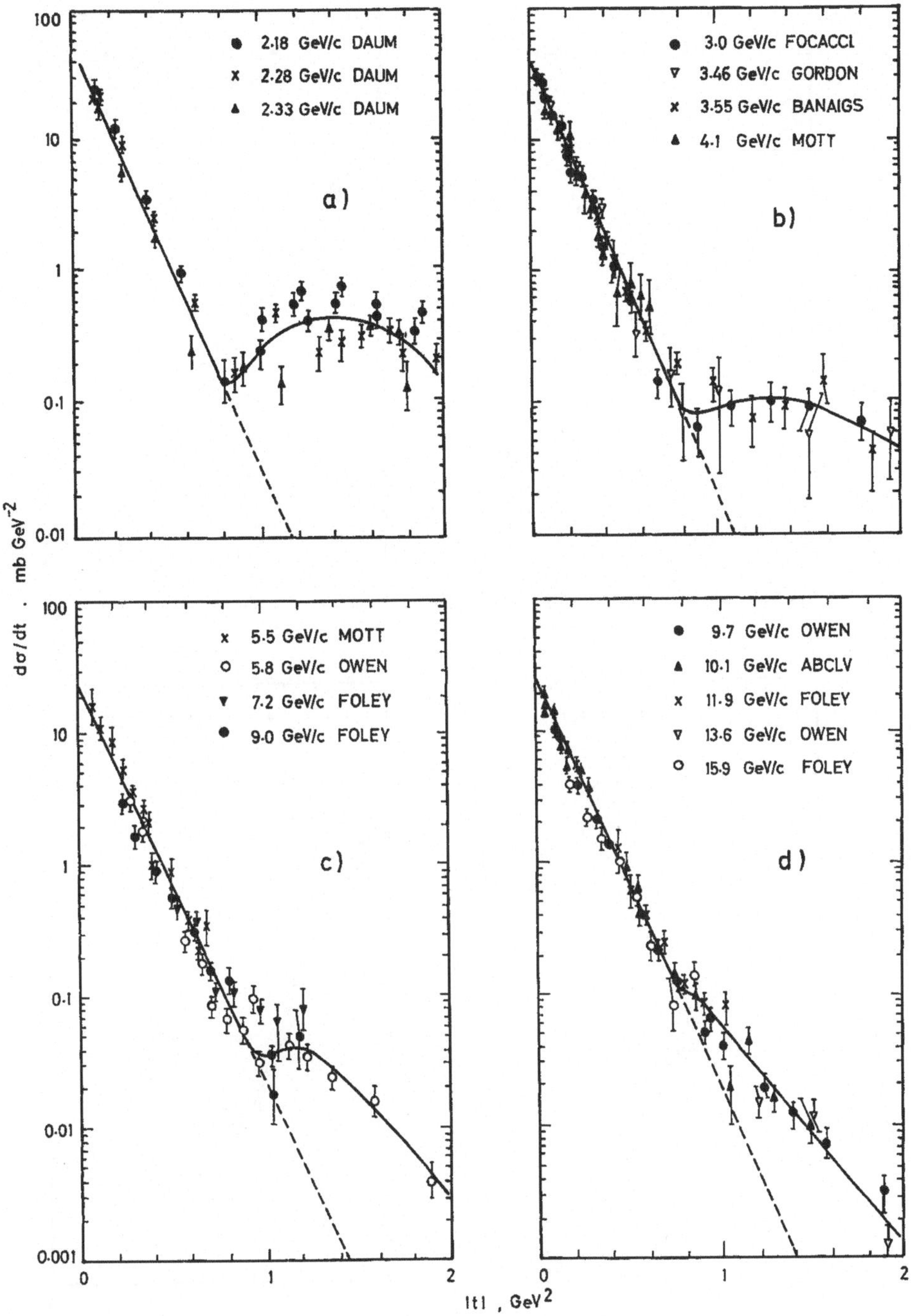

2.18 GeV/c DAUM
2.28 GeV/c DAUM
2.33 GeV/c DAUM
a)
3.0 GeV/c FOCACCI
3.46 GeV/c GORDON
3.55 GeV/c BANAIGS
4.1 GeV/c MOTT
b)
5.5 GeV/c MOTT
5.8 GeV/c OWEN
7.2 GeV/c FOLEY
9.0 GeV/c FOLEY
c)
9.7 GeV/c OWEN
10.1 GeV/c ABCLV
11.9 GeV/c FOLEY
13.6 GeV/c OWEN
15.9 GeV/c FOLEY
d)
dσ/dt , mb GeV⁻²
|t| , GeV²

Fig. 16

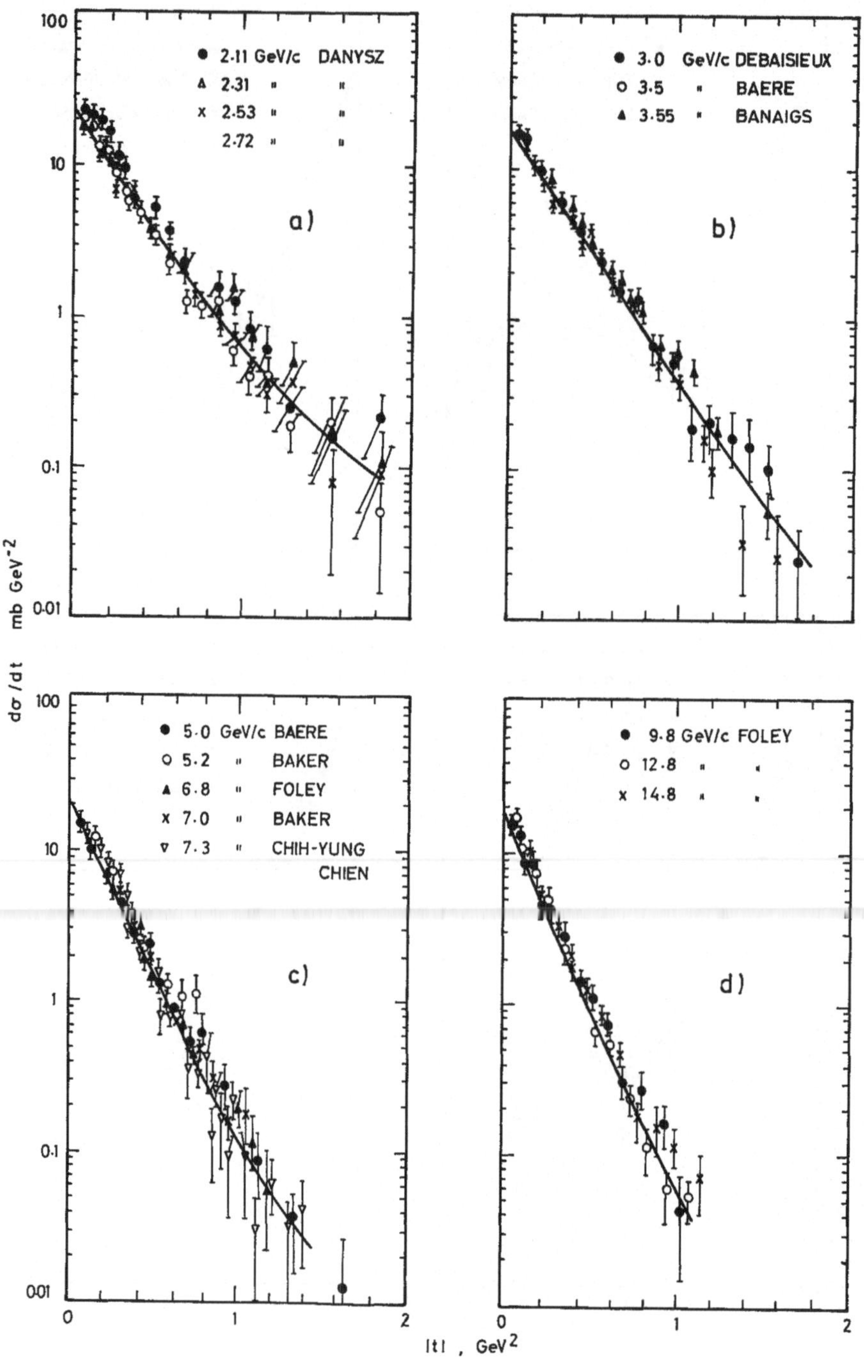

2·11 GeV/c DANYSZ
2·31 " "
2·53 " "
2·72 " "
a)
3·0 GeV/c DEBAISIEUX
3·5 " BAERE
3·55 " BANAIGS
b)
5·0 GeV/c BAERE
5·2 " BAKER
6·8 " FOLEY
7·0 " BAKER
7·3 " CHIH-YUNG CHIEN
c)
9·8 GeV/c FOLEY
12·8 " "
14·8 " "
d)
dσ /dt mb GeV^{-2}
100
10
1
0·1
0·01
0
1
2
|t| , GeV^2

Fig. 17

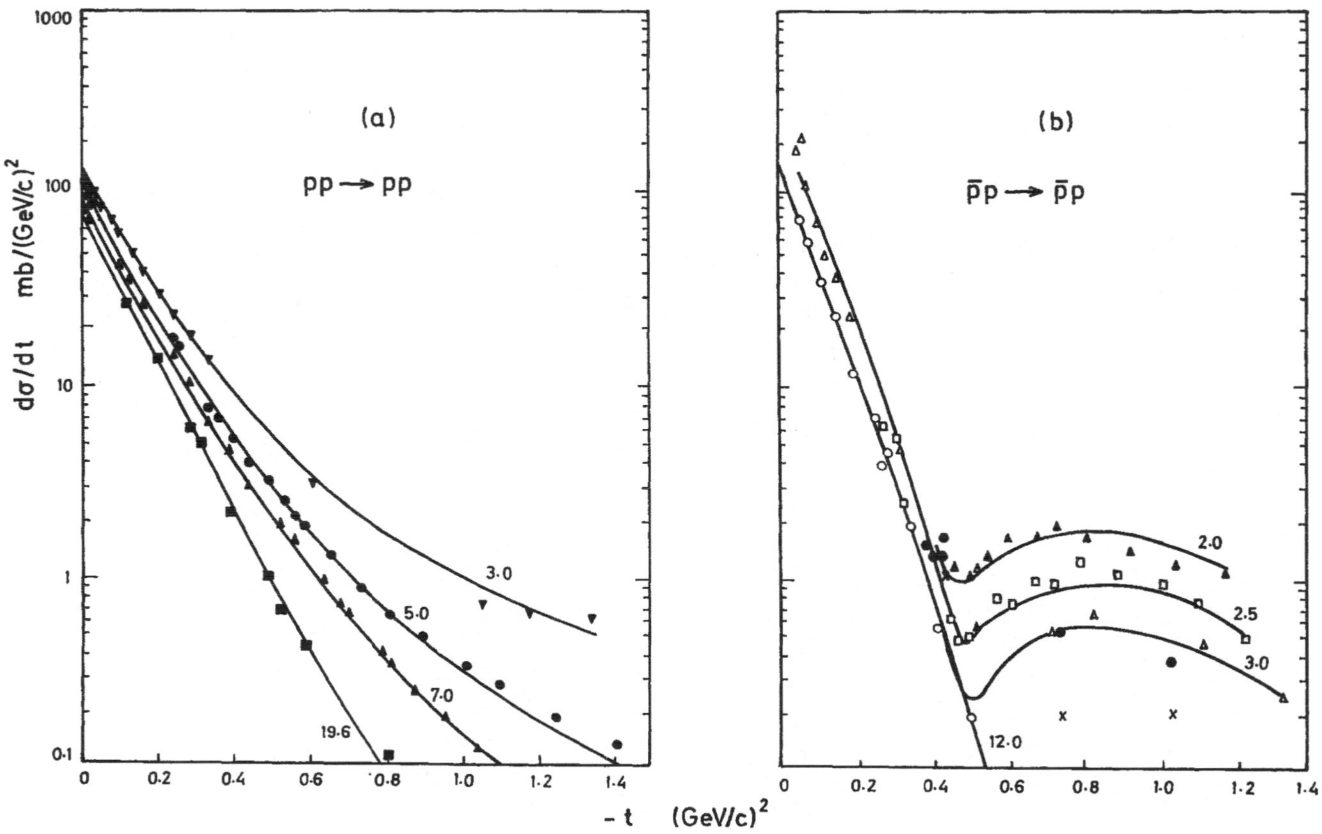
(a)
pp → pp
(b)
p̄p → p̄p
dσ/dt mb/(GeV/c)²
1000
100
10
1
0.1
3.0
5.0
7.0
19.6
2.0
2.5
3.0
12.0
0
0.2
0.4
0.6
0.8
1.0
1.2
1.4
- t (GeV/c)²

Fig. 18

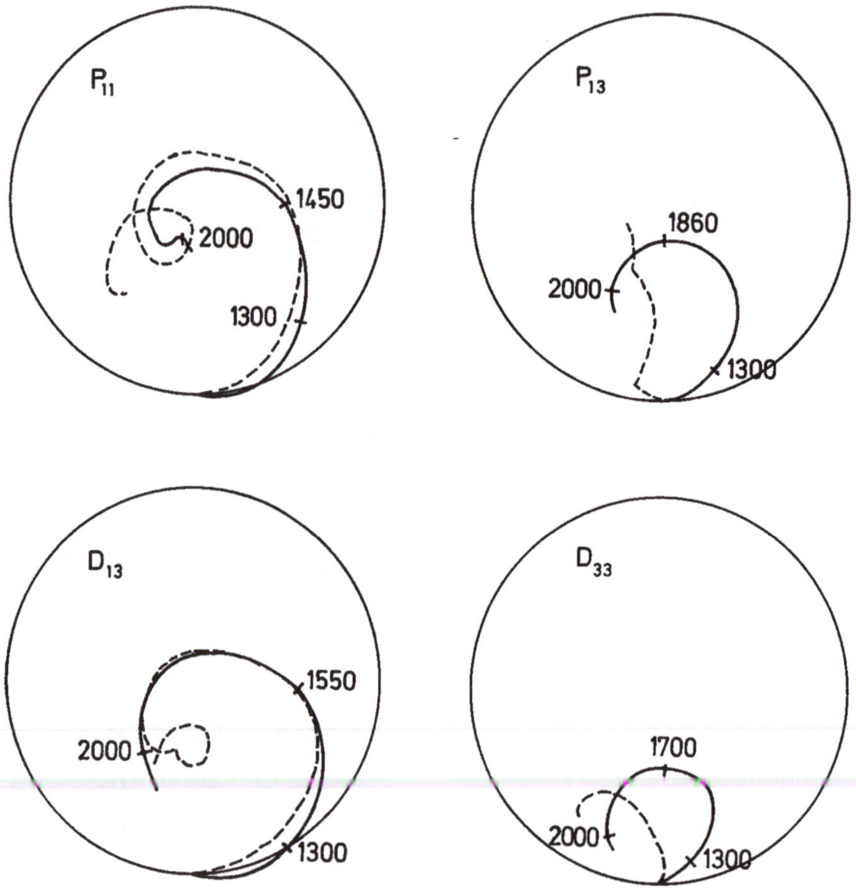
P11
1450
2000
1300
P13
1860
2000
1300
D13
1550
2000
1300
D33
1700
2000
1300

Fig. 19

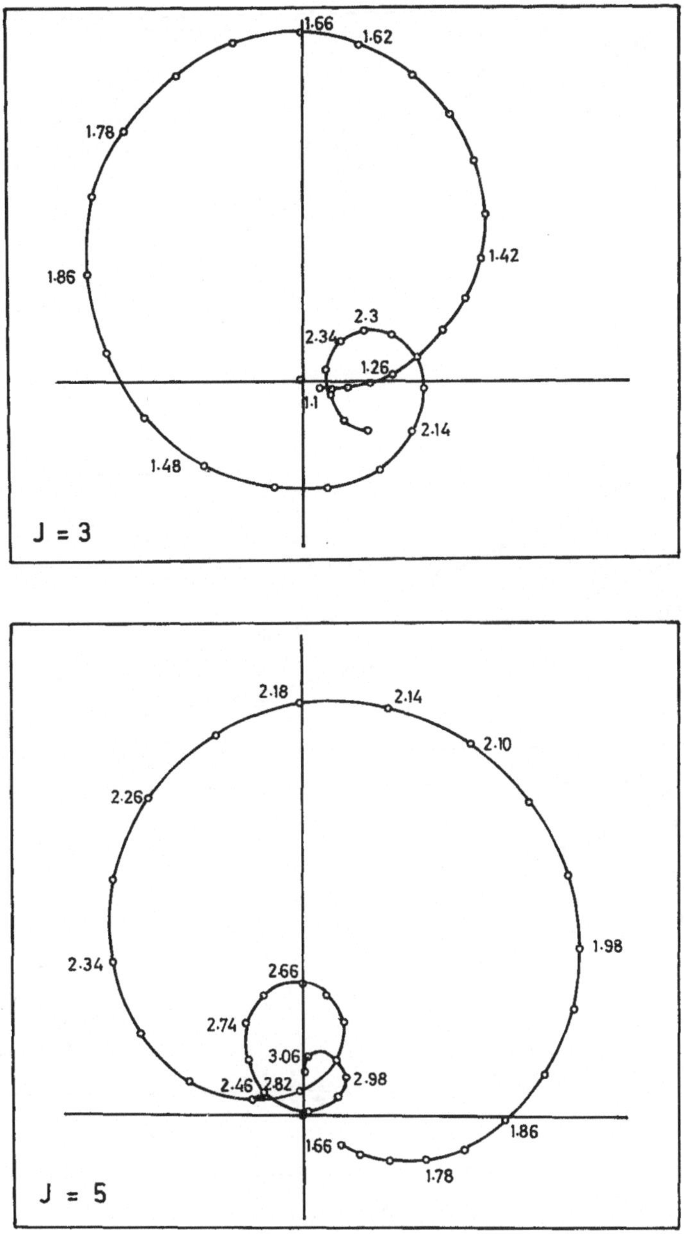
1.66
1.62
1.78
1.42
1.86
2.3
2.34
1.26
1.1
2.14
1.48
J = 3
2.18
2.14
2.10
2.26
1.98
2.34
2.66
2.74
3.06
2.98
2.46
2.82
1.86
1.66
1.78
J = 5

Fig. 2o

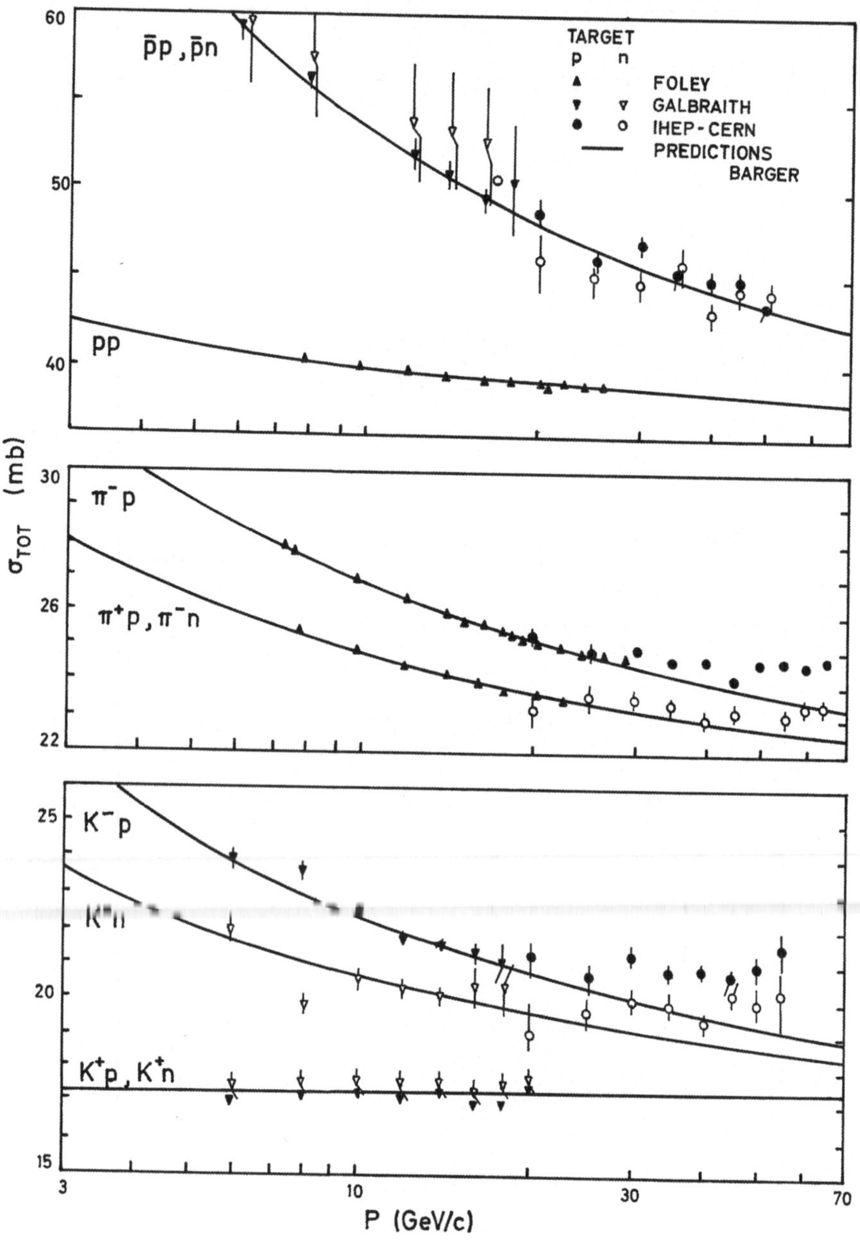
p̄p , p̄n
TARGET
p n
FOLEY
GALBRAITH
IHEP-CERN
PREDICTIONS
BARGER
pp
π⁻p
π⁺p , π⁻n
K⁻p
K⁻n
K⁺p , K⁺n
σTOT (mb)
P (GeV/c)
60
50
40
30
26
22
25
20
15
3
10
30
70

Fig. 21

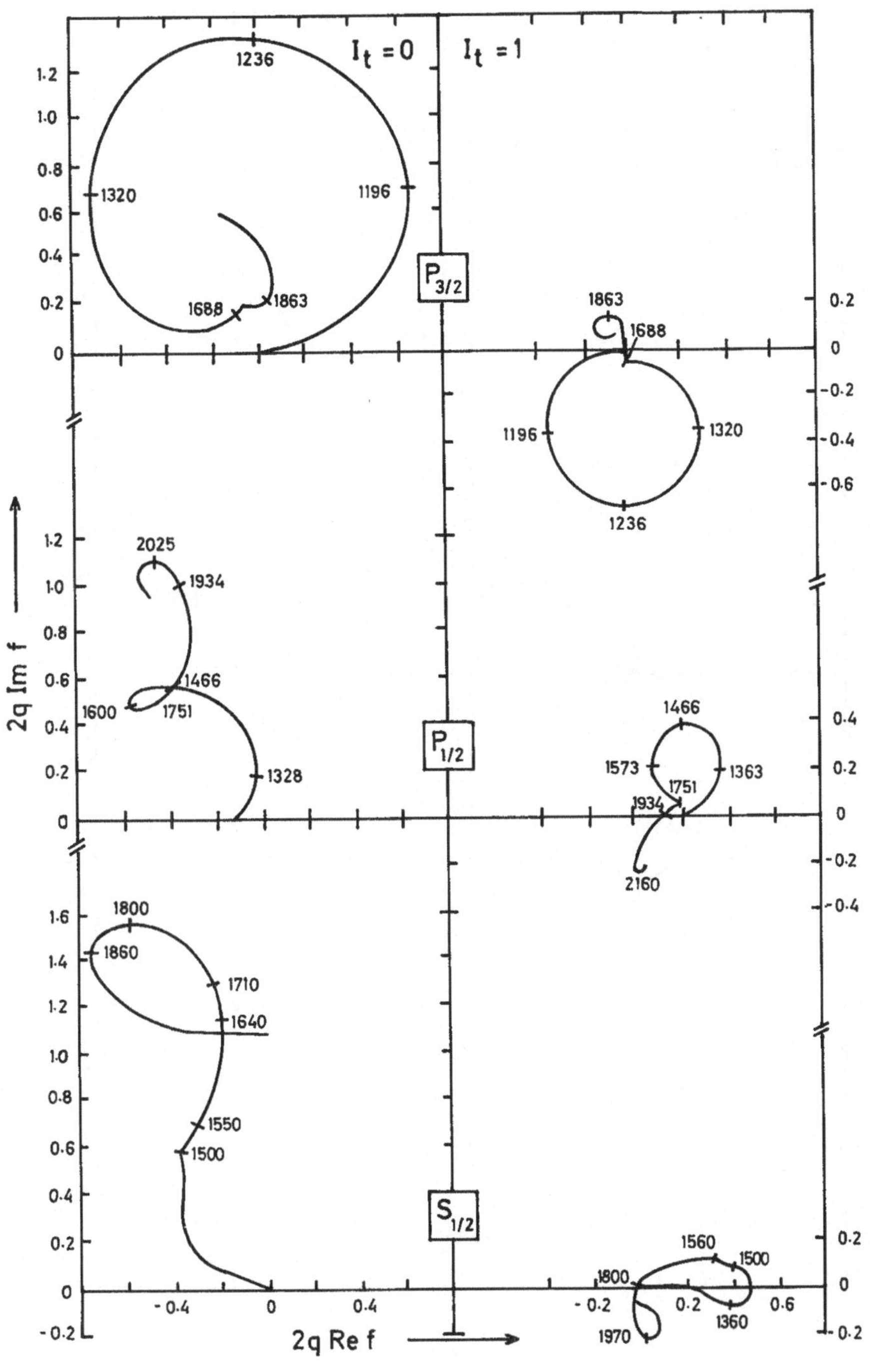

It = 0
It = 1
P3/2
P1/2
S1/2
2q Im f
2q Re f
1236
1320
1196
1688
1863
2025
1934
1466
1600
1751
1328
1573
1363
2160
1800
1860
1710
1640
1550
1500
1560
1360
1970

Fig. 22

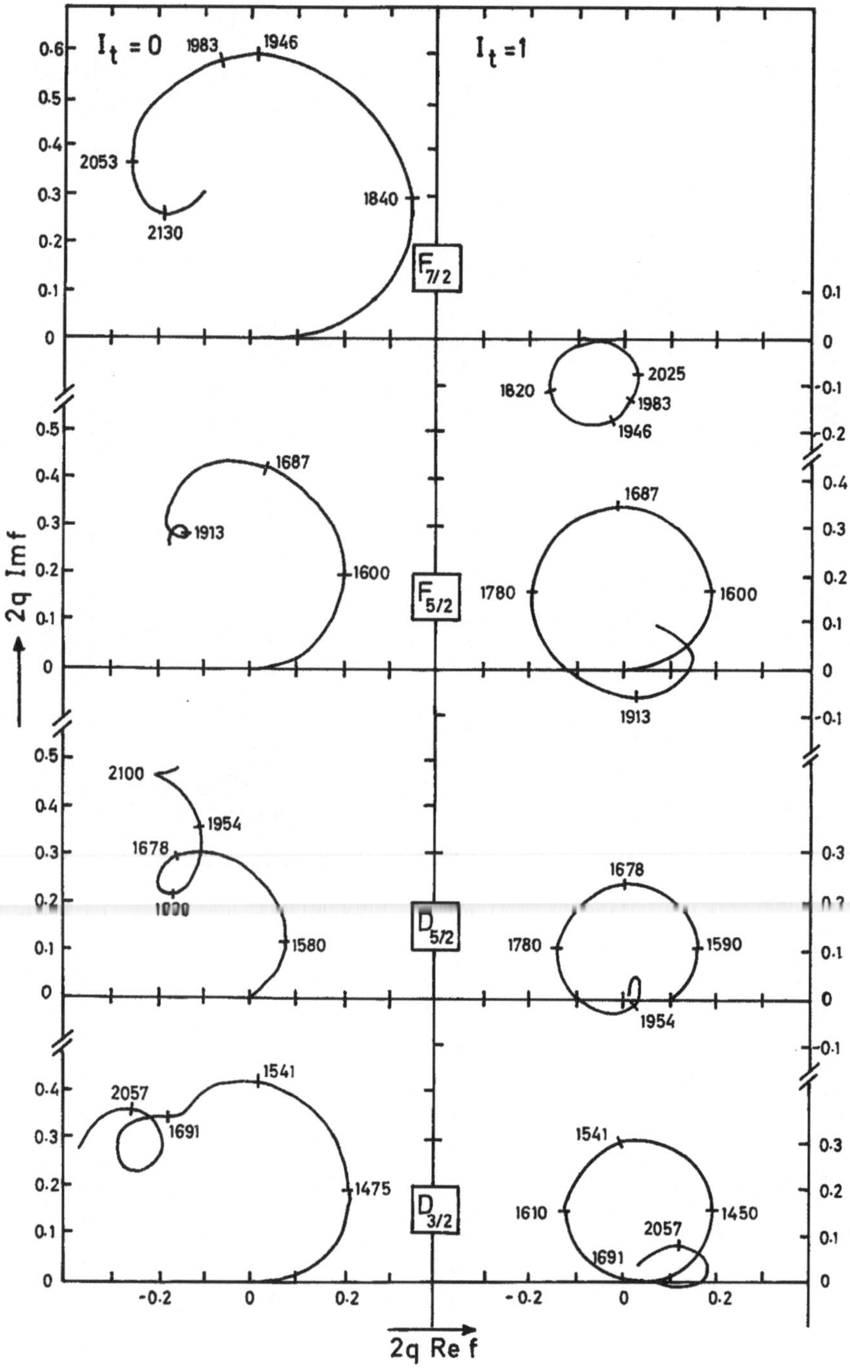

I_t = 0
I_t = 1
F7/2
F5/2
D5/2
D3/2
1983
1946
2053
1840
2130
2025
1820
1983
1946
1687
1913
1600
1780
2100
1954
1678
1580
1590
2057
1541
1691
1475
1610
1450
2q Im f
2q Re f

Fig. 23

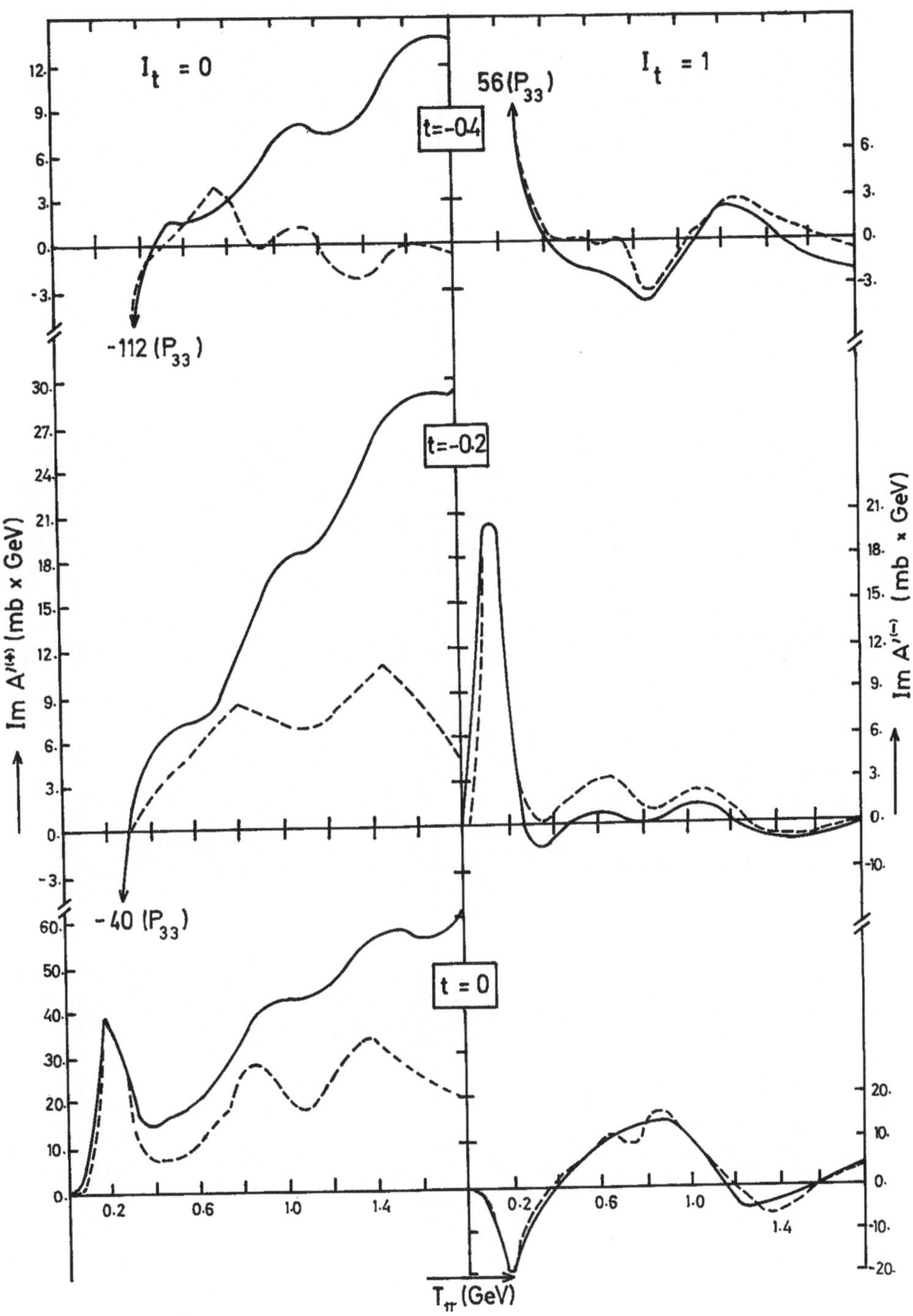

$I_t = 0$
$I_t = 1$
56 (P_{33})
-112 (P_{33})
-40 (P_{33})
t=-0.4
t=-0.2
t = 0
Im $A'^{(+)}$ (mb x GeV)
Im $A'^{(-)}$ (mb x GeV)
T_π (GeV)

Fig. 24

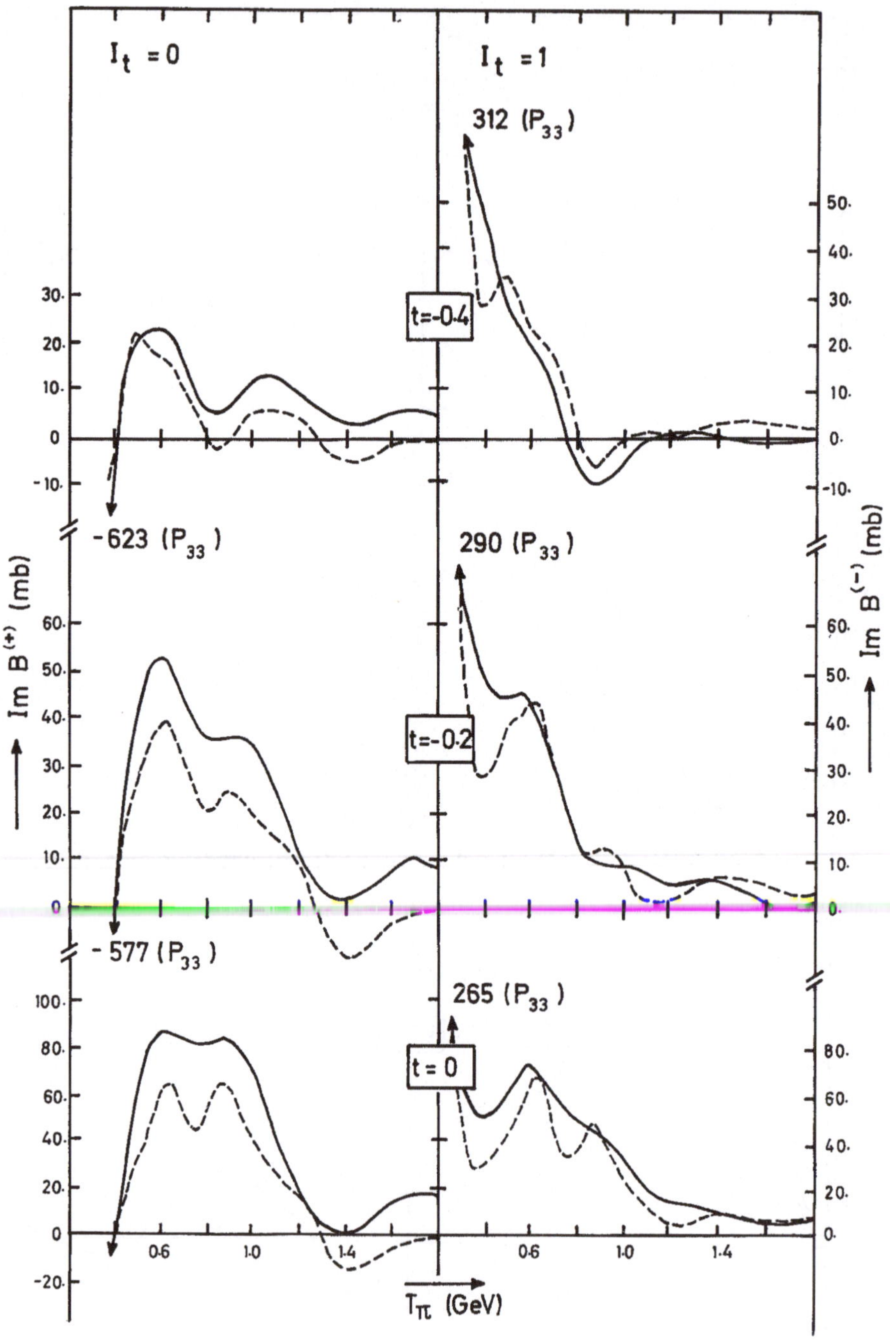

I_t = 0
I_t = 1
312 (P33)
t=-0.4
-623 (P33)
290 (P33)
t=-0.2
-577 (P33)
265 (P33)
t = 0
Im B(+) (mb)
Im B(-) (mb)
Tπ (GeV)

Fig. 25

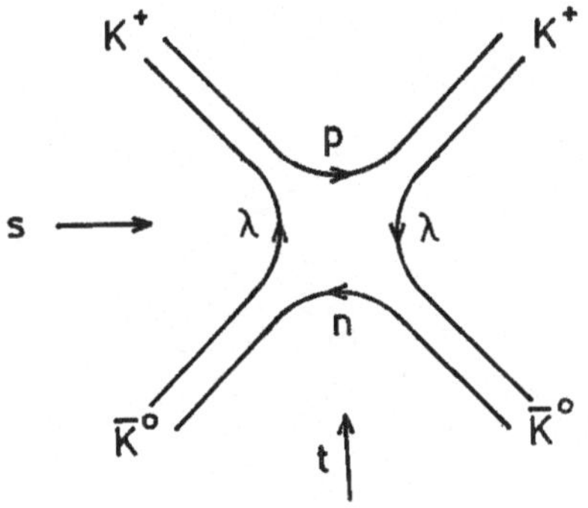

Fig. 26

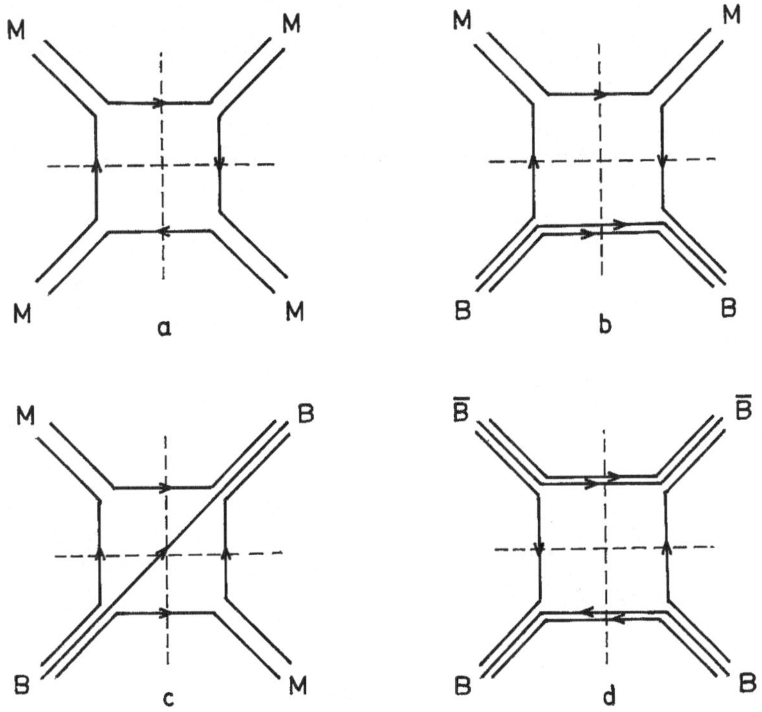

Fig. 27

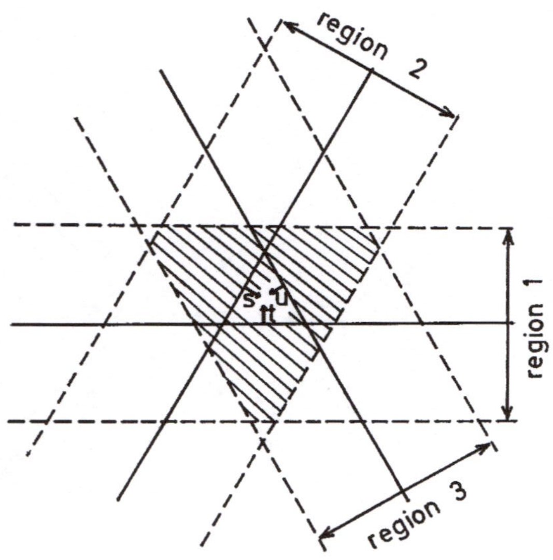
region 2
region 1
region 3
s
u
t

Fig. 28

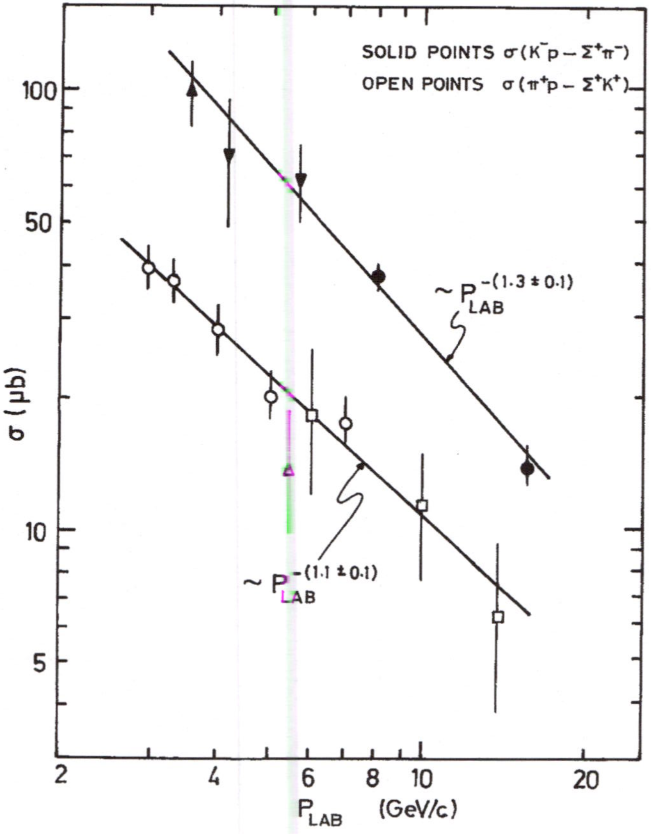
SOLID POINTS σ(K⁻p – Σ⁺π⁻)
OPEN POINTS σ(π⁺p – Σ⁺K⁺)
~P_LAB^-(1.3±0.1)
~P_LAB^-(1.1±0.1)
σ (μb)
100
50
10
5
2
4
6
8
10
20
P_LAB (GeV/c)

Fig. 29

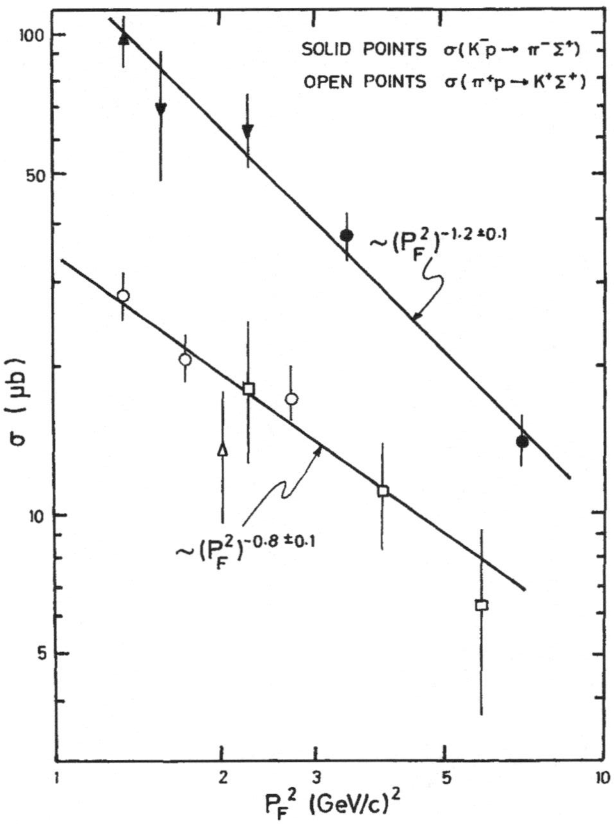
SOLID POINTS σ(K⁻p → π⁻Σ⁺)
OPEN POINTS σ(π⁺p → K⁺Σ⁺)
~(P_F²)^(-1.2±0.1)
~(P_F²)^(-0.8±0.1)
σ (μb)
100
50
10
5
1
2
3
5
10
P_F² (GeV/c)²

Fig. 3o

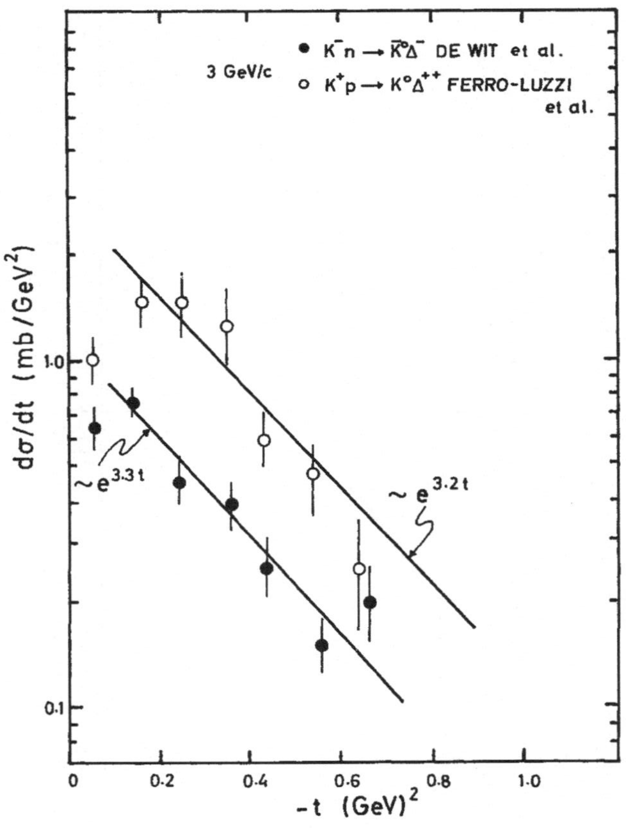
3 GeV/c
● K⁻n → K̄°Δ⁻ DE WIT et al.
○ K⁺p → K°Δ⁺⁺ FERRO-LUZZI et al.
~e^(3.3t)
~e^(3.2t)
dσ/dt (mb/GeV²)
1.0
0.1
0
0.2
0.4
0.6
0.8
1.0
-t (GeV)²

Fig. 31

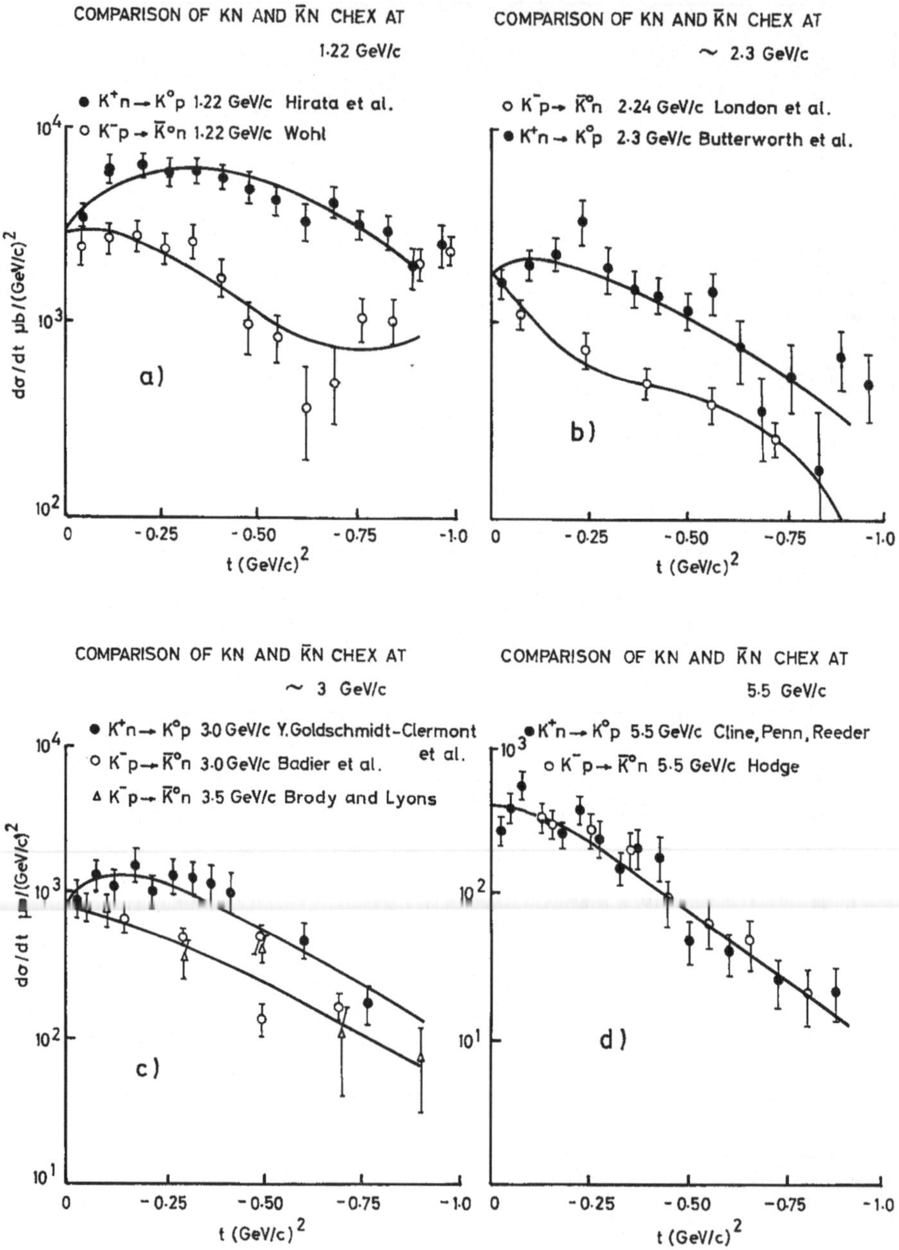
COMPARISON OF KN AND K̄N CHEX AT 1.22 GeV/c
● K⁺n → K°p 1.22 GeV/c Hirata et al.
○ K⁻p → K̄°n 1.22 GeV/c Wohl
dσ/dt μb/(GeV/c)²
a)
t (GeV/c)²
COMPARISON OF KN AND K̄N CHEX AT ~ 2.3 GeV/c
○ K⁻p → K̄°n 2.24 GeV/c London et al.
● K⁺n → K°p 2.3 GeV/c Butterworth et al.
b)
t (GeV/c)²
COMPARISON OF KN AND K̄N CHEX AT ~ 3 GeV/c
● K⁺n → K°p 3.0 GeV/c Y. Goldschmidt-Clermont et al.
○ K⁻p → K̄°n 3.0 GeV/c Badier et al.
△ K⁻p → K̄°n 3.5 GeV/c Brody and Lyons
c)
t (GeV/c)²
COMPARISON OF KN AND K̄N CHEX AT 5.5 GeV/c
● K⁺n → K°p 5.5 GeV/c Cline, Penn, Reeder
○ K⁻p → K̄°n 5.5 GeV/c Hodge
d)
t (GeV/c)²

Fig. 32

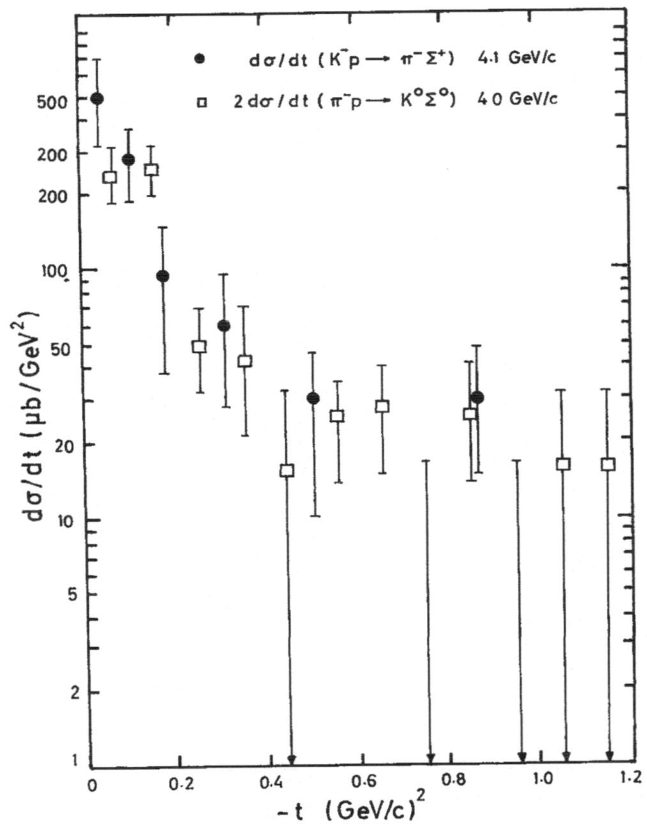
dσ/dt (K⁻p → π⁻Σ⁺) 4.1 GeV/c
2dσ/dt (π⁻p → K°Σ°) 4.0 GeV/c
dσ/dt (μb/GeV²)
-t (GeV/c)²

Fig. 33

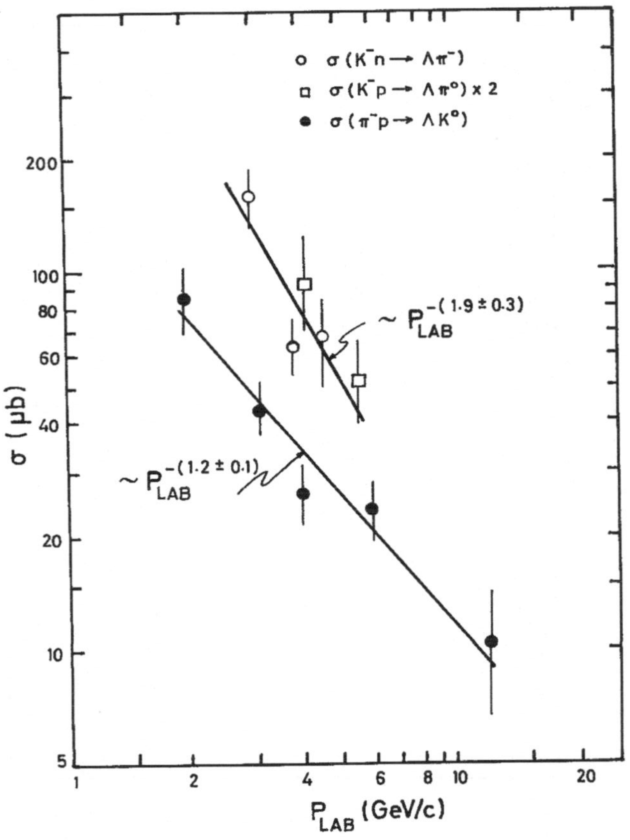
σ(K⁻n → Λπ⁻)
σ(K⁻p → Λπ°) x 2
σ(π⁻p → ΛK°)
~ P_LAB^-(1.9 ± 0.3)
~ P_LAB^-(1.2 ± 0.1)
σ (μb)
P_LAB (GeV/c)

Fig. 34

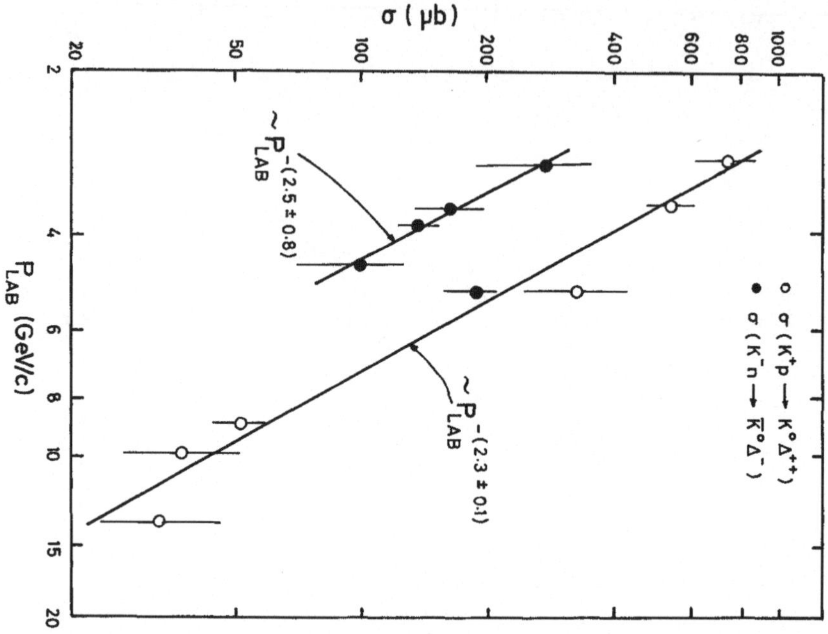
σ (μb)
P_{LAB} (GeV/c)
o σ ($K^+p \to K^o\Delta^{++}$)
• σ ($K^-n \to \bar{K}^o\Delta^-$)
$\sim P_{LAB}^{-(2.5 \pm 0.8)}$
$\sim P_{LAB}^{-(2.3 \pm 0.1)}$

Fig. 35

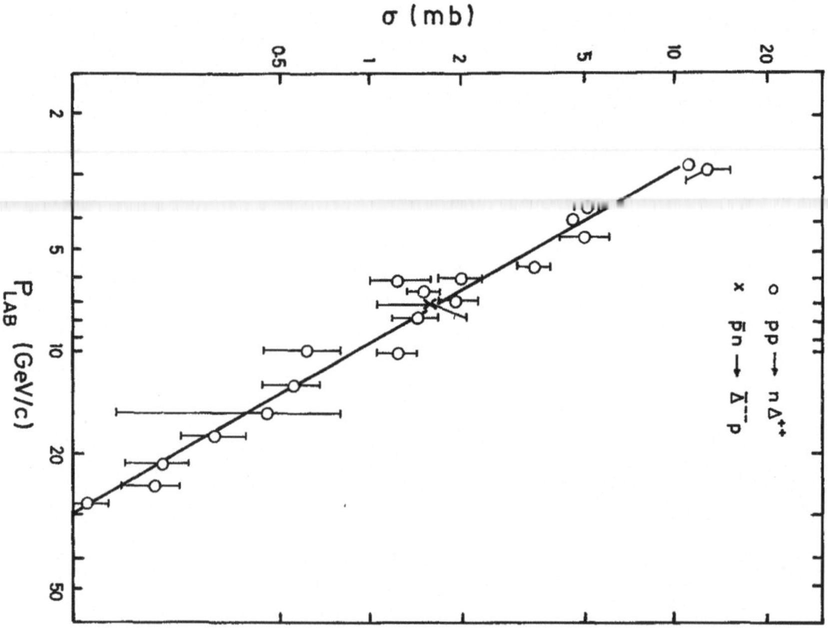
σ (mb)
P_{LAB} (GeV/c)
o $pp \to n\Delta^{++}$
x $\bar{p}n \to \bar{\Delta}^{--}p$

Fig. 36

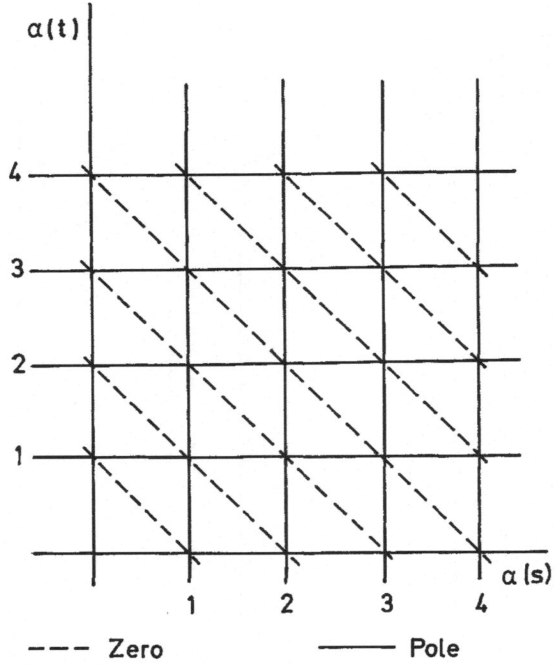

Fig. 37

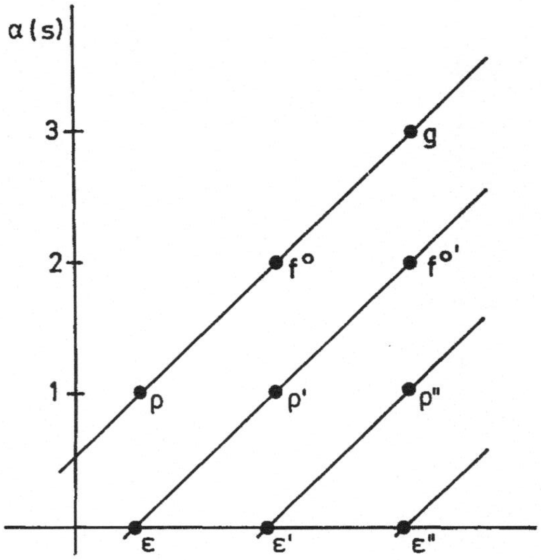

Fig. 38

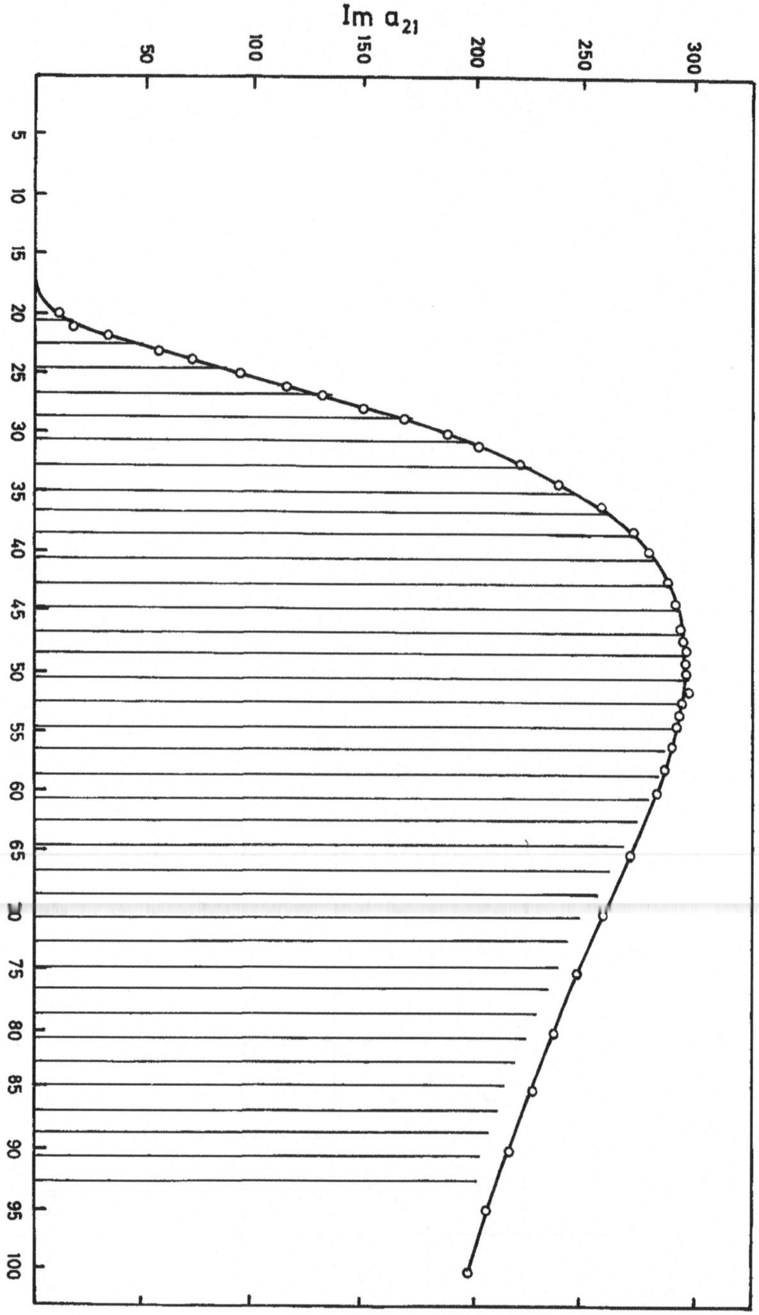

Im a_{21}
50
100
150
200
250
300
5
10
15
20
25
30
35
40
45
50
55
60
65
70
75
80
85
90
95
100
s

Fig. 39

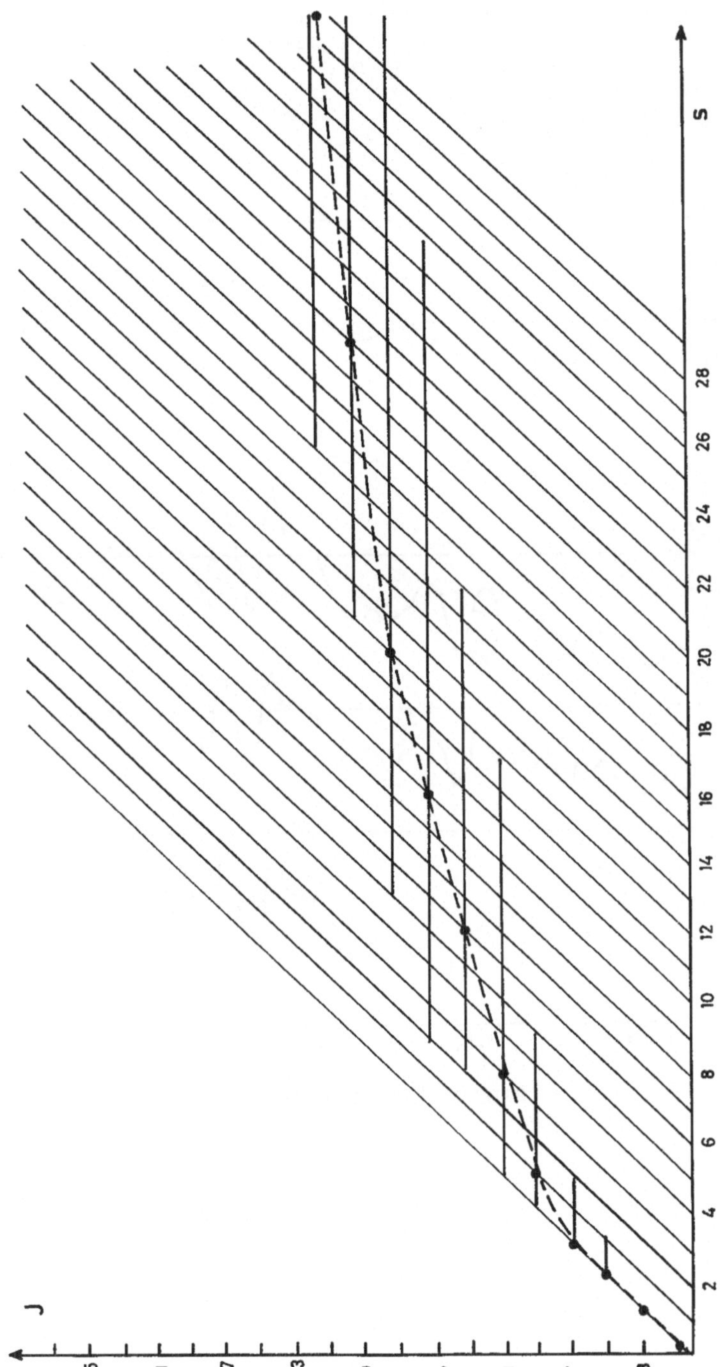

Fig. 40

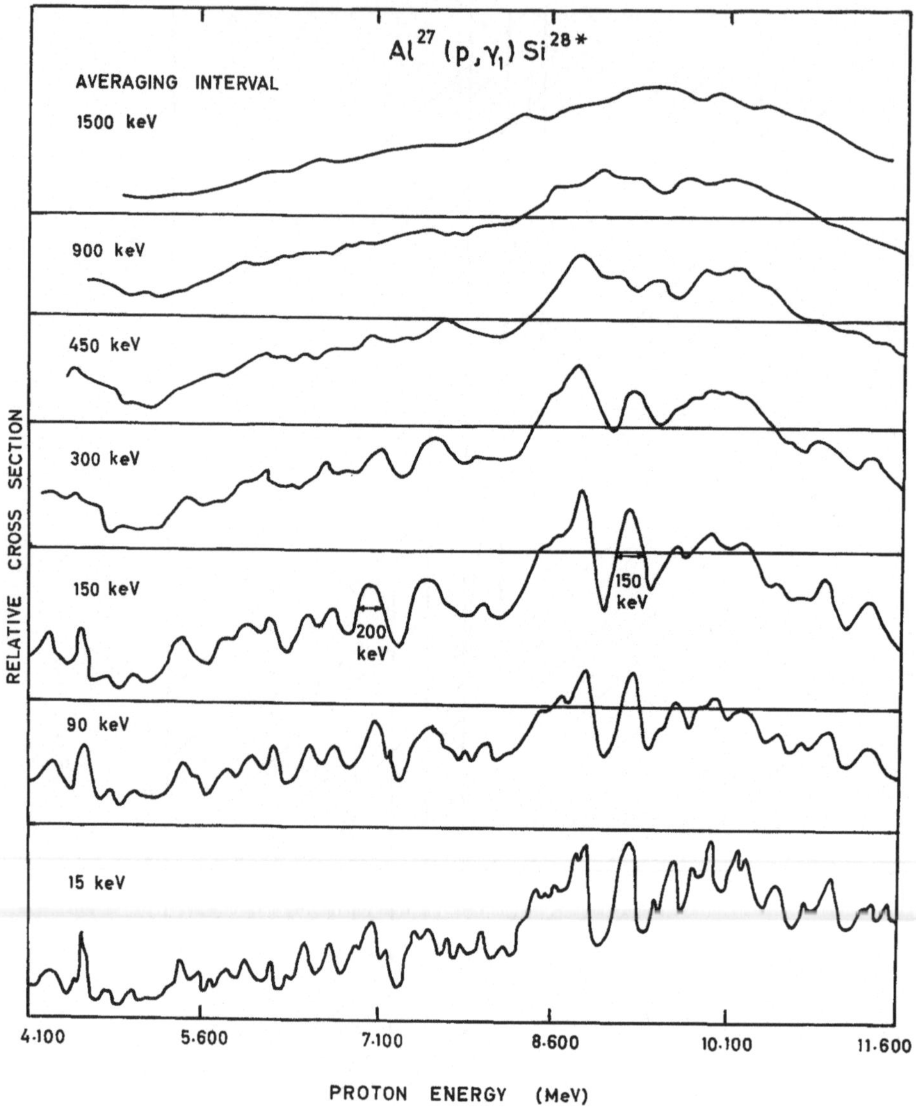
$Al^{27}(p,\gamma_1)Si^{28*}$
AVERAGING INTERVAL
1500 keV
900 keV
450 keV
300 keV
150 keV
200 keV
150 keV
90 keV
15 keV
RELATIVE CROSS SECTION
4·100
5·600
7·100
8·600
10·100
11·600
PROTON ENERGY (MeV)

Fig. 41

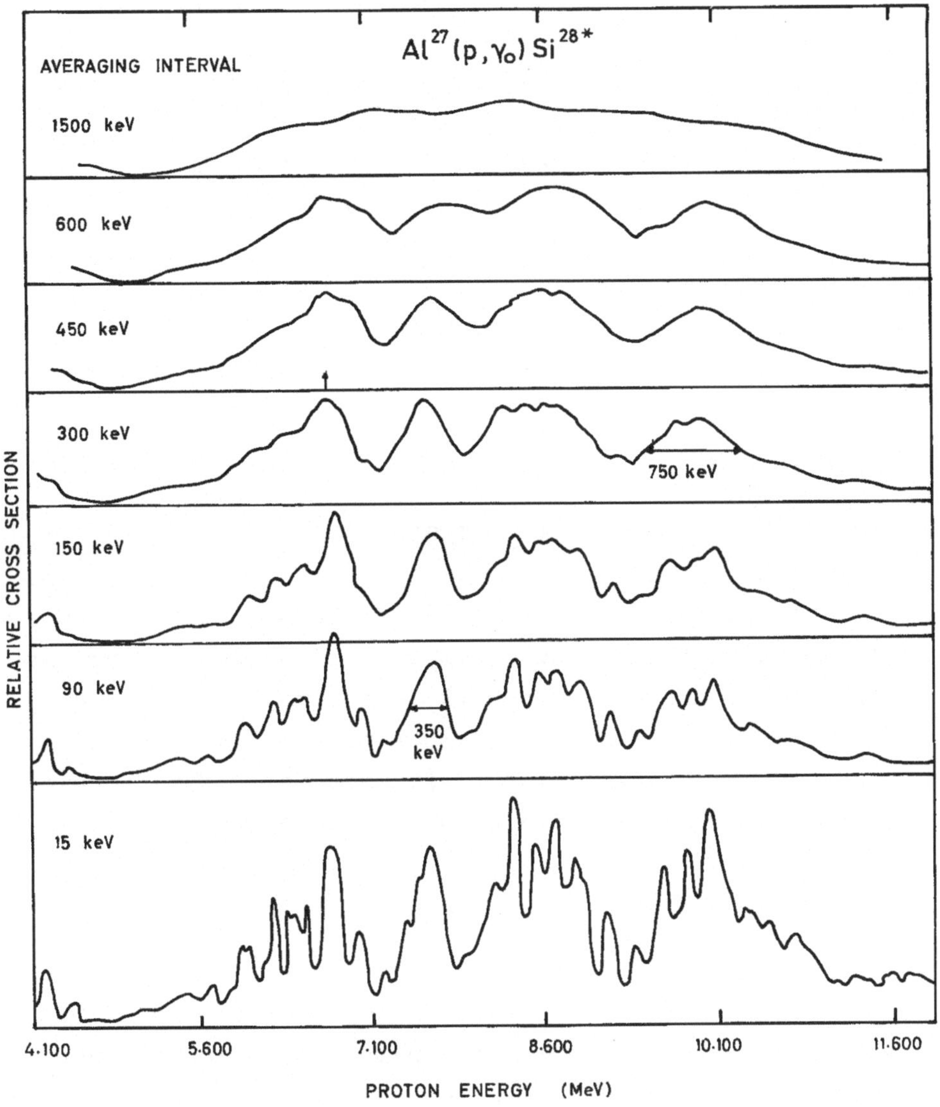
$Al^{27}(p,\gamma_0)Si^{28*}$
AVERAGING INTERVAL
1500 keV
600 keV
450 keV
300 keV
750 keV
150 keV
90 keV
350
keV
15 keV
RELATIVE CROSS SECTION
4.100
5.600
7.100
8.600
10.100
11.600
PROTON ENERGY (MeV)

Fig. 42

Acta Physica Austriaca, Suppl. VII, 548–566 (1970)

SOME REMARKS ON THE QUARK MODEL*

BY

D. FLAMM
Institut für Hochenergiephysik d. Österr. Akademie d. Wissenschaften, Vienna, Austria

1. INTRODUCTION

It is now six years that quarks were first proposed [1] and with the exception of five cloud chamber tracks reported by a Sydney group [2] which have been disputed a lot [3] and one bubble chamber picture [37] there is no experimental evidence for their existence. The trouble is that we know very little about physical quarks.

A. The Quark Mass

Some people believe that the quark mass m_q is about one third of the proton mass m_P ($m_q \approx m_P/3 \approx 336$ MeV/c^2) and that there is a selection principle which allows quarks and antiquarks to exist only in clusters of integral charge and baryon number [4]. Under this assumption the normal magnetic moment of the quarks $\mu_q = e_q\hbar/2m_qc$ is sufficient to explain the large anomalous magnetic moment of the nucleon. If one does not demand such a selection principle one has to assume that the quark mass and the binding- energy of quarks in hadrons are sufficiently high

* Seminar given at IX. Internationale Universitätswochen für Kernphysik, Schladming, February 23 - March 7,1970.

to prevent the present day accelerators from producing a quark pair or knocking one quark out of a hadron. The lower limit for the quark mass, which one obtains from the upper limit of the quark production cross section in accelerator experiments depends very much which of the various models (peripheral or statistical) one uses to estimate the quark production cross section. Recent Serpuchov data give for the total production cross section of quarks with a mass of 4.5 - 5 GeV/c^2 and charge 1/3 [5]

$$\sigma < 3\cdot 10^{-39}\ \mathrm{cm}^2$$

while for this mass range Hagedorns statistical theory [6] would, for instance, give

$$\sigma \sim 10^{-41}\ \mathrm{cm}^2$$

We might thus expect the quark mass somewhere between 4 and 10 GeV/c^2 or probably even higher. As a consequence any result which depends on the quark mass is suspicious.

B. The Quark Statistics

The most natural assumption for spin $\frac{1}{2}$ quarks$^+$ is Fermi-Dirac statistics. In this case it is difficult to understand why the symmetric s-wave bound state of the three quarks which is the 56-dimensional representation of SU(6) is lowest in energy [8]. One would also expect a zero in the proton form factor [9]. In order to avoid these difficulties it was proposed that quarks are parafermions of order 3 [10]. For parafermion creation a_k^+ and annihilation operators a_ℓ, where the indices k,ℓ denote all single particle quantum numbers like momentum, spin, isospin, etc., commutation or anticommutation relations do not hold. It is only required [11] that

---

$^+$ Also higher half-integral spin assignments for quarks have been proposed. See for instance [7].

$$((a_k^\dagger, a_\ell)_-, a_m)_- = -2\delta_{km}\, a_\ell \tag{1}$$

$$((a_k^\dagger, a_\ell^\dagger)_-, a_m^\dagger)_- = 0 \tag{2}$$

One can thus form products of parafermion operators which are symmetric, antisymmetric or of mixed symmetry. For parafermions of order three at most three identical particles may occur in a symmetric state. The operator

$$f_{k\ell m}^{(s)\dagger} = ((a_k^\dagger, a_\ell^\dagger)_+, a_m^\dagger)_+ \tag{3}$$

creates a symmetrical state $f_{k\ell m}^\dagger|0\rangle$ of three parafermions, which is a fermion since for order 3

$$(f_{k\ell m}^{(s)\dagger}, a_n^\dagger)_+ = 0 \quad \text{and thus} \quad (f_{k\ell m}^\dagger, f_{nop}^\dagger) = 0 \,.$$

This is just what we need to form the 56-dimensional ground state wave function of SU(6) from an s-wave bound state of three para-quarks. Due to eq. (2) an antisymmetric s-wave bound state of three paraquarks does not exist. On the other hand, a three para-quark state with mixed symmetry

$$f_{k\ell m}^{(m)\dagger}|0\rangle \quad \text{with} \quad f_{k\ell m}^{(m)\dagger} = ((a_k^\dagger, a_\ell^\dagger)_+, a_m)_- \tag{4}$$

can well exist. This composite state is a parafermion. Due to the selection rules which follow from the paralocality of the interaction Hamiltonian [12] such a parafermion could not be produced in a reaction where all other particles are ordinary bosons and fermions. Nothing prevents however the creation of pairs of paraparticles and an s-wave bound state of the type (4) should presumably not be higher in energy than the observed p-wave state of three paraquarks with mixed symmetry. Similarly from paraquarks and their antiparticles ($b_\ell^\dagger$) we may form antisymmetric states

$$d_{k\ell}^{(a)\dagger}|0\rangle \quad \text{with} \quad d_{k\ell}^{(a)\dagger} = (a_k^\dagger, b_\ell^\dagger)_- \tag{5}$$

which are the ordinary mesons[+] and a symmetric state

$$d_{k\ell}^{(s)\dagger}|0\rangle \quad \text{with} \quad d_{k\ell}^{(s)\dagger} = (a_k^\dagger, b_\ell^\dagger)_+ \tag{6}$$

which are parabosons. Green [13] has shown that calculations with parafermion operators of order p can be made easier by the ansatz

$$a_\ell^\dagger = \sum_{\alpha=1}^{p} a_\ell^{(\alpha)\dagger} \tag{7}$$

where the creation operators $a_\ell^{(\alpha)\dagger}$ satisfy anticommutation relations for the same α

$$(a_k^{(\alpha)}, a_\ell^{(\alpha)\dagger})_+ = \delta_{k\ell}, \ (a_k^{(\alpha)\dagger}, a_\ell^{(\alpha)\dagger})_+ = 0 \tag{8}$$

but commute for different α and β

$$(a_k^{(\alpha)}, a_\ell^{(\beta)\dagger})_- = (a_k^{(\alpha)\dagger}, a_\ell^{(\beta)\dagger})_- = 0 \quad \alpha \neq \beta \tag{9}$$

In our case p is equal to three and eqs. (7) - (9) mean that para-quarks of order 3 are equivalent to three indistinguishable triplets of quarks. The $a_k^{(\alpha)}$ obey abnormal commutation relations since they are spin $\frac{1}{2}$ particles and nevertheless eq. (9) holds.

There exists, however, a theorem which says [14]: In any field theory with abnormal commutation relations there always exists an irreducible set of fields with normal commutation relations obtained from the original set by a so-called Kleintransformation. Indeed if the Green components of the para-quark model (eq. (7)) are taken to be independent fields and are Kleintransformed, then, as far as the known hadrons are concerned the paraquark model can be considered a special case [15] of the three triplet model of Han and Nambu [16]. In this model one assumes nine fundamental fermions of spin $\frac{1}{2}$ which may be grouped in three triplets of integral charge and hypercharge but baryon number $\frac{1}{3}$. These three triplets transform

---

[+] This follows from eq. (2) .

as the representation (3', $\bar{3}$") of the group SU(3)'×SU(3)". Mesons are triplet-antitriplet states (T$\bar{T}$) and baryons are three triplet states (TTT), as usual.

The presently known baryons are placed in an SU(3)" singlet, which is antisymmetric. Fermi-Dirac statistics then allows the wave function to be symmetric with respect to the interchange of all other degrees of freedom of the nonet. The success of the symmetric quark model [10],[17] and in particular of the harmonic oscillator model [18] in predicting the energy levels of baryons and baryon resonances is an argument in favour of paraquarks or Han-Nambu triplets. Again results which do not depend on the quark statistics and these are at the same time all results which do not depend on a detailed assumption on the spatial wave function of the hadrons are more trustworth than others.

C. The Quark Charge

A third point is the charge of the fundamental particles. The Gell-Mann Zweig quarks p,n,λ have charge $\frac{2}{3}$e, $-\frac{1}{3}$e, and $-\frac{1}{3}$ e, respectively, as is well known. In 1964 and 1965 people were worried by the fractional charge of quarks and devised a lot of models to make the quark charge integer. All of them need more than one triplet and involve some higher symmetry group than SU(3) [19]. The most appealing one of these alternatives is the Han-Nambu triplet model [16].

It avoids observable fractional charges, reconciles the totally symmetric SU(6) baryon wave function with Fermi statistics and gives a plausible explanation for the saturation of the triplet triplet forces by the observed hadron states [15]. Integral charges and hypercharges are achieved by adding the fractional values of the individual SU(3) groups.

This amounts to adding an isoscalar with respect to SU(3)' in the Gell-Mann Nishijima relation. While the center of charge within all observed SU(3) multiplets is zero (zero-triality) this is not the case for the fundamental triplets. If we assume a very strong Coulomb type interaction coupling to the triality it is plausible that the triality zero states ($T\bar{T}$) and (TTT) are lowest in energy.

As far as the static properties of hadrons are concerned it is very hard to distinguish between ordinary quarks and Han-Nambu triplets. The latter give the same results for almost every calculation. This is also the reason why one usually calculates with ordinary quarks which after all are simpler. Okubo [20] and Adler [21] have, however, pointed out that the decay $\pi^0 \to 2\gamma$ and leptonic vectormeson decays depend on the triality of the fundamental triplet.

D. A Magnetic Model of Matter

One may ask about the physical significance of the second SU(3) of Han and Nambu. An interesting suggestion based on an idea of Schwinger [22] has recently been made by Han and Biedenharn [23]. In all experiments quarks have proved to be very elusive objects and so are magnetic monopoles. So why not combine them? In fact Schwinger showed that dually charged objects (i. e. objects which possess both electric and magnetic charge) allow charge quantization in units of e/3 for the electric charge, and g/3 for the magnetic charge with the dimensionless interaction strength $g^2/\hbar c \sim 36 \times 137 \cong 5000$. The magnetic charges interact via longrange, superstrong, forces. Since it is customary in physics to give every new concept an appealing name, Schwinger called these new objects dyons. It is now tempting to re-interpret the Han-

Nambu model by identifying SU(3)' with the SU(3) of electric charge and SU(3)" with an SU(3) of magnetic charge.

An attractive feature of this model is, that we might eventually understand the many kinds of interactions between quarks in terms of the superstrong magnetic force. At present one distinguishes at least three kinds of strong quark-quark interactions [24] : superstrong ones which allow for SU(6)-symmetry, strong ones which violate SU(6)-symmetry, strong ones which violate SU(6) but conserve SU(3), and moderately strong interactions which break SU(3) but conserve SU(2). From experiment we know very little about the nature of the superstrong forces which bind the quarks because the present hadron spectrum represents only the lowest bound states of quarks. The high mass meson resonances which will be looked for in a joint CERN-Serpukhov experiment may provide valuable information in this respect.

Schwinger guesses a mass of about 6 GeV/c^2 for the dyons. Due to their superstrong electromagnetic interactions the creation of a dyon-pair may require energies considerably above threshold. In most cases the pair will remain in a bound state and re-annihilate emitting an energetic photon shower. Some high energy cosmic ray events may be interpreted in this way [25]. Using the superstrong magnetic interaction Han and Biedenharn infer that exotic states lie reasonably high in mass. Even CP violation is no problem in the dyon theory because dyons themselves violate CP invariance. It rather is a problem to exclude unwanted CP violating effects like electric dipole moments of baryons. For this purpose one has to introduce a sufficiently strong magnetic-exchange interaction. Let us now list a few achievements of the quark picture. We shall in general use Gell-Mann Zweig quarks, which are the simplest model, but keep the alternatives in mind.

2. THE HADRON MASS SPECTRUM

A. Mesons

The simple assumption of $(q\bar{q})$ structure for mesons immediately gives selection rules:

a) From a quark and an antiquark one can only form a singlet and an octet of SU(3). This is usually referred to as: absence of exotic states of the first kind (No meson states with isospin $I>1$ and hypercharge $|Y|>1$ exist)

b) What J^{PC} combinations can we form? Here J is the total angular momentum, P the parity

$$P = (-1)^{L+1} \qquad (L = \text{orbital angular momentum})$$

and C the charge conjugation quantum number:

$$C = (-1)^{L+S} \qquad (S = \text{total } q\bar{q} \text{ spin})$$

which is only defined for neutral particles. We have to distinguish two cases:

1. $J = L \;;\quad P = (-1)^{J+1}$

$$CP = -1 \qquad \text{for } S = 0$$
$$CP = +1 \qquad \text{for } S = 1$$

An exceptional case is L=0 where only S=0 is allowed. As a consequence 0^{--} is forbidden.

2. $S = 1,\; J = L\pm1 \;;\quad P = (-1)^J$ (natural parity)

$$CP = (-1)^{2J} = +1$$
$$CP = -1 \text{ is forbidden } (0^{+-},\ 1^{-+},\ 2^{+-}\ldots)$$

Thus we have

absence of exotic states of the second kind (These are 0^{--} and all states with natural parity and odd CP).

In table 1 the allowed meson states are listed up to L=8 .

L	symbol	J^{PC} values for the various $^{2S+1}L_J$: S=1, $^3L_{L+1}$	S=1, ^3L_L	S=1, $^3L_{L-1}$	S=0, ^1L_L	tentative I=1 $^3L_{L+1}$ state (mass MeV)
0	S	1^{--}	-	-	0^{-+}	ρ(768)
1	P	2^{++}	1^{++}	0^{++}	1^{+-}	A_{2H}(1315), A_{2L}(1269)
2	D	3^{--}	2^{--}	1^{--}	2^{-+}	R(1630) or g(1630)[+]
3	F	4^{++}	3^{++}	2^{++}	3^{+-}	S(1929) ?
4	G	5^{--}	4^{--}	3^{--}	4^{-+}	T(2195) ?
5	H	6^{++}	5^{++}	4^{++}	5^{+-}	U(2382) ?
6	I	7^{--}	6^{--}	5^{--}	6^{-+}	X(2620) ?
7	K	8^{++}	7^{++}	6^{++}	7^{+-}	X(2810) ?
8	L	9^{--}	8^{--}	7^{--}	8^{-+}	X(3000) ?

Table 1 - Meson states allowed by the quark model

To each entry in the table corresponds a nonet of mesons. In addition there may be radial excitations of every state. In the last column a tentative assignment for the I=1 member of the $^3L_{L+1}$ nonet is given. ? means J^{PC} is not yet known. We see that the J^{PC} with natural parity and even CP (1^{--}, 2^{++}, 3^{--} ...) appear twice, all other allowed combinations only once. The A_2 splitting might come from an accidental degeneracy of the corresponding L=1 and L=3 states. Such a degeneracy is, however, not very plausible [26] and the problem is not settled.

It should be stressed that <u>every state</u> which appears in table 1 may have <u>a large probably even infinite number</u>

[+] R is the name in missing mass spectrometer experiments, g the one in bubble chamber experiments.

<u>of radial excitations</u> which form a sequence of higher mass states with the same quantum numbers. A candidate for such an excitation is the E(1422). If its J^{PC} assignment is really 0^{-+} it must be a radial excitation of η or X and we would expect the other members of the pseudoscalar nonet in the same mass region. On the other hand $J^{P}=1^{+}$ is not yet excluded for E and there is still a vacant place in the 1^{++} nonet.

What happens in the alternative quark models? In the <u>paraquark model</u> exotic states of the first kind contain more than one quark and one antiquark and thus if they exist at all lie higher in mass. Exotic states of the second kind may appear. They are the parabosons of eq. (6) and thus cannot decay in ordinary bosons which would explain why we have not yet seen them.

The <u>Han-Nambu triplet theory</u> reproduces for the low mass states just the ordinary quark results.

$$T\bar{T} : (3',\bar{3}'') \times (\bar{3}',3'') = (1',1'')+(8',1'')+(1',8'')+(8',8'')$$

Only the first two states which are SU(3)" singlets appear at low energy. If the energy is, however, large enough also states with nonzero triality may show up.

$$T\bar{T} : 1 + 8 + 8 + 1 + 8 + 8 + 10 + \overline{10} + 27$$

Thus even in the $T\bar{T}$ system a limited number of exotic states of the first kind may appear at high mass. The observed SU(3) group is the diagonal subgroup of SU(3)'×SU(3)". If we drop the distinction between the two SU(3) we may group the nine Han-Nambu particles into a nonet, which we may label by the names of the ordinary baryons [27]. Then $T\bar{T}$ contains for I=0

$$(\Lambda_1 \bar{\Lambda}_8 \pm \Lambda_8 \bar{\Lambda}_1)_{L,S}$$

and for I=1 [+] $(\Lambda_1 \bar{\Sigma}^o \pm \Sigma^o \bar{\Lambda}_1)_{L,S}$

which has both even and odd C for any L and S. This means that at high mass all exotic states of the second kind may be present. It should be pointed out that there is also a branch of solutions of the Bethe-Salpeter equation of the $q\bar{q}$ system which is exotic of the second kind. This essentially comes about because in the Bethe-Salpeter wave function the quark field operators appear at different times [8]

$$\sum_{\alpha,\beta} \int d^4x \; f_{\alpha\beta}(\vec{x}, t_1 - t_2) \; \psi_\alpha(\vec{x}, t_1) \; \psi^+_\beta(-\vec{x}, t_2)|0\rangle$$

and $(t = t_1 - t_2)$ $f_{\alpha\beta}(\vec{x}, t)$ may be a symmetric or an antisymmetric function of t. The solutions for which $f_{\alpha\beta}(\vec{x}, t)$ is an even function of t include the ordinary mesons. The solutions with $f_{\alpha\beta}(\vec{x}, -t) = -f_{\alpha\beta}(\vec{x}, t)$ would be exotic states. In the absence of any detailed dynamical calculation one can, however, not say whether bound states of this kind do exist. In any case such states have no vertex function coupling them to quarks[*] and thus even if they were bound states they were irrelevant to hadron physics.

Let us return to the nonrelativistic quark model and adopt a specific potential in order to obtain more information about the level spacing of mesons. From the Bethe-Salpeter equation for the $q\bar{q}$ system without spin one can derive in the nonrelativistic limit of strong binding the so-called Blankenbecler-Sugar equation

$$(E^2 - \frac{3}{2} p^2 - 4M^2) \; \Phi = V\Phi \tag{10}$$

If we use a harmonic oscillator potential

$$V = V_o + \frac{1}{\sqrt{6}} a^2 r^2 \tag{11}$$

and weak spin orbit coupling

$$V_{LS} = b \; \vec{L}\vec{S} \tag{12}$$

[+] Here Λ_1 denotes a unitary singlet with the quantum numbers of Λ while Λ_8 is the octet component.

[*] I thank Dr. Kugler for this remark.

we obtain the mass formula [26]

$$E^2(n,L,S) = E_o^2 + a(L+2n-2) + b\langle\vec{L}\vec{S}\rangle \qquad (13)$$

where $E_o^2 = 4M^2+V_o \sim m_\rho^2$ and n denotes the principal quantum number. The parameters a and b may be fitted to the L=1 mesons shown in table 2 .

J^{PC}	meson	mass squared
2^{++}	$A_2(1300)$	$E_o^2 + a + b$
1^{+-}	B(1220)	$E_o^2 + a$
1^{++}	A(1080)	$E_o^2 + a - b$
0^{++}	$\delta(965)$	$E_o^2 + a - 2b$

Table 2 - Spacing of the L=1 Mesons

One obtains $a = 0.84(\text{GeV})^2$ and $b=0.25(\text{GeV})^2$. It should, however, be mentioned that the simple spin orbit coupling of eq. (12) seems not to work for higher-L mesons. Eq. (13) means equal spacing for increasing L and radial excitation. The linear dependence of the meson mass squared on L which also implies a linear dependence on J is exhibited in fig. 1 .

Figure 1 is reminiscent of two linear rising Regge trajectories which are degenerate. Linear rising trajectories appear in the Veneziano model [28]. Potential models for such trajectories have been considered [29].

To obtain eq. (13) we have used a very specific model namely a harmonic oscillator potential. Is this really the physical situation? Shouldn't we rather use a Coulomb or a Yukawa potential which is singular at the origin? In the latter case the energy levels become more and more closely spaced as the principal quantum number increases. In the CERN boson spectrometer experiment an

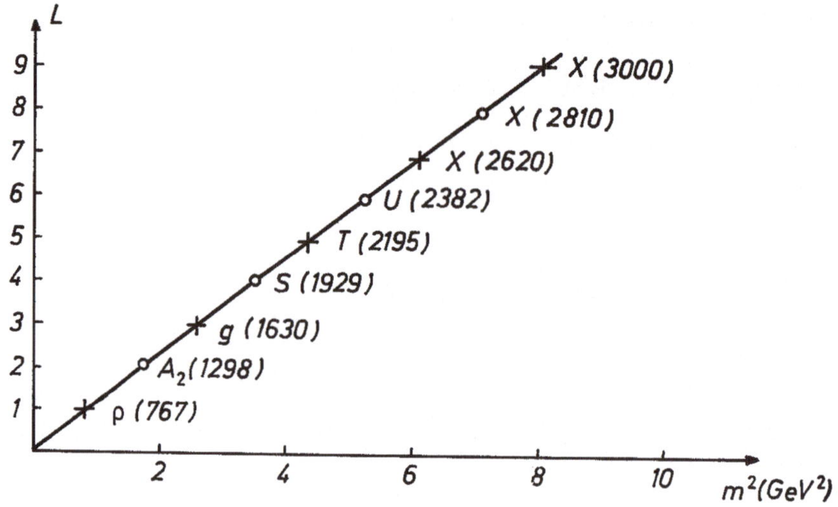

Figure 1 - A few I=1 mesons arranged according to eq. (13). The J^{PC} assignment are only known for the first three mesons .

accumulation of mesons near 4 GeV has been found [38]. A detailed knowledge of the high mass meson spectrum will help to decide on the character of the superstrong quark-quark force.

Now let us compare the mass spectrum of the quark model (table 1 and eq. (13)) with the Veneziano model [28]. For $\pi\pi$ scattering for instance the ρ and the f trajectory are exchange degenerate. These two parent trajectories are accompanied by a set of (Veneziano) daughter trajectories. $\varepsilon(700)$ is a daughter of ρ and ρ' a daughter of f^0 as is shown in fig. 2 .

The G parity of all trajectories is even to allow them to couple to the $\pi\pi$ system. The isospin is 1 for odd signature and 0 for even signature. Both the parity and the charge conjugation quantum number change with the signature of the Regge-pole. Similarly to the radial excitations in the quark model the Veneziano daughters form a sequence of higher mass states. In this framework exotic

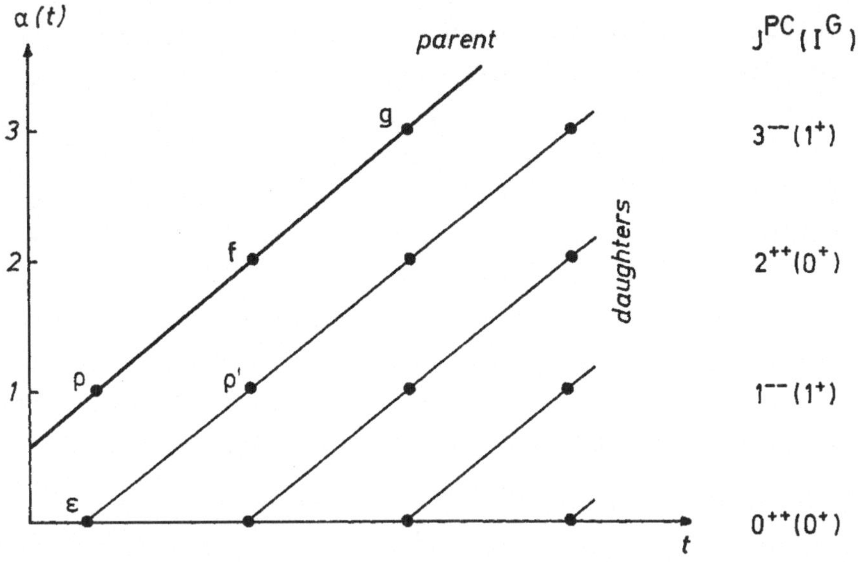

Figure 2 - Trajectories which contribute to the Veneziano amplitude for $\pi\pi$ scattering.

states of the second kind do not appear. The Veneziano daughter trajectories must not be confused with earlier daughter trajectories [30],[31] introduced to retain the correct Regge s behaviour at small t for external particles with unequal mass. Here the first daughter has the same isospin and C-parity as the parent but opposite space parity. A quark model which can accommodate such exotic daughters has also been devised [32] -[34].

A relativistic quark model for mesons based on the Veneziano representation has been proposed [35],[36]. It uses a multichannel bootstrap calculation. Any leading Regge pole which can be exchanged in the s or t channel of the meson-meson scattering amplitude may also appear as an external particle. A further ingredient is factorization which yields equations for the residua. Linear rising trajectories are used and the remaining free parameters are adjusted in such a way as to kill unwanted states. To incorporate SU(3) instead of the four point Veneziano amplitude for mesons the calculation is carried out with the eight point Veneziano amplitude for quarks. The trajectories obtained in this way correspond to representations of SU(12)×O(3) but the particle spectrum is classified according to SU(6)×O(3).

To conclude with mesons I want to mention that exchange degeneracy in the quark model is a simple consequence of the absence of quark-quark bound states. The t-channel exchange is allowed (fig. 3).

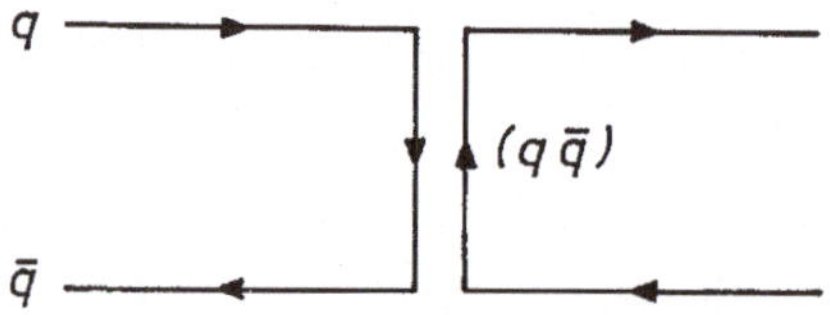

Figure 3 - The t-channel force, which binds a quark and an anti-quark to form a meson is again a meson.

The u-channel exchange, however, is exotic and thus contributes very little (fig. 4)

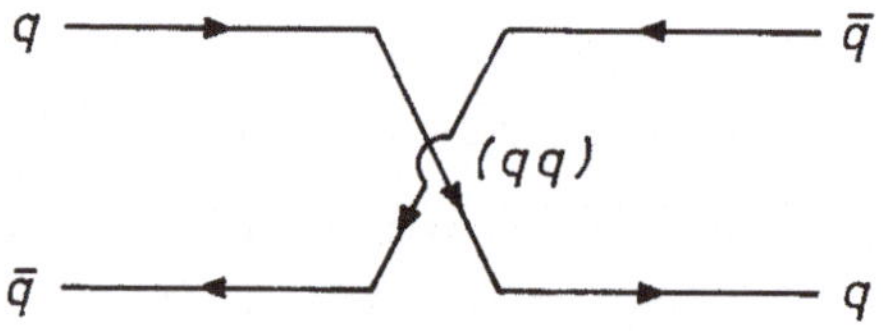

Figure 4 - The u-channel forces stem from two-quark exchange and are thus negligible.

If you remember that the signatured amplitude is defined by

$$f^{\pm}(\cos\delta, t) = \sum_j (F_D^j \pm F_E^j) \; P_j(\cos\delta)$$

where F_D^j is the direct (t-channel amplitude) and F_E^j is the exchange (u-channel amplitude). $F_E^j \sim 0$ gives immediately $f^+ \approx f^-$, i. e. mesons with opposite signature are degenerate.

B. Baryons

For baryons the shell model [10] with a symmetric overall wave function has been successful in deriving the mass spectrum [17]. From the Schrödinger equation one obtains with the shell model Hamiltonian [18]

$$H_{SM} = \sum_i \frac{\vec{p}_i^{\,2}}{2M} + \frac{1}{2} M \omega^2 \sum_i \vec{r}_i^{\,2} \tag{14}$$

a spin orbit coupling and other symmetry breaking terms a mass formula which is linear in L

$$E = (L+2k)\omega + \text{const.} + b\langle\vec{L}\vec{S}\rangle + \tag{15}$$

$$+ \text{SU(3) breaking and other terms.}$$

For higher orbital angular momenta mixing between various states complicates the situation.

Let me just mention what the quark model says concerning parity doubling.For the L=0 ground state there are certainly no parity doublets because they would require a spatial wave function of the form

$$(\vec{r}_1 \times \vec{r}_2) \cdot \vec{r}_3 = 0$$

where $\vec{r}_1$, $\vec{r}_2$ and $\vec{r}_3$ are the position vectors of the three quarks. For higher L there are a lot of positive and a lot of negative parity states, but it would be a mere coincidence should the energy levels of some of them coincide.

Concluding we may say that there are still a lot of open questions which leave the quark model rather flexible. As we learn more about the hadron spectrum, the quark model will be modified. Presumably the final theory of hadrons will still allow some kind of quark model. Lets hope that this model will be simple enough to be useful.

REFERENCES

1. M. Gell-Mann, Phys. Lett. 8, 214 (1964); G. Zweig, CERN Preprints TH 401, 402 (1964) (unpublished).
2. C. B. A. McCusker and I. Cairns, Phys. Rev. Lett. 23, 658 (1969) and I. Cairns, C. B. A. McCusker, L. S. Peak and R. L. S. Woolcott, Phys. Rev. 186, 1394 (1969).
3. R. K. Adair and H. Kaska, Phys. Rev. Lett. 23, 1355 (1969), H. Frauenfelder, V. E. Kruse and R. D. Sard, ibid. 24, 33 (1970) and D. C. Rahm and R. I. Loutit, ibid. 24, 279 (1970).
4. L. I. Schiff, Phys. Lett. 31B, 79 (1970).
5. Yu. A. Antipov et al.: Further search for quarks at 70 GeV proton synchrotron of IHEP, preprint IHEP Serpuchov (1969).
6. R. Hagedorn, Suppl. Nuovo Cim. 6, 311 (1968) and private communication.
7. J. Franklin, Phys. Rev. 181, 1984 (1969).
8. W. Thirring, Lectures held at the International School of Physics "Ettore Majorana" Erice, 1969.
9. R. E. Kreps and J. J. De Swart, Phys. Rev. 162, 1729 (1967). A.N.Mitra and R.Majundar,Phys.Rev.150,1194(1966).
10. O. W. Greenberg, Phys. Rev. Lett. 13, 598 (1964).
11. O. W. Greenberg, in Mathematical Theory of Elementary Particles, edited by R. Goodman and I. Segal, M.I.T. Press 1966, p.29.
12. O. W. Greenberg and A. M. L. Messiah, Phys. Rev. 138, B1155 (1965).
13. H. S. Green, Phys. Rev. 90, 270 (1953).

14. R. F. Streater and A. S. Wightman "PCT, spin, statistics and all that", W. A. Benjamin, Inc., New York 1964, p. 154 .
15. O. W. Greenberg and D. Zwanziger,Phys. Rev. 150, 1177 (1966).
16. M. Y. Han and Y. Nambu, Phys. Rev. 139, B1006 (1965).
17. O. W. Greenberg and M. Resnikoff, Phys. Rev. 163, 1844 (1967). A.N.Mitra, Nuovo Cim. 56, 1164 (1968).
18. D. Faiman and A. W. Hendry, Phys. Rev. 180, 1609 (1969).
19. T. D. Lee, Nuovo Cim. 35, 933 (1965).
20. S. Okubo: Test of Quark Models and Asymptotic Symmetry, University of Rochester, preprint 1969.
21. S. L. Adler, Phys. Rev. 177, 2426 (1969).
22. S. L. Adler, J. Schwinger, Science 165, 757 (1969), and 166, 690 (1969).
23. M. Y. Han and L. C. Biedenharn, Phys. Rev. Lett. 24, 118 (1970).
24. R. H. Dalitz, "Hadron Spectroscopy" in the Proceedings of the Haway Topical Conference in Particle Physics" (1967) edited by S. Pakavasa and F. S. Tuan, University of Hawaii Press; Honolulu 1968, p.363.
25. M. A. Ruderman and D. Zwanziger, Phys. Rev. Lett. 22, 146 (1969).
26. R. H. Dalitz, Lecture at the International Conference on Symmetries and Quark Models, Detroit 1969, Oxford University preprint.
27. O. W. Greenberg, Proceedings of the Lund International Conference on Elementary Particles, 1969, p.385.
28. G. Veneziano, Nuovo Cimento 57A, 190 (1968).
29. G. Tiktopoulos, Phys. Lett. 29B, 185 (1969).
30. D. Z. Freedman and J. M. Wang, Phys. Rev. 153, 1596 (1967).
31. L. Durand, Phys. Rev. 154, 1537 (1967).
32. M. Gell-Mann and G. Zweig, unpublished.

33. H. Harari, Proceedings of the 14th International Conference on High Energy Physics, Vienna 1968, p.197.

34. D. Horn, in: Particle Physics, edited by P. Urban, Springer Verlag Wien/New York, 1969, p.157.

35. S. Mandelstam, Phys. Rev. 184, 1625 (1969).

36. J. Mandula, J. Weyers and G. Zweig, Phys. Rev. Lett. 23, 627 (1969).

37. W. T. Chu, Young S. Kim, W. J. Beam, and Nowhan Kwak, Phys. Rev. Letters 24, 917 (1970).

38) R. Baud et al., Phys. Letters 31B, 549 (1970).

Acta Physica Austriaca, Suppl. VII, 567—587 (1970)

ANOMALIES IN WARD IDENTITIES?*

BY

W. KUMMER
Institut für Theoretische Physik der
Technischen Hochschule Wien
and
Institut für Hochenergiephysik der
Österreichischen Akademie der Wissenschaften
Wien

ABSTRACT

In certain generalized Ward-identities perturbation theory can lead to deviations from the relations, derived by canonical field theory. Such a modification caused by the fermion triangle graph in the axial-photon-vertex, was discussed in several papers during the last year. We try to show, also by means of other simple field theoretical models, that the problem of such "anomalies" is by no means solved, since the interpretation of the mathematical equations in terms of experimentally measurable ("renormalized") quantities is still very arbitrary. E.g. sometimes only by abolishing the normal-ordering prescription "anomalies" become "normal" tadpole-contributions.

---

* Lecture given at IX. Internationale Universitätswochen für Kernphysik, Schladming, February 23 - March 7,1970.

1. INTRODUCTION

One year ago the papers of Adler [1] and of Bell and Jackiw [2] have created renewed interest in problems, connected with the discrepancy of perturbation theory in vertices involving the axial vector current of fermions on one side and "canonical" relations on the other.

Taking not into consideration an old result of Schwinger [3], one had always tacitly assumed that no essential difference exists between Ward-Takahashi-identities [4] of "proper" n-point functions, where one corner is occupied by a "foreign" operator. With the latter expression, we mean a combination of field operators which does not appear as a current, mass-term etc. of the renormalizable theory. The most prominent example was discussed in ref. [1] namely the axialvector

$$j_\mu^5 = :\bar{\psi}\,\gamma_\mu\gamma_5\,\psi: \qquad (1.1)$$

in spinor-electrodynamics. It should have been clear, though that new problems might arise from a quantity like (1.1) in perturbation theory, because the axial vector current can never participate in any renormalizable field theory. This is also probably the basic reason for the infinities encountered in perturbation theory, even if j_μ^5 is considered to be coupled to an external field and only radiative corrections are calculated [5].

Any discrepancies between the formal canonical theory and perturbation theory within the context of a renormalizable theory have been settled long ago by appropriate "regularization" techniques - despite contrary statements in the recent literature [6]. One famous example is the Schwinger-term in the $[j_o, j_i]$-commutator which takes in spinor-electrodynamics its canonical value zero only if properly regularized. Two years ago we heard here a splendid lecture by the late G. Källén on this question

[7]. He stressed the close connection of the r.h.s. of this commutator with a contribution to the photon mass. A priori any cut-off procedure is allowed within a renormalizable theory from a mathematical point of view, the physical facts alone (e.g. the vanishing of the photon mass) support one way of handling the divergent integrals as compared to the others. In our opinion this point of view is even more appropriate for relations involving "foreign" operators. We shall emphasize that really at the moment no experimental evidence is in conflict with some "oldfashioned" approach.

We do not subscribe therefore to the ideas, as e. g. expanded in ref. [6], to change the canonical commutators, to abandon the Feynman conjecture of cancellation of seagull and Schwinger terms etc. Because even from a purely mathematical point of view, the new scheme does not yet seem to be consistent: anomalies in commutators as required in one case destroy the agreement in others [6].

In order to see the problems in their relation to the classical difficulty of the photon mass, we recall shortly the situation there (section 2) and turn then to the Adler-triangle in section 3. In the last section two other models are considered providing examples of even more troublesome but perhaps more clarifying cases.

2. PHOTON MASS

Twenty years ago the problem of the photon mass has been solved. We here only re-express it in a way which resembles our discussions below [8]. The divergence of the current

$$j_\mu = :\bar{\psi}\gamma_\mu\psi: \tag{2.1}$$

when calculated with the field equations of spinor-electrodynamics, becomes

$$\partial^{\mu} j_{\mu} = ie_o : \bar{\psi} \not{A} \psi - \bar{\psi} \not{A} \psi : \tag{2.2}$$

Its matrix element vanishes therefore certainly as long as the matrix element of the r.h.s. does not diverge. In the vacuum polarization $\Pi_{\mu\nu}$

$$\Pi_{\mu\sigma} D'_{\sigma\nu}(k) = -i \int e^{-ikx} \langle 0 | T(j_{\mu}(0) A_{\nu}(x)) | 0 \rangle d^4x \tag{2.3}$$

therefore to lowest order α (cf. fig. 1)

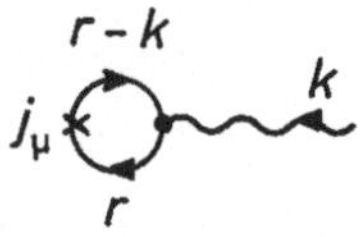

Figure 1 - Vacuum Polarization

$$k^{\mu} \Pi_{\mu\nu} = - \frac{e_o^2}{(2\pi)^4} \int d^4r \, \mathrm{Tr}\left(\frac{1}{\not{r} - \not{k} - m_o} \gamma_{\nu} - \frac{1}{\not{r} - m_o} \gamma_{\nu} \right) \tag{2.4}$$

We note that the r.h.s. represents really the difference of the two "tadpoles" of fig. 2 giving separately diverging contributions, which had been suppressed by the normal ordering prescription in (2.1) or (2.2).

Figure 2 - "Tadpoles" in $k^{\mu} \Pi_{\mu\nu}$

We know that the r.h.s. of (2.4) must vanish, otherwise the photon would acquire a mass. Also the W.I. (Ward identity) (2.2) with zero on the r.h.s. would not be fulfilled. So we conclude that perturbation theory must be done in a certain way, the regularization of $\Pi_{\mu\nu}$ must be such that $k^{\mu} \Pi^{reg}_{\mu\nu} = 0$. This is achieved here with the Pauli-Villars-regularization [9]

$$\Pi_{\mu\nu}(m_o,k^2) \to \Pi^{reg}_{\mu\nu}(m_o,M,k^2) = \Pi_{\mu\nu}(m_o,k^2) - \Pi_{\mu\nu}(M,k^2) =$$

$$= \frac{2i\alpha_o}{\pi}\int_0^1 dz\ z(1-z)\log\left(\frac{M^2}{m^2-k^2z(1-z)}\right)(k_\mu k_\nu - k^2 g_{\mu\nu}) \tag{2.5}$$

Let me stress that the physical facts - conserved current , vanishing photon mass - are the only guide to find the correct way, how to treat the problem. Consequently also for W.I.'s with "foreign" vertices only physical arguments - when they exist! - provide a legitimate starting point for any "regularization" in the most general sense.

3. AXIAL VECTOR CURRENT - TWO PHOTON - W.I. IN QED

The field equations of spinor electrodynamics let the divergence of the axial vector current (1.1) appear in the simple form

$$\partial^\mu j^5_\mu = 2i\ m_o\ j^5 \tag{3.1}$$

where

$$j^5 = :\bar{\psi}\gamma^5\psi: \tag{3.2}$$

All quantities are unrenormalized, j^5_μ and j^5 are the "foreigners". On the r.h.s. of (3.1) two terms $:\bar{\psi}\gamma_5 \not{A}\psi:$ have cancelled, which again is not necessarily true in diverging matrix elements. Moreover, gauge invariance is not manifest in (3.1). This could have been achieved by the usual limiting procedure [10]

$$j^5_\mu = \lim_{\varepsilon\to 0} j^5_\mu(\varepsilon,x) = \lim_{\varepsilon\to 0} :\bar{\psi}(x+\tfrac{\varepsilon}{2})\gamma_\mu\gamma^5\psi(x-\tfrac{\varepsilon}{2}) \times$$

$$\times\ \exp\{ie\int_{x-\varepsilon/2}^{x+\varepsilon/2} A_\alpha(y)\ dy^\alpha\}:$$

yielding on the r.h.s. of (3.1) the additional term

$$ie\ j^\mu_5\varepsilon^\alpha\ [\partial_\alpha A_\mu - \partial_\mu A_\alpha] + O(\varepsilon)$$

which may or may not for $\varepsilon \to o$ give zero in a given matrix element. We continue here with the simple expression (3.1). By a partial integration one verifies easily that the axial vector (or pseudoscalar) - two photon vertices (k_1, k_2 are the photon momenta, ρ and σ their polarization indices, respectively)

$$R_{\rho\sigma(\mu)}(k_1,k_2)D_F'(k_1)D_F'(k_2) = \tag{3.3}$$

$$= -\int d^4x\, d^4y\, \exp\{i(k_1x+k_2y)\}\langle O|T(A_\rho(x)A_\sigma(y)j^5_{(\mu)}(O))|O\rangle$$

obey the W.I. ($q=k_1+k_2$)

$$q^\mu R_{\rho\sigma\mu} = 2m_o R_{\rho\sigma} \tag{3.4}$$

and, of course, also the gauge condition

$$k_1^\rho R_{\rho\sigma(\mu)} = k_2^\sigma R_{\rho\sigma(\mu)} = O \tag{3.5}$$

By T we mean already the covariantized time ordered product, adapted for perturbation theory with Feynman rules (usually called T^*).

Similarly for the axial vector - fermion vertex

$$S_F'(-p)\Gamma^5_{(\mu)}S_F'(p') = \tag{3.6}$$

$$= -\int d^4x\, d^4y \langle O|T(\psi(x)j_{(\mu)}(O)\bar{\psi}(y))|O\rangle\, e^{i(px+p'y)}$$

one arrives at

$$(p+p')^\mu\Gamma^5_\mu = 2m_o\,\Gamma^5 + S_F'^{-1}(-p)\gamma_5+\gamma_5 S_F'^{-1}(p) \tag{3.7}$$

This resembles more the original W.I. in QED, the two propagator terms on the r.h.s. stemming from canonical equal time commutators of fermion fields which do not vanish in this case. The lowest order contribution to the quantities in (3.4) are the (superficially linearly divergent) triangle graphs of fig. 3 . Using an old result by Rosenberg [11] for $R_{\rho\sigma(\mu)}$ it turns out immediately that the

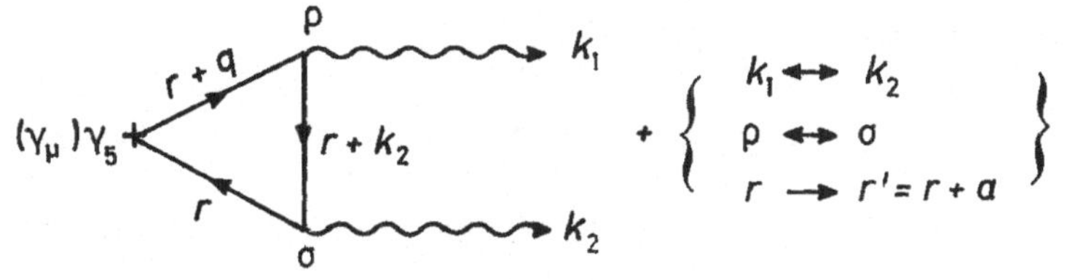

Figure 3 - Triangle graphs for $R_{\rho\sigma(\mu)}$

regularized quantity

$$R^{reg}_{\rho\sigma(\mu)} = R_{\rho\sigma(\mu)}(q^2,m_o) - R_{\rho\sigma(\mu)}(q^2,M) \tag{3.8}$$

is in fact finite even in the limit $M\to\infty$, gauge invariant and without arbitrary terms, whereas $R_{\rho\sigma}$ is finite altogether. Inserting into (3.4), however, shows disagreement [1], [2]. We prefer a very similar but perhaps more pedagogical approach in which explicit calculations of amplitudes are not necessary at all: The identity

$$\frac{1}{\not{r}+\not{q}-m_o}\not{q}\gamma_5\frac{1}{\not{r}-m_o} = \frac{1}{\not{r}+\not{q}-m_o}2m_o\gamma_5\frac{1}{\not{r}-m_o} + \frac{1}{\not{r}+\not{q}-m_o}\gamma_5 + \\ + \gamma_5\frac{1}{\not{r}-m_o} \tag{3.9}$$

in the triangle-contribution (where we have left free explicitly a shift r'=r+a in the integration of the crossed diagram!)

$$R_{\rho\sigma(\mu)} = \frac{-ie_o^2}{(2\pi)^4}\int d^4r\ Tr[\gamma_\mu\gamma_5(\not{r}-\not{k}_2-m_o)^{-1}\gamma_\sigma(\not{r}-m_o)^{-1} \times \\ \times \gamma_\rho(\not{r}+\not{k}_1-m_o)^{-1} + (r\to r+a, k_1\leftrightarrow k_2, \rho\leftrightarrow\sigma)] \tag{3.10}$$

gives in (3.4)

$$q^\mu R_{\rho\sigma\mu} - 2m R_{\rho\sigma} = -\frac{ie_o^2}{(2\pi)^4} Z_{\rho\sigma} \tag{3.11}$$

where $Z_{\rho\sigma}$ is the difference of linearly divergent integrals, differing by a shift of the integration variable only:

$$Z_{\rho\sigma} = \int d^4r \, \mathrm{Tr}[\gamma_5\gamma_\sigma(\not{r}-m_o)^{-1}\gamma_\rho(\not{r}+\not{k}_1-m_o)^{-1} - (r\to r'-k_1) -$$
$$- \gamma_5\gamma_\rho(\not{r}-\not{k}_2-m_o)^{-1}\gamma_\sigma(\not{r}-m_o)^{-1} + (r\to r'+k_2)]$$

E.g. the third and fourth term is easily shown to be a finite "surface contribution"

$$Z^{(1)}_{\rho\sigma} = \int_0^1 dx \int d^4r \, \frac{\partial}{\partial x} \mathrm{Tr}\{\gamma_5\gamma_\rho(\not{r}-\not{k}_2+x(\not{a}+\not{k}_2)-m_o)^{-1}\gamma_5(\not{r}+x(\not{a}+\not{k}_2)- -m_o)^{-1}\} =$$

$$= \int_0^1 dx(a_\gamma+k_{2\gamma}) \int d^4r \, \frac{\partial}{\partial r_\gamma} \, 4i\varepsilon_{\rho\sigma\alpha\beta}\left(\frac{r^\alpha k_2^\beta}{r^4} + O(1/r^4)\right)$$

evaluated with

$$\int df^\gamma \frac{r^\alpha}{r^4} = -i\,\frac{\pi^2}{2}\,g^{\gamma\alpha} \tag{3.12}$$

There i and (-1) come from the transition to an Euclidean R_4, $2\pi^2$ is the surface of the four-dimensional unit sphere, 1/4 the average over the four dimensions. The total $Z_{\rho\sigma}$ is then for

$$a = \lambda k_1 + \mu k_2 \tag{3.13}$$

equal to

$$Z_{\rho\sigma} = 2\pi^2 \,\varepsilon_{\rho\sigma\alpha\beta}\, k_1^\alpha\, k_2^\beta(\lambda-\mu) \tag{3.14}$$

In a completely analogous way with the identity

$$\frac{1}{\not{r}+\not{k}_2-m_o}\,\not{k}_1\,\frac{1}{\not{r}+\not{q}-m_o} = \frac{1}{\not{r}+\not{k}_2-m_o} - \frac{1}{\not{r}+\not{q}-m_o} \tag{3.15}$$

the deviations from the two gauge conditions (3.5) are calculated and found to vanish only for $\lambda=-\mu = 2$, $R^{g.i.}_{\rho\sigma\mu}$ is gauge invariant for this choice of a and the same is true for $R_{\rho\sigma}$, because there application of (3.15) gives a trace of γ_5 with at most three γ-matrices. Therefore

$$q^\mu R^{g.i.}_{\rho\sigma\mu} - 2m_o R_{\rho\sigma} = -\frac{ie_o^2}{(2\pi)^4}\,8\pi^2\,\varepsilon_{\rho\sigma\alpha\beta}\,k_1^\alpha\,k_2^\beta \tag{3.16}$$

is obtained. We note that in our derivation the deviation had its origin in the difference of integration which would have vanished for less than linearly divergent integrals. Any loop with more photons is more convergent, so that no deviation comes from those functions. On the other hand it is plausible that the insertion of all possible radiative corrections in the triangle just produces a replacement $e_o \to e$ the renormalized charge [12].

The relation to the Adler-Rosenberg calculation is clarified by using (3.8). In the difference all mass independent (surface) terms $Z_{\rho\sigma}$ cancel, thus

$$q^\mu R^{g.i.}_{\rho\sigma\mu} = \lim_{M\to\infty} q^\mu R^{reg}_{\rho\sigma\mu} =$$

$$= 2m_o\, R_{\rho\sigma}(q^2,m) - \lim_{M\to\infty} 2M\, R_{\rho\sigma}(q^2,M) \tag{3.17}$$

The regularization of $R_{\rho\sigma\mu}$ entails therefore automatically a "regularized" $[m_o R_{\rho\sigma}]$ which is something rather unusual. In fact from the explicit form of $R_{\rho\sigma}$ for $k_1^2 = k_2^2 = 0$,

$$R_{\rho\sigma} = -\frac{ie_o^2}{(2\pi)^4}\,\varepsilon_{\rho\sigma\alpha\beta}\, k_1^\alpha\, k_2^\beta\, 8\pi^2 m_o \times$$

$$\times \int_0^1 dx \int_0^{1-x} dy\, [xyq^2 - m_o^2]^{-1} \tag{3.18}$$

the limit is found to be finite and to coincide precisely with the "deviation" in (3.16). Apart from eq. (3.18), where it did not matter for $m_o \to \infty$, in all considerations up to now no use was made of $k_1^2 = k_2^2 = 0$, so everything was valid for virtual photons as well. For on-shell photons (3.16) is also understood from the most general form of $R_{\rho\sigma(\mu)}$ obeying Lorentz invariance and gauge invariance:

$$R_{\rho\sigma\mu} = q_\mu\, \varepsilon_{\rho\sigma\alpha\beta}\, k_1^\alpha\, k_2^\beta\, A(q^2)$$

$$R_{\rho\sigma} = \varepsilon_{\rho\sigma\alpha\beta}\, k_1^\alpha\, k_2^\beta\, B(q^2) \tag{3.19}$$

(3.16) or (3.17) with (3.18) just mean that

$$q^2A(q^2) = 2m_o[B(q^2) - B(0)] \tag{3.20}$$

providing thus the correct limit zero of both sides, despite $B(0) \neq 0$. This limit for the l.h.s of (3.20) is known as the Sutherland theorem: The $\pi^o \to 2\gamma$ rate vanishes for a soft pion [14], whereas $B(0) \neq 0$ and therefore both sides of (3.20) must be unequal without the additional term $-2m\,B(0)$ as noted first by Bell and Jackiw [2] in the σ-model [15]. Adler was led to modify (3.1) into

$$\partial_\mu j^{\mu 5} = 2\,i\,m_o\,j^5 + \frac{\alpha_o}{\pi} F^{\alpha\beta} \tilde{F}_{\alpha\beta}$$

$$\tilde{F}_{\alpha\beta} = \frac{1}{2} \varepsilon_{\alpha\beta\gamma\delta} F^{\gamma\delta} \tag{3.21}$$

Similarly, in the σ-model, where

$$j^5_\mu = \bar{\psi}\gamma_\mu\gamma_5\psi + 2(\sigma\partial_\mu\pi - \pi\partial_\mu\sigma) - f_\pi\,\partial_\mu\,\pi \tag{3.22}$$

with appropriate field equations, PCAC is supplemented

$$\partial_\mu j^{\mu 5} = f_{\pi^o}\, m^2_{\pi^o}\,\pi + \frac{\alpha_o}{\pi} F^{\alpha\beta} \tilde{F}_{\alpha\beta} \tag{3.23}$$

Adler uses the matrix element $\langle 2\gamma||\pi\rangle$ of (3.23) at $q^2=o$

$$0 = -\,i\,g_\pi\,f_{\pi_o} - \frac{2i\alpha_o}{\pi} \tag{3.24}$$

to predict the $\pi^o \to 2\gamma$ amplitude g_π

$$\langle 2\gamma|j_\pi(0)|0\rangle = -i\,g_\pi\,\varepsilon_{\rho\sigma\alpha\beta}\,k_1^\alpha\,k_2^\beta$$

in approximate agreement with the experimental lifetime of π^o, $\tau^{-1}_{\pi^o} = 7.4$ eV. This result is still considered to be about the only experimental prediction of W.I. - anomalies [6]. We remark:

1) The σ-model has but little relevance to physical problems of strong interactions where e.g. more (quark- or other) fermion loops change the prediction of g by a model dependent factor.

2) Any change of the approximate statement of PCAC in a certain application does not necessarily mean a modi-

fication of canonical field relations like (3.21) or (3.23). In fact (3.24) can as well be derived from the Sutherland theorem (the l.h.s. of (3.20)) alone.

This can be seen as follows. Take the renormalized axial vector current matrix element between the vacuum and two photons, (3.19) in an unsubtracted dispersion relation in q^2 with the pion pole at m_π^2 :

$$q^2 A(q^2) = \frac{f_\pi m_\pi^2 (-i) g_\pi}{m_\pi^2 - q^2} + \frac{1}{\pi} \int_{3m_\pi^2} \frac{x q(x)}{x - q^2} dx \qquad (3.25)$$

For the rest of the absorptive parts a = abs(A), b =abs(B)

$$q^2 a = b$$

holds, therefore the integral for a fermion-antifermion intermediate state equals simply 2m $B(q^2)$ if point-like fermions are assumed $\langle N\bar{N} | \partial_\mu j^{\mu 5} | O \rangle = 2m\, i\, \bar{u}\, \gamma^5 v$.
For $q^2 \to 0$

$$- i\, g_\pi = \frac{2m}{f_\pi} B(0) = \frac{2 i \alpha_0}{f_\pi \pi}$$

coincides with (3.24). It is, however, evident that strong interactions will certainly change this contribution entirely. In fact, with reasonable assumptions for form factors, coupling constants etc. the $N\bar{N}$-state is found not to be important at all. Furthermore, though (3.16) contains a finite modification from (3.21), this is no longer true in the modified axial vector two fermion W.I. (3.7):

$$(p'+p)^\mu \Gamma_\mu^5 = 2m_0 \Gamma^5 + S_F'^{-1}(-p)\gamma^5 + \gamma^5 S_F'^{-1}(p') + F \qquad (3.26)$$

$$S_F'(-p)\, F\, S_F'(p') =$$
$$= \frac{i\alpha_0}{4\pi} \int dx\, dy\, e^{i(px+p'y)} \langle O | T(\psi(x) :F^{\alpha\beta} \tilde{F}_{\alpha\beta}: \bar{\psi}(y)) | O \rangle \qquad (3.27)$$

Here something must be done with the additional term, because F is logarithmically divergent according to the graph of fig. 4. As shown in ref. [1] it is impossible,

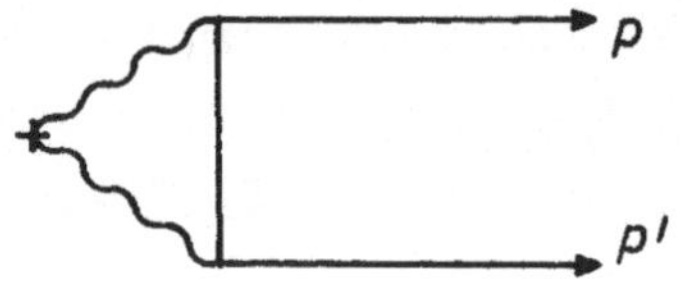

Figure 4 - Modification of the Axial-Vector-Fermion W.I.

because of the structure of F in terms of γ-matrices, to absorb the divergent part of F in a multiplicative renormalization of Γ_5. Admittedly Γ_5 is a "foreign" vertex function for spinor electrodynamics; if strong interactions are switched on, however, the (infinite) counter terms in the renormalization constant of this vertex are already completely determined just to make the measurable quantities of the renormalized pion-nucleon-photon system finite. Any additional counter terms for F would make the r.h.s. of (3.26) finite in its renormalized form but would at the same time destroy entirely the finiteness in the renormalized π-N-γ-theory.

As there is neither any conclusive experimentally nor theoretical necessity to maintain (3.21) we might as well rely more on the canonical result, the lesson learned from the photon mass problem, and modify the regularization, as it was so successfully done long ago in the case of the photon mass.

The first possibility is to define the properly regularized "foreign" $\tilde{R}_{\rho\sigma(\mu)}$ to be singular at $q^2 \to 0$ (destruction of Sutherland's theorem)

$$\tilde{R}_{\rho\sigma\mu} = R^{reg}_{\rho\sigma\mu} - \frac{2i\alpha}{\pi} \varepsilon_{\rho\sigma\alpha\beta} k_1^\alpha k_2^\beta \frac{q_\mu}{q^2} \tag{3.28}$$

which is somewhat ugly, although $\tilde{R}_{\rho\sigma(\mu)}$ is not accessible to direct observation anyhow.

Another way is suggested by (3.17) namely a "proper regularization" of $R_{\rho\sigma}$ as

$$\overbrace{(2m_o R_{\rho\sigma})} = \lim_{M\to\infty} (2m_o R_{\rho\sigma})^{reg} = 2m_o[R_{\rho\sigma}(q^2)-R_{\rho\sigma}(0)]$$

This amounts to fix the "renormalized" value of $R_{\rho\sigma}$ to zero. It might be cast in a form resembling somewhat the Pauli-Villars regularization by

$$\tilde{R}_{\rho\sigma} = \sum_i c_i R_{\rho\sigma}(m_i) \ , \ c_1 = 1 \ , \ \sum_i \frac{c_i}{m_i} = 0$$

Any of the two prescriptions is linear and guarantees therefore also the disappearance of terms like F in other W.I.'s related to the axial vector vertex. The first of the two possibilities has the further advantage to make $\tilde{\Gamma}^5_\mu$ finite. This is immediately seen because the loop contribution from $R_{\rho\sigma}$ in Γ^5 is finite. The logarithmically divergent is then taken together according to (3.28) with the equally divergent Γ^5_μ.

4. TWO EXAMPLES WITH TWO-CORNERED LOOPS [16]

For simplicity we assume in the first case a world of protons and π_o only (the generalization is obvious) with an interaction

$$L_{int} = i\, g_o :\bar{\psi}\gamma_5\psi\pi: \tag{4.1}$$

and therefore instead of (3.2)

$$\partial^\mu j^5_\mu = 2i\, m_o\, j^5 + 2g_o :\bar{\psi}\psi\pi: \tag{4.2}$$

From G-parity only the matrix element of j^5 into an odd number of pions can occur. The loop with three pions (fig.5) is even superficially only logarithmic divergent, so that no trouble can arise, which can be verified explicitly. We consider therefore

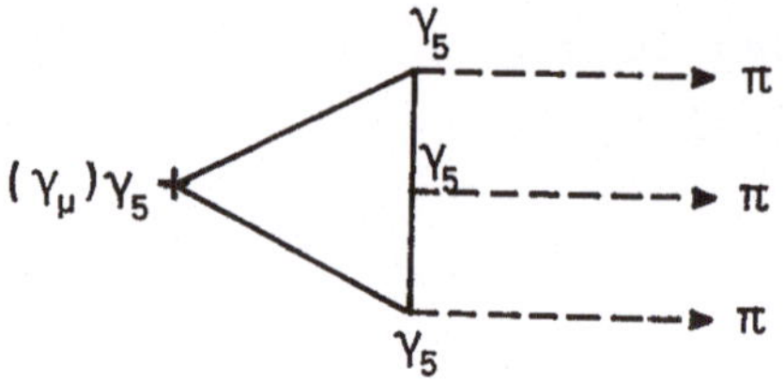

Figure 5 - 3-Pion Loop in π-N-Theory

$$\Delta'_\pi(q)\ F^5_{(\mu)} = -i\int d^4x\ e^{iqx} \langle 0|T(j^5_{(\mu)}\pi(x))|0\rangle \quad (4.3)$$

$$\Delta'_\pi(q)\ G = -i\int d^4x\ e^{iqx} \langle 0|T(:\bar{\psi}\psi\pi:\pi(x))|0\rangle \quad (4.4)$$

in the W.I.

$$q^\mu\ F^5_\mu = -2m_o\ F^5 + 2ig_o\ G \quad (4.5)$$

derived again with canonical commutators. To $O(g_o)$ no contribution comes from the G-term. Here we know that the "proper" F^5 obeys

$$\tilde{F}^{5reg}\Big|_{q^2=m^2_\pi} = 0 \quad (4.6)$$

because the renormalized

$$\tilde{F}^{5reg}\Big|_{q^2=m^2_\pi} = \langle 0|\tilde{j}^5|\tilde{\pi}\rangle = 0$$

This can be seen also from the fact that in the full pion propagator

$$\Delta'^{-1}_\pi = q^2 - m^2_\pi + \delta m^2_\pi - T(q^2, m_o, m_{\pi o})$$

g_oF^5 is the lowest order contribution of the proper self-energy T which appears always together with the mass counter term δm^2_π obeying

$$\delta m^2_\pi = T(m^2_\pi, m_o, m^2_\pi)$$

If we forget for the moment this δm^2_π and use (3.9) for the Feynman diagram of fig. 6 in (4.5), we obtain

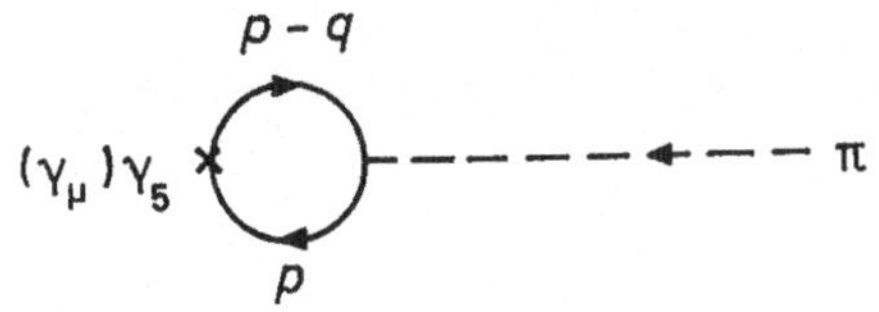

Figure 6 - Axialvector-Pion-Loop

$$q^\mu F^5_\mu = -\, 2m_o\, F^5 + R \tag{4.7}$$

$$R = -\frac{g_o}{(2\pi)^4}\int d^4r\, \mathrm{Tr}\left(\frac{1}{\not r - m_o} + \frac{1}{\not r + \not q - m_o}\right) \tag{4.8}$$

The deviation R is here: 1) divergent, 2) the sum of two "tadpoles" and hence certainly no surface term! The presence of those terms would be completely normal, if we would not have introduced our j^5_μ with the normal ordering prescription. We see that in general perturbation theory does not "respect" the normal ordering. If we regularize ($c_1=1$, $m_1=m_o$)

$$F^{5reg}_\mu = m_o \sum_i c_i \frac{F_\mu(m_i)}{m_i}$$

$$F^{5reg} = \sum_i c_i\, F^5(m_i)$$

$$R^{reg} = \sum_i c_i \frac{R(m_i)}{m_i} \tag{4.9}$$

in order to cut-off the divergence, we find

$$F^{5reg}_\mu = \phi(q^2)\, q_\mu$$

$$F^{5reg} = \alpha(q^2)$$

$$R^{reg} = q^2\phi + 2m_o\alpha \tag{4.10}$$

with the usual self-energy integrals

$$\phi = \sum_i c_i \frac{i m_o g_o}{4\pi^2}\int_0^1 dz \int_0^\infty \frac{1}{\lambda} \exp\{i\lambda(q^2 z(1-z)-m_i^2)\}d\lambda$$

$$\alpha = - \sum_i c_i \frac{ig_o}{4\pi^2} \int_0^1 dz[-3z(1-z)q^2+m_i^2] \int_0^\infty \frac{d\lambda}{\lambda} e^{i\lambda(q^2 z(1-z)-m_i^2)} \tag{4.11}$$

Including correctly the counter terms $\delta m_\pi^2/g_o$ in F^5 produces the replacement

$$F^{5reg} \to F^{5reg'} = \alpha(q^2,m_i)-\alpha(m_\pi^2,m_i) \tag{4.12}$$

This is not yet a finite quantity. Using the relation between T and the wave function renormalization of the π-field Z

$$Z^{-1} = 1- \left.\frac{\partial T}{i\partial q^2}\right|_{q^2=m_\pi^2} = 1 - \frac{g_o}{i} \left.\frac{\partial\alpha}{\partial q^2}\right|_{q^2=m_\pi^2} \tag{4.13}$$

we may express $F^{5reg'}$ by a cut-off independent $\bar{\alpha}$ and Z

$$F^{5reg'} = \bar{\alpha}(q^2) + \frac{Z-1}{g_o} i(q^2-m_\pi^2) + O(g_o^3) \tag{4.14}$$

where

$$\bar{\alpha}(q^2) = \alpha(q^2) - \alpha(m_\pi^2) - \left.\frac{\partial\alpha}{\partial q^2}\right|_{m_\pi^2} (q^2-m_\pi^2)$$

Since Z disappears only into the renormalized g if further graphs play together with (4.14), e,g. in an S-matrix element, F^5 alone will always depend on the cut-off M. But to include the counter-term $\alpha(m_\pi^2)$, already necessitates at least a corresponding change

$$R^{reg} \to R^{reg'} = q^2\phi + 2m_o[\alpha(q^2) - \alpha(m_\pi^2)] \tag{4.15}$$

in order to maintain the identity (4.7). If we invent also a similar correction to ϕ

$$\phi^{reg} \to \phi^{reg'} = \phi(q^2) - \phi(m_\pi^2) + f_\pi$$

with an arbitrary f_π, to be fixed by experiment ($\pi\to\mu\nu$) , $\phi^{reg'}$ becomes even cut-off independent. Again we might let disappear $R^{reg'}$ with an appropriate

$$\tilde{F}_\mu^5 = q_\mu(\phi^{reg} - \frac{R^{reg'}}{q^2}) = -2m_o F^{5reg'} \frac{q_\mu}{q^2} = q_\mu\tilde{\phi} \tag{4.16}$$

which has the disadvantage of making ϕ singular at $q^2=0$ but cancels the anomaly altogether. We note however again that the modification R is completely legitimate for the case in which no normal ordering is done in j^5_μ and j^5. An analogous example is the "tensor"-current in QED:

$$\ell_{\mu\nu} = :\bar{\psi}\,\sigma_{\mu\nu}\,\psi: \tag{4.17}$$

obeying

$$\partial_\mu \ell^{\mu\nu} = 2m_o :\bar{\psi}\gamma^\nu\psi: + 2e_o :\bar{\psi}\psi A^\nu: + i:[(\partial^\nu\bar{\psi})\psi - \bar{\psi}(\partial^\nu\psi)]: \tag{4.18}$$

Again we consider a two point function of the type

$$D'_{\rho\sigma}(q) H_{\mu\nu\sigma} = -i\int d^4x\, e^{-iqx} \langle 0|T(A_\rho(0), \ell_{\mu\nu}(x))|0\rangle \tag{4.19}$$

and similar definitions for $H_{\nu\rho}$, $L_{\nu\rho}$, $M_{\nu\rho}$ where $\ell_{\mu\nu}$ in (4.19) is replaced by $:\bar{\psi}\gamma_\nu\psi:$, $:\bar{\psi}\psi A^\nu:$, $:(\partial_\nu\bar{\psi})\psi - \bar{\psi}(\partial_\nu\psi):$, respectively. One easily derives the "W.I." .

$$q^\mu H_{\mu\nu\rho} = -\,2m_o\, i\, H_{\nu\rho} - 2\,i\,e_o\, L_{\nu\rho} + M_{\nu\rho} \tag{4.20}$$

$H_{\mu\nu\rho}$ and $H_{\nu\rho}$ are clearly gauge-invariant by themselves, but $M_{\nu\rho}$ and $L_{\nu\rho}$ are gauge invariant only <u>together</u>, due to the derivatives in $M_{\nu\rho}$. If we take the lowest order loop as in the example above, $L_{\nu\rho}$ will <u>not</u> contribute within usual Feynman rules, so (4.20) <u>must</u> obtain an additional term $Z_{\nu\rho}$ which <u>together</u> with $M_{\nu\rho}$ gives something gauge-invariant. $Z_{\nu\rho}$ is again the tadpole-contribution to $L_{\nu\rho}$, allowed only, if the normal ordering prescription is abolished. From Lorentz-invariance and gauge invariance and symmetry we expect

$$H_{\mu\nu\rho} = (g_{\mu\rho}\, q_\nu - g_{\nu\rho}\, q_\mu)\, A(q^2)$$

$$H_{\nu\rho} = (q_\nu\, q_\rho - g_{\nu\rho}\, q^2)\, B(q^2) \tag{4.21}$$

$L_{\nu\rho}$ and $M_{\nu\rho}$, being not gauge-invariant, consist in general of two parts, proportional to $q_\nu q_\rho$ and $g_{\nu\rho}$, respectively. We easily identify

$$- i\, e\, H_{\nu\rho} = \Pi_{\nu\rho} \qquad (4.22)$$

where $\Pi_{\nu\rho}$ is just the vacuum polarization (2.3), or (2.5) in its regularized form. Here the divergent part can be expressed by Z_3

$$\frac{\alpha_o}{3\pi} \log\left(\frac{M}{m}\right)^2 \cong 1 - Z_3$$

whereas the finite rest of $\Pi_{\nu\rho}$ vanishes at $q^2=0$. Again therefore no completely finite W.I. can be expected. The straight-forward calculation of $H_{\mu\nu\rho}$ shows that one must regularize as

$$H^{reg}_{\mu\nu\rho} = H_{\mu\nu\rho}(m_o) - \frac{m_o}{M} H_{\mu\nu\rho}(M)$$

so that

$$A^{reg}(q^2) = \frac{iem_o}{4\pi^2} \int_0^1 dz \log \frac{M^2}{m^2 - q^2 z(1-z)} \qquad (4.23)$$

since $H_{\mu\nu\rho}$ is only logarithmically divergent. The last bracket in the identity

$$\frac{1}{\not r + \not q - m_o} q^\mu \sigma_{\mu\nu} \frac{1}{\not r - m_o} = -2m \frac{i}{\not r + \not q - m_o} \gamma_\nu \frac{1}{\not r - m_o} +$$

$$+ \frac{i(2r+q)_\nu}{(\not r + \not q - m_o)(\not r - m_o)} - i\left(\gamma_\nu \frac{1}{\not r - m_o} \frac{1}{\not r + \not q - m_o} \gamma_\nu\right) \qquad (4.24)$$

yields the "deviation" $Z_{\rho\nu}$ (or the "tadpoles" from $L_{\nu\rho}$!) $M_{\nu\rho}$ is superficially quadratically divergent. A calculation analogous to the one in ref. [8] chapter 8.2 allows also to use identity (8.17) of the same ref. yielding

$$M^{reg}_{\nu\rho} = \frac{e_o m_o}{4\pi^2 i} \sum_i c_i \int_0^1 dz \int_0^\infty \frac{d\lambda}{\lambda} e^{i\lambda(q^2 z(1-z) - m_i^2)} \times$$

$$\times \left\{2g_{\nu\rho}[q^2 z(1-z) - m_i^2] - q_\nu q_\rho (2z-1)^2\right\} \qquad (4.25)$$

where $c_1 = 1$, $M_1 = M_o$. If at least i=1,2,3 with

$$\sum_{i=1}^3 c_i = \sum_{i=1}^3 c_i m_i^2 = 0$$

the λ-integration is finite. With a similar calculation we arrive for the deviation at

$$Z^{reg}_{\nu\rho} = - M^{reg}_{\nu\rho} + \frac{e_o m_o}{4\pi^2 i} (q^2 g_{\nu\rho} - q_\nu q_\rho) \int_0^1 dz \sum_i c_i \times$$

$$\times \int_0^\infty \frac{d\lambda}{\lambda} e^{i\lambda[q^2 z(1-z)-m_i^2]} (2z-1)^2 \qquad (4.26)$$

As we had expected, only the sum $M_{\nu\rho}+Z_{\nu\rho}$ is gauge-invariant. Again we are forced to introduce counterterms in the W.I., here in order to render all quantities to $O(e_o)$ gauge-invariant. This can be done e.g. replacing (4.20) by

$$q^\mu H_{\mu\nu\rho} = -2m_o \, i \, H_{\nu\rho} - 2i \, e_o \, L_{\nu\rho} + M^{g.i.}_{\nu\rho} \qquad (4.27)$$

where we make a prescription how to calculate $M^{g.i.}_{\nu\rho}$ to $O(e_o)$

$$M^{g.i.}_{\nu\rho} = M_{\nu\rho} + Z_{\nu\rho}$$

In higher orders, e.g. for $\sigma_{\mu\nu} \to 3$ photons, $Z_{\nu\rho}$ will no longer contribute, there $(M_{\nu\rho} - 2i \, e_o \, L_{\nu\rho})$ represents the gauge-invariant quantity.

A more detailed account of such models will be given elsewhere [16].

CONCLUSIONS

We summarize: Deviations from W.I.'s can have not only the form of a difference of "tadpole" contributions (surface-term) but can also be the (diverging) sum of such terms, making then a redefinition (counter-terms in the W.I.'s) an absolute necessity. The similarity to the mass renormalization let an additive "renormalization" be rather plausible. Sometimes the "anomalies" must be compensated entirely - as required by gauge-invariance in our second example of section 4 .

At the present time, in our view, neither experimental verifications nor theoretical consistency arguments require the introduction of anomalies. One should in this connection also keep in mind that in all two-cornered loops simply without normal ordering, all "anomalies" seem to become normal tadpole contributions, whereas the difference of "tadpoles" possessing more than one "tail" gives the deviation for the Adler triangle (cf. fig. 7). This

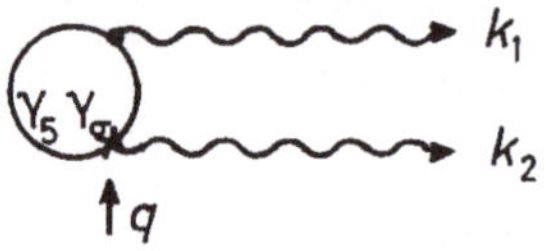

Figure 7 - "Tadpole" in $Z_{\rho\sigma}$ of the Adler-Relation.

would mean that W.I.'s are possibly only to be formulated in terms of "foreign operators" without normal ordering (situations as in section 4) or that appropriate "subtractions" must be made of diverging loop integrals (Adler triangle).

ACKNOWLEDGMENT

I am indebted to M. Schweda for much support in preparing this report.

REFERENCES

1. S. L. Adler, Phys. Rev. 177, 2426 (1969).
2. J. S. Bell and R. Jackiw, Nuovo Cimento 60, 47 (1969).
3. J. Schwinger, Phys. Rev. 82, 664 (1951).
4. Y. Takahashi, Nuovo Cimento 6, 370 (1957).
5. This is essentially the situation in weak interactions. For very valuable presentation cf. A. Sirlin, Particles, Currents, Symmetries (editor P. Urban), p.353,Springer

Wien-New York 1968.

6. cf. e.g. statements concerning eq. (21b) in R. Jackiw "Noncanonical Behaviour in Canonical Theories" CERN prepr. 1065 (August 1969).
7. G. Källén, in "Particles, Currents, Symmetries" (ed. P. Urban) p. 268, Springer Wien-New York 1968.
8. Our conventions and notations coincide with those of J. D. Bjorken and S. D. Drell, Relativistic Quantum Fields, McGraw-Hill, New York, 1965.
9. W. Pauli and F. Villars, Rev. Mod. Phys. 21, 434 (1949).
10. R. Jackiw and K. Johnson, Phys. Rev. 182, 1459 (1969).
11. R. A. Brandt, Phys. Rev. 180, 1490 (1969).
13. J. Steinberger, Phys. Rev. 76, 1180 (1949).
14. D. G. Sutherland, Nuclear Physics B2, 433 (1967).
15. M. Gell-Mann and M.Lévy, Nuovo Cimento 16, 705 (1960).
16. This and other models with deviations from W.I.'s have been worked out in collaboration with M. Schweda (to be published).

Acta Physica Austriaca, Suppl. VII, 588—596 (1970)

STRUCTURE EFFECTS IN WEAK INTERACTIONS[*] [†]

BY

H. PIETSCHMANN
Institute for Theoretical Physics
University of Vienna, Austria

1. A NON-LOCAL WEAK COUPLING

The standard way to formulate the theory (or model?) of weak interactions is to write down the "effective Lagrangian"

$$L_W = \frac{G}{\sqrt{2}} J_\lambda^\dagger(x)\ J^\lambda(x) \tag{1}$$

with

$$J_\lambda(x) = \ell_\lambda(x) + \cos\theta\ j_\lambda^{(\pi)}(x) + \sin\theta j_\lambda^{(K)}(x) \tag{2}$$

Here, θ is the Cabibbo angle and $j_\lambda^{(i)}(x)$ are the hypercharge conserving and hypercharge changing hadron currents. The lepton current $\ell_\lambda(x)$ is given by [1]

$$\ell_\lambda(x) = \sum_\ell \bar{\Psi}_\ell(x)\ \gamma_\lambda(1+\gamma_5)\ \Psi_{\nu_\ell}(x) \qquad \ell=e,\mu \tag{3}$$

The - by now fashionable - phrase "effective Lagrangian" which originated long ago in weak interaction physics, simply means that one uses the effective S-operator

$$S = I - \frac{iG}{\sqrt{2}} \int d^4x\ J_\lambda^\dagger(x)\ J^\lambda(x) + O(G^2) \tag{4}$$

---

[*] Seminar given at IX. Internationale Universitätswochen für Kernphysik, Schladming, February 23 - March 7, 1970.

[†] Supported by "Fonds zur Förderung der wiss. Forschung"

where nothing is said about $O(G^2)$.

For weak decay processes with small Q-value and for weak scattering processes with low momentum transfer, the S-operator (4) gives a very good phenomenological description. At higher Q-values or momentum transfers, i.e. at small distances, the question of the nature of the weak vertex comes up. Usually, the intermediate vector boson (IVB) is taken as an alternative to the four fermion interaction (1). In the absence of any experimental knowledge as to the existence of this particle, yet another alternative has recently been suggested [2]:

$$S = I - \frac{iG}{\sqrt{2}} \int d^4x\, d^4y\, J_\lambda(x)\, K^{\lambda\nu}(x-y)\, J^\dagger_\nu(y) + O(G^2) \qquad (5)$$

where $K^{\lambda\nu}$ is a c-number structure tensor whose invariant decomposition is given by

$$K_{\lambda\nu}(x) = M^4\{g_{\lambda\nu}\, F_1(x^2) + \frac{1}{M^2} \frac{\partial^2}{\partial x^\lambda \partial x^\nu} F_2(x^2)\} \qquad (6)$$

M is some characteristic parameter of the dimension of a mass. Note that $F_i(x^2)$ are dimensionless functions.

The structure tensor $K_{\lambda\nu}$ introduces an effective non-locality in the interaction without changing the local structure of the currents. Many theorems following from "local action of the lepton current" have been experimentally tested with success. Results obtained from the S-operator (5) can, of course, be immediately adjusted to yield IVB-predictions by simply setting

$$F_1(k^2) = F_2(k^2) = [\frac{k^2}{M^2} - 1 + i\varepsilon]^{-1} \qquad (7)$$

$$M = M_W$$

where the Fourier transform of the structure functions is defined by

$$F_i(x^2) = (2\pi M)^{-4} \int d^4k\, e^{ikx}\, F_i(k^2) \qquad i=1,2 \qquad (8)$$

The basic idea of a non-locality in weak interactions goes back to a paper of Lee and Yang in 1957 [3]. Even earlier, similar ideas have been tried to get rid of infinities in meson theories [4]. The basic difference to our approach lies in the fact that we keep the local structure of the currents and assume some phenomenological structure in the 4-fermion-vertex which may be caused by a highly unstable intermediate object just as well as by an intrinsic non-locality in the coupling of the two currents to each other.

2. EFFECTS TO LOWEST ORDER IN G^2

A list of the changes in semi-leptonic decays due to the structure tensor has been given in I. We may thus restrict ourselves in this paper to purely leptonic phenomena. The decay of the muon provides the classical example of a purely leptonic process and it has therefore been measured with excellent accuracy [5]. The energy spectrum of electrons from the decay of muons with polarization P is predicted by the V-A theory to be

$$N(x,\cos\theta) = \frac{G^2 m_\mu^5}{192\pi^3} x^2 \{(3-2x)+P\cos\theta(1-2x)\} + O(m_e^2/m_\mu^2) \qquad (9)$$

where x is the dimensionless electron energy, varying in the interval [0,1]

$$x = 2E_e/m_\mu \qquad (10)$$

and θ is the angle between the muon polarization vector and the direction of the electron momentum. Using the non-local S-operator (5), this spectrum becomes [6]

$$N(x,\cos\theta) = \frac{G^2\ m_\mu^5}{32\pi^3} \int_0^x d\sigma[F_1(\sigma)]^2(x-\sigma) \cdot$$

$$\cdot [(1+\sigma-x)-P\cos\theta(\frac{2\sigma}{x}+x-1-\sigma)] + O(m_e^2/m_\mu^2) \qquad (11)$$

where

$$\sigma = (p_e + p_{\nu_e})^2/m_\mu^2 \tag{12}$$

In principle, eq. (11) allows for a determination of the first structure function from muon decay. In practice, however, only the first coefficient of an expansion can be determined, because the structure functions are believed to vary little for $0 \leq \sigma \leq 1$. Hence we expand

$$F_1(\sigma) = 1 + \zeta\sigma + O(\sigma^2) \tag{13}$$

and the spectrum takes the form

$$N(x,\cos\theta) \doteq \frac{G^2 \; m_\mu^5}{192\pi^3} \; x^2\{(3-2x)+\zeta x(2-x)+P\cos\theta\,(1-2x-\zeta x^2)\} \tag{14}$$

It is immediately seen, that the spectrum is no longer of the Michèl type. Therefore, we urge all experimentalists to fit their data with the spectrum (14) rather than the standard Michèl parametrization. In so doing, they lose the power of testing the V-A nature (or, in muon decay, rather the two component neutrino assumption) but they gain the power of testing the structure of weak interactions at small distances. It is our parameter ζ which is most sensitive to this structure.

Next we turn to the "theoretical laboratory" of weak interactions, namely, $\ell\nu_\ell$- scattering. If the structure tensor is depicted by a blob the process can, to first order in G, be depicted as in fig. 1 . A standard

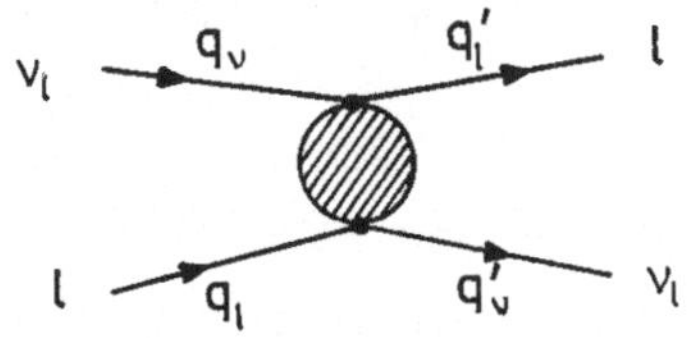

Figure 1 - $\ell\nu_\ell$ -scattering to first order in G .

calculation leads from the S-operator (5) to the sum over spins of the square of the T-matrix element for $\ell\nu_\ell$-scattering

$$\sum_{\text{Spins}} |T_{\ell\nu}|^2 = 32G^2\{[F_1(u)]^2 2(q_\ell q_\nu)(q'_\ell q'_\nu) - \frac{m^2}{M^2} F_1(u)F_2(u)(q_\nu q'_\nu) + \frac{m^4}{M^4}[F_2(u)]^2(q_\ell q'_\nu)(q_\nu q'_\ell)\} \tag{15}$$

where

$$u = (q_\ell - q'_\nu)^2 = m^2 - 2q(E+q\cos\theta) \tag{16}$$

and m is the lepton mass. q is the CMS-momentum and E the lepton energy. It is seen, that even for vanishing lepton mass, i.e. m=0, the coupling constant is effectively altered and becomes a function of u .

$$G_{\ell\nu} = G\,F_1(u) \tag{17}$$

The cross section calculated from eq. (15) is

$$\frac{d\sigma}{d\Omega} = \frac{G^2}{8\pi^2}\frac{1}{(E+q)^2}\{[F_1(u)]^2\, 4q^2(E+q)^2 - \frac{2m^4}{M^2} F_1(u)\, F_2(u)\, q^2(1-\cos\theta) + \frac{m^4}{M^4}[F_2(u)]^2 q^2(E+q\cos\theta)^2\} \tag{18}$$

3. HIGHER ORDER WEAK INTERACTIONS

Just as in section 2, we restrict ourselves to purely leptonic phenomena. The question we pose in this section is how to generalize the S-operator (5) to include higher orders. One possibility would be to simply exponentiate it, similar to the model of reference [7]. However, this possibility is immediately rejected for it introduces an extra non-causality which is <u>not</u> governed by the structure tensor and we do not want to accept this. Let us rewrite the S-operator (5) by a trivial change of integration variables

$$S = I - \frac{iG}{\sqrt{2}}\int d^4x\, d^4y\, \ell_\lambda^\dagger(x+\tfrac{1}{2}y)K^{\lambda\nu}(y)\ell_\nu(x-\tfrac{1}{2}y) + O(G^2) \tag{19}$$

In order not to introduce a new non-causal behaviour by higher orders, we now define a "gothic time ordering operator" $\mathfrak{T}$ by the requirement that it ignores the relative variables y but time-orders the "blobs" with respect to their "center of mass variable" x . Hence the full S-operator becomes

$$S = \mathfrak{T}\exp[-\frac{iG}{\sqrt{2}}\int d^4x\, d^4y\, \ell_\lambda^\dagger(x+\tfrac{1}{2}y)K^{\lambda\nu}(y)\ell_\nu(x-\tfrac{1}{2}y)] \tag{20}$$

The first question that arises is what happens with single particle matrix elements of the S-operator, i.e. to weak self-masses. Such a matrix element is graphically depicted in fig. 2.The dotted line does not represent a causal

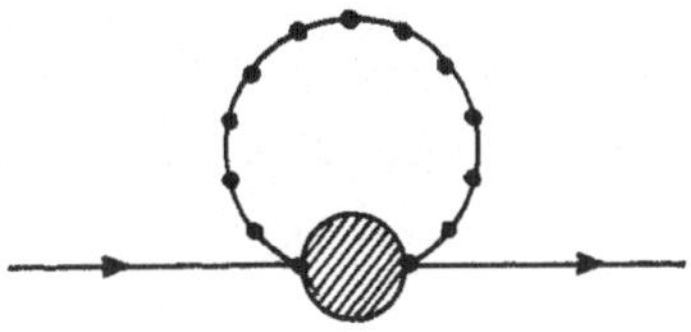

Figure 2 - Weak self -masses .

Feynman-propagator but stands for an S^+ function. The single particle matrix element can then be calculated to be

$$\langle p'|S|p\rangle = \delta^{(3)}(p'-p) - \frac{iG\sqrt{2}}{16\pi m^2 p_o}\,\delta^{(4)}(p'-p)\,\times$$

$$\times \int ds\, \varepsilon(m^2-s)(m^2-s)^2[2F_1(s) - \frac{m^2}{M^2}F_2(s)] \tag{21}$$

If the structure functions are Gaussian, the integral is perfectly finite. An inverse power behaviour would require at least s^{-4} to insure convergence.

Let us now use the "theoretical laboratory" again and calculate $\ell\nu_\ell$-scattering to order G^2. Forgetting about reducible diagrams, there are two contributions. They are shown in fig. 3 . In our non-local model they are of intrinsically different nature. Any Feynman diagram carries

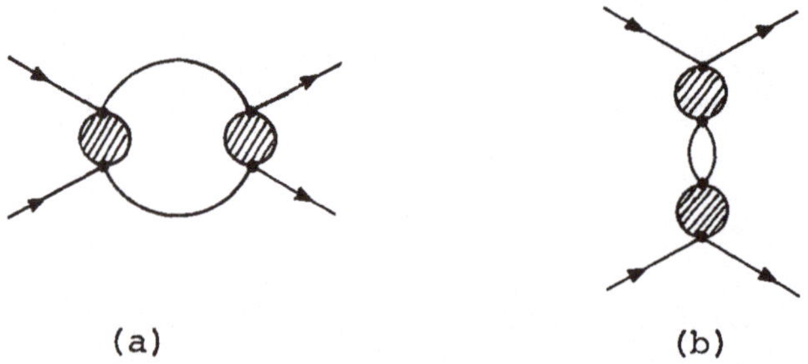

Figure 3 - G^2 contributions to $\ell\nu_\ell$-scattering .

integrations over closed loops. If we follow the particle lines along the loop, we meet "structure blobs" in the case of the graph of fig. 3a, while these blobs are left aside, i.e. do not interrupt the lines of fig. 3b. Therefore, the matrix element corresponding to fig. 3a has the structure functions under the integrations and the result is finite due to the damping caused by the F_i (provided they vanish rapidly enough at infinity). This is not the case for the matrix element corresponding to fig. 3b which reads

$$\langle q'_\ell q'_\nu | T_{(3b)} | q_\ell q_\nu \rangle = -i\ 4G^2\ \bar{u}_\nu(q'_\nu)\gamma_\lambda(1+\gamma_5)u_\ell(q_\ell)\ \cdot$$

$$\cdot\ \bar{u}_\ell(q'_\ell)\gamma_\sigma(1+\gamma_5)u_\nu(q_\nu)\cdot\tilde{K}^{\lambda\nu}(Q)\tilde{K}^{\mu\sigma}(Q)\cdot D_{\mu\nu}(Q) \qquad (22)$$

where $\tilde{K}_{\lambda\nu}$ is the Fourier transform of the structure tensor (6) and

$$Q = q_\nu - q'_\ell \tag{23}$$

The tensor $D_{\mu\nu}(Q)$ is the ordinary Feynman integral for the loop of fig. 3b, i.e.

$$D_{\mu\nu}(Q) = \int \frac{d^4k}{(2\pi)^4} \frac{1}{k^2-m^2} \frac{1}{(k+Q)^2}[2k_\mu k_\nu + Q_\mu k_\nu + Q_\nu k_\mu - (k^2+kQ)g_{\mu\nu}] \tag{24}$$

As in the conventional theory, $D_{\mu\nu}$ diverges badly, as had to be expected. This is the price we pay for keeping the local nature of the current.

A very radical way out would be to recall the early days of quantum electrodynamics and to simply drop "by request" all graphs containing uninterrupted closed loops. However, this would mean to repeat barking up the wrong tree which has once already turned out to be a dead end [8].

The possibility we choose is to contemplate the starting point of our model (see I) and to say that the difficulty stems from the fact that we integrate over energies which are larger than the total energy of the universe, in other words that we again go to distances that are too small to be physically meaningful (cf.I.). In the spirit of our model, we have to turn to a phenomenological description in such a case. Therefore we write

$$D_{\mu\nu}(Q) = \frac{1}{M^2}[g_{\mu\nu}\,\Phi_1(Q^2) - \frac{Q_\mu Q_\nu}{M^2}\,\Phi_2(Q^2)] \tag{25}$$

where Φ_i are again structure functions. If they vanish rapidly enough at infinity, we have succeeded to calculate $\ell\nu_\ell$-scattering to order G^2 without infinities. The price we pay is that we now have four phenomenological structure functions which can, however, be determined by experiment. It is the belief, that they will eventually be calculable from a future theory which will probably combine general relativity into elementary particle physics.

REFERENCES AND FOOTNOTES

1. The notation is that of J. Bjorken and S. Drell, Relativistic Quantum Fields, McGraw-Hill 1965, except for a change of sign in γ_5.
2. H. Pietschmann, Springer Tracts in Modern Phys. 52, 193 (1970). This paper will be referred to as I.
3. T. D. Lee and C. N. Yang, Phys. Rev. 108, 1611 (1957).
4. P. Kristenson and C. Møller, Dan.Mat.Fys.Medd.27/7 (1952).
5. See for example C. Rubbia in "The growth points in physics", Proceedings of the EPS Conference in Florence, April 1969.
6. H. Pietschmann and H. Stremnitzer, to be published.
7. H. Pietschmann and J. Nilsson, Phys. Rev. 142, 1173 (1966).
8. This is why we call it a "radical point of view" because repeating historical mistakes seems to be the defining quality of radicalism.

Acta Physica Austriaca, Suppl. VII, 597—609 (1970)

REALIZATIONS OF LIE ALGEBRAS THROUGH RATIONAL FUNCTIONS OF CANONICAL VARIABLES[*]

BY

H. D. DOEBNER

Institute for Theoretical Physics (I),
University of Marburg, Fed. Rep. Germany

and

Institute for Theoretical Physics
Technical University of Clausthal
Fed. Rep. Germany

AND BY

T. D. PALEV

Institute for Physics, Bulgarian Academy of
Sciences, Sofia, Bulgaria

1. INTRODUCTION

1. The group theoretical approach to relativistic and non-relativistic quantum mechanics is usually based on integrable representations of Lie-algebras G being considered as dynamical algebras, non-invariance algebras or spectrum-generating algebras. Because the generators of G are, in general, physical observables, it is reasonable to assume that they can be expressed as functions of creation and annihilation operators A_i, A_i^* or of momentum and posi-

[*] Seminar given at IX. Internationale Universitätswochen für Kernphysik, Schladming, February 23 - March 7, 1970.

tion operators P_i, Q_i, $i=1,\ldots,n$. If such functions exist for a given algebra G , we call this a canonical realization of G . The two sets of 2n non-commuting variables (being different only if an involution is introduced) are denoted as canonical variables written as $r_-^{2n} = \{a_1,\ldots,a_n, b_1,\ldots,b_n\}$ and with commutation relations (Heisenberg-Lie algebra H_{2n})

$$[a_i,b_j] = \delta_{ij} \qquad\qquad (1)$$
$$[a_i,a_j] = [b_i,b_j] = 0 \quad i,j=1,\ldots,n$$

So the various types of dynamical algebras are directly related to canonical realizations and a closer study of canonical realizations is necessary. For a representation theory of G, the realizations are also useful. If a realization of G is known, an irreducible representation of H_{2n} acting on a representation space $\mathcal{G}$ can be inserted and one has already a representation of G on $\mathcal{G}$. Or one can use a representation space spanned by all polynomials of a_i (or b_i) $i=1,\ldots,n$, in which representations of the realization can easily be found. A further application of this technique is the derivation of equations being invariant under a given group, e.g., infinite- and finite-component field equations of various types.

2. For a construction of canonical realizations, first a function space S_- over V_-^{2n} is needed, with V_-^{2n} being a 2n-dimensional vector space with basis elements $a_1,\ldots,b_n$. Then one has to calculate commutators between different elements of S_- from those of the canonical variables and to look for Lie subalgebras in S_- being isomorphic to G .

Because of the non-commutative variables, the construction of a function-space S_- is difficult. There are two possibilities. One can use purely algebraic methods, without any reference to a representation. Or one already starts with a representation of H_{2n} and has to apply methods from functional analysis, say the spectral theorem or the Fourier transform. The first approach is restricted to function spaces of a special type. However, it turns out that this type is suitable for physical applications and so we use the algebraic approach. It should be mentioned that all the results can also be derived by the second method (operator approach).

For Fermi creation and annihilation operators, written as $r_+^{2n} = \{a_1,\ldots,a_n,b_1,\ldots,b_n\}$, and with anti-commutation relations (Clifford-Jordan algebra Cl_{2n})

$$[a_i,b_j]_+ = \delta_{ij} \qquad (1')$$

$$[a_i,a_j]_+ = [b_i,b_j]_+ = 0 \quad i,j=1,..,n .$$

the same problem can be formulated and canonical realizations in a function space S_+ over V_+^{2n} are of some physical interest. Therefore we treat both problems along parallel lines.

Concerning the Lie algebras G , we consider the most general case, i.e. arbitrary G over a field ϕ, which includes the real and complex case.

3. Some necessary definitions and theorems are collected in Section II. The canonical algebra $F_\mp$ over ϕ , g generated from $r_\mp$ is defined in Section III.1 as the set of all (finite) polynomials in $r_\mp$, containing the commutation and anti-commutation relations as a sort of subsidiary condition. Then we enlarge F_- such that quotients of polynomials will also exist, we embed F_- into the canonical quotient division ring Δ_-, which is impossible for F_+. Commutators between any two second-order polynomials are calculated in Section III.2 using a trace formula (lemma 2). This formula leads to canonical realizations for any G through second-order polynomials

(theorem 2) and using so-called canonical isomorphisms to a realization in terms of rational functions (Section III.3).

4. Most of the results presented here are collected from [1], [2], [3]. For applications we refer to [4], [2] (representations) and to [5] (invariant equations). A more systematic treatment of the operator approach can be found in [6]. A list of some significant references concerning canonical realizations in particle physics and in quantum mechanics is contained in [2]. The realization problem for function spaces with variables taken from para-field operators is discussed in [7], [8]. Extremely useful applications to differential equations via a canonical embedding of differential operators, considered as functions of canonical variables in the operator approach, in a suitably chosen Lie algebra are described in [9].

2. SOME PROPERTIES OF ASSOCIATIVE RINGS R

1. Definitions

a) Let R be an associative ring with unity. An element $r \in R$ is called left (right) zero divisor if there exists $s \in R$, $s \neq 0$ such that $rs=0$ ($sr=0$). A commutative ring without non-zero divisors of zero is an integral domain. R is a division ring if for any $r \neq 0$ an element r^{-1} exists in R with $rr^{-1}=1=r^{-1}r$. If, furthermore, any element of R can be written as left (right) quotient, i.e. $r^{-1}s$ (sr^{-1})r, $s \in R$, then R is called left (right) quotient division ring Δ. A commutative division ring is a field. An isomorphism mapping a ring R_1 into a ring R_2 is called an embedding of R_1 in R_2.

b) The intersection of all (both sided) ideals in R containing a given set K is again an ideal, the so-called ideal J_K generated from K. All elements of J_K can be represented as $\sum_{i\varepsilon I} r_i x_i t_i$, with r_i, $t_i \varepsilon R$, $x_i \varepsilon K$ and with I being a finite index set.

c) R satisfies the common left (right) multiple condition, known also as left (right) Ore-condition, if any pair of non-zero elements r, r' has a common non-zero left (right) multiple $m\varepsilon R$, $m\neq 0$, i.e. two non-zero elements s, s' exist such that sr=s'r'=m (rs=r's'=m).

d) Let $L=\{\ell_1,\ldots,\ell_N\}$ be a set of indeterminates. Then $F(L,\phi)$ denotes the free (associative) algebra over ϕ generated by L. An isomorphic copy of F is a tensor algebra constructed from a vector space V_N over ϕ with basis L

$$F = \phi 1 \oplus V_N \oplus V_N \otimes V_N \oplus \ldots$$

The unity in F is 1 .

2. Embedding of R in a quotient division ring

a) Considered as a function space, $F(L,\phi)$ is spanned by all (finite) polynomials in $\ell_1,\ldots,\ell_N$ (non-commutative variables) with coefficients taken from ϕ . To enlarge F to the space of rational functions it is necessary to define for any non-zero $f\varepsilon F$ an element f^{-1} such that $f^{-1}g$ exists for all $g\varepsilon F$ and $f^{-1}f=1=ff^{-1}$ holds, i.e., F has to be embedded in a quotient division ring. This is not possible in general. The following embedding theorem holds [10]: A given ring R can be embedded in a left quotient division ring Δ if and only if R satisfies the left common multiple condition.

b) We sketch a proof in one direction by constructing Δ , i.e., we introduce quantities of the form $(r^{-1}s)$ and we define addition and multiplication such that the set

$$\Delta(R) = (r^{-1}s) \mid r,s\ \varepsilon R, r\neq 0$$

is a left quotient division ring containing an isomorphic copy of R. We start defining the equality (as equivalence relation) for elements from $\Delta(R)$:

$$(r^{-1}s) = (r_1^{-1}s_1) \quad , \quad r,\ r_1, s, s_1 \varepsilon R$$

if t, $t_1 \varepsilon R$ exist with $tr=t_1r_1$ and $ts=t_1s_1$. From the left common multiple condition we infer the existence of t,t_1, $u,u_1 \varepsilon R$ in the following definition for addition and multiplication:

$$(r^{-1}s) \pm (r_1^{-1}s_1) = ((t.r)^{-1}(t.s \pm t_1.s_1)) \qquad tr=t_1r_1,$$

$$(r^{-1}s)\cdot(r_1^{-1}s_1) = ((u\cdot r)^{-1}(u_1\cdot s_1)) \qquad u\ s=u_1r_1 \quad .$$

One can check that $\Delta(R)$ is indeed a division ring. $(s^{-1}r)$ is the inverse of $(r^{-1}s)$. Zero and unity are given by $(r^{-1}0)$ and (r^{-1},r), $r \neq 0$, respectively. The set $\{(1^{-1},r) | \ | r \varepsilon R\}$ is isomorphic to R. To simplify the notation we write r and r^{-1} instead of $(1^{-1}r)$ and $(r^{-1}1)$, respectively.

3. MAPPING OF LIE ALGEBRAS INTO THE CANONICAL QUOTIENT DIVISION RING

1. The Canonical Quotient Division Ring

a) To define functions of canonical variables r_α, α=+ or α= − , a suitable function space at hand is the free algebra $F_{\mp}$ over ϕ generated by $r_{\mp}$. However, we have to assure the validity of the commutation and anti-commutation relations in F_- and in F_+, respectively. To do this we impose them as a subsidiary condition. Take for example $[a_i,b_j] = \delta_{ij}1$. It implies in F_- that all expressions like $f(a_ib_j-b_ja_i-\delta_{ij}1)g$ with $f,g \varepsilon F_-$ vanish, i.e., elements in F_- are compared modulo $f(a_ib_j-b_ja_i-\delta_{ij}1)g$. For a rigorous formulation consider the sets K_- and K_+

$$K_{\mp} = \{a_i b_j \mp b_j a_i - \delta_{ij} 1,\ a_i a_j \mp a_j a_i,\ b_i b_j \mp b_j b_i\ ,$$
$$i,j=1,\ldots,n\}$$

Then the factor space

$$E_\alpha = F_\alpha / J_\alpha$$

with J_α being the ideal generated from K_α, possesses the necessary algebraic structure and we can say (although improperly) that (1) and (1') are valid in E_- and in E_+, respectively. E_α is referred to as canonical associative algebra or as associative algebra generated from canonical variables. E_- and E_+ are isomorphic to the quasienveloping algebra of H_{2n} and to the enveloping algebra of Cl_{2n}.

b) E_α can be embedded in a quotient division ring Δ_α if the left common multiple condition is valid. This is not the case for ε_+. Suppose Δ_+ exists, take $a_1 \neq 0$ and calculate $a_1^{-1} \cdot (a_1 \cdot a_1) = 0 = (a_1^{-1} \cdot a_1) \cdot a_1$ (Δ_+ is associative) then $a_1 = 0$ holds in contradiction to $a_1 \neq 0$, because all elements of the generating set r_+ are already non-zero divisors of zero $a_i^2 = b_i^2 = 0$. The dimension of E_+ considered as a vector space is finite.

c) The situation for E_- is rather different. Here the canonical quotient division ring exists. We formulate this as:

Lemma_1: The associate algebra E_- generated by canonical variables of the Bose type defined over a field ϕ satisfies the common left multiple condition and can be embedded into a left quotient division ring Δ_- .

For a proof we refer to [11] or to [2]. Hence, the space of rational functions with variables from r_- is a well-defined algebra. The vector space E_- is infinite dimensional.

2. Mapping of Lie Algebras into Second-Order Polynomials of Canonical Variables

a) Now we are prepared to construct canonical realizations for any G over ϕ. Because Δ_- and E_+ can be considered both as Lie algebras with a Lie bracket $[f,g]=fg-gf$, $f,g\in\Delta_-$ or $f,g\in\varepsilon_+$ any realization of G, in Δ_-, E_+ corresponds to an isomorphism Θ_-, Θ_+ mapping G into Δ_-, E_+

$$G \xrightarrow[\Theta_-]{} \Delta_- \qquad G \xrightarrow[\Theta_+]{} E_+$$

and any such Θ_α produces a canonical realization. The isomorphisms exist if and only if Δ_-, E_+ contains a Lie subalgebra isomorphic to G .

b) For a first simple construction of Θ_α we remember that all second-order polynomials $E_\alpha^{(2)} \subset E_\alpha$ already form a Lie algebra. It is convenient to introduce in $E_-^{(2)}$ a completely symmetrized basis [12] and in $E_+^{(2)}$ a completely antisymmetrized one, i.e.,

$$E_{\mp}^{(2)} = f_{\mp} | f_{\mp} = \frac{1}{2} \sum_{i,j=1}^{n} [(a_i b_j \pm b_j a_i) Z_{ij} \mp \frac{1}{2}(a_i a_j \pm a_j a_i) S_{ij}^{\mp} + \\ + \frac{1}{2}(b_i b_j \pm b_j b_i) T_{ij}^{\mp}]$$

with Z, $S^{\mp}$, $T^{\mp}$ being $n\times n$ matrices over ϕ and with $S_{ij}^{\mp}=\pm S_{ji}^{\mp}$, $T_{ij}^{\mp}=\pm T_{ji}^{\mp}$. With a pair of n-component vectors $A=(a_1,..,a_n)$, $B=(b_1,..,b_n)$ and a further 2n-dimensional pair $\psi_{\mp}=(B,\mp A)$, $\phi_{\mp}=(A,B)^t$ (the index t means transposed) any f_α can be written as

$$f_\alpha = \frac{1}{2}\, \psi_\alpha\, C_\alpha^{2n}\, \phi_\alpha$$

with C_α^{2n} being a $2n\times 2n$ matrix from the set

$$C_\alpha^{2n} = \left\{ C_\alpha^{2n} \middle| C_\alpha^{2n} = \begin{pmatrix} Z & T^\alpha \\ S^\alpha & -Z^t \end{pmatrix} ;\; S_{ij}^{\mp} = \pm S_{ji}^{\mp},\; T_{ij}^{\mp} = \pm T_{ji}^{\mp}, Z_{ij} \right\}$$

c) The calculation of commutators in $E_\alpha^{(2)}$ is made using:

<u>Lemma 2</u>: Let X_α, Y_α be two matrices from C_α^{2n}. Let ψ_α, Φ_α be 2n-component vectors defined as above. Then the trace formula

$$[\tfrac{1}{2}\psi_\alpha X_\alpha \Phi_\alpha, \tfrac{1}{2}\psi_\alpha Y_\alpha \Phi_\alpha] = \tfrac{1}{2}\psi_\alpha [X_\alpha, Y_\alpha]\, \Phi_\alpha \qquad (2)$$

holds in $E_\alpha^{(2)}$. For a proof we refer to [1].

Consider now a d-dimensional faithful matrix representation $D_d(G)$ of G, which exists by the Ado-Iwasawa theorem (d finite). If it is possible to put all the matrices $D_d(g_k)$ k=1,...,m, with $g_1,\ldots,g_m$ being a basis of G, in the form C_α^{2n}, d=2n, then the mapping Θ defined via

$$g_k \xrightarrow[\Theta_\alpha]{} \tfrac{1}{2}\psi_\alpha D_d(g_k)\Phi_\alpha \ , \quad D_d(g_k) \in C_\alpha^{2n} \ , \quad d=2n,\ k=1,\ldots,m \qquad (3)$$

is an isomorphic of G into the set $E_\alpha^{(2)}$. It can even be shown that all possible isomorphisms can be constructed with formula (2).

The Lie algebra $E_\alpha^{(2)}$ is isomorphic to the Lie algebra formed by C_α^{2n}. The identification of $E_-^{(2)}$ as sp(2n,C) or sp(2n,R) is obvious if the field of complex or real numbers is inserted for ϕ. By putting $S^\alpha=T^\alpha=0$ we have, furthermore, that $g\ell(n,C)$ (or $g\ell(n,R)$) is contained in $E_-^{(2)}$ and in $E_+^{(2)}$.

d) We are left with the problem of constructing finite-dimensional representations $D_d(G)$ of the form C_α^{2n}. Because of $D_d(G) \subset g\ell(d,\phi) \subset E_\alpha^{(2)}$ it is easy to do this. Hence the mapping

$$g_k \xrightarrow[\Theta_\alpha]{} \frac{1}{2}\,\psi_\alpha \begin{pmatrix} D_d(g_k) & 0 \\ 0 & -D_d(g_k)^t \end{pmatrix} \Phi_\alpha$$

exists and is a mapping into the set of bilinear polynomials in a_i, b_j $(S^\alpha = T^\alpha = 0)$. A complete discussion of all isomorphisms can be found in [2].

We collect the result in

Theorem 1: Let G be any Lie algebra over ϕ and let d' be the dimension of the smallest representation space carrying a faithful representation of G. Then isomorphisms Θ exist mapping G into $E_\alpha^{(2)}$, i.e., into the set of second-order polynomials of 2n canonical variables of Bose or Fermi type, for some n with $2n \geq d'$.

There are Lie algebras with isomorphisms for 2n=d, e.g. $sp(2n,\phi)$.

3. Mapping of Lie Algebras into Higher-Order Polynomials and into the Canonical Quotient Division Ring

a) A construction of more complicated realizations is interesting only for Δ_- because E_+ is finite dimensional and contains only polynomials linear in all canonical variables. We use, following [3], so-called canonical isomorphisms τ, defined as mappings of H_{2n} into Δ_-. τ is supposed to be different from the trivial isomorphisms. Consider $\tilde{a}_i, \tilde{b}_i$ obtained via $\tilde{a}_i = \tau a_i$, $\tilde{b}_i = \tau b_i$ and apply formula (2) with vectors $\tilde{\psi} = (\tilde{B}, -\tilde{A})$, $\tilde{\Phi} = (\tilde{A}, \tilde{B})$, $\tilde{A} = (\tilde{a}_1, \ldots, \tilde{a}_n)$, $\tilde{B} = (\tilde{b}_1, \ldots, \tilde{b}_n)$. Then the mapping $\Theta\tau = \rho$

$$g_k \xrightarrow[\rho]{} \frac{1}{2} \tilde{\psi} D_d(g_k) \tilde{\Phi} \in \Delta_-, \quad D_d(g_k) \in C_-^{2n},$$

$$k = 1, \ldots, m$$

is again an isomorphism sending G into higher-order polynomials or into rational functions.

b) A class of this isomorphism is available from:

Theorem 2: Let h be any non-zero element in Δ_- and let $f_i(b_i) \in \Delta_-$ be a function of b_i only. Then the following mappings of type τ_1, τ_2 are canonical isomorphisms:

$$\text{Type } \tau_1: \quad \tau_1 a_i = h\, a_i\, h^{-1} \;, \quad \tau_1 b_i = h\, b_i h^{-1} \;, \quad i=1,..,n$$

$$\text{Type } \tau_2: \quad \tau_2 a_i = ([a_i, f_i(b_i)])^{-1} \cdot a_i \;, \quad \tau_2 b_i = f_i(b_i) \quad i=1,..,n \;.$$

Let $K_{1,2}$ be the set of all isomorphisms of type τ_1, τ_2, then the intersection of K_1 and K_2 contains only the trivial isomorphism.

The theorem guarantees canonical realizations for any G such that at least some of the generators are quotients of (finite) polynomials and opens the possibility to treat a large class of Hamiltonians and differential operators. Isomorphisms of type τ_1 are related to unitary transformations if the operator approach is used.

4. CONCLUDING REMARKS

1. We have shown how functions of canonical variables, i.e. momentum- and position- or creation- and annihilation operators, can be used to realize the generators of any Lie algebra G by purely algebraic methods, such that the canonical commutation relations will imply the commutation relation for the Lie algebra G in question. The largest function space available for us was the space of rational functions, i.e. quotients of (finite) polynomials. However, in this space we construct only some of the canonical realizations. A complete list of all canonical isomorphisms for Δ_- is not known. There are only some results for E_-, n=1 .

2. To enlarge the function space further, one has to equip Δ_- or E_- with a suitable topology and to complete Δ_- or E_- in respect of this topology. This is possible for E_- and so, presumably, square roots, trigonometric functions, etc., can be treated.

3. For an application to differential equations or to spectrum-generating algebras, i.e. with a given Hamiltonian as one of their generators, one has to insert for $a_i=P_i, b_i=Q_i$ in a next step the irreducible unitary representation $D_1(H_{2n})$ of H_{2n} (unique up to unitary equivalence) in $\mathcal{G}$. Then a representation of G in $\mathcal{G}$ is obtained. Because the generators of G are assumed to be physical observables, only such canonical isomorphisms are physically admissible which map the generators into a set of operators in $\mathcal{G}$ being essentially self-adjoint on a common dense invariant domain θ. If, furthermore, the Nelson operator of G is also essentially self-adjoint on θ the representation can be integrated to a representation of the corresponding group, which is of some physical relevance. The integration problem can be solved for realizations in $E_-^{(2)}$. The sp(2n,R) representation obtained from (3) with $D_1(H_{2n})$ is integrable and hence also the representation of its subalgebras. For a treatment of differential equations, nonintegrable representations are in general also useful.

REFERENCES

1. H. D. Doebner, T. D. Palev, "The trace formula for canonical realizations of Lie algebras", preprint, University of Marburg, 1970.
2. H. D. Doebner, T. D. Palev, "On canonical realizations of Lie algebras" (in preparation).

3. H. D. Doebner, T. D. Palev, "On Lie subalgebras and isomorphisms of the canonical quotient division ring" (in preparation).
4. T. D. Palev, ICTP Trieste, preprint IC/68/23 (1968).
5. T. D. Palev, Nuovo Cim. 62, 585 (1969).
6. H. D. Doebner, B. Pirrung, "On the construction of spectrum-generating algebras by embedding methods", preprint, University of Marburg, 1970.
7. K. Kademova, T. D. Palev, "Second-order realizations of Lie algebras with parafield operators", preprint, Bulgarian Academy of Sciences, 1970.
8. K. Kademova, ICTP Trieste, preprint, IC/69/108 (1969) (to appear in Nuclear Physics).
9. W. Miller, "Lie Theory and Special Functions" (Academic Press, New York 1968).
W. Miller, "On Lie Theory and some Special Functions in Mathematical Physics" (American Mathematical Society, Providence, R.I. 1964).
10. N. Jacobson, "Lecture on Abstract Algebra, I" (Van Nostrand, Princeton 1951).
N. Jacobson, "Structure of Rings" (American Mathematical Society, Providence, R.I. 1964).
11. I. M. Gel'fand, A. A. Kirillov, "On the division rings connected with enveloping algebras of Lie algebras", preprint, Steklov Institute of Mathematics, Moscow (1965).
12. S. Helgason, "Differential Geometry and Symmetric Spaces" (Academic Press, New York 1962).

Acta Physica Austriaca, Suppl. VII, 610–619 (1970)

MASS OPERATOR IN THE SU(3)-SYMMETRIC STRONG COUPLING THEORY WITH PSEUDOSCALAR MESONS*

BY

J. TOLAR**
Institut für Theoretische Physik der Universität
Marburg, BRD

1. INTRODUCTION

The strong coupling approximation scheme has been used in the quantum field theory almost only for the meson field interacting with a static fermion source. The first works in this direction considered the πN system [1]. At the end of the 40's the interest decreased, but later the excited isobaric states of the nucleon predicted by the strong coupling theory were found experimentally as nucleon resonances. New interest arose after the discovery of the SU(3) symmetry of strongly interacting particles [2] with the aim to connect the properties of different SU(3)-multiplets by a dynamical theory.

The subject of this lecture is a consistent formulation of the strong coupling approximation scheme for the meson-baryon system exhibiting SU(3) symmetry along the lines suggested by Jahn [3] for the SU(2)-symmetric system. First it will be shown how the Hamiltonian de-

---

* Seminar given at IX. Internationale Universitätswochen für Kernphysik, Schladming, February 23 - March 7,1970.

** On leave from the Faculty of Nuclear Science, Czech Technical University, Prague

scribing an octet of pseudoscalar mesons (π, η, $K,\bar{K}$) coupled to a bare baryon source (N,Σ,Λ,Ξ) is transformed to a form allowing perturbation theory in the inverse powers of the coupling constant. Then the basic properties of the lowest order approximation will be considered: the wave functions and the energies of the isobaric states (i.e. the mass formula).

2. TRANSFORMATION OF THE HAMILTONIAN

In the usual weak coupling perturbation approach the zeroth order approximation is given by the Hamiltonian of the free fields, the energies and wave functions being expanded around g=0. In the strong coupling approximation one would like to make the perturbation expansion around $g \to \infty$; the first step is to fix the starting point, i.e. a suitable splitting of the Hamiltonian in two parts

$$H = H^{O} + H^{W}$$

so that the perturbation series in H^{W} represents an expansion in the powers of 1/g. It is not certain a priori that such a splitting exists.

Let us formulate the strong coupling condition [3].

Let

$$H^{O}\theta_n = E_n^{O}\,\theta_n\,, \quad H\chi_n = E_n\,\chi_n$$

The perturbed wave function is expanded according to

$$\chi_n = \sum_{n'} f_n(n')\theta_{n'}$$

where the expansion coefficients are

$$f_n(n') = \delta_{nn'} + \frac{(n'|H^{W}|n)}{E_n - E^{O}_{n'}} + \dots$$

Then the strong coupling condition is

$$\lim_{g\to\infty} f_n(n') = \delta_{nn'}$$

For the model with one fixed baryon source (non-relativistic, without baryon pair creation) the SU(3)-symmetric Hamiltonian has the form

$$H = H_{mes}(q_{i\vec{k}},\ p_{i\vec{k}}) + H_{int}$$

where

$$H_{mes}(q_{i\vec{k}},p_{i\vec{k}}) = \frac{1}{2}\int (p^+_{i\vec{k}}p_{i\vec{k}} + \varepsilon^2 q^+_{i\vec{k}}q_{i\vec{k}})\, d^3k\ I\ ,$$

$$H_{int} = \frac{g}{2\pi\mu}\ \Lambda_i \otimes \sigma_a \int v_a(k)\ q_{i\vec{k}}\ d^3k\ .$$

Here the constant baryon mass term has been neglected; the mesons of mass μ are treated relativistically but with a cut-off function $v_a(k) = -ik_a v(k)$,

$$v(k) = \int K(r)\ e^{i\vec{k}\vec{r}}\ d^3r \qquad ,\ v(0) = 1\ ,$$

$K(r)$ describing the extension of the baryon source. Fourier transforms $q_{i\vec{k}}$, $p_{i\vec{k}}$ of the meson field operators $\phi_i(\vec{r})$, $\pi_i(\vec{r})$, $i=1,2,\ldots,8$ satisfy the commutation relations

$$[q_{i\vec{k}},\ p_{j\vec{k}'}] = i\ \delta_{ij}\ \delta^3(\vec{k}-\vec{k}')$$

and further

$$q^+_{i\vec{k}} = q_{i,-\vec{k}}\ ,\ p_{i\vec{k}} = p_{i,-\vec{k}} \qquad ;\ \varepsilon^2 = k^2+\mu^2.$$

The interaction Hamiltonian is a 16×16 matrix acting in the space of bare baryon states with

$$\Lambda_i = \alpha F_i + (1-\alpha)D_i$$

describing antisymmetric and symmetric couplings of SU(3) octets (α is the mixing parameter); I is the 16-dimensional unit matrix and the summation over repeated indices a=1,2,3 and i=1,2,...,8 is understood.

The splitting of the Hamiltonian is attained by a suitable splitting of the meson field operators in "bound" and "free" parts (denoted by superscripts b and f, respectively), the bare baryon and the bound mesons forming a "compound baryon" (superscript c). The conditions to be satisfied by this splitting are:

(1) The splitting of the integrals of motion for all values of g

$$G_i \equiv F_i \otimes \sigma_o + f_{ij\ell}\int q_{j\vec{k}}\, p_{\ell\vec{k}}\, d^3k\, I =$$

$$= F_i \otimes \sigma_o + F_i^b\, I + G_i^f\, I \quad ,$$

$$J_a \equiv I_8 \otimes \frac{\sigma_a}{2} - \varepsilon_{abc}\int p_{i\vec{k}} k_b \frac{\partial}{\partial k_c} q_{i\vec{k}}\, d^3k\, I =$$

$$= I_8 \otimes \frac{\sigma_a}{2} + L_a^b\, I + J_a^f\, I \quad ,$$

where σ_o and I_8 are 2- and 8-dimensional unit matrices.

(2) The operators of bound and free mesons commute with one another so that they correspond to independent subsystems;

(3) The splitting transformation is invertible;

(4) The strong coupling condition for $H = (H^{oc}+H^{of})+H^W$.

The conditions (1) - (4) can be satisfied by a rather complicated splitting of the meson field. First split

$$q_{i\vec{k}} = u_{a\vec{k}}\, \overset{o}{q}_{ai} + \overset{1}{q}_{i\vec{k}} \quad ,$$

$$p_{i\vec{k}} = v_{a\vec{k}}\, \overset{o}{p}_{ai} + \overset{1}{p}_{i\vec{k}} \quad ,$$

where

$$\int u_{a\vec{k}}\, v_{b\vec{k}}\, d^3k = \delta_{ab}$$

so that

$$[\overset{o}{q}_{ai}, \overset{o}{p}_{bj}] = i\, \delta_{ab}\, \delta_{ij} \quad .$$

Now $\overset{o}{q}_{ai}$ transforms as the adjoint representation ($\underline{8}$,J=1) of $SU(3) \times SU(2)_J$. The strong coupling condition (4) can be satisfied [4] (or the strong coupling limit exists) if $\overset{o}{q}_{ai}$ is constrained to a certain $SU(3) \times SU(2)$ - orbit in $\underline{q}$-space, namely of the type $SU(3) \times SU(2)_J \,/\, SU(2)_V \times U(1)_Y$, where $v=\vec{I}+\vec{J}$, Y is the hypercharge.There is a one-parameter family of these orbits Ω

$$\overset{o}{q}_{ai} = q\, S_{ai}$$

where S_{ai} depends on 7 internal coordinates in the orbit. The splitting of momenta

$$\overset{o}{p}_{ai} = \overset{o}{p}{}^{(int)}_{ai} + \overset{o}{p}{}^{(rad)}_{ai}$$

with respect to Ω will be explained in Sect.4 .

Our final splitting is therefore

$$q_{i\vec{k}} = u_{a\vec{k}}\, S_{ai} + q'_{i\vec{k}} \;, \qquad p_{i\vec{k}} = v_{a\vec{k}}\, \overset{o}{p}{}^{(int)}_{ai} + p'_{i\vec{k}} \;,$$

where $q'_{i\vec{k}}$, $p'_{i\vec{k}}$ are connected with the free meson field operators $a^{+}_{i_o\vec{k}_o}$, $a_{i_o\vec{k}_o}$. Further define as usual

$$H^{of} = \int \varepsilon_o \langle a^{+}_{i_o\vec{k}_o}\, a_{i_o\vec{k}_o} \rangle\, d^3k_o\, I \;,$$

the brackets denoting symmetrization. The compound-baryon Hamiltonian defined as

$$H^{oc} \equiv H(q^{b}_{i\vec{k}}=u_{a\vec{k}}\, S_{ai},\; p^{b}_{i\vec{k}}=v_{a\vec{k}}\, p^{(int)}_{ai}) = T + V =$$

$$= \frac{1}{2} A\, p^{(int)}_{ai} p^{(int)}_{ai}\, I + [\, \frac{1}{2} C\, S_{ai} S_{ai}\, I +$$

$$+ \frac{gD}{2\pi\mu}\, \Lambda_i \otimes \sigma_a\, S_{ai}\,]$$

is an operator acting only on functions defined on Ω with $S_{ai}S_{ai} = 3$,

$$A = \frac{1}{3}\int v^{*}_{a\vec{k}}\, v_{a\vec{k}}\, d^3k \;, \qquad C = \frac{1}{3}\int \varepsilon^2 u^{*}_{a\vec{k}}\, u_{a\vec{k}}\, d^3k \;,$$

$$D = \frac{1}{3}\int v_a(k)\, u_{a\vec{k}}\, d^3k$$

(note that $\overset{o}{p}{}_{ai}^{(int)} \neq p_{ai}^{(int)}$, for details see [3], [4]).

Let us look more closely at the strong coupling condition; the transformed Hamiltonian takes the form

$$H = (H^{oc}+H^{of}) + \{[H_{mes}(q'_{i\vec{k}},p'_{i\vec{k}})-H^{of}]+\int[-\varepsilon^2 u_{a\vec{k}} S_{ai}\, I + \frac{g}{2\pi\mu}\Lambda_i \otimes \sigma_a v_a(k)]d^3k - \langle \overset{o}{p}{}_{ai}^{(int)} \int v_{a\vec{k}} p'_{i\vec{k}} d^3k\rangle\, I + \frac{1}{2}A(\overset{o}{p}{}_{ai}^{(int)}\overset{o}{p}{}_{ai}^{(int)} - p_{ai}^{(int)}p_{ai}^{(int)})\, I\} \quad .$$

The compensation of terms of order g^1 and g^o in $H^W = \{\dots\}$ leads to the condition that

$$u_{a\vec{k}} = -\frac{1}{2}\frac{g}{2\pi\mu}\frac{v_a(k)}{\varepsilon^2}$$

is proportional to g so that $v_{a\vec{k}} \sim g^{-1}$, and another condition

$$H_{mes}(q'_{i\vec{k}},p'_{i\vec{k}})-H^{of} = \frac{1}{2d}\sum_{bj}(\Lambda_{bj,ai}\int v_{a\vec{k}} q'_{i\vec{k}} d^3k)^2 \quad ,$$

where

$$d = \frac{1}{3}\int[\frac{k}{\varepsilon}v(k)]^2 d^3k \qquad \text{and } \Lambda_{bj,ai}$$

is defined in Sect. 4

3. THE DYNAMICAL GROUP AND EIGENSTATES OF V

The first task is the determination of the eigenstates of H^{oc} for $g\to\infty$. In this limit the potential term V (proportional to g^2) prevails in H^{oc}. It can be diagonalized with the lowest eigenvalue $-\frac{3}{2}(\frac{g}{4\pi\mu})^2 d$ belonging to the eigenvector labelled by the quantum numbers v=0 and Y=1 of the stability group $SU(2)_v \times U(1)_Y$ of Ω . (This holds for $0<\alpha<0.725$; for other values of α one gets v=0,Y=-1 and $v=\frac{1}{2}$,Y=0, but they do not lead to results conform to reality [2] .)

The quantum numbers v and Y specify a unitary irreducible representation of the dynamical group $T_{24}\otimes[SU(3)\times SU(2)_J]$ defined as the symmetry group of V. Its generators are S_{ai}, $G_i^c = F_i^b I + F_i \otimes\sigma_o$ and $J_a^c = L_a^b I + I_8 \otimes \frac{\sigma_a}{2}$. The unitary irreducible representation of the dynamical group labelled by v=0, Y=1 contains an infinite series of SU(3)× SU(2)-multiplets beginning with $[\underline{8},\frac{1}{2}]$, $[\underline{10},\frac{3}{2}]$, $[10^*,\frac{1}{2}]$, $[27,\frac{1}{2}]$ etc; the first two multiplets have been observed experimentally.

The normalized eigenstates of H^{oc} for $g\to\infty$ have the form [2],[4]

$$|\Theta_\rho> \equiv |\lambda\ J,\nu\ M, I_o>_{(v,Y_o)} =$$

$$= \sqrt{N_\lambda(2J+1)} \int_\Omega d\underline{q} \sum_{I_{3o},M_o} (I_o I_{3o} JM_o|v\ v_3) D^{(\lambda)*}_{\nu_o\nu} D^{(J)}_{M_oM} |\underline{q}> \quad ,$$

where $\rho\equiv[\lambda,J]$, $\nu\equiv(II_3Y)$, N_λ = dimension of the SU(3) representation λ.

4. THE MASS OPERATOR

The eigenstates of V are infinitely degenerate. Their splitting to the order g^{-2} is due to the kinetic term $T\sim g^{-2}$ in H^{oc}; H^W brings about only higher order corrections. We consider now the energies of isobars $|\Theta_\rho>$ taking T as first order perturbation [4].

The orbit Ω can be looked upon as embedded in the Cartesian space of coordinates q_{ai}. In this space the kinetic energy is given by

$$\frac{1}{2}\Lambda\, p_{ai}\, p_{ai} \;, \quad (p_{ai} = -i\frac{\partial}{\partial S_{ai}}) \quad ,$$

but we are interested in the part of it confined to Ω.

In the work in collaboration with L. Kanthack [5] it is shown that the splitting of momenta (or kinetic energy) into internal and radial parts with respect to a given hypersurface in $\underline{q}$-space can be easily achieved by means of a perpendicular projection matrix Λ_{AB} ($A\equiv ai$, $B\equiv bj$) projecting on the tangent linear space in each point of the hypersurface:

$$p_A^{(int)} = \Lambda_{AB}\, p_B \quad , \quad p_A^{(rad)} = (\delta_{AB}-\Lambda_{AB})p_B \ .$$

The basic properties of Λ_{AB} are:

$$\Lambda_{AB}\Lambda_{BC} = \Lambda_{AC} \quad , \quad \Lambda_{AB} = \Lambda_{BA} \quad , \quad \mathrm{Tr}\Lambda = \text{dimension of the hypersurface} \ .$$

Then

$$\frac{1}{2}\, A\, p_A\, p_A = \frac{1}{2}\, \Lambda\, p_A\, \Lambda_{AB}\, p_B + \frac{1}{2}\, A\, p_A (\delta_{AB}-\Lambda_{AB}) p_B$$

where the first term is T and can be shown to be expressible in terms of the second order Casimir operators of $SU(3)\times SU(2)$:

$$T = \frac{1}{2}\, A\, p_A^{(int)} p_A^{(int)} I = -\frac{1}{2}\, A(\frac{4}{3}F^{b2} - \frac{5}{6}L^{b2}) I \ .$$

This leads to a change of energy of isobars $|\Theta_\rho\rangle$ proportional to g^{-2}

$$\Delta E_\rho = \langle\Theta_\rho|(-\frac{A}{2})(\frac{4}{3}\,F^{b2} - \frac{5}{6}\,L^{b2}) I|\Theta_\rho\rangle$$

Using a unitary transformation diagonalizing V [6] we get ΔE_ρ expressed in terms of compound-baryon operators

$$\Delta E_\rho = \langle\Theta_\rho|(-\frac{A}{2})(\frac{4}{3}\,G^{c2} - \frac{5}{6}\,J^{c2})|\Theta_\rho\rangle + \text{const.}$$

Hence the mass operator to the order g^{-2} is given by the expression

$$-\frac{1}{2}\, A(\frac{4}{3}\, G^{c2} - \frac{5}{6}\, J^{c2}) + \text{const} \ .$$

with

$$G^{c2}|\Theta_\rho\rangle = [\frac{1}{3}(p^2+pq+q^2)+p+q]|\Theta_\rho\rangle \quad ,$$

$$J^{c2}|\Theta_\rho\rangle = J(J+1)|\Theta_\rho\rangle \quad ,$$

where $\rho\equiv[(p,q),J]$.

<u>Summarizing</u>: the lowest order approximation is given by

$$H^{o} = H^{oc} + H^{of}$$

with the eigenstates

$$\theta^{\vec{k}_{o1} \dots , \vec{k}_{on}}_{i_{o1}, \dots , i_{on}} , \rho$$

where

$$a_{i_o \vec{k}_o} \theta_{o\rho} = 0 \quad , \quad \theta_{o\rho} \equiv \theta_{\rho} \; ,$$

and n is the number of free mesons present in this state.

ACKNOWLEDGMENTS

The author is greatly indebted to Professor H. D. Doebner who stimulated this research. He is thankful to Professor H. Jahn for many discussions and comments regarding this work.

His thanks belong also to the Alexander von Humboldt Foundation under the support of which most of this work was carried out.

REFERENCES

1. The early works are summarized in G. Wentzel, Rev. Mod. Phys. 19, 1 (1947).
2. T. Cook, C. J. Goebel and B. Sakita, Phys. Rev. Letters 15, 35 (1965); Y. Dothan and Y. Ne'eman, Proc. 2nd Topical Conf. on resonant particles, Athens, Ohio, 1965; T. Cook and B. Sakita, J. Math. Phys. 8, 708 (1967); C. Dullemond and F. J. M. von der Linden, Ann. of Phys. 41, 372 (1967); M. Bednar and J. Tolar, Nucl. Phys. B5, 255 (1968) .

3. H. Jahn, Nucl. Phys. 26, 353 (1961); "Die Näherungen der starken Kopplung in der Mesonentheorie", Oberwolfach (1959), mimeographed lecture notes; see also: Topical Conf. on dynamical groups and infinite multiplets, ICTP report IC/69/54, Miramare - Trieste, 1969.
4. J. Tolar, to be published
5. L. Kanthack and J. Tolar, to be published
6. P. Exner and J. Tolar, Czech J. Phys. B19 (1969).

Acta Physica Austriaca, Suppl. VII, 620–628 (1970)

SUMMARY - FIRST WEEK[*]

BY

D. ATKINSON
CERN - Geneva

It is my pleasure to attempt an impossible task, that of providing a coherent summary of the first week's lectures at this, the ninth, and my first Schladming school.

I would like first to commend most highly the organizers of the course, and the noble authorities of the town of Schladming, who have appreciated the fact that the rarefied thought-processes of a theoretical physicist proceed best at an altitude above 1000 metres, and at a relative velocity of around 80 km per hour with respect to the snow.

Let me start with Salam's masterly treatment of non-polynomial Lagrangian theories. The idea is to try to treat, for example, Lagrangians that are rational fractions, rather than polynomials in the fields. The hope is to find such Lagrangians that are (a) physically interesting, and (b) which either have no ultraviolet divergences, or which can be renormalized in the usual way by a finite number of counter terms.

The basic method is to expand the denominator in the Lagrangian as a power series, and get expressions for

---

[*] Summary given at IX. Internationale Universitätswochen für Kernphysik, Schladming, February 23 - March 7,1970.

expectation values as usual (except that they are horribly divergent), and then to resum the divergent series of divergent terms by the Borel method. This is done in the unphysical Symanzik region, and 3+1-dimensional integrals are converted into Euclidean ones by a Wick rotation. At the end, the momentum space quantities have to be analytically continued out of the Symanzik region, and it appears that they can be made to have the correct cut structure. The unfortunate thing is that the continued matrix elements turn out to have an exponentially exploding behaviour as $p^2 \to \infty$. It was suggested that a possible way out of the impasse is to pretend that one was calculating K matrix elements all along (in the Symanzik region), and then when you construct the T matrix from this, you have one exponential upstairs, and one downstairs, and everyone should be happy.

It seems to me that the method holds great promise. I don't think one needs to worry about trying to provide an a priori justification for the resummation technique, nor for the Wick rotation. This can be done a posteriori, by showing, if one can, that the final T-matrix has the correct analyticity, decent asymptotics, etc. In fact, since the Borel resummation is but one of an infinite number of possible ways to make sense of a divergent series, it is tempting to think that someone might be able to construct a resummation technique which fits a particular non-linear Lagrangian, as a glove fits a hand, and gives a glowingly healthy S-matrix. Of course, quite a different problem is that the connection between the quantities that appear in the S-matrix and those that appear in the Lagrangian must surely be much more tenuous than in a normal theory.

These ruminations apply directly to the first order in the major coupling constant, and I refer you to Salam's notes for his discussion of what happens in higher orders, where we meet super-graphs, Salam rules for writing them down, and starfish diagrams to enrol in the field-theoretical bestiary.

We had a talk by Kummer on certain discrepancies between the perturbation treatment of generalized Ward-identities, and that provided by more general field-theoretical considerations. He maintained that the discrepancies can be traced to ill-defined differences of divergent quantities, and he persuaded us all that there exist values, κ, such that $\kappa - \kappa \neq 0$. This was done by introducing tadpoles with more than one tail, a mutant form which can be traced to the action of foreign operators that do not respect the normal order.

Next, let me sketch the outline of Rühl's lectures. We first looked at the representations of SL(2,C), the group that covers O(3,1) twice. Rühl concentrated on the principal series, defining a suitable L^2 Hilbert space of homogeneous functions in which to discuss the representation theory. A specific problem of physical interest is the reduction of a current density between single-particle states into an integral of irreducible covariant operators. Asymptotic expansions of form factors were obtained; and the matrix elements of current commutators were examined in two ways: first, with fixed photon mass, and, secondly, as the photon mass tends to infinity.

I come now to Landshoff's lectures on perturbation theory-inspired ideas about Regge cuts. We were reminded about the Amati-Fubini-Stanghellini diagram, shown where its angular-momentum branch-point would be, and then instructed how to wash away the discontinuity across the cut by making certain assumptions about the analytic and asymptotic behaviour of the particle-particle-Reggeon coupling functions. In the case of the Mandelstam double-cross, which has also been described as the picnic-table diagram, the AFS cut really is there, because the coupling functions have both left-hand and right-hand cuts, and the simple closure of the contour of the loop-integration can no longer be done.

We were then presented with thoughts, along the same lines, about the absorption model and about Glauber theory. The question as to when Regge cuts are there, and when they are not, appears to depend upon scholastic distinctions as to when a pole contribution cancels,or does not cancel, an unknowable cut contribution, or when the cut cancels itself (as in the simple AFS case), or cannot do so.

It was suggested that sums of Regge poles and cuts, in series, should lead to a cut and a new pole, with a "renormalized" location. We were presented throughout with simplified demonstrations, that were doubtless considered to be suitable for skiers, and for this I am sure everyone was grateful.

Next, let me summarize Renner's admirable introduction to chiral symmetry. We started with a derivation of the Goldberger-Treiman relation, from a momentum-transfer dispersion relation for the form factor of the weak axial divergence between nucleon states, and the assumption of pion pole dominance. We were then led to criticize this latter approximation by considering in detail the $\rho\pi$ contribution to the cut-discontinuity. The problem is that, whereas the pion pole contribution is suppressed by the factor m_π^2 , it is not apparent why the $\rho\pi$ continuum states, for example, should contribute so much less than the pole. However, the Goldberger-Treiman relation is good experimentally, so we have to hunt for a mechanism to suppress the cut. What one needs in fact is that all matrix elements of the axial current be of order m_π^2 . In the limit of vanishing pion mass, we would then have a conserved axial current. With a small pion mass, one is led to expect a nearly, or partially conserved axial current. This is the first step.

Renner then considered tests of pion pole dominance, and hence low energy theorems, in particular the existence of Adler zeros. In taking two pions to zero energy, we needed to compute several commutators of axial currents and their divergences. It was suggested that the commutator of an axial current with the axial charge should be identified with the isospin current, which is a true vector. The coefficient of proportionality, which fixes the scale of the weak currents, is set at unity, by the fiat of Gell-Mann, in his hypothesis of the SU(2) × SU(2) structure of the algebra of weak currents.

We then looked at the Ward identities that were implied by current algebra, and were drawn into the problem of constructing chiral Lagrangians. One can fabricate such a Lagrangian by first cooking up a piece that is completely invariant under SU(2) × SU(2). This piece alone would give conserved currents, and an SU(2) × SU(2) current algebra. One now adds on a term to the Lagrangian that depends only on the fields - the pion field, for example - which breaks chiral symmetry, and the conservation of the axial current, but which leaves the current algebra inviolate.

Renner then went back a step and showed how in fact the consistent exploitation of Adler zeros, in a tree-graph model with only pions, forces one to the current-algebra expression for the axial-current and axial-charge commutator, except that the proportionality factor is not determined, either in magnitude or in sign. It was suggested that the Adler-zeros, that have been read into the Veneziano model, may not necessarily argue a deep connection between current algebra and duality. An extension of these ideas to strangeness-changing currents was presented finally, in which an $SU(3) \times SU(3)$ current algebra was proposed. However, the present position is unclear, since it is not sure whether the K-meson Goldberger-Treiman relation is experimentally good. It could be that SU(3) symmetry breaking is not part of a chiral $SU(3) \times SU(3)$ breaking, but is of quite a different origin.

I should mention also that Landshoff presented an interesting Veneziano-like formula for a four-point function with two currents. Although there is a certain amount of freedom in the model, it turned out that, once the moving ρ-pole, and the required fixed pole at $j=-1$ were incorporated, then the Bjorken high-energy scaling law was automatically satisfied.

To conclude, I will discuss the fascinating talks on gravitational collapse given by Sexl, and the backing up course on differentiable manifolds provided by Hlawka. Sexl started with a welcome revision of standard general relativity, which included an up-to-the-minute report of experimental tests, in particular a modern light-deflection experiment, where you time radio waves that graze the sun and bounce back from Venus. The existence of a coordinate-dependent proof of the non-existence of God, attributable to Fock, was touched upon, but regrettably no details were given. The detection of gravitational waves by Weber was mentioned, and we learned the surprising fact that we are continually bathed in gravitational galaxy-shine that could be a thousand times as bright as sun-shine at the earth's surface. The exact figure is uncertain, since Weber's measurements cover only a very narrow wave-band (at a frequency incidentally that is well within the capabilities of a good Austrian Jodler).

Weber considers that if his government would lend

him two battleships and two strong hooks from which to suspend them, he could do a gravitational Herz experiment.

Next, as an exercise, preparatory to collapsing, we saw how to calculate curvature tensors in an arbitrary base, and then this was specialized to show how the new techniques, based on the general theory of differential forms, simplify enormously the calculation of the familiar spherically-symmetric exterior Schwarzschild metric.

Let me see if I can provide a few land-marks to help you read through Hlawka's notes. It is only to be regretted that the future reader will necessarily miss the instruction that we foreigners had in the German pronunciation of the names of the letters of the alphabet. We were first introduced to a chart on a general set, which is just a subset of the set, together with a one-to-one mapping of the subset onto an open set in n-dimensional Euclidean space. Two charts are said to be compatible, roughly speaking, if the direct mapping of the Euclidean image of the intersection of the two sets by one of the chart mappigs, into the image by the other chart mapping, is r-times differentiable, in the usual sense, and the converse. A set of charts, the subsets of which cover the whole set, is very reasonably called an atlas. Two atlases are equivalent if their intersection is an atlas, and a differentiable manifold is then the original set, together with an equivalence class of atlas.

Hlawka defined for us a differentiable mapping between two different manifolds, and then we had an abstract definition of a Lie group. Next we learned about the tangent-vector at a point of the basic set. The whole set of all tangent-vectors constitutes the tangential space. Sexl used an orthonormal base of this space in his pedagogic exercise on the exterior Schwarzschild solution. The cotangential space is the dual of the tangential space; and we finally learned what a tensor is, namely a function on a suitable product of tangential and cotangential spaces. Finally,we learned about vector- and tensor-fields, and what is termed connexions, or covariant differentiation. This, of course, is what allows you to displace a vector parallel to itself, in infinitesimal steps, and it is an indispensable concept in relativity.

After this digression, I return to Sexl. We learned about the Tolman-Oppenheimer-Volkoff equation for the dynamics of a spherical object in the static approximation. Due to the pressure terms in this equation,it follows, at least for incompressible matter, that when a shrinking object gets to 9/8-ths of its Schwarzschild radius, collapse to, and beyond this radius becomes, sadly, inevitable. During the process of collapse, it seems that about 30 % of the mass, in the form of gravitational self-energy, can be given up, and radiated, for example electromagnetically.

It was interesting to hear that Kruskal, as recently as 1960, found the maximal analytic extension of the Schwarzschild manifold, by a simple transformation involving the t and r coordinates in the Schwarzschild metric, and that this shows that the Schwarzschild limit is not singular. One can use Kruskal space to analyze the experience of any space-time observer of a collapsing object. Suppose, for example, that your lecturer, as a result of the hideous stress of delivering this talk, were to collapse towards his own Schwarzschild radius. So far as I am concerned, I could quite happily pass my own Schwarzschild radius in a finite time with only the minor discomfort of being compressed, and stretched, and perhaps torn apart, and in fact I could reach zero radius in a finite time, if not in one piece, then at least in several pieces,or should it be no pieces? But from your point of view, I would never seem to shrink beyond my Schwarzschild radius, but I would simply become redder and redder, and even dimmer than before, and in fact I would begin to look more and more like an elementary particle, since eventually all you would be able to see of me would be my mass, charge, and other quantum numbers.

Acta Physica Austriaca, Suppl. VII, 629—633 (1970)

SUMMARY - SECOND WEEK*

BY

H. PIETSCHMANN
Institute for Theoretical Physics,
University of Vienna, Austria

Once again it is my privilege and pleasure to summarize the second week of this wonderful Winterschool on High Energy Physics. Because of the bad weather this year, the spirit was particularly serious and dedicated to hard work. This spirit was anticipated by the invitation of a pure mathematician who succeeded in strengthening our understanding of the fundaments on which the delicate building of theoretical physics is erected. Although it was a series of lectures in the first week, let me briefly turn back to it and show you on one particular example where the complex terminology of the theory of differentiable manifolds stems from:

Let W be a set

Take a subset

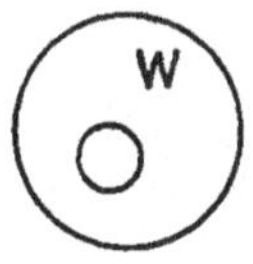

* Summary given at IX. Internationale Universitätswochen für Kernphysik, Schladming, February 23–March 7, 1970.

Now define a mapping of this subset into an open set.

If you add another mapping and various cross-mappings you end up with what is called "Atlas"

Atlas

This "introductory joke" leads us smoothly into a problem which - though seemingly of entertaining nature - certainly is worth some serious considerations: The problem of communication in high energy physics. To begin with, we note in passing that only a small fraction of our colleagues speaks our working language as mother tongue. Therefore, a "pure mathematician" very often becomes a "poor mathematician", a "rational number" very often becomes a "russianal number", an "angle" becomes an "angel", an "arrow" becomes an "error", "heuristic" becomes "juristic" (and, to our deepest regret, "polology" very often becomes "polution"). I think we all enjoy these little mistakes and there is no need of complaining. What should concern us much more seriously is the circumstance that oral communication as opposed to written papers becomes increasingly important in the rapidly changing field of high energy physics, but it seems that physicists have not yet adapted to the new way of doing research in which oral presentation plays a major role. Giving talks - short contributions or lectures of one to several hours - requires a special skill and technique which few colleagues are really trained in. Consequently, the majority of lectures are extremely tiring for the audience if not very difficult to follow at all. Serious thought should be given to this situation by all of us. Each of us should try to improve his lectures and it would be a good advice if a lecturer spends about half or a third of the time he uses for preparing the content of his lecture, on the very way of presenting this content.

The reason, why written communication becomes less important is of course the flood of publications which, in turn, is due to a theorem*:

To each N, there exists an $\varepsilon(N)$ such that to each nonsense bigger than N, there exists a physicist whose importance is less than ε, who publishes this nonsense.

Let us now turn to the main subject of this years second week. There were two concentration points: "Regge phenomenology" and "Rigorous methods". Both fields are not new and have been discussed in Schladming before. The new developments mainly lie in branches which are being developed, or in different approaches which were recently opened. It seems that the theory (of strong interactions) is right now going through a state of confusion. I say this with a positive emphasis; i.e. I mean to say that this is a good sign, because after we have gone through a state of confusion, we certainly have learned something. Right now, we are still waiting for the break-through which will end our journey through this confusion.

M. Kugler's lectures on "Duality" gave a very clear review of the subject which has became one of the main concentration points of current efforts in strong interaction physics. It was fascinating to see, how in a very phenomenological way the vast material of data can be ordered, handled, and finally even modelled into a very new physical concept: Duality.

---

* Private communication

R. Phillips and G. Hite talked about Regge theory which had once already been buried but - like a phoenix - returned to live again and now forms the basis of almost all new models in hadron physics.

D. Atkinson showed us how he got to make use of non-linear techniques by studying theorems on Banach spaces. It was very interesting to be informed that one can actually extract physically useful information from very fundamental mathematical considerations. Along similar lines, A. Martin lectured on his results for scattering amplitudes. The efforts in field theory, which right now seem to advance at a faster pace, were excellently presented by L. Streit.

Let me characterize some of the phenomenological approaches by a picture: A physicists proof of the theorem, that all odd numbers are prime numbers. Draw a graph

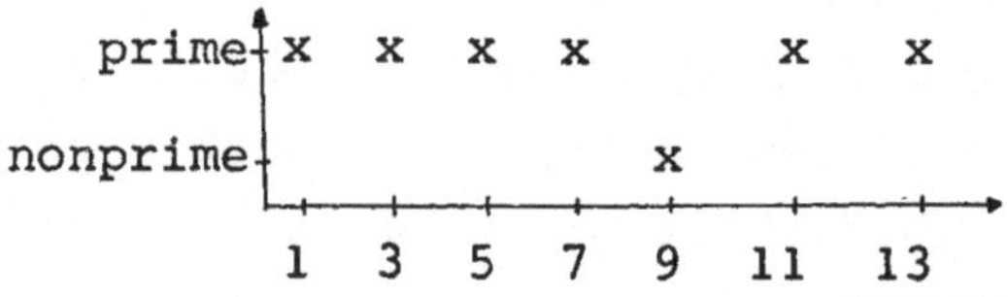

With such an overwhelming evidence no real phenomenologist would hesitate to draw a straight line through 6 out of the 7 points and extrapolate it to infinity, which proves the theorem.

It is elucidating to try a backview over the last seven Schladming Winterschools. (The first two were needed before the Schladming Winterschool was established beyond doubt as a regular event on every high energy physicists calendar). In the following table the number of contributions in each of the six fields "Mathematics, Weak Interactions, Quantum Electrodynamics, Strong Interactions, Experiments and Miscellaneous" are listed (for 1970, the seminars have not been included).

Year	Math.	W.I.	QED	Str.I.	Exp.	Misc.
1964	xx	xxxxxx xxxxx		xxx	x	
1965	x		xxxxxx xx	xx		x
1966	xxxxxx	xx		xxxxxx xxxxxx		
1967	xx	xx	x	xxxxx		xxx
1968	x	xxx	x	xxxxxx x	x	xx
1969	xxx	xx	x	xxxxxx x	x	x
1970	xxxx			xxxxxx x		x

It is noticable that a marked concentration of talks in one particular field is caused by a new idea in this field. 1964 was the year following "Peratization". 1965 was the year of the historical confrontation of Källén's and Johnson's views in Quantum Electrodynamics. 1966 was the year of Higher Symmetries and afterwards Chiral Lagrangians came into play. Later on it becomes increasingly difficult to point to real new inventions that have reshaped our fundamental understanding.

I am authorized by Prof. Urban to announce to you that next years 10th Winterschool in Schladming will have the title: "The growth points of high energy physics". It shall be a resumee of the development and the status of our branch of physics. After this, a new idea in our field is urgently needed and I hope that you will all do your home work quite diligently, so that one of you will eventually come up with this new idea.

It is the honour of the last speaker to express our sincere thanks to Prof. Urban for the fine organization of this Winterschool. We shall also not forget his diligent staff who worked in the organizing committee.